WOMEN SCIENTISTS IN AMERICA

WOMEN SCIENTISTS IN AMERICA

Forging a New World since 1972

MARGARET W. ROSSITER

The Johns Hopkins University Press
Baltimore

Printed in the United States of America on acid-free paper
2 4 6 8 9 7 5 1

The Johns Hopkins University Press
2715 North Charles Street
Baltimore, Maryland 21218-4363
www.press.jhu.edu

Library of Congress Cataloging-in-Publication Data

Rossiter, Margaret W.
Women scientists in America : forging a new world since 1972 / Margaret W. Rossiter.
p. cm.
Includes bibliographical references and index.
ISBN-13: 978-1-4214-0233-8 (hardcover : alk. paper)
ISBN-10: 1-4214-0233-5 (hardcover : alk. paper)
ISBN-13: 978-1-4214-0363-2 (pbk. : alk. paper)
ISBN-10: 1-4214-0363-3 (pbk. : alk. paper)
1. Women scientists—United States—History—20th century. 2. Women in science—United States—History—20th century. 3. Women scientists—United States—History—21st century. 4. Women in science—United States—History—21st century. I. Title.
Q130.R6835 2011
509.2′273—dc23 2011016268
[B]

A catalog record for this book is available from the British Library.

Special discounts are available for bulk purchases of this book. For more information, please contact Special Sales at 410-516-6936 or specialsales@press.jhu.edu.

The Johns Hopkins University Press uses environmentally friendly book materials, including recycled text paper that is composed of at least 30 percent post-consumer waste, whenever possible.

CONTENTS

ILLUSTRATIONS

PLATES

Following page 212:

FIGURES

TABLES

ACKNOWLEDGMENTS

Once again it is a great pleasure at the end of a long project to look back and try to thank the many people and organizations who helped and made a difference. First there were the many librarians and archivists at Cornell University and elsewhere, who are too numerous to name individually but include Patricia Albright, Tanya Zarnish Belcher, Leonard Bruno, Peter Campbell, David Caruso, Elaine Engst, Sheridan Harvey, Pamela Henson, Robin Rider, Wilma Slaight, Sharon Gibbs Thibodeau, and Pat Viele. They were masters of their material, printed and electronic. At the Johns Hopkins University Press Henry Tom ushered the manuscript through the selection process, as he had the two previous volumes, before his retirement, and Joanne Allen brought the text and notes to their high level of accuracy, completeness, and consistency.

Valuable financial support for part of a sabbatical year and several research trips was provided by a grant from the Science and Technology Studies Program of the National Science Foundation, directed by Ronald Rainger and later Frederick Kronz. Cumulatively as important, and immeasurably more convenient, have been the annual research support of the Marie Underhill Noll Professorship at Cornell and the sabbaticals and study leaves of Cornell's College of Arts and Sciences. Deborah Van Galder handled the administrative side throughout. At the end Mark Collins helped considerably with the illustrations, permissions, and some tables, Honghong Tinn performed some timely cross-checking, Sara Cleary generously volunteered to reformat the figures, and Joy Harvey constructed the fine index. The views expressed here are my own and do not represent those of any agency or institution.

Throughout, there was the continuing encouragement and support of my fellow scholars and friends, Sally Gregory Kohlstedt, Michele and Mark Aldrich, Joy Harvey, Clark Elliott, Ronald Kline, Mary Oates, Peggy Kidwell, and Pnina Abir-Am.

Last but far from least are the three "without whoms," those without whom this book and much more—a life and career—would have been very different: my late mother, Mary Madden Rossiter (1913–2005), who was always there, always understood, and always supported me, even and especially in tough times; Sally Gregory Kohlstedt, who for more than forty years has believed in me and this long project, which she suggested in a Simmons College lunchroom in the early 1970s and of which she has since read every page, some more than once; and Patricia Carey Stewart, longtime vice chairman of the Cornell University Board of Trustees, who made sure I did not leave Cornell in 1991, when a temporary post ended, even if it meant that the College of Arts and Sciences had to create a new department, its first in twenty-five years. I hope this third (and final?) volume is worthy of their sustained support.

INTRODUCTION: THE NEW ERA

We live in historic times. This is especially true for American women in science and engineering. Opportunities have greatly expanded since the early 1970s because of a variety of factors, starting with but extending beyond two landmark laws passed in 1972—new expectations, new energy, a growing economy, new technical industries and opportunities, battles won and programs instituted. The women's liberation movement of the late 1960s and early 1970s inspired many women scientists and their supporters to new levels of activism, and legislators passed and President Richard M. Nixon signed significant legislation that greatly affected traditional patterns in academia.

In a sense it started quietly. In the spring of 1972 the 92nd U.S. Congress, the most liberal of the twentieth century, passed historic legislation on equal employment opportunity for women in government and universities, the Equal Employment Opportunities Act and the controversial Education Amendments Act.[1] Before long the conservative but opportunistic President Richard Nixon signed both into law. But few among the public were aware of its passage and import, for the publicity in the *New York Times,* the *Washington Post,* and elsewhere paid more attention to Nixon's accompanying remarks about a divisive upcoming election issue, school busing of white children. Only the *Wall Street Journal,* a month later, pointed out that Title IX, as the second bill came to be known, could have far-reaching implications at colleges and universities, affecting such internal practices as admissions, stipend levels, and other opportunities of all kinds.[2] The *Wall Street Journal* was right, and wrangles over enforcement have persisted ever since. But the enactment of the new laws was a godsend to those few already trying to sue universities (and discovering the many existing loopholes). The 1970s saw many epic lawsuits, including a few that went to the Supreme Court, which itself was changing and included one woman justice after 1981 and two for twelve years after 1993. The new legislation affected women scientists and engineers in several ways, for it outlawed the rampant discriminatory behavior that riddled academia, government agencies, industry, and private clubs, which all too many scientific societies were at the time.

But passage was one thing. How was this new legislation to be enforced, and if it was, how vigorously? Or would it just fall by the wayside, bypassed and ignored, a kind of paper tiger? What difference would it make? Or were the new laws to be just one part of a larger movement already under way? And what was to come of all the new energies released by a plethora of opportunities—or were they, too, mere rhetorical illusions?

As a graduate student at Yale in the early 1970s, I was a small part of this. All of it was in the air when I asked my Yale professors of the history of science whether there had ever been any women scientists. The emphatic answer was assuredly not. (Even Madame Curie with her two Nobel prizes was discounted as a mere drudge who helped her husband Pierre.)[3] Years later, as a postdoctoral fellow at the Charles Warren Center at Harvard University and having found several hundred women in the early editions of the *American Men of Science,* I pressed on with this topic, determined to write a one-volume survey of women scientists from the beginning, whenever that was, to the recent past. In 1982 I published a volume that got the story to 1940, as it did not seem reasonable to start World War II on page 400.[4] A few years later I decided to take on the postwar period, and once again I found far more archival material than anyone had imagined was available. The more I looked, the more I found. That led to another lengthy volume, whose end point was 1972, the year when presumably everything changed.[5] In fact, I was only able to complete that volume by putting aside everything after 1972, the beginning of a deluge. At that point, dealing with the later period seemed an impossible or at least Herculean task. But as time passed, as a few (immense and unprocessed) archives became available, and as colleagues began to ask whether I was going to do a third volume, it seemed that I ought to try, although it would be difficult, since the cast was large, the topic wide-ranging, and everything seemed to have happened at once. Women scientists and, increasingly, engineers, often obscure and not the usual characters in the history of science, had come out of "nowhere" and done things women had never done before. For a while the topic threatened to diverge endlessly, but of course one does not need to find and read everything by everybody; one only needs enough to get the story nearly right, but not so complete that the publisher rebels and the reader gives up. Eventually the project reached its, and my, limits, and for those who wish I had done more, I urge them to take up their pen and dig in, for the topics abound, and the materials available continue to mount.

The chapters are arranged topically and somewhat chronologically. The first five deal with topics that were most prominent in the 1970s—the formation of the women's caucuses and their first challenges; events leading up to the passage of the Equal Opportunity Act in Science and Technology in 1980; the major academic lawsuits of the 1970s; the various programs to strengthen precollege training, especially in mathematics, for college-bound girls; and the major restructuring of undergraduate opportunities for women at the schools of engineering and agriculture, the newly coeducational men's colleges, the traditionally coeducational colleges, the historically black colleges and universities, and the women's colleges. Then the focus shifts to graduate and postdoctoral work, as the numbers of women doing such work grew dramatically in the 1970s through the 1990s in most but not all fields for reasons that are not entirely clear but seem to be responses to higher expectations, expanded opportunity, and reduced discouragement. In the second half of the book the emphasis is on the several sectors of employment—first in industry, the federal government, and the nonprofit institutions, where by the 1990s and the first decade of the twenty-first century a few women were rising to top levels, and then in academia (especially the research universities), where too little was happening. The final chapter has to do with "recognition," or women's expanding role and visibility in the scientific societies, their winning of such top honors as election to presidencies, membership in the national academies, and, not least, nine Nobel prizes. The volume might have ended there had biologist Nancy Hopkins's revelations about conditions at MIT in 1999 and Harvard president Lawrence Summers's remarks in January 2005 not put the subject on the front pages of newspapers across the country. Thus, I added an epilogue on the major events after 1999, when the whole topic entered a new era of institutional contrition and perhaps the start of institutional transformation. Thus, I can finally claim to have reached the goal I set almost four decades ago of reaching the "recent past." Of course, the story has not reached its end there; in fact, for many younger readers it may be just beginning.

The overall tale in the years 1970–2010 is one of a small number (50–100) of marginal individuals finding one another, forming organizations and electing leaders, starting and running grass-roots campaigns, and raising funds for projects that, while not centrally directed, collectively made certain inroads into a sexist and elitist system over several decades. It is rather amazing how much was accomplished by a relatively few clever, astute, hardworking, and determined individuals, some of whom set up offices in Washington, DC (called "shops" inside the Beltway). There were many legal battles and numerous firsts at each step, for everything needed to be changed at nearly every institution. All this occurred amidst significant national political swings, mostly to the right, numerous lurches in government organization and funding, and uncertain (increasing but also decreasing) funding from foundations. But certain beachheads were established by women, such as the posts of dean of a graduate school or executive officer of a scientific society, from which perch individuals might network with others and move ahead in a kind of upward migration. There were numerous individuals with particular passions and programs who worked on one but rarely two projects, followed by others who took on related tasks in other guises. The decades might also be viewed as a series of meetings or mini-events on various issues, each with a certain rationale and a changing cast of participants but without much overall central direction, as local (or national) groups identified a key issue, got a small grant to work on it, perhaps held a meeting and disseminated the results, which others copied or built upon. Linked by a network of newsletters, a new medium in the 1970s, and occasional items in mainstream science journalism, as well as new federations or organizations, together they accomplished much on many fronts. There was no written master plan, but many interacting pieces that resulted in a new vocabulary (*math anxiety, chilly climate, biological clock, Mommy track, glass ceiling, spousal hires,* etc.) and a new career support system. Individuals created many new events, traditions, awards, that increase the positive feedback and cultural atmosphere that a woman entering science or engineering might encounter. How all this happened in thirty-odd years is the subject of this book.

ABBREVIATIONS

AAAS	American Association for the Advancement of Science
AAUP	American Association of University Professors
AAUW	American Association of University Women
ACS	American Chemical Society
AEC	Atomic Energy Commission
AMS	American Mathematical Society
AMWS	*American Men and Women of Science*
APA	American Psychological Association
APS	American Physical Society
ASCB	American Society for Cell Biology
AWG	Association for Women Geoscientists
AWIS	Association for Women in Science
AWM	Association for Women in Mathematics
C&EN	*Chemical and Engineering News*
CHE	*Chronicle of Higher Education*
CIW	Carnegie Institution of Washington
COACh	Committee for the Advancement of Women in Chemistry
COSEPUP	Committee on Science, Engineering, and Public Policy
CSWP	Committee on the Status of Women in Physics
CWSE	Committee on Women in Science and Engineering
DOE	Department of Energy
EEOC	Equal Employment Opportunity Commission
ERA	Equal Rights Amendment
FOPW	Federation of Organizations of Professional Women
HBCUs	Historically black colleges and universities
HEW	Department of Health, Education, and Welfare
HHMI	Howard Hughes Medical Institute
HHS	Department of Health and Human Services
IEEE	Institute of Electrical and Electronic Engineers
IOM	Institute of Medicine
LBL	Lawrence Berkeley Laboratory
LLNL	Lawrence Livermore National Laboratory

MAA	Mathematical Association of America
NAE	National Academy of Engineering
NARA	National Archives and Records Administration
NAS	National Academy of Sciences
NASA	National Aeronautics and Space Administration
NIH	National Institutes of Health
NIST	National Institute of Science and Technology
NOAA	National Oceanic and Atmospheric Administration
NOW	National Organization for Women
NRC	National Research Council
NSB	National Science Board
NSF	National Science Foundation
NYAS	New York Academy of Sciences
NYT	*New York Times*
OOS	Office of Opportunities in Science
PI	Principal investigator
POWRE	Program of Opportunities for Women in Research and Education
RIAS	Radcliffe Institute for Advanced Study
SHER	Self Help for Equal Rights
SWE	Society of Women Engineers
USFS	U.S. Forest Service
USGS	U.S. Geological Survey
VPW	Visiting Professorships for Women program; also a woman taking part in the program
WCC	Women Chemists Committee
WEAL	Women's Equity Action League
WEPAN	Women in Engineering Programs and Advocates Network
WHOI	Woods Hole Oceanographic Institution
WOTY	Woman-of-the-year award

WOMEN SCIENTISTS IN AMERICA

From "Sisterhood" to Interest Group

Learning to Lobby

In the heady atmosphere of the early 1970s, many previously isolated, silent, and invisible women scientists began to find both their voices and one another. Almost immediately they formed organizations, within a few years they had a movement, and by the end of the decade there was a new area of science policy. Once their "consciousness" was raised, young and not so young women in science found that they had certain interests and problems in common. Collectively they had access to considerable resources, they liked working together to reach solutions, and over time they obtained greater access to the inner workings of government and scientific institutions. A major result of this more assertive "sisterhood" was the creation of several independent organizations and new units within existing institutions that would represent the interests of women scientists and work for their emerging goals. Prominent among these were the Association for Women in Science, a society of their own (to paraphrase the English author Virginia Woolf), a women's committee at the New York Academy of Sciences, the Office of Opportunities in Science at the American Association for the Advancement of Science, and a committee of the National Science Board. By 1980 they had taken advantage of some grant money, a lot of volunteers, a pioneering lawsuit, and a growing political savvy to achieve some spectacular successes and some more modest toeholds for future growth.

The Association for Women in Science

It all started in a rather quiet, mundane, even ladylike way. On April 13, 1971, a group of twenty-seven women, mostly biologists, attended an evening "champagne mixer" at the annual meeting of the Federation of American Societies of Experimental Biology (FASEB), held that year, significantly, in Chicago rather than in its usual locale, Atlantic City, New Jersey. Since 1967 the physiologist G. Virginia Upton, of the Veterans Administration Hospital in West Haven, Connecticut, had been organizing this annual reception, funded by laboratory supply houses, as a way to introduce the women attendees to one another. She had been hoping that such a get-together would kindle some sort of women's group across the FASEB member organizations, some of which, such as the American Physiological Society, had (for the time) many women members and participants, and there had been some interest at the 1970 reception. The 1971 mixer, which she was unable to attend because of the recent death of her husband, is now recalled as the first meeting of the Association of [later "for"] Women in Science (AWIS). Upton was later included in the list of forty-two "charter members."[1]

Yet this gathering was not a routine event. The attendees had come prepared for an organizational meeting, and the "members," as yet without a constitution or by-laws, elected two copresidents for 1971–72, Judith Pool, of the Stanford University Medical School, who had done landmark work in hemophilia research, and Neena Schwartz, then at the University of Illinois College of Medicine in Chicago, who was an important researcher in reproductive physiology.[2] Subsequently, as surviving correspondence shows, Pool wrote to foundations and sought outside financing, while Schwartz coordinated the spawning energies of the executive and steering committees.[3] Particularly important was the biochemist Anne Briscoe, of Harlem Hospital Center and Columbia University's College of Physicians and Surgeons in New York City, who volunteered to put out the newsletter, at first a quarterly of eight pages, presenting the group, its goals ("to promote equal opportunities for women to enter the professions and to achieve their career goals"), and its desired image to members, recruits, and other interested persons.[4] Before long this fledgling organization had some specific projects and audacious ambitions.

One of the association's first projects was to establish a roster of names and credentials of women scientists who might be considered for the new jobs and advisory posts that were opening up. If jobs were to be advertised more often or even if women were to be considered for those that were still unadvertised, it would help both employers and applicants to have master lists of the many prospects available. Yet this goal, the perfect roster, which also occurred to many other emergent groups in the early 1970s—including the Office of Opportunities in Science at the American Association for the Advancement of Science, the American Council on Education, the Federation of Organizations for Professional Women, and the American Society of Biological Chemists (as it was then named)—proved both difficult to design, since it should not duplicate anyone else's roster, and expensive to set up and maintain. Funding such a project (one estimate was for sixty thousand vitae) had a catch-22: it would require getting a grant, a step that required a preexisting stable organization, preferably one with a staff, a good track record, and an audited financial statement, which generally ruled out new groups of volunteers like the AWIS. For a while, the anatomist Fann Harding, of the National Institutes of Health (NIH), was an advocate of a registry, but it was the physicist

Natalia Meshkov, then of George Mason University, who ran the AWIS's fee-charging registry, from her office, in the mid-1970s.[5]

Another early, related activity that the new group wished to undertake, and one that grew out of a pressing problem that was common among women biologists at a time of reductions in federal budgets for research, was finding some way to increase the paltry number and percentage of women on the NIH technical panels and study sections that evaluated (and all too often rejected) their grant applications. Successful within a year beyond anyone's wildest dreams was a project spearheaded by the ophthalmologist turned biophysicist Julia Apter, from Chicago, whose research revealed that on average only 2 percent of the members of the NIH study sections were women, and most sections, including that on breast cancer, had no women. What was to be done? United in their venom against the NIH, Apter and a delegation of others met with the NIH director and an assistant to Health, Education, and Welfare Secretary Elliot Richardson in November 1971.

Dissatisfied with their response, these angry women did not give up but instead planned the unthinkable: they determined to sue the federal Department of Health, Education, and Welfare (DHEW, referred to as HEW), which oversaw the NIH. Finding a feminist lawyer to take the case was one early hurdle, and before long they had three—Marguerite Rawalt, of Arlington, Virginia; Sylvia Roberts, of Louisiana; and Helen Hunt Jones, of Chicago and the National Organization for Women's Legal Defense Fund. Shortly thereafter, the AWIS became the lead plaintiff in a suit filed in March 1972 along with several individuals and other newly established women's organizations (including, among others, the vision psychologist Naomi Weisstein, the National Organization for Women, the Association for Women in Mathematics, and the Association for Women in Psychology) that became known as *AWIS, et al. v. Elliot Richardson.* Although the government's lawyers retaliated by urging the male judge to dismiss the case and filed for summary judgment, he did not; instead he denied the NIH motion in August 1972, thereby awarding the AWIS the legal standing to sue the federal agency, in itself a historic step. This meant that for the first time a women's group had the authority to sue the federal government for discrimination and could enter the discovery phase, in which it would find out even more about how and why women were not being appointed to various advisory NIH committees. For example, when Apter scanned the latest list of NIH panelists, she found that several of the men on them were from the same (or branches of the same) universities, thus violating federal conflict-of-interest rules. She later discovered even more irregularities and cronyism in that there were no clear selection criteria for training grants, for which 80 percent of the applications arrived after the deadline, and committee members were routinely reappointed. Finally, in May 1973 the judge heard oral arguments and once again ruled in the

AWIS's favor. Muting the rejoicing somewhat was the lawyers' bill, which came to $6,904.[6]

Despite protests from NIH officials—such as that they were already making substantial efforts to identify qualified women and that there was not enough turnover on its committees for there to be much change in one year—within months of the initial filing the proportion of women on NIH study sections jumped from 2 percent to 20 percent. (Among the new members were the distinguished virologists Charlotte Friend and Gertrude Henle, the well-regarded biochemist Mary Ellen Jones, and the emerging chemist Marjorie Caserio.) In 1975 the women's percentage jumped to 34 percent (254 of 747), yet by 1979, the year in which the feisty Julia Apter passed away, then DHEW secretary Patricia Roberts Harris, herself an African American attorney, reported that women made up only 24 percent (and minorities, 15%) of members on the DHEW advisory committees.[7]

This dramatic campaign against the DHEW, reported in the newspapers and in the AWIS newsletter, attracted many new members—more than seven hundred by May 1972. But this sudden growth created administrative problems, as Copresident Schwartz was about to move to Northwestern University. Governing by correspondence and group consensus was becoming difficult, even with a personal secretary.[8] No one was quite sure how the next president, for 1973–74, the flamboyant physiologist Estelle Ramey, of Georgetown University Medical School in Washington, DC, elected under the new constitution (drawn up by the African American physiologist Eleanor Franklin, of Howard University) over the biologist W. Ann Reynolds, would be able to cope. But Ramey, who was witty—the *Washington Post* called her "the Mort Sahl of Women Scientists," likening her to a popular stand-up comic of the time—and a darling of the media in the early 1970s, had her own way of getting things done. She had come to national attention in 1970 when she took on the sexist Democratic political adviser Edgar Berman, MD, in the press and in a debate at the Washington National Press Club over the role of women's hormones in behavior and public life. Then in 1972 she was successful in taking on the textbook publisher Williams and Wilkins for selling an offensive (because pornographic) medical textbook. Her other advantage was that she was well connected, especially in Washington, DC, where her husband was a prominent lawyer. Since she knew and had entertained many lawyers, she persuaded one to get the AWIS tax-exempt status and to establish the AWIS Foundation, which would use her donated lecture fees to pay a part-time staff person to do what Neena Schwartz and her secretary had been struggling to do themselves.[9] AWIS was growing rapidly and needed a paid employee and some space, preferably in Washington, DC. Fortunately, another new group already had both and was willing to share.

When in 1974 the AWIS sought affiliation with the large American Association for the Advancement of Science (AAAS), its Committee on Council Affairs rejected its application on the grounds that its name seemed to discriminate against men, although it had some male members. This potentially crippling difficulty (itself a product of the new stress on equal opportunity) was, however, eventually resolved by changing the preposition in the AWIS's name from *of* to *for,* which additionally opened the door to nonscientists who supported greater opportunities for women in science.[10]

Thus by 1974 a strong AWIS had in just a few years crystallized out of isolated bits of marginality. It was off to a vigorous start and quickly showed itself to be a forceful presence on the legal as well as the professional scene. It had a remarkable group of energetic and resourceful leaders, a nationwide membership that was starting to form regional chapters, a part-time staff member in the nation's capital, access to top-notch feminist legal assistance, effective publicity, the ability to raise funds, a self-supporting registry, a tax-exempt educational foundation, and a newsletter capable of mobilizing the members. In short, about a thousand women in science had a society of their own that stood ready to respond in whatever way it could to appeals for information and assistance from suddenly aroused women scientists around the country. Before long it would be asked to help in various ways.

Other Groups

Meanwhile other women scientists and social scientists had also begun to meet and organize committees and caucuses within and outside their professional associations. One of the earliest was the independent Association for Women in Psychology (AWP), formed at the annual meeting of the American Psychological Association (APA) in August 1969.[11] Another early women scientists' group was the Association for Women in Mathematics (AWM), founded by Mary Gray, of American University (later called affectionately "the mother of us all"), and five other women at a meeting of the American Mathematical Society (AMS) in January 1971. It was deliberately independent of the major mathematical associations, reflecting Gray's direct, even confrontational style, for she was impatient with and dismissive of the sexist and patronizing practices of the existing mathematical organizations. Previously active in campaigns for social justice in Kansas and California, Gray kept up with current legal rulings, of which she informed the AWM's burgeoning membership through its bimonthly newsletter, and she earned a law degree in 1979. She was ideally suited for the emerging role of feminist activist professor; located in Washington, DC, she could testify before Congress on a number of issues on short notice.[12] Other early independent organizations were the Sociologists for Women in Society (SWS), the women's caucus in biophysics, and the Women in Cell Biology, formed by the Yale biologists Mary Clutter and Virginia Walbot at a meeting of the American Society for Cell Biology (ASCB) in 1971. It produced a newsletter notable for its many job advertisements and occasional special issues on such practical topics as "How to Get a Job," "How to Get a Postdoc," and "How to Hold onto Your Job," that is, how to get tenure. Extra copies of these became bestsellers (at $1 each), and the revenue gained helped support the newsletter when the council of the ASCB, disturbed by the newsletter's critical editorial tone, cut off its initial subsidy. Even so, additional contributions were collected at the annual meeting.[13]

Most caucuses were, however, simply women's committees formed within existing scientific societies that held national meetings at least annually, such as the venerable Women's Service Committee of the American Chemical Society, which dated from the 1920s. One of the most energetic of the new women's committees was that in the American Physical Society (APS), chaired by Vera Kistiakowsky, of MIT, in 1971–72. Kistiakowsky obtained a $10,000 grant from the Alfred P. Sloan Foundation, and in one year her committee produced an extensive report and roster of nearly all U.S. women in physics. The APS Council then voted to establish a standing committee on the status of women in physics, which still exists. In 1973 the American Psychological Association, which already had a committee on women, went even further and approved (not without controversy) the formation of Division 35 on the psychology of women. Before long it had 804 charter members, started a journal (the *Psychology of Women Quarterly*), formed numerous task forces, created several awards, and was well on its way to fostering a new specialty. By 1978 there were at least seventy such committees and caucuses, and many women scientists and engineers belonged to more than one.[14] Typically, members volunteered to collect data on the number and proportion of women in their fields, started newsletters and prepared rosters, and strategized to get more women elected to office (including some of the first women presidents of major scientific societies). Their leaders also began to think of affiliating or federating with other women's groups, but this usually required asking their parent groups for approval and for money to pay any dues, since they had no resources of their own.

Another burst of energy in the early 1970s that increased consciousness nationwide was the spring 1972 conference hosted by the New York Academy of Sciences in New York City. In 1971, as the women's liberation movement was gathering energy, Ruth Kundsin and other middle-aged medical scientists in Boston started to plan a conference that would feature or showcase the many, they hoped, successful women in the sciences. They defined successful women as those who held a professional position, were married, and had children. The widowed sociologist Jessie Bernard, whom they invited twice, criticized the seeming elitism of this emphasis on success. Wasn't it enough to have survived? By May 1972, when the meeting was held at the New York Academy of Sciences in Manhattan, it featured a program of twelve women in science, broadly interpreted to include architecture, horticulture, electrical engineering, and pediatrics, who were genuinely successful in some sense. There was great interest and some press coverage.

In fact the meeting served several purposes. It brought together many women in or near science, some for the first time, and demonstrated how very difficult it had been simultaneously to have a successful marriage and a career in science. The meeting also opened a new genre of self-revelatory writing by and about women in science—up close and personal, as television's *ABC's Wide World of Sports* would say about sportscaster Jim McKay's interviews with Olympic hopefuls about the same time. Autobiographical, nearly confessional, only a bit sugar coated, their stories attracted readers. Many speakers admitted, for example, that they had been considered freaks at some point in their career. Gone was the stonewalling message of the 1950s and 1960s that there was no problem and it was all up to the individual. Now one could say, as these "successful" women did repeatedly, that it had been difficult and continued to be so. Sisterhood had come to science. Within a year the academy's volume of proceedings had sold so well that a commercial press published another version.[15]

The volume also included contributions from about thirty other early experts on women in science, including the endocrinologist Estelle Ramey, but most were social scientists, such as the sociologists Cynthia Fuchs Epstein and Arlie Hochschild, then a junior faculty member at Berkeley. There was a heavy contingent from Cambridge, including Radcliffe College president Mary Bunting, the psychologists Matina Horner and Hilda Kahne, of the Radcliffe Institute, the sociologist Dorothy Zinberg, and the historian Barbara Solomon. The agenda was wide-ranging, analytical, largely psychosocial, and apolitical, making no demands for or reference to federal funding, new legislation, or expanded enforcement of existing legislation (despite Nixon's signing epic legislation in this area in March and June). The sociologist Joan Huber, in a review in the *Chronicle of Higher Education*, complained of its focus on the individual motivation of relatively few individuals—a "pseudo-problem" that gave the volume "a built-in conservative bias"—which grew out of current interest in Matina Horner's "fear of success."[16] Yet the wide interest in the conference led to the creation of an energetic women's committee at the New York Academy that would hold a second major symposium on women in science in 1978, that had more impact on proposed legislation, and a third in 1998.

The Office of Opportunities in Science

An even more effective organizational innovation of the early 1970s was the Office of Opportunities in Science (OOS), established at the headquarters of the American Association for the Advancement of Science in Washington, DC, in response to the "non-negotiable demands" by women protestors. With its long history of representing scientists' concerns in the nation's capital, its considerable political voice, with more than 100,000 members (subscribers to its journal *Science*) nationwide, and its vast network of contacts with federal agencies, professional societies, and even Congress, the AAAS offered a key vantage point from which to affect almost all the sciences across the nation. If it could establish innovative and effective programs, even a small operation (or "shop," as they like to say in Washington) could have large repercussions nationwide.

At the December 1971 annual meeting of the AAAS in Philadelphia, a session titled "Women in Science" was held as part of the association's new Youth Council, an initiative designed to attract more young members. At the session, a women's caucus was formed that drew up a proposal for an "office for women's equality." Shortly thereafter, Mary Clutter and Virginia Walbot presented the proposal to Hazel Fox, a professor of home economics at the University of Nebraska, the only woman on the AAAS Council at the time (representing Sigma Delta Epsilon, an affiliate society). Later there was some dispute as to the legality of the council's voting on this proposal, which had not been presented a month before the meeting as the AAAS statutes required, but the council discussed it and passed it on to the executive officer for consideration by the board of directors. At about the same time, the board heard from its recently formed Committee on Minorities that it would like the AAAS to set up an office for ethnic affairs. Because the board (then chaired by Mina Rees, the AAAS's first woman president) did not feel that the association could afford two new offices, it recommended the creation of a single combined Office of Opportunities in Science with one advisory committee comprising both women and minorities (which was chaired by Claire Nader, of Oak Ridge National Laboratory).[17]

In July 1972 the board approved the creation of one combined office, and in December 1972 it appointed as its director the political scientist Janet Welsh Brown, formerly on the faculties at Sarah Lawrence College and Howard University but then at Federal City College in Washington, D.C. (since renamed the University of the District of Columbia). She quickly calmed the embattled atmosphere and took advantage of the AAAS's location in Washington, DC, where there were by then a sizeable number of women scientists. Some were career or short-term civil servants, others were former academics now at nonprofit institutions, and the rest were victims of the new "PhD glut," who worked at desk or policy jobs rather than in laboratories or classrooms. There were many biologists among them. Collectively they had their fingers on many levers of influence, if not yet power. Brown and her small staff "networked" with them, to use a new term of the time, and over the next several years worked to form an emerging movement of women and minority scientists.[18]

One of the biennial tasks of the leaders of the AAAS, the center of the American scientific community, was to suggest names for appointment to the National Science Board (NSB), which watches over the National Science Foundation (NSF), the source of a sizeable portion of the federal government's support of academic science. Brown immediately took advantage of this opportunity to nominate Jewel Plummer Cobb, an African American biologist who also had taught at Sarah Lawrence College and who had since been a dean at Connecticut College, another former women's college that had recently gone coed. Perhaps surprisingly, in 1974 President Richard Nixon appointed Cobb to a full, six-year term on the NSB, the first black woman and one of the few young persons ever to serve on the board. There she soon started an ad hoc and later (in 1976) standing committee on minorities and women, which guaranteed that the subjects were on the agenda for every board meeting. Perhaps to mollify the growing number of minority members of Congress who voted on the foundation's appropriations, the NSF staff reallocated within its 1974–75 (FY 1975) budget a paltry $390,000 for special studies of, and experimental programs for, women and minorities. (One, for reentry women chemists at Rosemont College, in Bryn Mawr, Pennsylvania, would be widely copied.) This tiny appropriation became a toehold from which larger initiatives could emerge later, and it gave Cobb's committee a budget entry, if not yet a program to oversee.[19]

In October 1975, as part of a reorganization of the NSF's Education Directorate, the NSB, emboldened by President Gerald Ford's endorsement of the United Nations proclamation on behalf of the International Women's Year and assured by the foundation's lawyers of the legality of programs specifically for one sex, recast this small effort as the Women in Science Program (WIS), to be directed by the former chemist M. Joan Callanan. Congress increased its budget to $1 million, of which it awarded the bulk of its funds ($950,000) to twenty-two institutions for science career workshops (one- to two-day events for undergraduate women) and eleven others for "career facilitation workshops" (multiweek reentry training programs) modeled on the one at Rosemont College. By 1981 it had supported fifty-one reentry programs, largely in engineering, chemistry, and computer sciences. These were run by, among others, the engineers Bonita Campbell, of California State University at Northridge, and Carol Shaw, of the University of Dayton, chemists Nina

Roscher, of American University, and Ann Benham, of the University of Texas at Arlington, and the computer scientist Nell Dale, at the University of Texas at Austin. In 1977–79 the WIS also awarded a contract to the Research Triangle Institute, in North Carolina, to administer a visiting-women-scientists program for high-school students. It chose thirty women scientists (from six hundred volunteers!) to visit 120 high schools and colleges across the nation in order to interest young women in careers in science. Although all accounts indicate that these visits and workshops filled a significant need at a modest cost, and a 1980 evaluation by the social scientist Alma Lantz, of the Denver Research Institute, confirmed that the reentry programs were successfully meeting the foundation's goal of having an immediate or short-term impact, the initially projected budgetary growth of the program to $8 million by FY 1982 did not materialize. In fact, the budget never got beyond $1.4 million, and in 1981 the program was discontinued by the incoming Reagan administration, which dismantled the whole Education Directorate at the NSF primarily because of its ideological opposition to federal support of education.[20]

Raising Expectations

But even before the Women in Science Program was terminated, it was being criticized by well-placed women scientists for what it was not. Each year when its selection panel assembled to rank that year's proposals, there were discussions on how the money (or even more of it) might better be spent. One panel included the former chemist Lilli Hornig, of the Higher Education Resource Service (HERS), a placement service located first at Brown University and then at Wellesley College; more importantly politically, she was the wife of the former science adviser to Democratic president Lyndon Baines Johnson, Donald Hornig, a prominent insider in science policy. She and others were dissatisfied that the NSF was making no effort to help women scientists at elite institutions, whose problems were being chronicled in study after study, including some by a National Research Council (NRC) committee on women that Hornig herself had headed since 1975. Why should the NSF's actions be limited to remedial programs at the bachelor's- and master's-degree levels in the Education Directorate, while the more prestigious research directorates were doing nothing for women postdoctoral fellows and faculty. Meanwhile Senator Edward Kennedy (D-MA) began in the spring of 1975 to take an interest in what the NSF was doing for women scientists. He alerted the male heads of both the NSF and the NSB that he would like to hear their suggestions for innovative programs. As a result, the NSF asked Janet Welsh Brown to run a meeting of about sixty "young" women scientists and ask them for ideas for remedies to their career difficulties that Kennedy might later use in crafting legislation. This meeting, initially set for May 1976, was eventually postponed until October 1977 "because," as the *Chemical and Engineering News* put it, "the NSF approval process hadn't moved along far enough."[21]

Hornig's forceful letter to Callanan in February 1977 (with carbon copies to the NSF director and four male top officials at the National Academy of Sciences and the NRC) led to two important top-level meetings, one in April 1977 with officials at the NSF and a second in May with then presidential science adviser Frank Press, a former Harvard geophysicist, at the Old Executive Office Building. This was attended by an emerging power group that included Jewel Plummer Cobb, Janet Welsh Brown, Vera Kistiakowsky, Lilli Hornig, Shirley Malcom, Anne Briscoe, Marian Koshland (by then on the NSB) and Julia Lear, a former physicist at the Aerospace Corporation and now a social scientist at the National Planning Association, who sought to bring a variety of women-in-science issues to the attention of newly elected President Jimmy Carter. Unfortunately, Carter, trained as an engineer and a bit of a rival of Senator Kennedy's, did not embrace the cause, perhaps because the latter already had.[22]

A few months later, Frank Press's Office of Science and Technology Policy gave Lilli Hornig's Committee on the Education and Employment of Women in Science and Engineering, at the National Research Council, which had been established in 1975 and had floundered for lack of funds and lack of a role, $57,000 toward the preparation of two reports, *Climbing the Academic Ladder: Doctoral Women Scientists in Academe* and *Women Scientists in Industry and Government: How Much Progress in the 1970s?* These were admittedly not "original research" but timely and authoritative studies that used matched pairs from the National Academy's database on earned doctorates to show that five years after Nixon had signed the historic legislation, women science doctorates were not advancing equally. Although both reports, especially the first one, on academia, had to be successively revised and toned down to get through the academy's vetting process, a preliminary version of the second appeared in time to be included in the published version of Senator Kennedy's hearings on the Women in Science Bill. Upon their completion the committee was discontinued, although a later version resumed in 1983.[23]

The Federation of Organizations for Professional Women: Learning to Lobby

Janet Welsh Brown was also instrumental in establishing and fostering another uniquely Washington coalition, the Federation of Organizations for Professional Women (FOPW), which in the late 1970s grew to represent more than a hundred such groups in Washington, DC (it faded away in the early 1990s). The goal was to have one central organization represent all the small, scattered women's groups that were springing up around the country. Thus, rather than having each women's caucus or committee try to follow something as Byzantine as pending federal legislation, these organizations would pay a modest fee to one comprehensive federation that would maintain a Washington headquarters, testify and lobby on behalf of all the members, and supply information and feedback to the caucus leaders and newsletters. This arrangement could be particularly useful to those women's committees whose parent organizations' tax-exempt status prevented them from lobbying themselves. The FOPW held its first meeting in November 1972 at Marymount College in Arlington, Virginia. Reportedly the best part of the meeting was the introductions and presentations by women from thirty-nine affiliated organizations. The biochemist Anne Briscoe, representing the AWIS, and Elizabeth Baranger, of the newly formed Committee on Women (as it was temporarily called) of the APS, found it electrifying to meet so many dynamic women from diverse fields such as history, communications, and law and "church employed women." Some formed lasting friendships.

Yet the implementation of the vision floundered. A central registry seemed an obvious goal, yet it would be expensive to run and keep current. No foundation was interested in supporting it. Most of the potential constituent organizations were focused on their own immediate needs and run by volunteers who could see little need for continuing representation in Washington. If a particular issue did arise, most had a member in the area who could attend a meeting, take notes, and report back to her group. But Janet Welsh Brown, a political scientist who saw the value of a forceful presence in Washington, persevered, even running an emergency fund drive in 1973 to pay the staff's overdue salary, and by the mid-1970s the organization began to find a niche for itself. Starting in 1975 it held an annual cocktail party and awards ceremony to introduce and meet the growing number of women in Congress and to recognize those who had made notable contributions, and in 1978 it published *Washington Women: A Directory of Women and Women's Organizations in the National Capitol*, an invaluable guide that included the phone numbers of congressional representatives and high-ranking women whom Presidents Gerald Ford and Jimmy Carter were appointing to federal office. From time to time it also put together roundtables on issues affecting women and health or women and public policy, as well as providing expert testimony for the myriad congressional initiatives in these years that affected professional women. Thus, in the late 1970s there was beginning to be a women's network in Washington, and the FOPW and the women scientists it represented were becoming a full-fledged part of it.[24]

One particularly useful set of workshops run by Brown and the FOPW in 1975–76, the year she was president, was on economic equity and funded by the Ford Foundation. Brown, who was herself learning how to work the levers of science policy in Washington, invited both activists and researchers from the by then eighty member organizations. Thinking that once they met, productive communication would follow, she was surprised at the many frictions and suspicions between those who were at some level working for the same goals. One major conclusion of her final report was that most researchers, which would include most women scientists, lacked even the basic knowledge and skills necessary for political lobbying on their own behalf. This insight would underlie the big push of her final years at the AAAS.[25]

The Equal Rights Amendment: A Test of Muscle

Brown's insight was immediately relevant to the central issue of feminist politics in the late 1970s and early 1980s, the movement to ratify the Equal Rights Amendment (ERA). Introduced into every U.S. Congress since 1923, the ERA finally passed both the House of Representatives and the Senate in 1972. Very quickly thirty-five states ratified it, but at that point, a tantalizing three states short of the goal of three-fourths of the fifty states, the momentum stalled, as an anti-ERA movement led by the conservative activist Phyllis Schlafly, of Illinois, rose up to stop it. As state after state voted no, supporters of the ERA began to worry and regretted that in their early optimism they had allowed a time limit to be placed on the ratification process. Accordingly, in 1977 Eleanor Smeal, the newly elected president of the National Organization for Women (NOW), which had made ratification of the ERA its top priority, asked NOW members to pressure other groups to which they belonged to refuse to hold large conventions in the states who had voted against the ERA. This included such prime sites as Las Vegas, Nevada; Orlando and Miami, Florida; Chicago, Illinois; Atlanta, Georgia; New Orleans, Louisiana; and St. Louis and Kansas City, Missouri. Thus, between 1977 and 1982, when the boards and councils of many scientific societies—groups to which women were increasingly being elected—debated and voted on the heretofore nonpolitical issue of where to meet in the future, there were often heightened tensions.

In August 1977, for example, the council of representatives of the large and liberal American Psychological Association was among the first scientific-organization governing boards to adopt the NOW boycott policy (confirmed later that month when its more fiscally cautious board of directors agreed), and by 1979 about twenty-five scientific organizations (but not the American Chemical Society) had followed suit. Yet, voting on such a divisive issue was a wrenching experience for the officers and council members of some societies. It thrust them, including some of the first women presidents of major organizations, such as Anna Jane Harrison, of the American Chemical Society, C. S. Wu, of the American Physical Society, and E. Margaret Burbridge, of the American Astronomical Society, uncomfortably into the political limelight. Perhaps the most dramatic episode of this campaign was the turnaround at the AAAS in February 1978, when its board of directors reversed a previous decision to meet in Chicago in March 1979. William Carey, its executive officer, and Janet Brown had received many letters and telephone calls from members threatening not to attend a meeting there. Fearing financial loss whatever they did, since they had a contract with the Statler Hotel in Chicago, the board members voted to meet in Houston instead. More than a year later, Carey explained to Harvey Brooks, then president of the APS, that there had been nothing "furtive" about the decision: The AAAS had passed many resolutions affirming equal opportunity over the years, and it had recently held a meeting on the barriers to women's advancement; thus when the time had come to stand up for the principle, the board had voted to change the meeting site. Undoubtedly he reflected Brown's adamancy on the issue. But in the end neither this tactic nor any other was successful in getting even one state legislature to ratify the ERA, which therefore failed to be added to the U.S. Constitution in 1982.[26]

Innovative Projects: Broadening the Constituency

Aside from these external activities, in her six years at the OOS Brown focused on making equal opportunity a serious concern for scientific organizations. Her immediate problem was money, for although the AAAS provided some space and support, the OOS was strapped for funds. An active program would require outside grant support, which was unusual at the time for a scientific society. Yet Brown readily raised and juggled modest grant money from the Ford Foundation in 1974 and from the NSF and other organizations thereafter. Surprisingly, in most years her OOS brought in more outside support than any other AAAS program. She used the funds to hire a young staff that worked on a series of extended projects from which came a veritable cascade of useful written products—rosters, inventories, bibliographies, reports, and lists of women and minority women scientists.[27] Along with these ongoing projects, Brown and her staff also held events at which well-chosen participants brainstormed about what more was needed, and then she and these *rapporteurs* refined these nuggets into a final published report with a list of possibly fundable recommendations. Thus at almost no cost to the AAAS, Brown ran a pathbreaking operation on issues of importance to women and minorities.

For example, sensing that many groups were interested in but undecided about rosters, Brown obtained a small grant from Mariam Chamberlain, at the Ford Foundation, to run a meeting in Washington, DC, in February 1974 at which, despite a major snowstorm, about fifty interested persons talked about their experiences with and needs in rosters. Among the final report's recommendations was that while any one comprehensive roster was too expensive to contemplate, a clearinghouse or central coordinating bureau whose small staff would help those starting new rosters, experiment with improved techniques, and avoid duplication by keeping roster makers informed of one another's efforts would be useful, but it would need foundation support.[28]

Another of Janet Welsh Brown's innovative events, which the NSF supported, was a weekend meeting at Airlie House, in Warrenton, Virginia, in December 1975 of thirty minority women scientists. Locating thirty African American, Mexican, Puerto Rican, and Native American women in a variety of scientific fields who would come on short notice was a difficult task in 1975, but out of the busy weekend came the AAAS's pioneering report *The Double Bind: The Price of Being a Minority Woman in Science.* At the top of its list of many recommendations for improvements was a "communications network" among minority women, who were all too often isolated at their workplaces. By the next June the OOS had drafted a proposal for a two-year grant to prepare a directory of minority women, start a newsletter, and add a staff member who would help the network members to organize local events and invite them to the AAAS annual meeting. Although this proposal went unfunded, seventy-seven participants at a second conference, at the 1978 AAAS annual meeting in Washington, DC, went ahead and formed the National Network of Minority Women in Science, chaired by Yolanda George, of the Lawrence Livermore National Laboratory. This was the start of several future AAAS programs directed at minority women in science.[29]

After the AAAS added the needs of the physically handicapped to Brown's domain in 1975, she and her staff hosted, with the support of Exxon, DuPont, and the DHEW, a meeting of representatives from twenty-five other scientific associations who were willing to share their experiences in running meetings that were accessible to the handicapped. Out of this came another booklet, *Barrier Free Meetings*, which sold thousands of copies.[30]

Climax or Anti-Climax? The Women in Science Legislation

The big push of Brown's final years at the AAAS was the steps leading up to pioneering legislation, the Women in Science and Technology Equal Opportunity Act of 1980. It grew out of the 1975–76 FOPW workshops and a request from NSF officials in April 1977 that Brown run a conference of young women in science. Initially set for June 1977 but postponed until October—the week, it turned out, after Rosalyn Yalow won the Nobel Prize—the conference was aimed at collecting the women's thoughts on what more the NSF might do to help them. With Cobb as the meeting's honorary chairperson, the OOS brought sixty women aged 27 to 47 to Washington for several days of small-group discussions, a speech by Margaret Mead, past president of the AAAS, and introductions of many NSF officials (including some of its first high-ranking women) and other federal and industrial representatives. One goal was to get the young women to articulate their needs and dilemmas in such a way that NSF or congressional staff could formulate programs that would solve or alleviate them.[31] The conference ended with a trip to Capitol Hill for a briefing by Senator Edward Kennedy on the current contents of his proposed "Women in Science and Technology Equal Opportunity Act." If the young women scientists at the meeting took the opportunity to visit their senators and push for this bill or talk about the Equal Rights Amendment, that was their choice. They were learning to lobby on their own behalf, something women scientists had never done before.

The proposed bill (S. 2550) had been prepared by members of Kennedy's staff, who had been in touch with the MIT physicist Vera Kistiakowsky, Brown's network, and others since 1975. It accompanied the annual NSF authorization bill (S. 2549, for $934 million) and initially proposed support for a comprehensive set of innovative programs for ten years, at an annual level of $25 million, a substantial sum. The programs, which largely built upon efforts already under way at the NSF, were designed to "encourage" and "increase the participation of women in" science and technology in a variety of ways. These included improving science teaching in grades 7–12, providing college traineeships and fellowships, training and retraining faculty and counselors, starting a continuing-education program for women already trained in science, establishing a clearinghouse for information on women in science, sponsoring relevant media and museum projects, supporting research and books on women in science, establishing several awards for women's contributions to science and technology, and supporting at least thirty female scientists who would visit high schools and colleges to encourage young women to become scientists.[32]

Yet the groundwork for this ambitious set of programs was not well laid. At the bill's Senate hearings in April 1978 only one person, the sociologist Lucy Sells, then at the University of Maryland, testified about women's issues, specifically "math filters," or female students' tendency not to acquire sufficient mathematical background in high school to pursue advanced careers in science. The chemist George Pimentel, then an official at the NSF, called attention to the fact that at Berkeley and at the University of Delaware there had recently been large, sustained increases in the number of women undergraduates in engineering. They might constitute, he estimated, as much as 15 percent of the engineering school's undergraduate enrollment. But despite the support of the American Chemical Society, whose president that year was its first woman president in one hundred years—Anna Jane Harrison, of Mount Holyoke College—the bill died in committee.[33]

Meanwhile, back in New York City several members of the AWIS, which had formed a local chapter in 1974, and the New York Academy of Sciences (NYAS), which had formed a committee on women in 1977, got approval to hold a second conference to assess what had or had not happened to women in science and engineering since 1972. Originally scheduled for 1977, it was for various reasons postponed until March 1978. By then it could relate to several ongoing issues. There were the gloomy ones of the lack of progress in ratifying the ERA and the recent loss of the Sharon Johnson lawsuit at the University of Pittsburgh, on which much hope had been placed (see chapter 2). But the NYAS speakers also called attention to certain bright spots amidst the gloom: female admissions and enrollment had increased considerably in just six years at medical as well as engineering schools, and the NSF's newly established Women in Science Program was beginning to sponsor certain activities to encourage more women to become scientists. Seventeen corporations and foundations, as well as the AWIS Educational Foundation, sponsored the meeting, which was attended by about five hundred persons and even attracted some local television coverage. Senator Kennedy's staff used the meeting's published proceedings to modify their revised "women in science" bill.[34]

Then in 1979 a sour note was cast when the sociologist Jonathan Cole's book *Fair Science: Women in the Scientific Community,* started years before, was published. It seemed to conclude, from a variety of data samples of thousands of men and women who had earned doctorates from 1911 to 1970 in the physical, biological, and social sciences, that women deserved their lesser place in American science because of the quality of their work, as measured by the number of times it was cited by other scientists. Although the book, which was published by the Free Press of New York City, was widely and favorably publicized by persons who had not read it carefully, knowledgeable insiders such as the manpower analyst Betty Vetter, the Berkeley statistician Elizabeth Scott, and the administrator and adviser Lilli Hornig privately feared that it would trigger a backlash. Yet Cole eventually reversed his publicists and in a letter to the editor of the *Chronicle of Higher Education* claimed that his book "underscores the strong evidence for substantial gender-based discrimination in promotion to high rank, even after taking into account a variety of factors including substantial productivity differentials . . . at this point, there is only one strong conclusion that can be made from the data on academic rank: there is substantial evidence of gender discrimination in promotion of women to high-ranking academic positions."[35]

Undeterred, Senator Kennedy submitted in the next congressional session (1979–80) a revised "women in science" bill (S. 568), which promised to set up an advisory committee on women and minorities (the Committee on Equal Opportunities in Science and Technology, or CEOST) for the director of the NSF (then the African American engineer John Slaughter, a Carter appointee) and provide support for a new program of visiting professorships for women (VPW) and other educational and media ventures. (Initially it had also provided reimbursement for legal expenses incurred in filing a lawsuit against an employer and reduced the overhead on federal grants to those employers—including the national laboratories—that did not employ enough women scientists, but these provisions were dropped along the way.) This time the Senate hearings lasted two days and displayed a better orchestrated set of presentations by several spokeswomen for the emerging women scientists' interest group. In August 1979 the biochemist Anne Briscoe, representing the AWIS Educational Foundation, and Eleanor Smeal, representing NOW, testified in favor of the bill, and in March 1980 Betty Vetter, of the Scientific Manpower Commission (affiliated with the AAAS), and a panel of women scientists (including the biologist W. Ann Reynolds, provost at Ohio State University, perhaps because Senator Howard Metzenbaum [D-OH] was on the Senate committee and involved in the hearings; Margaret Dunkle, of the FOPW; Shirley Malcom, of the AAAS/OOS; and Mary Kostalos, of Chatham College, which had run a reentry program for women chemists) provided further evidence of the need for the bill. In addition, Alice Schafer, of Wellesley College, representing the joint AMS–Mathematical Association of America–National Council of Teachers of Mathematics–Society for Industrial and Applied Mathematics Committee on Women in Mathematics, also sent in a statement, and Jewel Plummer Cobb, by then dean of Douglass College in New Jersey, sent a supportive telegram to Senator Harrison Williams Jr. (D-NJ), chairman of the Senate Committee on Labor and Human Resources. There was also more publicity this time, including editorials in *Science* and *Physics Today* and progress accounts in *Science*, the *New York Times*, and the *Chronicle of Higher Education*.[36]

But not all the comment was favorable. *Science* reported an unfortunate remark by the Berkeley immunologist Marian Koshland about the bill's proposed visiting professorships for women, to be funded through the research directorates. She reportedly said that "many women would prefer not to have this award, because it could be viewed as a second-class citizenship award." And one letter writer to *Physics Today* said that special fellowships for one sex were "reprehensible" and illegal. Providing written testimony (which unfortunately was not published), this time on behalf of Committee W (on women) of the American Association of University Professors, the hard-hitting mathematician and lawyer Mary Gray raised several relevant points: the funding had been reduced to a pitifully small amount; because of its own "poor record of employment of women in high-level positions," the NSF might not be the best agency to run women-in-science programs; it was wasteful and dispiriting to train more women in science if they could not rise at the nation's top universities; more women scientists should be included in setting up any new programs; and the penalties for contractors who did not practice equal employment opportunity were far too weak. In fact the deleted provision for paying the legal expenses of female litigants was extremely important, easily the best part of the bill, for the AAUP already had lots of data on faculty salary disparities. It was time for action, and not just more data collection and subsequent hand-wringing. She also felt that the VPW program was misconceived, for it would reward the most prestigious departments, the worst offenders of equality for women faculty, rather than penalizing them. It would be far better to cut off their federal funds, which would save money. The nation's wealthiest and most prestigious universities could certainly afford to hire women science faculty without financial help from the NSF.[37]

Eventually, revised versions of this weakened bill passed both chambers of Congress, and a compromise conference bill was settled upon, but with the addition of minorities, a particular concern of House members, and drastically reduced levels of funding. Although Senator Kennedy usually is given credit for this bill, it was the crowning achievement of Cobb's term on the NSB and Brown's six years at the OOS. (Brown had left in the spring of 1979 to become the executive director of the Environmental Defense Fund, which would soon be battling President Ronald Reagan on other issues.) Meanwhile, President Jimmy Carter signed the bill in November 1980, just weeks after Ronald Reagan had defeated him handily in his bid for reelection. Officials of NSF, which fluctuated uncertainly between a role limited to funding research and a larger one offering outreach and educational programs, were testy about being required to venture into new territory,[38] and the incoming budget director, David Stockman, threatened to eliminate all new programs ("no new starts"). Nevertheless, the program not only survived but was in fact the beginning of what became after 1983 an area of considerable growth for the NSF. Cobb's term on the NSB had ended, and it would be a while before another minority woman was appointed, but later in the 1980s the NSF would take on most of the equal-opportunity and science-education agenda proposed by Janet Welsh Brown and others at the AAAS in the 1970s.

In the early 1980s, with the coming of President Ronald Reagan and all his new appointees, his rhetoric about family values, David Stockman's cuts to the federal budget, particularly to special programs for women and minorities, the failure to ratify the ERA, as well as the loss of the Sharon Johnson case (to be addressed shortly), the future looked decidedly uncertain for women scientists. The promise unleashed by the reforms and laws of the early 1970s might easily be rolled back and come to nothing. Not much of substance had really happened as yet. A lot—perhaps too much—depended on Congress, the courts, and the president, which until then had been largely dominated by Democrats and liberal Republicans.

Conclusion

In the 1970s women scientists in Washington and elsewhere had forged a framework of organizations from which they might launch future projects. Scientists and social scientists, who had been marginal and invisible at Sarah Lawrence College and elsewhere in the 1960s, had emerged as national leaders on a number of fronts, supported by cadres of others running meetings, putting out newsletters, publicizing events, and writing congressmen. Collectively they had overcome the financial strains of the early days and successfully made the transition from all-volunteer organizations to an organized interest group with a paid staff in Washington, DC. Thus they were linked in a network to women scientists across the nation, to most of the nation's major scientific and professional societies, to some federal agencies, and to the halls of Congress. In one very busy decade women scientists (and to a lesser extent engineers) had organized themselves into an efficient political constituency that had gotten increased membership on NIH panels, established new programs and an advisory committee to the director at the NSF, weathered the attacks in Jonathan Cole's book, and survived the nonratification of the ERA. They could count the Women in Science and Technology Opportunity Act, modest as it turned out to be, as a significant triumph.

This good news was all the more welcome because the atmosphere on campuses, where by 1980 several of the long-running lawsuits of the 1970s had finally been settled, was mixed at best. Several women scientists there had spent much of the same decade playing major roles in landmark cases. Alas, only a few had been successful.

CHAPTER TWO

Taking On Academia

Tokenism, "Revolving Doors," and Lawsuits to 1985

As soon as President Richard Nixon signed the Equal Opportunity Act in March 1972, a few women scientists were ready to take advantage of its provisions and do what had never been possible before—take a university to court for discriminatory hiring, denial of tenure, unequal pay, or other disparate treatment. Finally the victims of such pervasive practices had the right and a route, albeit cumbersome and costly, to corrective action. With enough money, determination, and stamina, they could have their days (or months or years) in court. Many universities, however, stood ready to fight back. Although the 1970s were a time of economic hardship for many of them—enrollments had reached a steady state, the faculty was highly tenured, hiring was cut back, energy costs were skyrocketing, interest rates were in double digits, and one book about the period was entitled the "academic depression"[1]—major universities managed to find the funds to mount long and strong legal defenses. Nevertheless, against high odds over more than a decade several women scientists fought key courtroom battles, including some epoch-making, transformative "class action" lawsuits (a new concept then), that slowly clarified just what equal employment opportunity might mean in academia. The costs and carnage were considerable.

False Hope: Executive Orders 11246 and 11375

Women's prospects in academia had already entered a new phase back in January 1970, when the clinical psychologist Bernice "Bunny" Sandler filed a protest with the Department of Labor (DOL). Having been denied several jobs at the University of Maryland because "she came on too strong for a woman," she turned to the only existing legal remedy of the time, Executive Order 11246, which prohibited sexual discrimination by federal contractors. (An executive order is a proclamation by the president that has what is called "the force of law." Several already applied to conditions surrounding federal contracts.) Claiming that discriminatory practices were rampant "industry-wide," Sandler eventually filed complaints against 260 universities. This set off a series of skirmishes between universities and federal contract officials, who, unprepared for such an onslaught and amidst much publicity, began ever so slowly to investigate the complaints. Meanwhile, groups of angry women coalesced on many campuses, including initially and especially the University of Pittsburgh, the University of Wisconsin, and the University of Hawaii.[2] Among these women's first actions was to hold an event, invite a speaker who would attract an audience, and see who came. Often they invited Bernice Sandler, who was becoming a Washington insider: In 1970 she worked for a few months for Congresswoman Edith Green (D-OR) and the Special Subcommittee on Education of the House Committee on Education and Labor, which held hearings on what became the Education Amendments Act of 1972. Later Sandler was briefly deputy director of the Women's Action Program at HEW. Starting in 1971 she headed what became the influential Project on Status and Education of Women at the Association of American Colleges, which was funded by grants from the Carnegie Corporation of New York and the Ford Foundation for twenty years.[3]

Over the years Sandler visited many campuses, performing a catalytic function. Like Estelle Ramey, of the AWIS, she was informed and witty and revved up the audiences, but she also offered shrewd tactical advice to those who had invited her. She urged them to file grievances immediately with the DHEW Office of Civil Rights (or the DOL Office of Federal Contract Compliance Programs), for the university would respond to the current uproar by setting up a commission on women of its own choosing. This group of loyal yes-women would then spend a year or more talking, but little would result. But it would also take almost that long for the federal investigators to come to campus. When they did come, their visit and the continuing low status of women on campus would be highly publicized once again, thus providing a good prod to the dilatory commission, which by then could be revealed to have accomplished next to nothing. Also by then there would be more potential grievants who might be willing to come forward to talk about their discontent, adding fresh fuel to the fire. It was an effective strategy for keeping the cause alive and in the campus newspapers throughout the early 1970s.[4]

Over the years, however, the executive order proved a big disappointment to women on most campuses. The investigators came, first to the University of Michigan and later to perhaps ten other campuses, went away, and later sent letters saying that the local "affirmative action plan" was incomplete and inaccurate, as they often were, since campus records on employment were often decentralized and unsystematic. Then many months or perhaps a year later, almost out of the blue, the grossly understaffed regional office of the DHEW (or the DOL) would send a report stating that it had accepted the university's affirmative-action plan, which provided little more than promises of future compliance with some set of numbers. The editors of the campus paper would be astonished at this mockery—how, knowing what everyone on campus knew by then of the prevailing employment practices, could the university possibly be "in compliance"?[5]

In 1977–78, as President Jimmy Carter prepared to issue an executive order that would reorganize the several enforcement agencies, their level of activity intensified, and the DHEW Office of Civil Rights even forced four large midwestern universities—Ohio State, Purdue, Michigan, and Wisconsin—to compensate victims for past discriminatory practices. In 1978 the Department of Labor's Office of Federal Contract Compliance Programs took over for the DHEW, which was in the process of losing several agencies to the new and separate Department of Education. Eventually the expanded and better-funded Equal Employment Opportunity Commission (EEOC), headed by the African American lawyer Eleanor Holmes Norton, took over enforcement in academia. Under her directorship in the late 1970s and early 1980s some cases were settled, and a few academic women (such as the botanist Shirley Graham, at Kent State University, who had filed her initial complaint back in 1974) even got their jobs back.[6] The revamped EEOC also threatened to cut off funding in thirty days at just a few places but never really did, as the affected universities quickly signed a settlement instead.

But with the coming of Reagan appointees in early 1981 the EEOC, headed by Clarence Thomas, lessened enforcement considerably. It reduced the number of its visits to universities, summarily disposed of its monumental backlog of incomplete cases, and was satisfied simply with data or even promises from the universities' central administration of better data in the future. By then the Title IX in the Education Amendments Act of 1972, which had seemed to apply to whole educational institutions, was being interpreted by conservative judges and government officials as applying only to "educational programs" or students and especially to athletics, a highly controversial area. Increasingly Title IX, like the executive orders before it, was seen as a false hope. Government agencies, often staffed by former academic administrators and accustomed to assisting educational institutions, and courts manned by conservative judges all proved unwilling to coerce powerful organizations that were well represented in Washington both individually and collectively (via various organizations) to change their practices. For that, still other measures were needed.[7]

Token Voluntary Measures

In the atmosphere of a potential federal investigation of their hiring practices and the abysmally low numbers of female professors reported in their own recent status-of-women reports, several major universities took voluntary steps in the 1970s to appoint a few senior women to their faculties. The easiest way to do this was to move a few of their long-time marginalized research associates into full professorships. This was possible in part because of the rapid waning of the antinepotism rules that had formerly justified either not employing or underemploying wives of faculty members.

Taking a leading role in this change was the newly revived Committee W of the American Association of University Professors. In February 1970, prodded by the sociologist Alice Rossi, the AAUP reactivated its Committee W, on the status of women, which had existed during the association's first decade but had been discontinued back in 1928, when the problems of women faculty had seemed solved. As one of its early activities, Committee W, together with Committee A, on academic freedom and tenure, prepared an official policy statement, "Faculty Appointments and Family Relationship," which the AAUP as a whole endorsed at its next general meeting, in April 1971, and the Association of American Colleges concurred in June 1971. Endorsement of the statement was not binding on any college but did indicate that blanket denials of faculty employment to academic spouses were no longer acceptable. Thus, this serious obstacle to generations of academic women was rather quickly pushed aside, though it lingered in various forms at some universities.[8] (In the next several decades the resuscitated Committee W took on the myriad other issues affecting women and men in academia, most notably tenure clocks, the status of women's studies, sexual harassment, and TIAA-CREF pensions. Prominent committee members were the economist Barbara Bergmann, the statistician Elizabeth Scott, and especially the mathematician and after 1979 lawyer Mary Gray, who, as mentioned in chapter 1, was a founder of the Association for Women in Mathematics.)[9]

Among the women scientists affected by these voluntary in-house rectifications of status (which often involved a local letter-writing campaign) were the psychologist Margaret Kuenne Harlow and the mathematician Mary Ellen Rudin at the University of Wisconsin,[10] the physiologist Lee Tidball at George Washington University, the physicist Vera Kistiakowsky at MIT,[11] the political scientist Judith Shklar and the biologists Ruth Hubbard and Ruth Dixon Turner at Harvard,[12] the physiologist Judith Pool and the botanist Isabella Abbott at Stanford,[13] the biochemist Joanne Ravel at the University of Texas, the physicist Phyllis Freier at the University of Minnesota, the psychologist Lois Hoffman and the demographer Beverly Duncan at the University of Michigan,[14] the mathematician Julia Robinson at Berkeley (in 1975, immediately *after* she was elected to the National Academy of Sciences),[15] the nutritionist Daphne Roe, the neurobiologist Miriam Salpeter, and the botanist Elizabeth Earle at Cornell,[16] and the space physicist Margaret Kivelson and the electrical engineer Thelma Estrin at UCLA in 1980.[17] Other women scientists who had been employed in auxiliary units were moved into the main departments; for example, the psycholinguist Susan Ervin-Tripp, long in the undergraduate-only rhetoric department at Berkeley, was now welcomed into its prestigious graduate psychology department.[18] Some other women already on the faculty, such as the physicist C. S. Wu, at Columbia University, were given substantial pay raises when in-house surveys revealed that they were grossly underpaid.[19] Thus the new atmosphere of voluntary compliance in the 1970s offered universities a window of opportunity during which several found it possible to make some long-overdue adjustments regarding senior personnel. Most of these women were loyal faculty spouses who, except for Ruth Hubbard at Harvard, Susan Ervin-Tripp at Berkeley, and possibly Wu at Columbia, were grateful and quiet and did not press for further changes for women students or faculty.

Some other distinguished senior women took advantage of this window of opportunity to leave employers that did not, or under new budget realities could not, support their research adequately. For example, the physiologist Anne Marie Weber left St. Louis University for the University of Pennsylvania in 1972, and the physicist Fay Ajzenberg-Selove went to Penn from nearby Haverford College a year later, after some procedural difficulties.[20] Similarly, in 1973 the dissatisfied psychologist Naomi Weisstein left Loyola University in Chicago for a full professorship at SUNY-Buffalo, and in 1975 the biologist Ruth Sager, disappointed with Hunter College's waning commitment to research, moved to Harvard Medical School's then new Dana-Farber Cancer Institute, where she founded and headed its Division of Cancer Genetics. Likewise, in 1976 the demographer Judith Blake left Berkeley's School of Public Health, which had terminated her program, for better prospects at UCLA.[21]

Meanwhile, other female long-term research associates who were not grandfathered (or here, grandmothered) into faculty positions on campus left for such posts at less prestigious universities, which in the new era were also seeking more senior women faculty. Among these were the biochemist Carolyn Cohen, who, since earning her PhD at MIT in 1954, had held a series of research posts at Boston cancer centers and accepted a full professorship at nearby Brandeis University in 1972; the immunochemist Justine Garvey, of Caltech, who became an associate professor at Syracuse University in 1974, then a full professor in 1980, and retired in 1989 (whereupon she returned to Caltech as a visiting researcher); the physicist Elsa Garmire, an unhappy research associate at Caltech, who in 1974 moved to the University of Southern California, where she directed a project on laser studies and in 1981 became a full professor of electrical engineering; the botanist Ruth Satter, who left Yale for a faculty position at the University of Connecticut; and the geneticist Joan Stadler, who after earning her PhD in 1955 had been a postdoctoral fellow, research assistant, and research associate at Yale and left in 1972 for an assistant professorship at Iowa State. For Stadler it may have felt like starting over at the bottom, but by 1976 she was tenured, and by 1981 she was a full professor. The longest-running saga may have been that of Anne Cowley, an astronomer and faculty wife at the University of Michigan, who in 1983, after ten years, finally obtained a professorship at the rising Arizona State University. Some other research associates left for positions in the federal government.[22]

More often, and with more long-term consequences, universities also began to hire younger (and so cheaper) women into what were coming to be called "tenure-track" assistant professorships. These appointments also had to follow new affirmative-action procedures, which required, for the first time, that jobs be advertised, searches monitored, and data collected on the numbers and proportions of women earning doctorates in each field so as to have some idea of the size of the pool of potential candidates. These new appointees were novel not only at the formerly all-male institutions such as Caltech, Yale, Columbia, and Princeton, which had had nearly no women faculty in 1970, but also at century-old coeducational universities, which made up most of the land-grant system. Among this first wave (pre-1973) of young, newly hired women scientists at prestigious institutions were the biochemist Joan Steitz, at Yale; the molecular biologist Nancy Hopkins, at MIT; and the computer scientist Susan Graham, at Berkeley. Graham's appointment was initially disapproved by the dean in June 1971 when he learned that she would soon be marrying Michael Harrison, who was already on the faculty. But the very next month the rule was changed, and shortly thereafter she was appointed.[23] Thus in the 1970s a slow trickle of firsts began to occur at the major universities—the first woman hired in any science or engineering college, the first woman in a particular department, the first woman tenured, the first woman full professor, and even the first woman chair or assistant dean. But some universities were slow and needed an outside prod. Cornell University, for example, was still making such firsts as late as 1980, when it hired the assistant professors Barbara Baird in chemistry, Birgit Speh in mathematics, and Rosemary Loria in plant pathology, and even into the early 1980s (i.e., during the Cornell Eleven lawsuit, discussed below), with Martha Haynes in astronomy, Teresa Jordan in earth sciences, and Barbara Cooper in physics.[24]

One university that seemed to be making a notable attempt to hire and promote women faculty in many of its departments in the 1970s was the Massachusetts Institute of Technology (MIT). Early on, its administrators realized that they had little endowment and were almost totally dependent on federal grants. As such, they needed to make more than a minimum effort to hire women. They hired the social scientist Mary Potter Rowe to do what was necessary to restructure the institution. She did a lot with the staff to change what would later be called the "micro inequities," or collectively the "climate," of the institution for both women and minorities, and under pressure from the institute's top administrators and the vocal physicist Vera Kistiakowsky, even the faculty began to change. In 1974 the African American economist Phyllis Wallace, who as deputy director of research at the EEOC had worked on the impact of racial and sexual discrimination, joined the faculty after a year as a visiting professor. Also the alumna and aeronautical engineer Sheila Widnall rose through the ranks to become a full professor, and in 1979 she became chair of the MIT faculty. By the late 1970s, MIT's very large physics department had seven women faculty, the most in the nation, and by 1980 its biology department had that many as well.[25] Its chemistry and mathematics departments, however, did not.

Outside Pressure and Publicity

But these modest and in some cases long-overdue changes receive little publicity and appeared to those aware of them to be mere "tokenism." What remained striking on a national scale was how *few* women were being hired, let alone retained or tenured, by the major departments in their field. Some women's caucuses and committees prepared and distributed charts that emphasized where the women were *not.* In this tactic the most vocal group was the longstanding but newly energized Women's Service Committee (renamed Women Chemists Committee) of the American Chemical Society, for which Sister Agnes Ann Green, IHM, of Los Angeles, on more than one occasion prepared a list of all the doctoral departments in the field, with the number of faculty by sex and sometimes by rank. Of the 172 chemistry departments offering the doctorate in 1971, for example, only 5 had three or more women faculty, only 13 had two (including, significantly, the University of Wyoming), 41 had just one, and fully 113 had no women faculty at all.[26] The *Chemical and Engineering News* occasionally published articles on the slow changes in this dismal situation.[27]

Similarly, in 1979 Vera Kistiakowsky, of MIT, counted the number of women faculty in all 941 physics departments in the nation that were listed in the most recent annual directory of the American Institute of Physics. She found a total of just 218 women full, associate, and assistant professors. At the top ten departments there were just 11 women in these three ranks—of whom, as mentioned above, 7 were right there at MIT.[28] In 1980–81 Laura Eisenstein, of the University of Illinois, and Elizabeth Baranger, of the University of Pittsburgh, prepared for the Committee on the Status of Women in Physics (CSWP) a count and list of women faculty at the 171 departments offering a PhD in physics. This came to be called "the list of zeroes," for out of a total of 4,176 faculty members in these departments only 79 (fewer than 2%) were women. Committee members then pressured the council of the American Physical Society to direct the society's president, Arthur Schawlow, of Stanford, to send letters to the presidents of those universities whose departments had no women to inform them that they needed to change.[29] In 1981 Mary Gaillard, formerly of the Centre National de la Recherche Scientifique (CNRS) and the Conseil Européen pour la Recherche Nucléaire (CERN), became the first woman in the physics department at Berkeley.[30]

Other groups, where they existed, could have done the same; women engineers, for example, could have produced charts with even more zeroes, but the exercise must have struck them as futile. Similarly, women in mathematics and the geosciences knew only too well that there were next to no female faculty in the major geology departments.[31] Because "feminine consciousness" was still low in the agricultural sciences—there were no women's groups in the agricultural sciences at all until 1980, when the (male) president of the American Phytopathology Society finally started a committee on women— in 1985 two female assistant professors conducted their own survey, which showed that out of 11,069 faculty members at seventy land-grant colleges in agronomy, animal science, entomology, agricultural economics, forestry, food science, horticulture, plant pathology, and other agricultural fields (veterinary medicine was not included), only 514 were women. Agricultural engineering was the worst, with just 6 women faculty, or 0.8 percent of all faculty in that field, in the entire land-grant system. The largest number of women was 27, at the University of California at Davis, followed closely by the University of Wisconsin at Madison, the University of Florida, Cornell, and Rutgers, but thirty-nine other institutions had fewer than 10 women, and twelve had none at all.[32]

Besides these charts, one particularly noteworthy attempt to change hiring practices at prestigious places in the 1970s was that of the American Anthropological Association. In 1971, even before the new federal laws had gone into effect, the association's council passed a resolution presented by its energetic committee on the status of women that any department that was not hiring a proportionate number of women five years hence would be censured. The council adopted this recommendation and then charged the committee with finding a way to implement it, as results would vary depending on which data were used. The Berkeley statistician Elizabeth Scott served as a consultant. In April 1979 the three-person committee presented its report, which recommended censuring five departments and, most surprisingly, dropped for no apparent reason the University of Chicago, whose department had ranked as the worst employer of women. The association's leadership and its paid executive officer were evidently under pressure not to censure these departments. The matter might have ended there, except that the voting members, informed by a mail campaign by several activists, rose up at the next annual meeting and voted that they be censured. But the board stalled further by insisting on a mail vote, which also (but narrowly) supported the censure resolution. A year later a resolution was passed extending this scrutiny to tenure awards and equal pay, as well as initial hires, for the next five years. Perhaps as a result, during the 1980s the proportion of assistant professors who were women approached 50 percent (48.1% in 1987).[33]

Lawsuits: "The Strategy of Last Resort"[34]

Whereas universities had once claimed that academic freedom exempted them from governmental intrusions and that they should be allowed to function as they wished, federal legislation in the 1960s had begun to change all that. In fact, the field of college and university law started to grow explosively after the passage of the Buckley student privacy law in 1974.[35] Also starting in the 1970s, the publication of statistics on the lack of equal employment opportunity for women and minorities in academic hiring and promotion, as documented in universities' own status-of-women reports, and the tales of discrimination reported frequently in the national, local, and campus media steadily undercut universities' credibility still further.

Thus the real implementation of the equal-employment-opportunity legislation of the 1970s into the 1990s, and the place where the toughest battles were fought, was in the lawsuits. Under Title VII of the Civil Rights Act of 1964 (as amended by the Equal Employment Opportunity Act of 1972), individuals could, after receiving a letter from the EEOC, take a university to court to hold onto a job, protest a denial of employment or tenure, or sue for equal pay. (In some states one could also ask the state human-relations commission to determine "probable cause" by filing a complaint and letting the commission's staff investigate. Often this was a more user-friendly process, but even a finding in one's favor got few results. A lawsuit was still necessary.) Although initiating an action could be easy, most plaintiffs lost or dropped their cases, for the preponderance of evidence showed that it was costly and futile. But experience was also beginning to show that if one persisted over the several years, the "named plaintiff" could accomplish several things in the process. She might not "win" in the conventional sense, but she could inflict considerable damage on the reputations of the institution and of those individuals who had ruined her career.

In a more positive sense, the plaintiff might rouse others still at the institution to change it from within. First, for example, during the discovery stage she could gain access to many old university files of internal memos and confidential correspondence. If the files contained damaging, demeaning, and possibly incriminating informal remarks, they might be leaked to and published in the campus newspaper, creating a sensation. This could damage the university's reputation as a high-minded institution that used only the most enlightened and dispassionate personnel practices and so win a few points in the arena of public opinion. Second, with the information gained by discovery one could "depose" the university's faculty members and administrators, that is, force them to respond in excruciating detail to questions about the full circumstances of the case while the plaintiff's lawyer badgered them relentlessly. (Of course, the defense lawyers would do the same to plaintiffs and their principal supporters.) Third, if the case went to court and professors were forced to testify against one another, this could polarize a department's atmosphere for years to come. In addition, if the plaintiff's credentials were compared with those of men tenured before or after her, their reputations could also be tainted by public disclosure even though they had done nothing wrong themselves. Fourth, any mention of or publicity about the lawsuit, including the very fact of having been sued, was a public-relations nightmare for a major university, for it could appear to be guilty whatever the complicated circumstances around the actual case. The institution's reputation would be damaged not only on campus but also among alumni, donors, students' parents, and the public, who might write letters to the president or ask questions at alumni functions. Fifth, a major lawsuit would cost the university a significant amount of money, ranging from $1 million to $4 million, which might have been spent on other things, though universities often have insurance to cover expenses and settlements. Sixth, the faculty's dislike of being in the midst of a long-term, heavily publicized lawsuit might make some of them work to transform the university's grievance and appeals procedures into meaningful avenues of redress in the future, something they might not have bothered to do without a lawsuit. Seventh, one might actually win, for, as was noted early on by the feminist lawyer Sylvia Roberts, everything depended on the judge; even when there was strong evidence, two judges could rule differently.[36] Finally, a victory would be widely noted and have repercussions elsewhere. This motivated many women plaintiffs who wanted to reform the system and not just get a monetary settlement for themselves.

If one actually won a class-action suit or even settled out of court with a consent decree, she would win something not only for herself but for all past (since 1972), present, and future women employees of the institution, which might be put under a court order and forced to change its ways. Moreover, as news of the victory spread rapidly to other institutions, it might encourage others to start their own lawsuits or to work to change practices at their institutions as well. If one won the case in court (and did not settle out of court with a consent decree), she could also set a legal precedent that would broaden women's rights across the nation. This was heady stuff for bright, angry, and idealistic women who felt betrayed. All the considerable personal costs—financial, emotional, and even physiological—might be worth it if a plaintiff could improve academic careers for future women.

A whole book could be written on the pioneering academic-discrimination cases of the 1970s, and a few have been.[37] An important trio of key cases were fought by desperate and intrepid women scientists. An early, highly publicized lawsuit started at the University of Pittsburgh Medical School in 1973, when the biochemist Sharon L. Johnson, who earned a PhD from MIT in 1959 and who had been hired as an assistant professor in 1967, was coming up for tenure. She expected no difficulty, as she had been told when she was hired that tenure would be no problem. But by the time of the actual decision, those that had appointed her had moved on, and the verdict was negative. There was no appeals process; she was supposed to accept the decision and leave within eighteen months. Aghast but clear-headed, she sought legal advice within weeks of the new law's coming into force and sued the University of Pittsburgh in February 1973. In May a federal judge awarded her an injunction against the university, forcing it to keep her until her NIH grants ran out. This decision startled the university's lawyers, who had expected the plea to be thrown out of court, as all previous cases had been. It was a historic first and attracted a lot of press attention in *Science* and the *Chemical and Engineering News.*[38]

Johnson's subsequent suit for the award of tenure became the centerpiece of a NOW-supported campaign against selected universities. Situated in Pittsburgh, the home of NOW organizer Eleanor Smeal and several active chapters, employing as counsel (pro bono) the prominent feminist lawyer Sylvia Roberts, who had already won other key victories (including that for AWIS over the NIH), and obtaining modest financial support from the NOW Legal Defense and Education Fund, the members of the then new AWIS, and a fundraiser by the feminist comedian Lily Tomlin, the case featured testimony by many male and female scientists nationwide, including a few social scientists who were experts in the then new area of citation analysis. At the outset the case looked strong and winnable, and it attracted national attention, including occasional updates in the *New York Times.* Everyone seemed to expect that a victory would open doors for women in universities nationwide, and that would be that. But the University of Pittsburgh fought back, and its six lawyers, who had initially underestimated the case, got many delays, charged Roberts with unprofessional conduct and possible disbarment (for signing a fund-raising flyer), and spent a very large sum (possibly $1 million) to defend their client, despite a deficit in the university's budget that required a tuition increase. In addition to many interruptions and postponements, the case eventually consumed seventy-four days of courtroom procedures over four long years, from 1973 to 1977. In time the judge, perhaps overwhelmed by the number of persons involved in academic decision making, grew cautious about intervening, even though he agreed that statistically the university employed discriminatory practices. In the end the judge refused to award Johnson tenure or even to decide on the merits of her case, a cautious practice later called "judicial abstention." In August 1977, when the university threatened her with a bill for its legal expenses but also offered her another year of lab space, Johnson decided not to appeal the case, thus ending the ordeal.[39] This loss was devastating to her and to the many nationwide who had pinned so much hope on her epic lawsuit. For a while it cast an additional pall over the women's movement, which was already facing the faltering fortunes of the campaign to ratify the ERA.

But all was not lost. In September 1977, a month after Johnson's case was finally settled and she decided not to appeal, another woman scientist, the anthropologist Louise Lamphere, inexpensively and relatively quietly settled her more ambitious class-action lawsuit—one of the first of that genre—against Brown University in an innovative way that suited all parties. Like many other female assistant professors encountering what came to be called the "revolving door," she had been hired in the early 1970s but then denied tenure a few years later. But unlike many others, she did not go quietly but filed suit against the university in 1975. In response, Howard Swearer, the new president of Brown, realistically feared how much a possibly lengthy court case might cost the university, which had only recently come out of a period of financial difficulty. He dreaded the public disclosure of sexist and disparaging faculty comments and wished to put the matter behind him as quickly as possible so that he could get on with a major fund drive. Accordingly, he did the unthinkable: without going to court at all, he signed a consent decree under which the university would be monitored for a decade or more—until its faculty employed at least sixty-seven tenured women (it had twelve at the time) and until its academic-personnel practices could pass judicial scrutiny. Three women—Lamphere, the biologist Helen Cserr, and the Slavic scholar Claude Carey—got tenure retroactively, and a fourth, a former instructor in German, received a substantial monetary settlement. Other universities were horrified that a major university would allow its hiring practices to be overseen by a court. They were equally displeased that Brown agreed to pay the women's legal fees (just over $250,000) and estimated its own as about $700,000, for a total settlement of nearly $1.1 million. But women faculty elsewhere who were not getting tenure in the late 1970s saw the Lamphere case as a precedent for a speedy, low-cost, relatively unpublicized settlement on campus that would allow them to keep on with their research and careers. Compared with all that Sharon Johnson's case against Pitt had entailed, the Lamphere settlement, unthinkable as it seemed at first, offered both sides many advantages.[40]

In the later 1970s it became clear that the number of women on the faculties of prestigious universities was actually dropping. Some left voluntarily, such as the geophysicist Tanya Atwater, who left MIT in 1980 for a full professorship at the University of California–Santa Barbara, and the physicist Pamela Surko, who left Princeton for Bell Labs.[41] Many others, perhaps even most, of the newly hired young women like those at Brown were not receiving tenure. This phenomenon soon was labeled cynically the "revolving door," as all too often these appointments had been made as temporary evils, never intended to be permanent or to bring about lasting change.[42] When each spring word spread about the latest women denied tenure, there was a spate of publicity and talk of more lawsuits. For example, when in 1979 eight women did not receive tenure at Princeton, this was reported and discussed in the *Washington Post* (but curiously not in the *New York Times* until eleven months later in a letter to the editor). One of those terminated, the psychologist Diane Ruble, who had published thirty articles, including one in *Science,* and whose department had voted in her favor 9–1, filed a complaint with the EEOC, but she dropped it later after she got a tenured position at New York University and Princeton promised to enlarge its important Committee of Three to include a woman or minority-group member.[43] Similarly, the geologist Judith Moody, at the University of North Carolina–Chapel Hill, one of eleven special hires there in 1974, was astonished not to receive tenure in 1979. She protested through all means open to her, even finding some that had never been used before, but eventually she decided not to file a lawsuit, as the chance of victory was so remote in the South (Fourth Circuit) at the time.[44] Likewise, in the early 1980s the highly regarded Ellen Daniell, the first female assistant professor in the molecular biology department at Berkeley, and Sandra Panem, in pathology at the University of Chicago, were stunned when they were denied tenure.[45] Such was the state of affairs—bad-faith hiring, "revolving doors," and a lack of accountability—when the remarkable Rajender decision of 1980 riveted everyone's attention.

In 1971 the Indian-born chemist Shyamala Rajender, who had received her PhD in 1965 from the University of Wyoming, where she had learned a lot about academic discrimination from two women chemists there, was upset to discover that despite a series of one-year positions as a research associate and instructor at the University of Minnesota, she was not being interviewed for tenure-track posts. Being familiar with the ACS's Women's Service Committee's chart of zeroes, which she had once helped prepare, and with the case of Sharon Johnson, with whom she corresponded, she knew there was a pattern bigger than her own case, and in 1973 she sued. When the university's central administration, which presumably had vowed to the DHEW that it supported equal employment opportunity, decided instead to stand behind the outraged professors in the chemistry department and to fight the case in the courts, Rajender, dissatisfied with her first lawyer, hired on a contingency basis the best lawyer she could find in the Twin Cities, Paul Sprenger, who had recently won a large settlement ($4.5 million) in a sex-discrimination suit against the Minnesota Mining and Manufacturing Corporation. The case encountered postponements, and the university, perhaps emboldened by the University of Pittsburgh's eventual victory, unwisely rejected two offers of settlement (for $50,000 and $35,000). Sprenger then broadened and strengthened the case into a class-action one, to include all those women discriminated against by the University of Minnesota, systemwide, since March 1972. When in April 1980 Judge Miles Lord oversaw a consent decree, an out-of-court settlement, in Rajender's favor, not only did he award her $100,000 but he set up a "special master" program, perhaps modeled on that recently established at Brown University, whereby a committee would oversee settlements and future hires at the university. In addition, he required that the university pay Rajender's legal fees, or about $2 million on top of the class members' settlements, which by 1983 amounted to another $2 million. The university had some insurance to pay part of the expenses, but even so, the costs certainly got attention in *Science,* the *Chronicle of Higher Education,* and elsewhere. Rajender's success heartened women plaintiffs at other universities, who dared to hope that federal judges would extend these precedents to them.[46]

In particular the Lamphere and Rajender victories inspired five women at Cornell University, including the psychologist Donna Zahorik and sociologist Judith Long Laws, to file a class-action suit against their institution in 1980, chiefly for denial of tenure. As the case was often delayed and costs mounted, a group known as the Friends of the Cornell Eleven raised money, including the first grant from what became the Legal Advocacy Fund of the mainstream even bourgeois American Association of University Women. In addition, the Friends, headed by the redoubtable Jennie Farley, a Cornell alumna, faculty wife, former journalist at *Redbook,* and reentry sociologist, got a great deal of publicity for the case in the alumni magazine and elsewhere. One clever gambit was having PEER (Project on Equal Education Rights), a group affiliated with the NOW Legal Defense Fund, award Cornell its Silver Snail Award for the least progress on hiring women faculty in the Ivy League. The university's board of trustees was incensed by an offer to settle out of court and had Cornell hire lawyers from Washington, DC, at an eventual cost of about $2.5 million. The case focused attention on the university's tenure procedures, which lacked clear criteria, varied by department and college, and often were not followed, and on the lack of a realistic appeals process. As the case continued, the faculty and administration revised the faculty handbook several times to provide for annual evaluations in writing, more feedback, and some redress on campus. Finally, the federal judge in Syracuse decided that the plaintiffs did not qualify as a "class" (several other women at Cornell had been promoted) and awarded no one tenure, but in 1984 he did award thirty-seven women a modest ($100,000) monetary settlement for past salary inequities.[47]

Internal Remedies

In 1980–81, in the early stages of the Cornell case, Jennie Farley took advantage of her position in Cornell's Institute of Labor Relations to run several symposia on recent and current cases elsewhere and alternative methods of settling tenure disputes. These included panels of feminist lawyers from major pending cases nationwide, other experienced activists, and the biologist Anne Fausto-Sterling, of Brown University, who spoke on the Lamphere settlement there. Farley published the proceedings and explored alternative methods of solving faculty labor grievances.[48]

When by the late 1970s and into the early 1980s, as some of these epic sagas were resolved and others persisted, many universities' central administrators began to feel that although they should defend their institution's authority as far as possible, many factors justified finding other ways to settle with individuals. These included the high costs of spending years in court, as in the Pittsburgh and Minnesota cases; the certainty of prolonged negative publicity, as at Pittsburgh and Cornell; the chance of losing and being subject to a court's special master, as at Brown and Minnesota; and the toll on them personally, as shown in many early retirements. Universities began to settle with disgruntled former employees quietly out of court for undisclosed sums of money. Since many post-1980 settlements required silence for a number of years, it is hard to know much about them, although the long list of cases followed by the Women's Equity Action League (WEAL) or supported by the AAUW's Legal Advocacy Fund indicates that many suits continued.[49]

At the same time, university personnel policies began to change. Departments had to advertise their positions, run reasonable searches, follow clear procedures, and document their decisions; faculty senates introduced more grievance measures and appeals procedures and generally allowed those denied tenure a chance to present their cases to a university body. Whereas formerly a departmental vote (if even something that formal occurred) was all that mattered, and the plaintiffs had no rights or recourse other than to engage a lawyer, it began to seem reasonable and prudent to give grievants or possible plaintiffs a chance to respond. Maybe with luck or more information a dean or a provost (or at Harvard the president) would overturn a negative decision. This rarely happened, but at least there began to be some accountability for what had been all too often collegiality contests not based on careful assessments of merit or accomplishment.

One example of this more moderate in-house appeals process was the unusual case of the sociologist Theda Skocpol at Harvard in the early 1980s. When Skocpol came up for tenure in 1980, the sociology department voted 5–5, which was not strong enough to justify tenure. Ordinarily the case would have ended there, but as permitted under grievance procedures adopted in 1978, Skocpol appealed to President Derek Bok, a labor lawyer and formerly dean of the Harvard Law School, saying that there had been procedural irregularities. A three-person panel voted 2–1 to recommend that she be given a second chance three years later. Meanwhile, Scokpol did everything she could to document her merit: she won a fellowship at the Institute for Advanced Study at Princeton, got several job offers, picked up some other prestigious awards, including a new prize from the American Sociological Association, and accepted a tenured position at the University of Chicago. When Harvard's President Bok reconsidered her case in 1985, he decided in her favor, and she returned to Harvard. Because the case attracted so much sympathetic local and national publicity (including several stories in the *New York Times* and the *New York Times Magazine*) and Harvard did reverse itself internally, the case showed the effectiveness of this new institutional responsiveness and flexibility. Had Skocpol gone to court in 1980, no such relatively amicable in-house settlement would have been possible, as Harvard's lawyers would have fought vigorously to defend the department's decision. Yet by using recently instituted internal processes and communications, sympathetic faculty got the upper-level administrators to admit that there had been irregularities and that given fresh evidence, they would reconsider her case later.[50]

Battles for Equal Pay

Meanwhile, many women scientists were discovering that they were often paid far less than men in comparable positions on the faculty. Some fought within the university, as did the paleoecologist Margaret Bryan Davis at the University of Michigan in the early 1970s, when she discovered that her salary, which had initially been based on her lowly status as a faculty wife, was still below the stated minimum for professors, even though she had become a full professor and was getting a divorce. (After winning her raise and some back pay, but not an admission that she had been discriminated against based on her sex, she left Michigan in 1973 for a major post at Yale. Three years later she left there to chair a department at the

University of Minnesota, whose salary she later discovered was well below that of her fellow [male] Regents Professors. After the Rajender decision, she filed for and got a 35% increase.)[51] Another faculty member, Elizabeth Scott, a professor of statistics at Berkeley, after hearing Martha Peterson, president of Barnard College, say at a meeting of the American Council on Education that unequal pay in academia was disgraceful, devoted several years to developing a salary evaluation kit for determining whether an individual was underpaid. In the late 1970s the AAUP and many universities adopted it, but it had limitations, as it omitted publications.[52]

Other academic women scientists joined class-action lawsuits at their universities, including the sociologist Ruth Coser at SUNY–Stony Brook and several at the City University of New York, who got a $7 million settlement in 1982, after seven years of litigation. A case at the University of Rhode Island started in 1977 by Lucy Peng-Fei Chang, an underpaid instructor in statistics, eventually won $1.24 million. Once the lawyers were paid and the sum was divided up in many ways, the individual settlements were not large, but it was hoped that the institutions' leaders would do better in the future.[53]

Results

One result of this embattled decade or more of legal action was that a few women scientists and social scientists began to have a chance at a close to normal academic career at more and especially major universities. The big hurdle was tenure, and if they did manage to get that at a major institution, they would then have a whole career ahead of them in which to make contributions. Formerly, tenure had not been a possibility for most women scientists at research universities. Those there at all had instead labored as lecturers or as research associates in someone else's shadow. Before March 1972 the most research-oriented among them would have fared better outside academia, in separate nonprofit institutions.

But turnover was low, and change was slow in coming. Whereas most science departments at major universities had had no women scientists at all about 1970, by the early 1980s the situation had changed a bit in some departments, as shown in table 2.1 for the College of Arts and Sciences at Berkeley, which presents an unduly optimistic picture, for it includes junior untenured women, several of whom did not get tenure. Overall there was more than a doubling of a small number of women faculty (tenured and untenured), but at the departmental level there was no pattern. Some large departments had made no changes, and were still stuck at zero. Five others had actually gotten worse since 1973. Interestingly, several fields that were to emerge nationally in the 1990s as particularly slow to hire women (chemistry, mathematics, physics, economics, and engineering) ranked in the

Table 2.1. Women faculty (ladder appointments) in science and engineering at the University of California–Berkeley, by department and rate of improvement, academic years 1972–73 and 1982–83

Department	1972–73 Total	1972–73 Women	1972–73 % Women	1982–83 Total	1982–83 Women	1982–83 % Women	Change in % Women
Big Improvement (more than 5% increase)							
Paleontology	8	0	0.00	7	2	28.57	+28.57
Linguistics	11	1	9.09	11	3	27.27	+18.18
Psychology	38	2	5.26	37	8	21.62	+16.36
Molecular Biology	15	0	0.00	14	2	14.29	+14.29
Public Health	31	2	6.45	42	8	19.05	+12.60
Plant and Soil Biology	9	0	0.00	9	1	11.11	+11.11
Economics	27	0	0.00	28	3	10.71	+10.71
Zoology	28	1	3.57	30	4	13.33	+9.76
Anthropology	33	4	12.12	33	7	21.21	+9.09
Agricultural and Resource Economics	13	0	0.00	12	1	8.50	+8.50
Entomology	26	0	0.00	26	2	7.69	+7.69
Political Science	41	1	2.44	44	4	9.09	+6.65
Little Improvement (less than 5% increase)							
Mathematics	68	0	0.00	69	3	4.35	+4.35
Physiology and Anatomy	13	2	15.38	16	3	18.75	+3.37
Chemistry	50	0	0.00	52	1	1.92	+1.92
Nutritional Sciences	11	6	54.55	16	9	56.25	+1.70
Physics	59	0	0.00	62	1	1.61	+1.61
Engineering	208	0	0.00	217	3	1.38	+1.38
Statistics	22	1	4.55	20	1	5.00	+0.45
No Change							
Chemical Engineering	20	0	0.00	21	0	0.00	0.00
Forestry and Resource Management	23	0	0.00	29	0	0.00	0.00
Plant Pathology	13	0	0.00	14	0	0.00	0.00
Biochemistry	15	0	0.00	16	0	0.00	0.00
Biophysics and Medical Physics	17	0	0.00	11	0	0.00	0.00
Botany	12	0	0.00	15	0	0.00	0.00
Astronomy	10	0	0.00	13	0	0.00	0.00
Geology and Geophysics	17	0	0.00	17	0	0.00	0.00
Got Worse (up to 10% decrease)							
Electrical Engineering and Computer Science	59	1	1.69	72	1	1.39	−0.30
Genetics	7	2	28.57	11	3	27.27	−1.30
Sociology	25	3	12.00	22	2	9.09	−2.91
Microbiology and Immunology	9	2	22.22	11	2	18.18	−4.04
Geography	10	1	10.00	12	0	0.00	−10.00
Total	948	29	3.06	1,005	74	7.33	−4.27

Source: Data on faculty women at Berkeley, carton 74, Elizabeth L. Scott Papers.

middle of departments showing little improvement (less than 5%) at Berkeley in the early 1980s.

Conclusion

Even this brief sketch shows that many academic women scientists fought a variety of battles at their workplaces to bring some modest but essential change to higher education in the years 1972–85. Using publicity and hiring the best lawyers they could afford, they fought for themselves, for their "class" of fellow female academics at their universities, for colleagues in their fields across the nation, and for future women scientists, who years later became the chief beneficiaries. These efforts consumed much time, money, energy, emotion, and health that in a more perfect universe might have gone into scientific teaching or research. Someone had to be the named plaintiff and try to bell the cat, lonely, unpopular, and expensive though the role was.

Some won and, once tenured, were able to continue in science, with more than a few rising to distinction. For example, by 2000 eighty or more women scientists had been elected to the National Academy of Sciences, and several had been elected to the National Academy of Engineering, and a long list of others held top posts in the Clinton administration in the 1990s. After 1987 a few even became university and foundation presidents or filled other academic leadership positions in which they might do even more. The women's numbers were growing, and a major logjam had been broken. Even so, there should have been more, far more of them. The high hopes of the early 1970s, codified in federal legislation as well as executive orders, had encountered much resistance from scornful faculty, hesitant academic administrators, and cautious judges. The women had fought back with help from one another, lawyers, and legal defense funds. Yet despite the setbacks, the few victories, such as the Lamphere and Rajender class-action cases and the Skocpol case at Harvard, had a disproportionate impact on the academic workplace, where practices were already beginning to change. Once administrators were forced to be legally accountable for their faculty's behavior and to pay multi-million-dollar settlements, they turned their attention to constructing other, less adversarial avenues and incentives for change. Among these would be new programs at the NSF and philanthropic efforts elsewhere beginning in the 1980s to create positive incentives for fostering the career development of women scientists in imaginative ways.

But even as lawyers at many universities resolutely fought these cases through the courts year after year, many administrators were working overtime to recruit women students to fill their classrooms. Their goals were to enroll enough female students to make their institutions attractive to male undergraduates, who no longer preferred the single-sex colleges of yesteryear and/or to maintain at all costs their total enrollments in a time of declining demand. Ironically, the feminization of the student body would in time transform academic institutions as much or even more as the fiercely fought lawsuits.

Taking Advantage of Undergraduate Openings

Impetus to Ever-Broadening Reforms

Potentially as transformative as the major lawsuits of the period 1970–85 was the tremendous increase in the number of women embarking on and successfully completing undergraduate degrees in the sciences, including the social sciences, mathematics, and engineering. Where before relatively few women had majored in the sciences, and many of them at the women's colleges, now much larger numbers completed their degrees in more fields and at a full range of institutions, including the engineering and agricultural colleges, the prestigious liberal arts schools (including the Ivy League and Jesuit-run institutions), and the historically black colleges and universities (HBCUs). Overall the total number of women completing bachelor's degrees in science and engineering more than doubled, from about eighty thousand per year in 1970 to more than two hundred thousand per year in 2000, as shown in figure 3.1. But the patterns varied greatly by field. A particular breakthrough occurred in engineering, where in 1970 women earned fewer than 1 percent, or about 350, of the BS degrees, but by 2000 they earned just over 20 percent, or about 13,000, annually. The earth and agricultural sciences also changed markedly, and computer science came into being. Yet these striking transformations did not just happen by themselves and were not lasting in all fields. Why and how so much change occurred so fast in so many fields, especially in comparison with the lackluster results of the "scientific womanpower" campaign of the federal government in the 1950s and 1960s,[1] is unknown and open to speculation. Similarly, why it was so slow and even collapsed in one field (computer science) after 1985 is under investigation.[2] Much of the change, however, was the result of several economic and social forces in addition to legislation and lawsuits.

About 1970 there were several structural shifts in the educational and professional opportunities open to women and of interest to men, as much of higher education faced a variety of challenges. In engineering and agricultural in particular, men's enrollments (and so degrees) were sagging, just when employers were feeling federal pressure to hire women and minorities. Thus the solution for those concerned with maintaining the size of the engineering and agricultural colleges and with training enough women and minority graduates to satisfy employers (and to win the Cold War) was to admit American women. Also, between 1968 and 1985

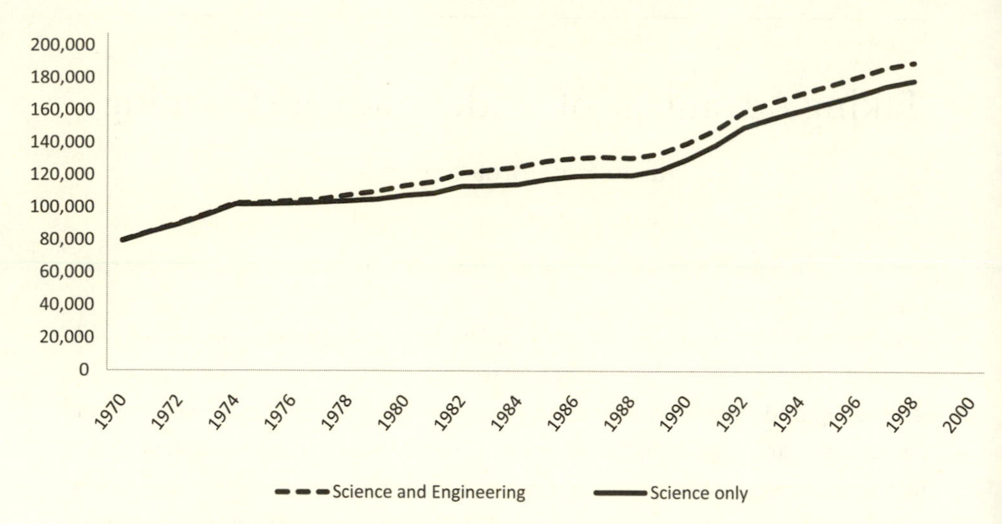

Fig. 3.1. Number of bachelor's degrees awarded to women in science and engineering, 1970–2000. Adapted from Susan T. Hill, *Science and Engineering Degrees: 1966–2000,* NSF 02-327 (Arlington, VA: National Science Foundation, Division of Science Resources Statistics, 2002).

many traditionally men's colleges began, for their own reasons, to recruit women. The trustees' goal was not primarily to train more women but simply to make their colleges more attractive to top male students and thus more competitive with their peer institutions, most of which were already coeducational. Only in this way could they maintain or increase their schools' reputation, enrollment, and income. This changeover had consequences for the traditionally coeducational institutions, which had to respond with innovations of their own. In addition, the HBCUs, with increased resources after about 1980, began to expand and attract more women students. Meanwhile, after 1971, when Congress passed and President Richard Nixon signed the Comprehensive Health Manpower Act, a federal ruling that prohibited discrimination in admissions, the nation's many medical, dental, and veterinary schools, which formerly had set low quotas for women applicants, changed their admissions practices. Immediately, vastly more women took up premedical, predental, and preveterinary studies at all institutions. (It also helped that after 1965 thirty-nine new medical schools were founded.) Thus the lowering of age-old barriers and the lure of greatly expanded professional opportunities attracted many women into formerly unwelcoming institutions and technical majors after 1970.

But the mere presence and passing through of so many women students was just the beginning of what turned into a complex story for these institutions as well. Over time the women students' experiences and needs changed the institutions (and later the graduate schools and employment sectors) in an ever-broadening process, as criticisms of the women's initial maltreatment surfaced and feminists, coining the phrase "chilly climate" in the early 1980s, identified further institutional changes that were needed. When, for example, in the mid-1980s data on current and former students indicated that women's participation and retention in science at some colleges were lower than men's and that their numbers had fallen off nationally in certain fields, administrators at many engineering schools and colleges responded by establishing special programs for women, which helped to maintain or even increase women's participation in the 1990s. The presence of women students in turn created pressure for more women faculty and administrators in the sciences, as advocates used the undergraduate women's growing (or even shrinking) "market share" of the student body to justify even broader and more thoroughgoing changes than initially envisioned. Thus, like a kind of cascade, one step led to others not envisioned at the outset (at least not publicly), and each in turn led to still others that might, it was hoped, in time transform the institutions into truly coeducational ones. By the year 2000, when many institutional differences had been reduced (though many remained), attention turned to the continuing disparities among the technical fields, and new programs were introduced (often with NSF support) to "increase participation" by members of all disadvantaged groups in lagging fields.

We will start by examining the responses of four, slightly overlapping types of institutions[3]—engineering schools, colleges of agriculture, former men's colleges, and traditionally coeducational institutions, including the historically black colleges and universities—and then focus on a few specific fields—geoscience, mathematics, and computer science. (The often innovative response of traditional women's colleges is considered in chapter 5.)

Engineering Schools: Recruiting Women

In the early 1970s, just when the legal status of women in higher education was changing so dramatically, engineering enrollments, harbinger of future BS degrees, dropped off, owing, it was said, to word about layoffs in the aerospace industry at the end of the Apollo space program. At this low point most engineering deans, concerned about their institutions' survival, began to do the unthinkable: they urged bright women, whom they had formerly ignored or even scorned, to apply. One reason for this drastic action was that corporations, required by the federal government to hire more women and minority engineers, had begun to pressure the engineering schools to train more such persons. The deans responded by sending out the message to high-school women with mathematical ability that engineering was now a career for women, and a well-paying one at that. They and leaders of the engineering societies published a torrent of articles in engineering and guidance journals disclaiming traditional stereotypes of engineering as masculine as a thing of the past, featuring upbeat messages about future opportunities, and welcoming women to specialties from aeronautical and agricultural to chemical, civil, electrical, and mechanical engineering. They also held a host of meetings at various locations, most notably two at Henniker, New Hampshire, in 1971 and 1973 but also at MIT and Cornell. Although initially the programs of these meetings featured autobiographies of middle-aged women engineers (the survivors or "exceptions" from past decades), by the mid-1970s they began to include data on the current surge of women in engineering, who were seemingly different in certain respects from earlier women.[4]

As early as 1968 Purdue University, with one of the largest engineering schools in the nation, started what later became its pioneering Women in Engineering Program (WIEP). Led by Jane Z. Daniels, its systematic efforts got such effective results that in just seven years women's enrollment in engineering at Purdue had jumped from 87 in 1972 to 1,143 in 1979, and their attrition rate had also dropped substantially, from more than 80 percent to less than 40 percent, which was about the same as that for men. Daniels was greatly helped, especially in recruitment and in job placement, by the voluntary efforts of the student members of the campus chapter of the Society of Women Engineers (SWE), who telephoned applicants, invited them to campus, and urged those admitted to accept.[5]

Elsewhere, such as at MIT, which, like Purdue, had a large engineering school and where women had been undergraduates since the 1870s but had not always felt very welcome, the top leadership took the new emphasis on equal educational opportunity for women and minorities quite seriously. Aware that they did not have a large endowment and were heavily dependent on federal funds, they hired the economic consultant Mary Potter Rowe in 1973 as "special assistant for women and work" and gave her full authority to make any changes that were needed at the institute to bring about equal opportunity, in areas ranging from admissions to athletics to housing to the health center to maternity leave (but not quite to sexual harassment, the curriculum, and faculty hiring). She emphasized more recruitment, mentoring, and networking and better communication and complaint resolution. Her view was that the many daily "micro-inequities" in the treatment of women at MIT aggregated into "Saturn's rings" of considerable impact. Yet she was hopeful in the face of this daunting challenge, claiming that if every person at MIT did one constructive thing per year, the cumulative result would be substantial change. Meanwhile, the physicist Margaret MacVicar, who had started her career in the 1960s as a tutor in McCormick Hall, MIT's then new dormitory for women, pioneered an innovative Undergraduate Research Opportunities Program (UROP) in the 1970s and became dean of undergraduate education, a new post created for her, in 1985. In 1980 the mathematician Shirley McBay, a former dean of science at Spelman College, in Atlanta, who had also spent five years at the NSF, became the dean for student affairs at MIT, where she was the first, and for many years only, African American on its top-level Administrative Council. By the fall of 1986 the freshman class was 38 percent female, a far higher percentage than at competing technical universities.[6]

Not to be outdone, the College of Engineering at the University of California at Berkeley pioneered several programs in the 1970s and 1980s, including one for reentry women that occasionally even offered financial aid. In 1979 Sheila Humphreys started a long career there as an "academic coordinator," designing and running workshops and programs with corporate and philanthropic support for the recruitment and retention of women, including those returning to prepare for graduate work. As word of her success spread, she, Daniels, and others became part of an emerging national movement whose participants began to report on their efforts at the annual meetings of the AAAS and elsewhere.[7]

The big surge in women's bachelor's degrees in engineering came in the decade 1977–87, as shown in figures 3.2 and 3.3. Because of the continuing drop in the number of men completing degrees, there was a sharp increase in the women's proportion, which shot up from less than 1 percent in 1970 to almost 15 percent in 1985. Contemporaries noticed and mentioned the upsurge. For example, in April 1978 an NSF official testifying at a congressional hearing on the pending Women in Science Bill (see chapter 1) called attention to the recent large and sustained increases in the number of women undergraduates in engineering. And a year later, at a New York Academy of Sciences symposium on women in science, Naomi

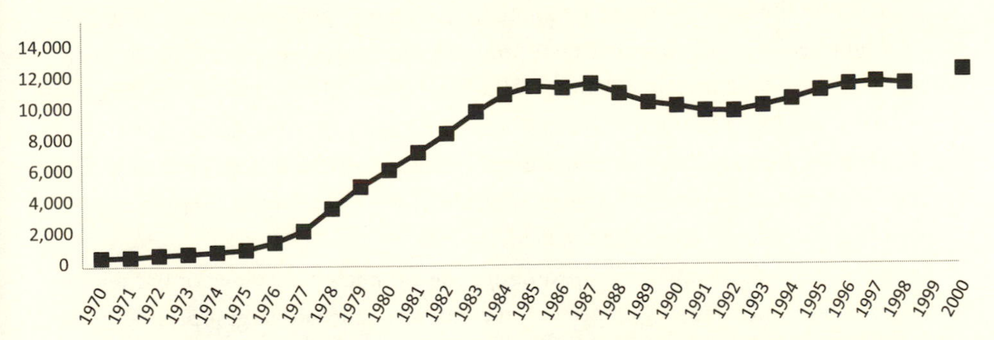

Fig. 3.2. Number of bachelor's degrees awarded to women in engineering, 1970–2000. Adapted from Susan T. Hill, *Science and Engineering Degrees: 1966–2000*, NSF 02-327 (Arlington, VA: National Science Foundation, Division of Science Resources Statistics, 2002).

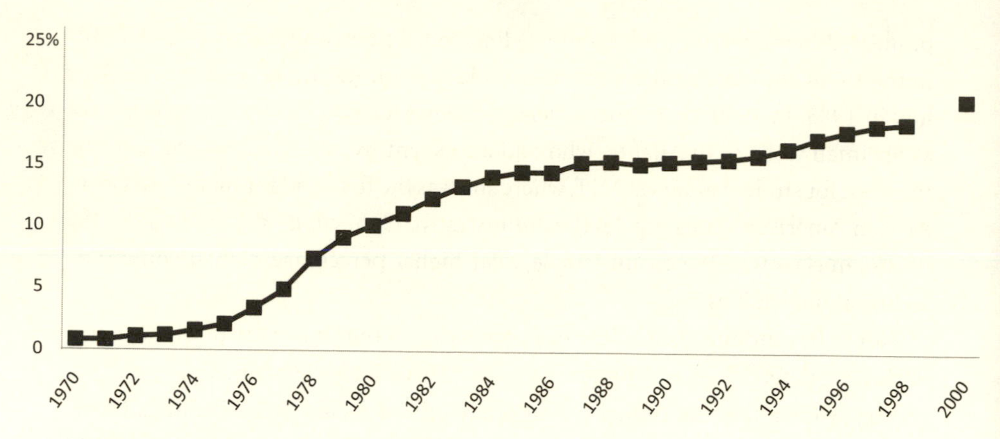

Fig. 3.3. Percentage of bachelor's degrees awarded to women in engineering, 1970–2000. Adapted from Susan T. Hill, *Science and Engineering Degrees: 1966–2000,* NSF 02-327 (Arlington, VA: National Science Foundation, Division of Science Resources Statistics, 2002).

McAfee, past president of the SWE, presented data on the burgeoning female presence in undergraduate engineering education.[8]

Yet the upward trend did not continue monotonically. After 1984 the women's enrollment and their proportion of BS degrees in engineering began to level off, as also shown in figures 3.2 and 3.3. Some alarmed engineering deans began to emulate Purdue and Berkeley, hiring a coordinator of women's programs to recruit more women and improve their retention rates. So many new coordinators were isolated and relatively inexperienced in this kind of work, however, that in 1990 Jane Z. Daniels, still at Purdue, Susan Staffin Metz, of the Stevens Institute of Technology, and Suzanne Gage Brainard, at the University of Washington, cofounded with the help of an NSF grant and other support the Women in Engineering Programs and Advocates Network (WEPAN), an organization of such coordinators from across the country, who met annually to listen to government and foundation officials, to report on their own programs and experiences, and to network and exchange information with one another. In 1991 Emily Wadsworth, of Purdue, started a "catalog of resource materials" to help program directors keep up with the burgeoning literature, and in 1997 Carol Muller, recently associate dean of Dartmouth's engineering school, founded WEPAN's nationwide online mentoring service, MentorNet, for women in the field.[9]

These and other remedies helped nudge the women's percentage of bachelor's degrees in engineering to 20.6 percent in 2000, when the two largest producers of women with BS degrees in engineering were the University of Michigan and the Georgia Institute of Technology (with 320 and 319, respectively), followed by Texas A&M, Purdue, and MIT.[10]

Yet WEPAN's efforts were usually seen as stopgap measures until more women engineering faculty could be hired; those women would, it was hoped, be young and enthusiastic and relate well to the women students and thus humanize the engineering school. The assumption was that if the women's percentage of the engineering student body was to stay at 20 percent or rise a little higher, as seemed desirable and even necessary in 2000, then the number and percentage of women on the engineering faculty (and in time even in the deanships) should also increase, if not to 20 percent or higher then at least to 10–15 percent. These goals were slow in being met, although occasionally a donor contributed funds for scholarships for women students and professorships for women faculty. A few women did become deans of engineering, starting with the electrical engineer Eleanor Baum at the Pratt Institute in New York City in 1985 and continuing with the electrical engineer Denice Denton at the University of Washington in 1996, the computer scientist Maria Klawe at Princeton in 2002, and the mechanical engineer Pamela Eibeck at Texas Tech in 2004.[11] Thus the initially expedient goal in the early 1970s of admitting some females to fill empty seats in the freshman class in order to help employers meet their goals had expanded incrementally into a movement that at some distant future point might transform the whole engineering college, the statistically lagging part of most university campuses.

Colleges of Agriculture: Silent Revolution

A basically similar but eerily different scenario unfolded at the colleges of agriculture at the nation's fifty-six traditional land-grant universities, which were facing their own set of challenges. In the 1960s these universities had worked to move beyond their traditional "cow college" image by diversifying their programs, increasing in size, accepting suburban high-school graduates, and building large residential complexes. Like Michigan State University, formerly the Michigan Agricultural College, they built up their libraries and graduate programs and tried to become prestigious enough to be elected to the Association of American Universities. About 1970 some of the colleges of agriculture, the traditional heart of such institutions, were on the brink of merger or extinction because of a drastic drop in the nation's farm population, which had traditionally provided most of their students. One response was to develop closer ties to the growing world of agribusiness, which employed few women at the time. Meanwhile, the faculties at colleges of agriculture also began to incorporate more basic science, such as genetics and molecular biology, into the traditional agricultural sciences of plant breeding, food science and engineering, entomology, horticulture, plant pathology, animal breeding and nutrition, dairy and poultry science, soil science, forestry, fish and wildlife management, and conservation and range management. Sometimes this "fundamentalization," as Stéphane Castonguay has called it, resulted in a name change, as, for example, at Cornell, where the College of Agriculture became the College of Agriculture and Life Sciences (CALS) in 1971. It may also have been one reason why once the federal Education Amendments Act of 1972 went into effect, the number of women enrolled in agricultural colleges increased dramatically; high-school senior girls may have been better qualified academically for spaces in the freshman class than their more traditional farmboy competitors. They may also have been spurred by the decision at the nation's veterinary schools, which had also been shrinking in size, to admit women students in order to retain their federal funding.[12]

Before long, the employers of the graduates, the large agrichemical and agribusiness firms, who played a strong role on campus in shaping the colleges' mission, felt themselves under pressure to hire women and minority graduates. The deans (all men then) and faculty (nearly so) complied quietly, but unlike their colleagues at the colleges of engineering, they did not undertake a campaign to recruit women students; in fact, the subject seemingly never came up in print, unlike the situation at the colleges of engineering discussed above, whose deans created a whole literature of recruitment and then offices of special programs. Nothing similar happened at the colleges of agriculture. Yet, even without such encouragement, the applicant pool and the entering class changed dramatically. By 2000 the number of women earning BS degrees in agricultural sciences had jumped almost twentyfold, from 458 in 1970 to 8,464, in roller-coaster style, increasing tenfold from 1970 to 1983, then sagging by 40 percent, and then nearly tripling again after 1990 (fig. 3.4). Yet the numbers could have been still higher, since many agricultural colleges retained a preference for those sons of agribusiness who would be returning to the farm. The deans did not remark publicly on this feminization of the classroom and the feedlot, but gender would become a salient factor when the time came to replace the aging agricultural faculty, and some of their former women students would be among the candidates. Occasionally a college of agriculture moving into the twenty-first century would select its first woman dean, such as the molecular biologist Susan Henry at Cornell in 1999 and the entomologist Wendy Wintersteen at Iowa State and the plant breeder Molly Jahn at Wisconsin, both in 2006.[13]

It is tempting to speculate why there was so little notice of or even comment on the changing student body at the agricultural colleges, the biggest demographic change since the first Morrill Land-Grant Act, in 1862. The schools' leaders did not feel the need to go out and persuade women to apply; evidently, as with the medical schools, but not the engineering schools, there were many—perhaps some felt too many—who were already eager to attend if given the chance. In any case, in 2000

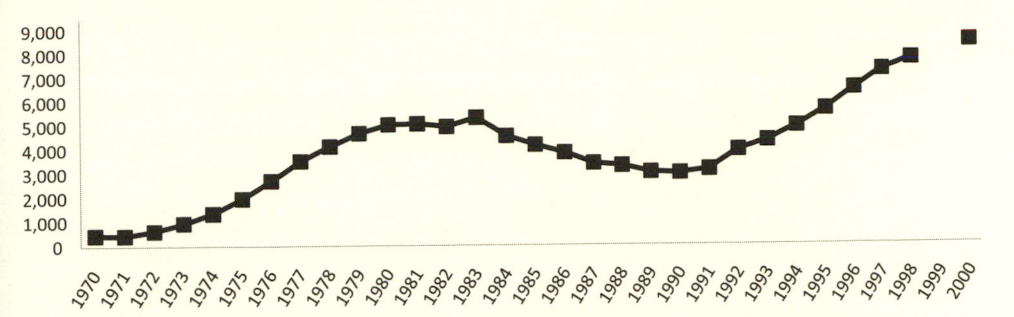

Fig. 3.4. Number of bachelor's degrees awarded to women in agricultural sciences, 1970–2000. Adapted from Susan T. Hill, *Science and Engineering Degrees: 1966–2000,* NSF 02-327 (Arlington, VA: National Science Foundation, Division of Science Resources Statistics, 2002).

the proportion of BS degrees in agriculture going to women reached 50 percent, and more than 70 percent of the DVMs at veterinary schools went to women.

Former Men's Colleges: "Going Coed"

The many historically men's colleges that began to admit women in the period 1968–85 are listed in table 3.1. As one by one the all-male boards of trustees of these wealthy, prestigious, and Roman Catholic colleges voted to admit women within the next year (or even six months, as at Princeton in 1969), certain events quickly unfolded. In fact their very rush to "go coed," as a recent book calls it, was a deliberate tactic to outmaneuver their opposition. Thus their presidents, far from being feminists, were adroit politicians who intentionally minimized the many issues that would be involved in this drastic institutional transformation, which usually provoked opposition from some quarters.[14] Admissions officers, including some newly hired women, at the formerly all-male colleges quickly began to reach out to women applicants, even offering significant amounts of financial aid.[15] Some colleges hired an older woman to oversee the process, such as the microbiologist Mary Bunting at Princeton and the former chemist Elga Wasserman at Yale. The latter later claimed that both her expected title (assistant dean) and her early public assurances of considerable authority were soon undercut by various elements within the college's own administration, as when she insisted that more women faculty be hired.[16] Thus at the outset the process was deliberately made to look deceptively simple, yet it was a major one, with many incremental steps, that would have justified a multiyear plan. (At Princeton coeducation was still considered a work in progress at its twentieth anniversary.)[17] For starters, what, for

Table 3.1. Major men's colleges admitting women, 1968–1985

Year	Colleges
1968	Regis (Denver), Seton Hall, Villanova, Wesleyan, Worcester Polytechnic
1969	Boston College, Franklin and Marshall, Georgetown, Princeton, University of the South, Trinity (CT), Yale
1970	Bowdoin, Caltech, Colgate, Johns Hopkins, Union, University of Virginia (by court order), Williams
1971	Brown (merged with Pembroke), Lehigh, Providence
1972	Dartmouth, Duke (merged with Trinity), Holy Cross, Kenyon, University of Notre Dame
1974	Fordham (merged with its women's college), Pennsylvania (merged with its women's college)
1976	Amherst, U.S. Air Force Academy (Colorado Springs), U.S. Military Academy (West Point), U.S. Naval Academy (Annapolis)
1977	Davidson, Harvard (merged with Radcliffe), Haverford (upperclassmen)
1978	Hamilton
1980	Haverford (all classes)
1983	Columbia
1985	Washington & Lee

Sources: Ohles and Ohles, *Private Colleges and Universities; NYT; CHE; Wikipedia.*

example, were the options in enrollment size and space needs? Princeton University enrolled so few women in the first year—148 on a campus of 3,400 men—that they were isolated and lonely, while Dartmouth doubled its total enrollment in one year by adding a summer semester. Would there be equal admissions (accepting only the best applicants, even if this meant a lot fewer men), or would there be quotas by sex with tougher standards for the women? How much money would be needed for additional housing, classrooms, and laboratories? Would women use laboratories?

Women students applied, and some, often those with alumni ties, were admitted, were housed on campus, and met a mixed reception. Some of the all-male faculty welcomed them, but others, perhaps alumni themselves, who had never taught women students before made belittling remarks and treated them awkwardly. Some longstanding traditions, such as the lyrics to campus songs, were changed, the health center added a gynecologist, and the counseling center added a female psychologist. But many other traditions and practices remained the same. For example, Yale's secret societies long refused to admit women, and Sally Franks, an undergraduate at Princeton, started what turned into a thirteen-year lawsuit (1979–92) against the private eating clubs near campus. Many women students quickly transferred to other institutions, and those who remained were pioneers who rarely felt equal. Twenty years after the coming of coeducation, some of Princeton's first alumnae printed a scathing report in a separate reunion book, and in 2005 a Dartmouth alumna (class of 1979) published a humorous account provocatively entitled *Babes in Boyland.*[18]

At Caltech, which went coed in 1970, women were still only 16 percent of the undergraduates a decade later. But their numbers began to grow in the 1980s, under President Thomas Eberhardt, and they constituted 37 percent of the freshman class in 2007. There, unhappy women undergraduate students could benefit from the organizational efforts of the even more disgruntled female graduate students, who in 1988 formed Women in Science and Engineering (WISE), a group that included all women students, postdoctoral fellows, technicians, and faculty members. Supported by the institute's Graduate Student Council, it sponsored brunches, picnics, social hours, a reading and discussion group, and featured the Undergraduate-Graduate Connection, whereby each undergraduate was assigned to a female graduate student, who was to serve as her mentor. The institute paid for them to go out for lunch together once each semester. Later, WISE members were consulted when the institute established a women's resource center, which offered a lending library, a monthly reading group, and advice and coaching, especially in communication skills and self-defense. These activities and facilities provided a certain space and refuge for relaxation and encouragement to Caltech women to tough it out and stay enrolled.[19]

At first, few of the women who stayed (except those at Caltech and MIT) majored in the physical sciences. For example, even those women who had been admitted to Yale with top test scores in mathematics did not major in a science after all. There was only one female senior in mathematics in 1986 and just one in physics in the class of 1991.[20] In 1986 the *New York Times* reported that since few women were majoring in the sciences or engineering at Princeton, President William Bowen proposed to recruit more women students in the humanities, which a letter to the editor promptly ridiculed as old-fashioned and sex-stereotyped. Shortly thereafter, the Bell Labs physicist Barbara Wilson, a frequent visitor at nearby Princeton, pointed out in the newsletter for women in physics that it might be more effective for Princeton to recruit more women science faculty, as it had none at all in physics and no tenured women in chemistry, mathematics, computer science, and several other departments.[21]

Yet over time some things began to change at the former men's colleges. Having learned from their female human guinea pigs, some institutions, often under the next president, began in the mid-1980s to take belated steps to remedy the problems with new programs and informal attempts to modify the culture. Illustrative of some of these new programs at the formerly men's colleges was the woman-in-science program started at Dartmouth College in 1990, by which time at least four circumstances prevailed: President James O. Freedman was determined to bring about greater gender equity; a report on the women graduates of 1985–89 indicated that only half as many women as men majored in science at Dartmouth; a woman faculty member had been appointed associate dean for science (not a coincidence, as she was given some responsibility for dealing with gender issues); and the NSF was advertising the availability of its new institutional multiyear Career Access awards for junior women faculty. Spearheaded by Carol Muller and Mary Pavone, the program got NSF funding and focused on first-year students, providing them with laboratory internships, sometimes with a stipend, and a tutoring room, which was open four nights a week to students of both sexes but staffed by upper-class female science majors. Added to this female-friendly study space were a faculty retreat and a summer seminar on science teaching funded by the Alfred P. Sloan Foundation. By 1995 the results were impressive: the proportion of women majoring in the sciences had doubled and was nearly equal to that for men. Word of the program spread and even influenced the decisions of some applicants to choose Dartmouth over other institutions.[22]

In time some women faculty were hired, and a few even tenured, in the science departments of these former men's colleges. Among these were the biologist Helen Robinson, at Dartmouth (who unfortunately died young); the astronomer Judith Cohen, the biologist Mary Kennedy, and the chemist Jacqueline Barton (hired first at Columbia), all at Caltech; and the anthropologists Kay Warren and Hildred Geertz and the psychologist Joan Girgus, at Princeton. Something of a turning point was reached at Princeton in the mid- to late 1980s, when its faculty added several senior women scientists: the chemist turned geophysicist Alexandra Navarotsky in 1985, the molecular biologist Shirley Tilghman in 1986, the psychologist Joan Girgus in 1987 (after a decade as dean of the college), and the astronomer Neta Bahcall in 1990. In the 1990s the Clare Boothe Luce Program provided faculty positions for numerous women scientists and scholarships for women science students at several formerly men's Catholic colleges.[23] Something could be written about each of these colleges, as has been done in occasional articles in the *Dartmouth Alumni Magazine* and more comprehensively for the College of the Holy Cross on the twenty-fifth anniversary of the coming of coeducation. In time it might be interesting to know about the first women hired, the first tenured, the first department chair, and on up the ladder, as well as all the other women faculty who left early by the "revolving door."

A little evidence indicates that at the higher ranks the battles were intense. The anthropologist Mary Catherine Bateson has told the tale of her unsettling years as a dean and then acting president in the early 1980s at the recently all-male Amherst College. When hired as dean in 1980, she had hoped to recruit more women faculty to make the place more female-friendly, but with the sudden death of the president who had just hired her she found herself instead abandoned, an outsider, even an intruder targeted for removal. The men's sense of entitlement and sex-solidarity resurfaced undiminished. She left quickly and spent the next few years trying to understand the experience.[24]

Coeducational Colleges: Warming the "Climate"

Although most of the nation's colleges and universities—large or small, public or private, religious or non-sectarian—were in 1972 and are now "coeducational," the term is deceptive, for within these institutions there had long been much internal segregation, as the number of status-of-women studies in the early 1970s documented all too well. Women students (and faculty) were then largely congregated in nursing, education, the arts and humanities, and "human ecology," the new name for home economics, while the men predominated in engineering, agriculture, and the physical sciences. This was the pattern until the early 1970s, when two things happened. First, these ostensibly coeducational colleges suddenly faced intense competition from Amherst, Williams, and the other formerly men's colleges for their traditional applicant pool, those smart men and women high-school seniors who wanted to attend a coeducational college.[25] Second, a few individuals at some of the these institutions began to take a new look at their colleges and attempted to reconceptualize "coeducation" as a broader, more pervasive integration of women into all parts of the institution's power structure rather than just having some women clustered in a few marginal places.

For example, a traditionally coeducational institution such as the University of Oklahoma could count few women science majors or faculty members in 1970, when Betty Pollak, an associate professor of physics there, applied for and got an NSF grant to design and teach a separate section of introductory physics for women students only. The section ran for two years (the academic years 1971–72 and 1972–73), enrolled about twenty-five students each year, and featured visits to class and campus by several nationally known women physicists, such as Melba Phillips, Fay Ajzenberg-Selove, and Johanna Sengers. The students also enjoyed special scholarships, girls-only time in the machine shop, the personal attention of a clinical psychologist attached to the course, and the friendly attention of the physics department personnel. But when the NSF funds ran out, Pollak apparently did not continue the course, and it is not known what happened to the students.[26]

Similarly, Oberlin College, historically the nation's first coeducational college, had almost no women on its science faculty in 1970. Encouraged by its liberal president (and his feminist wife) to scrutinize many of its practices and procedures, from faculty hiring and promotion to women's studies, student counseling, athletics, maternity leave, and faculty status for librarians, the biologist Anna Brummett, one of the few women scientists on the faculty at the time, took the lead on this in

1970. A doctoral student of the biologist Jane Oppenheimer at Bryn Mawr College in 1953, she had survived a lonely time at Carleton College, in Minnesota, before being hired as an assistant professor by Oberlin in 1961 to replace the longtime biology professor Hope Hibbard, the only female in the department at the time, who had insisted as a condition of her retirement that she be replaced by a woman. Brummett had carried on, earned tenure, and even served as acting chair of the department, rather unusual in those days, when she began reading about the women's liberation movement. With the help of others at Oberlin, she documented the school's shortcomings and then got the major recommendations passed by the faculty senate in 1972.[27]

Oberlin was one of about fifty small liberal arts colleges that statistics showed had been and continued to be notable for the high percentage of their men (and increasingly women) graduates who went on to earn PhDs in the sciences, especially in the physical sciences. Among the others were Amherst, Bates, Carleton, Grinnell, Harvey Mudd, Pomona, Reed, Wesleyan, and Williams, to name just a few. At those liberal arts colleges the enthusiastic and accessible faculty (including by the 1980s several females), high expectations, personal attention, small, sometimes individualized labs allowing independent research, frequent field trips (in geology), the presence of other women science majors, and what one informant called the lack of the unnecessary "general meanness" that often typified such colleges transformed a relatively high proportion of talented freshmen into successful science majors who went on to graduate schools.[28]

After 1970 the proportion of women among the scientist graduates of these traditionally coeducational institutions began to rise markedly. This was partly owing to the rise in the percentage of women students enrolled (up to 60% at some in the 1990s and early 2000s) and perhaps also to the rise in the number and percentage of women on the science faculties there. A 2001 study of 136 undergraduate institutions by the Research Corporation showed that whereas women had constituted 10 percent of the faculty hired at 136 undergraduate colleges in the 1970s, they made up 21 percent of faculty hired in the 1980s and 40 percent in the 1990s.[29] Some had come after not getting tenure at research universities, but others had come directly from graduate schools or postdoctoral fellowships. Receiving special attention in 2006 for the high proportion of its female graduates who went on to get PhDs in the physical sciences, especially geology, was Carleton College, thirty-two of whose female graduates earned PhDs between 1999 and 2003, more than from either Dartmouth or Princeton, which had had twice as many women students. Particularly notable at Carleton was that five of the nine faculty members in its geology department were women.[30]

Yet despite these notably successful producers of future women scientists, there was still plenty of work to be done at most coeducational institutions, including the large state universities. In the early 1980s Roberta Hall, Bernice Sandler's assistant at the Association of American Colleges, came up with the term *chilly climate* to refer to the host of behaviors and attitudes that still pervaded many campuses, but not including the newly identified "sexual harassment."[31] Once a practice or situation was identified as "chilly," that is, as having different impacts on men and women, some remedies for improvement could be proffered, thus allowing the typical woman at an institution to play a role. Ingenious, the concept raised awareness, multiplied the number of implementers, and broke down the immensity of the task into doable portions and possibly externally supported projects. Similarly, in the late 1990s the American Association of University Women published a report showing that male students were called on in class much more often than women students.[32] Identification of these pervasive phenomena and practices spurred many individuals at coeducational colleges across the nation to examine behavior at their institutions and, what was even more remarkable, to do something to change it.

One of the innovations at some coeducational universities in the 1980s was an on-campus women-in-science program, usually separate from the women-in-engineering one mentioned above. One of the largest was that at the University of Michigan, where in 1980 several groups, including the Academic Women's Caucus, the Center for Continuing Education of Women, and the university administration, started its Women in Science Program (WIS) to work to increase the number of women earning degrees in science. Cinda-Sue Davis, a PhD in chemistry, became its director in 1984 and for decades has overseen its extensive program, from providing information and counseling to running workshops and conferences, hosting the many Warner-Lambert Foundation visiting speakers, and being involved since its inception in 1993 with the residential program (in Mosher-Jordan Hall) for about 120 freshmen and sophomore female science and engineering students. In addition, in July 1987 the program undertook the immense task of hosting, in conjunction with the Office of Opportunities in Science of the AAAS, the fourth international GASAT (Girls and Science and Technology) congress, with participants from twenty-eight countries and whose proceedings, assembled by Jane Z. Daniels and Jane Butler Kahle, came to five volumes and thirteen hundred pages.[33] Many other WIS programs were started after the mid-1980s, when the NSF resumed support for educational projects.[34]

HBCUs

Another particularly bright spot for women in science at coeducational institutions were the historically black colleges and universities (HBCUs) and the University of

Puerto Rico. Although these colleges were floundering about 1970, as many black and other minority students and faculty left for the newly integrated white institutions, they began to grow in the period 1978–94, when federal support increased as a result of the growing political muscle of the largely Democratic Congressional Black Caucus and executive orders signed by Republican presidents. Although the nation's 103 HBCUs enrolled only 2 percent of the nation's college students in 1994, together they accounted for 28 percent of the bachelor degrees earned by African Americans (as well as several hundred whites, Hispanics, and others).[35] As enrollees at the HBCUs were usually well over 50 percent female, their resurgence, with its federally mandated strong emphasis on training future scientists and engineers, has meant that many of their women graduates have majored in science. For example, of the eighteen physics departments in the nation that graduated the most women with a BS in physics in 2005, seven were at HBCUs—Alabama A&M, Grambling State, Hampton, Jackson State, North Carolina A&T, North Carolina Central, and Xavier University of Louisiana. Similarly, a recent study of the undergraduate colleges of the African American women who later earned PhDs in science showed that almost half were from the HBCUs, especially the two historically black women's colleges, Spelman College and Bennett College.[36]

Thus the institutionally segregated educational world of 1970, with its engineering schools, agricultural colleges, men's colleges, coed schools, including the HBCUs (as well as the women's colleges to be considered in chapter 5), had, through a variety of largely self-imposed and voluntary changes fueled by competition and employer demands, become more homogeneous coeducational institutions, a process called "institutional isomorphism" by sociologists. Yet by 2000 they remained different in other, more subtle ways. The agricultural schools, the liberal arts colleges, and the HBCUs had 50 percent or more female students and were training many women scientists, while the engineering schools, the former men's colleges, and many large universities, which had initially recruited women students energetically, were still lagging, despite the help of special programs. Similarly, some disciplines were proving less successful than others in attracting female undergraduate majors.

Chilly Fields: A Bewailing Literature

Another way to look at divisions among the undergraduate population is to focus on their majors or fields from 1970 until 2000. Although there was a remarkable overall upsurge in the number of women completing bachelor's degrees in engineering and scientific fields from 1970 to 2000 (see fig. 3.1), this aggregate concealed sharp differences between fields, which began to be noticed. When the data were presented as percentages of the total (which was at times falling), these trends provoked a variety of reactions. Thus when the numbers of women earning bachelor's degrees in psychology, biological sciences, political science, and even economics continued to increase to record levels (figs. 3.5–3.8), this was greeted with relative silence and probably approval, though private fears must have been mounting. Instead, attention focused on the fields where the more publicly discussed question became, why so few? Why, despite all the activism in the 1970s and 1980s, did so many physical science departments at coeducational institutions remain decidedly "chilly" for female undergraduates? What might be done about it? Were new, more focused intervention programs needed? If so, who should be asked to pay for them? Then when such cautionary "plateaus" in the enrollment and degree

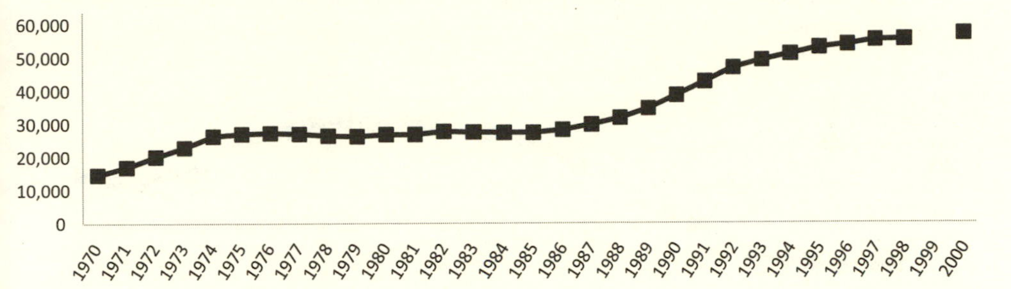

Fig. 3.5. Number of bachelor's degrees awarded to women in psychology, 1970–2000. Adapted from Susan T. Hill, *Science and Engineering Degrees: 1966–2000,* NSF 02-327 (Arlington, VA: National Science Foundation, Division of Science Resources Statistics, 2002).

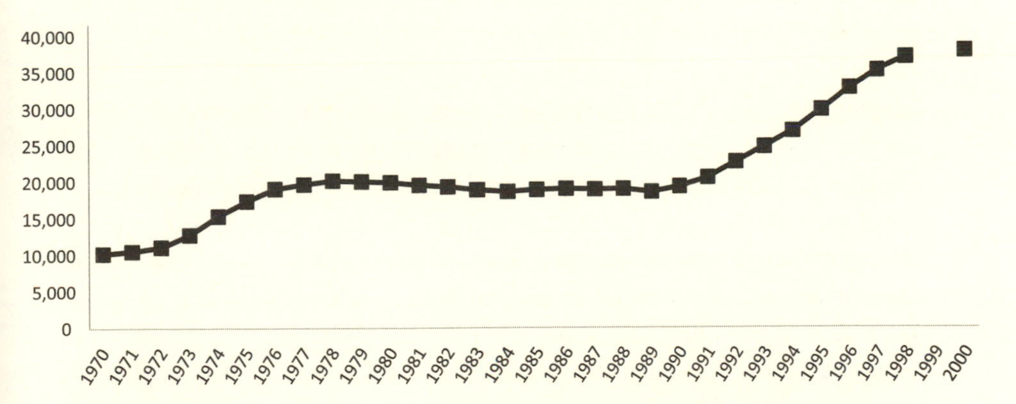

Fig. 3.6. Number of bachelor's degrees awarded to women in biological sciences, 1970–2000. Adapted from Susan T. Hill, *Science and Engineering Degrees: 1966–2000,* NSF 02-327 (Arlington, VA: National Science Foundation, Division of Science Resources Statistics, 2002).

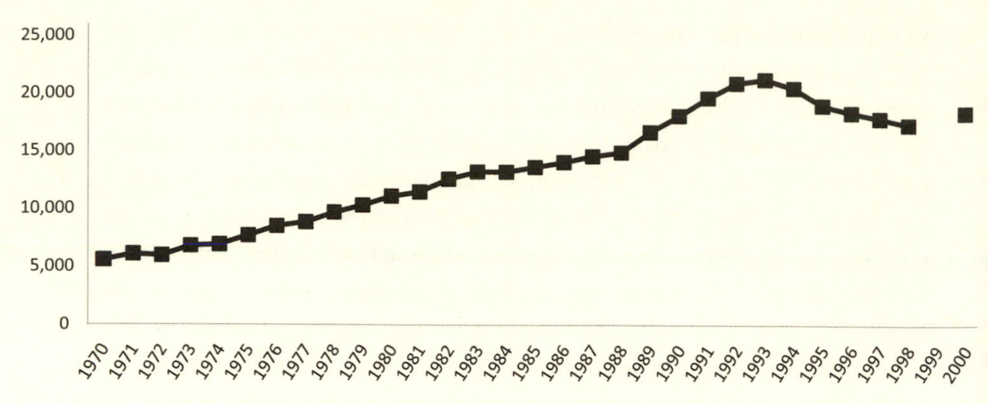

Fig. 3.7. Number of bachelor's degrees awarded to women in political science, 1970–2000. Adapted from Susan T. Hill, *Science and Engineering Degrees: 1966–2000,* NSF 02-327 (Arlington, VA: National Science Foundation, Division of Science Resources Statistics, 2002).

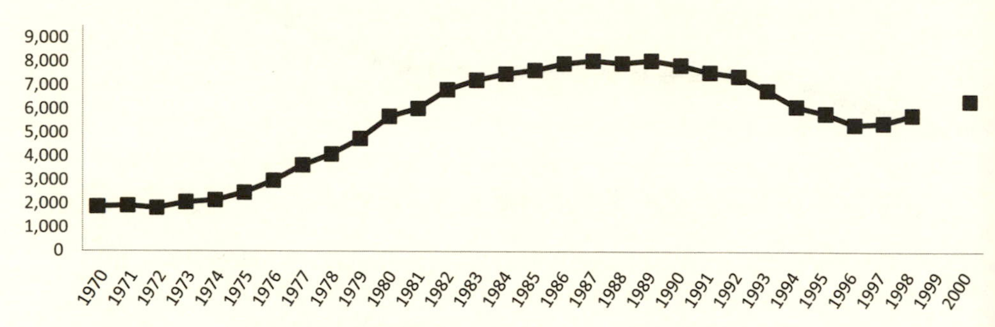

Fig. 3.8. Number of bachelor's degrees awarded to women in economics, 1970–2000. Adapted from Susan T. Hill, *Science and Engineering Degrees: 1966–2000,* NSF 02-327 (Arlington, VA: National Science Foundation, Division of Science Resources Statistics, 2002).

data not only continued but even gave way to actual drops in numbers and percentages in some fields (computer science especially) after 1985, this led to real concerns, almost a crisis, at the federal level. Several dramas were under way.

For example, the rapid rise in the numbers of women earning degrees in the geosciences (earth, atmospheric, and oceanographic sciences) in the 1970s led to attention and reforms at the undergraduate level. Kathleen Crane, later an oceanographer of note, recalled how in the early 1970s she was caught in the double bind not unusual for women students of the earth sciences at the time. To complete her major in geology at Oregon State University in 1973 she had to complete a field-camp component, but women were forbidden to take part in the outdoor program. Nevertheless, she persevered and improvised and was eventually allowed to meet the requirement with a summer course in the Virgin Islands. At the University of

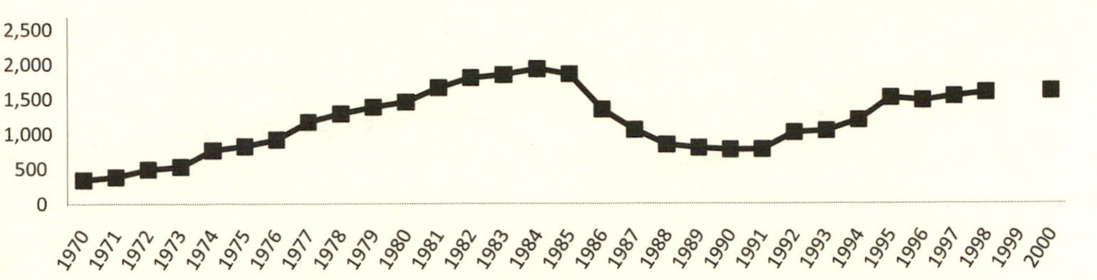

Fig. 3.9. Number of bachelor's degrees awarded to women in earth, atmospheric, and oceanographic sciences, 1970–2000. Adapted from Susan T. Hill, *Science and Engineering Degrees: 1966–2000,* NSF 02-327 (Arlington, VA: National Science Foundation, Division of Science Resources Statistics, 2002).

Michigan, the portion of women geology majors doubled from one-sixth to one-third between 1975 and 1980, leading the local paper to headline its story "The Women Are Coming!"[37] When the increase not only stopped but, as shown in figure 3.9, was followed by a steep decline in the 1980s, the old practices were not reinstated, but neither were new recruitment programs undertaken, as happened in engineering and would later happen in computer science. The rise and sudden drop were instead shrugged off as part of the traditional boom-and-bust cycle in the oil and energy industries, where the last hired (the young women) had been the first fired. With so much unemployment, it was not wise to train many more any-time soon (see also chapter 1). Before long the numbers were rising sharply again.

But two other fields—mathematics and computer science—attracted the most attention and spawned a special bewailing literature of their own. The refrain "why so few?" was invoked even when the numbers (and percentages) were not so low. In the field of mathematics, for example, the undying assumption that "girls can't do math" persisted even though they had constituted 40 percent of the majors before 1970 and still made up about 40 percent of a much lower total number in 2000 (fig. 3.10). The drop-off in the numbers of both men and women was noted, and efforts were made to publicize the downturn as a kind of national issue, but as in geology, it did not blossom into a full-blown crisis. Perhaps it was felt that all the potential math majors had gone into other technical fields (including possibly engineering). But what did bother at least one leading woman mathematician was that too few of these majors were going on to graduate school in mathematics. Disturbed by this situation, Karen Uhlenbeck, of the University of Texas, the most prominent woman in the field in the United States after the death of Julia Robinson, and others seeking to inspire undergraduate women to become full-fledged mathematicians secured funding to start a different sort of math program for women in 1993. A ten-day residential program of lectures and activities held each May at the famed School of Mathematics at the Institute for Advanced Study in

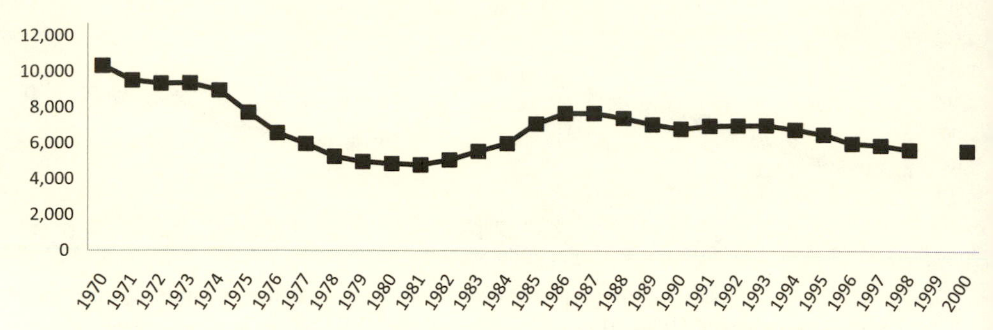

Fig. 3.10. Number of bachelor's degrees awarded to women in mathematics, 1970–2000. Adapted from Susan T. Hill, *Science and Engineering Degrees: 1966–2000,* NSF 02-327 (Arlington, VA: National Science Foundation, Division of Science Resources Statistics, 2002).

Princeton, New Jersey, it was attended by about sixty undergraduates, graduate students, and others from all over the country. This program, as well as one held annually in June at Park City, Utah, allowed potential math majors and future mathematicians to spend several weeks together getting acquainted, doing math problems, and preparing presentations for the larger, more male-dominated and intimidating audiences at traditional mathematics meetings.[38]

The field of computer science was in more serious difficulty starting in 1986. In 1988, after more than a decade of sharply rising enrollments and degrees to both men and women, Betty Vetter, who was making a career out of interpreting myriad manpower statistics, was among the first to point out that not only had the numbers of women earning bachelor's degrees in computer science leveled off, they had actually begun to drop (fig. 3.11). Shortly thereafter Erich Bloch, director of the National Science Foundation (and formerly of International Business Machines [IBM]), asked Nancy Leveson, then at the University of California–Irvine, who was also on an NSF advisory committee, to write a report including recommendations for NSF action. Since there was at the time no women's organization in computer science to ask for suggestions, she issued an open call for advice via a computer-science newsletter. With about fifty responses from men as well as women, she presented her findings.[39]

In her report Leveson mentioned briefly an array of projects already under way, focusing especially on the one best known to her respondents, that at UC–Berkeley, where Sheila Humphreys had expanded beyond electrical engineering into computer science in 1983. In particular, with the support of the dean of electrical engineering she had established and obtained corporate support for a reentry program for women and minorities in computer science and electrical engineering. Up to twenty students per year undertook personalized programs of at least two courses per semester to prepare them for graduate work in the field. In 1994, after a decade

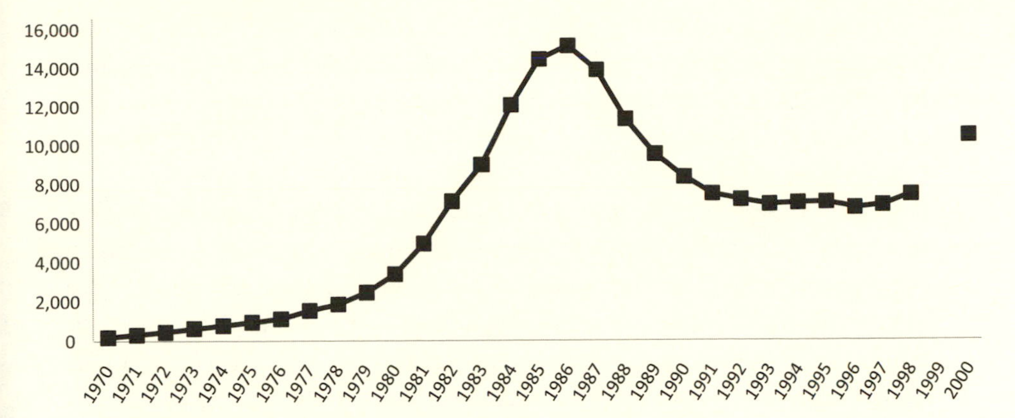

Fig. 3.11. Number of bachelor's degrees awarded to women in computer science, 1970–2000. Adapted from Susan T. Hill, *Science and Engineering Degrees: 1966–2000,* NSF 02-327 (Arlington, VA: National Science Foundation, Division of Science Resources Statistics, 2002).

in operation, about forty women had gone on to earn master's degrees in the field at Berkeley and elsewhere. At the undergraduate level, Humphreys had helped students form the Association of University Women in Computer Science and Electrical Engineering (AUWICSEE), which got some support from the student government on campus to hold events, help recruit more women, build solidarity, and raise morale.[40] But unlike the Society of Women Engineers, which was flourishing on more than a hundred campuses, the AUWICSEE did not become a national organization. On most campuses the women students in computer science were intimidated and lacked the solidarity and political skills to protest daily indignities. Rather than organizing and fighting back in a field requiring some militancy, they just gave up and transferred out.

Among Leveson's numerous recommendations for the NSF were the support of summer and after-school programs for precollege women; graduate fellowships specifically for women in computer science; more reentry programs; special grants for women faculty; funds for home computers; support of women speakers; local or regional workshops with women participants; and the formation of a women's group in computer science. Shortly after she submitted her report, the NSF supported the formation by the Computer Research Association of a committee on women (CRA-W); C. Diane Martin organized a workshop on gender and the teaching of computer science in June 1990; and Amy Pearl, of Sun Microsystems, edited a special issue on women and computing of the main journal in the field, the *Communications of the A[ssociation for] C[omputer] M[achinery],* in November 1990. This burst of interest and activity culminated in 1994 in the large Grace Hopper Celebration, held in Washington, DC, which brought large numbers of women in computer science together for the first time (see chapter 8).[41] Then in 1995 the biggest curriculum-reform project of the decade got under way at Carnegie Mellon University, one of the nation's top universities in computer science. With support from the Alfred P. Sloan Foundation, the NSF, and the Spencer Foundation, the psychologist Jane Margolis and the assistant dean, Alan Fisher, started an innovative four-year program to recruit and retain women in its computer-science major. One of their first steps was to persuade the faculty to drop the admission requirement of extensive previous experience in programming, which served as a kind of "filter" that kept most women out and yet over time turned out not to be critical. By 2000 the proportion of women majoring in computer science at Carnegie Mellon had grown from 7 percent to nearly 40 percent.[42]

Yet Carnegie Mellon remained an isolated case, and nationwide the numbers of women earning bachelor's degrees in computer science continued to drop. In fact, after the "dot-com bust" of 2000 and the wide publicity about the movement of programming jobs offshore, the numbers plummeted for both men and women. In 2004 the billionaire Bill Gates, of Microsoft, made a personal tour of five major campuses, imploring students to major in computer science and help with the challenges ahead. But still the numbers dropped. In fact the lack of women in computer science began to be something of a national crisis, and in the spring of 2005 the NSF started a new $14 million program—Broadening Participation in Computing—to tackle the problem. Headed by Janice Cluny, of the University of Oregon, a former chair of CRA-W, the program gave large grants to a variety of institutions to develop innovative curricula. One award went to the University of Maryland–Baltimore County, whose new Center for Women and Information Technology ran summer and after-school programs for middle-school girls and reserved space in a residential hall for female undergraduate computer majors. Another went to Georgia Tech, which in a joint program with Bryn Mawr College taught students how to program a robot. Some of the women got so interested that they did the unforeseen—stayed after class and figured out how to make the robot dance, a sign, perhaps, that if changes were made in standard class projects, more girls might come.[43]

Conclusion

The process of "going coed" at the undergraduate level in science and engineering education was not initially intended to do much for women students, let alone transform them or the science profession. It had the rather limited goal of helping to salvage certain traditional institutions that were facing dwindling enrollments, such as the engineering colleges, the schools of agriculture, the men's colleges, and the HBCUs. The coming of women accomplished this limited purpose and saved many such institutions from merger or extinction. But it also had important unintended consequences for women and for science and engineering. It opened the doors to greater opportunities for women and over time gave rise to efforts to change the colleges themselves. At first it seemed simple and easy to recruit and admit women students. But once the women came, their presence and awkward experiences demonstrated the necessity for further, even more fundamental changes within their institution, chiefly the appointing of junior women administrators, who might then begin to work from within the power structure, and in time the hiring and tenuring of more women faculty. Entrenched resistance and inertia—"institutional sexism"—oftentimes necessitated special programs and funds, which for a time were made available, and a stronger commitment to full coeducation by fresh administrators. Over time the results were substantial and would in turn justify even more changes at academic institutions in the 1990s.

In short, in the period 1970–2000 doors for students opened initially and voluntarily because administrators and boards of trustees believed such a drastic change to be in the institutions' overall best interest and because employers were under pressure as federal contractors. Record numbers of women applied, matriculated, and even graduated. But changing college admissions was just the beginning and perhaps the easiest step of a long process, for many faculty and institutions resisted further modifications. Many individuals devoted much effort to setting up programs that encouraged women students to persevere: they volunteered or got the programs funded by a variety of industrial, philanthropic, and, after the mid-1980s, government sponsors. The female students' presence and needs also justified making inroads into a hostile male culture or changing it into a more receptive one. It was an ever-broadening mission that stretched upward to the faculty, deans, and even presidents.

Making much of this continuing growth in female enrollments possible was a burgeoning new activity, if not actual industry. Many newly formed women's groups started outreach efforts and projects in the 1970s and after to get out the message to high-school girls, their parents, and their younger sisters that that it was now acceptable and prudent to study math in high school and even contemplate and complete a college major in a physical science—including chemistry, mathematics, geology, and to a lesser extent physics and astronomy—or even engineering, computer science, or agriculture. The federal government wanted more women trained, employers were looking to hire them, and most colleges had removed their most blatant barriers and were competing for good students. Yet tradition, stereotypes, and caution still deterred too many young women. The next target was girls, K–12.

CHAPTER FOUR

Innovative Outreach

Expanding Girls' Options and Opportunities

The remarkable increases in girls' enrollment in science and engineering in the 1970s and thereafter did not just happen by themselves. They grew out of the changed climate of the early 1970s, including the new federal equal-opportunity legislation, and reflected the increased opportunities for coeducation at so many former men's colleges. But putting this message of expanding options into practical everyday terms and providing help on how to position oneself to take advantage of them was the work of another branch of the women's movement. It involved new organizations and coalitions that were more akin to the collegiate women-in-engineering and women-in-science programs mentioned in chapter 3 than to the women's caucuses that, as discussed in chapter 1, were sprouting up in the 1970s and taking on the power structure of their disciplines. This new locus of feminist energy led to the spontaneous emergence of many innovative women-run organizations devoted to broadening girls' awareness of both their expanding career options and the mathematical preparation that they would need to fulfill them. With dedicated leadership, feminist insights, modest and intermittent corporate or governmental support, and the largely volunteer zeal of thousands of individuals, these groups formed a movement that reached out to young women, including those in the inner city, to alert them to the new possibilities in the world of science and engineering. Sensing a new era of opportunity, accepting the necessity for dual-career marriages and incomes, perhaps recalling their own early struggles and discouragements, and worrying that their daughters might have to face the same experiences, many women scientists and engineers found the energy to work together on a variety of projects. These programs varied from single presentations to annual day-long workshops such as career days and award ceremonies, to after-school, Saturday, and summer programs, to semester-long reentry programs, to major media and museum exhibits. Some were free, some cost money (especially the summer camps), and some paid a stipend (e.g., the reentry courses). They all got publicity, attracted girls, their parents, and their sisters, and urged them to consider and prepare for nontraditional careers. Several of these groups had ties to the women's colleges, which were facing their own set of challenges (see chapter 5).

These activities were largely altruistic, but they also held the women's groups together with a yearly round of meaningful activities that brought satisfactions and sustained bonds of friendship and affection. In short order they created a whole new realm of activity that had barely existed before. They started in the 1970s with modest foundation support, flourished in the late 1970s, when the NSF's Women in Science Program (WIS) funded several of them, and surprisingly managed to survive the abrupt termination of that program in 1981. They then expanded greatly with the revival in federal support of science education after the 1983 report *A Nation at Risk* and the 1986 Neal Report, of the National Science Board, which recommended major expansion of such programs. Under the rubric of summer camps and "informal education," a wide variety of organizations launched many activities aimed at young and high-school girls. Although it was hard to evaluate the long-term impact of any one program, in time the numbers of young women going to college and earning bachelor's degrees in science and engineering shot up dramatically. Something changed the prevailing "climate of unexpectation," as Mary Bunting had called it back in the 1960s. The proliferation of these new programs may have made the difference for many of these newly motivated young women who began to choose more math and science classes in high school and even to major in them in college. These programs varied in subject matter, but more mathematics was their main message.

Math Mobilization

The movement to recruit more women into science put particular emphasis on mathematics education, where there quickly emerged a nearly insatiable demand for encouragement and innovative teaching. Yet it all started quietly. In 1973 a graduate student in sociology, Lucy Sells, was working on her Berkeley dissertation, which highlighted *math filters,* or women students' practice of taking so few math courses in high school that most of the majors offered at Berkeley, with its large engineering college, were not open to them.[1] (Others would term this *math avoidance.*) As part of her effort to document this phenomenon, Sells went up to the Lawrence Hall of Science, a center for science learning dedicated to the memory of the physicist E. O. Lawrence situated high above the Berkeley campus, to collect more data on the participation of men and women in its math and science outreach activities. There she met the staff member Nancy Kreinberg, a former English major and political activist, who was not as fatalistic as many others about the inevitability of this prevailing pattern and even saw an opportunity in Sells's idea: if someone was seeking a way to broaden women's future prospects, teaching them math was a promising place to start (or "intervene"). Soon thereafter Kreinberg,

Diane Resek, and others on the staff of the Lawrence Hall of Science organized a program of afternoon and Saturday courses called Math for Girls for those aged 6–14. By 1977 this had expanded into a program for hundreds of local math and science teachers, called EQUALS, which still exists.[2]

At about the same time the mathematician Lenore Blum, on the faculty at Mills College, just a few miles away in Oakland, began a campaign to teach all of its 850 students precalculus and more. In 1975 she, Kreinberg, and others joined forces and formed the Bay Area Math/Science Network. With the help of Jean Fetter, a former AAUW fellow in astronomy and then head of the Danforth Center for Teaching and Learning at Stanford University, and a grant from the Syntex Corporation, they recruited enough local women in industry, government, and higher education to hold a career day at Mills for area high-school girls in April 1976. It included what became standard elements at such events—a plenary lecture, hands-on demonstrations, box lunches, practical career advice, and an evaluation form. This event was a great success, and they soon repeated it, calling it "Expanding Your Horizons." The next year they arranged to hold four such events simultaneously at different locations around the Bay Area, attracting a total of two thousand people, and by 1980 the events took place at up to fifteen locations, with four thousand students and two thousand parents participating. In the late 1970s they got grants from the Carnegie Corporation of New York, the San Francisco Foundation, the federal Women's Educational Equity Act Program, IBM, and other foundations and companies, which allowed them to set up a resource center at Mills and hire a staff person to enlarge the project. The Network experienced a low point in 1982, when the Carnegie grant was not renewed and the leaders had to ask for personal donations. After that, however, they developed a marketing plan that turned their soft-money operation into a trademarked national, even international, phenomenon—EYH—that still exists at 105 locations and has attracted more than 575,000 persons. It filled a need, generated positive publicity, and attracted volunteers, whose employers were often supportive. In short, it was innovative, efficient, effective, and organized so that even busy scientists could replicate it anywhere.[3]

Another large effort of the 1970s and after was the Women and Mathematics (WAM) initiative, run by members of the Mathematical Association of America (MAA), an organization composed primarily of college math educators. In 1974 the members of its committee on visiting secondary-school lecturers wished to expand its modest program, but the association had no funds for this. Meanwhile, officials at IBM had noticed that there were no girls at the USA Olympiad, a contest for high-school students started in 1972, which it supported. In 1975, convinced that despite the obstacles and inertia there was important work to be done, the ubiquitous Mary Gray, by then the past president of the Association for Women in

Mathematics, wrote a grant proposal that would support MAA outreach activities in a few urban areas, such as New York City, Chicago, and San Francisco. IBM awarded this project $7,500 and renewed it annually at various levels for nearly two decades. Several other foundations and corporations, including General Electric, John Hancock, GTE, and Hewlett-Packard, among others, supported WAM by providing visiting speakers to tenth-grade girls in urban and suburban high schools. The reasoning for focusing on tenth-graders was that the spring of the sophomore year was a key point, since too few girls were electing to take mathematics classes in their junior and senior years, thereby reducing or even eliminating their chances for further training or careers in many fields. A series of dynamic MAA members ran the program: Eileen Poiani, of St. Peter's College in Jersey City, New Jersey (a Douglass alumna and Rutgers PhD), Carole Lacampagne, of the University of Michigan at Flint (and a graduate student at Teachers College), and Alice J. Kelly, of the University of Santa Clara in California. Utilizing a large number of volunteers and broadening its coverage to other parts of the nation when it got enough support (about seventeen localities at its height in the late 1980s), the program's visiting speakers eventually reached more than twenty thousand tenth-grade girls and two thousand teachers and guidance counselors each year. At some point the MAA, which had initially been reluctant to start such a program, began to charge a substantial, 35 percent overhead, and so paid some of its own staff out of this new revenue stream. In the mid-1990s IBM decided not to continue its grant, and the once so successful program shrank considerably.[4] But by then the disparity in math preparation and achievement of high-school boys and girls had diminished to insignificance.

In addition, in 1978 a new term, *math anxiety,* entered the nation's lexicon when Sheila Tobias, a feminist writer and former associate provost at Wesleyan University, which went coed in 1968, published *Overcoming Math Anxiety,* an analysis of a condition she had been studying there ever since she first learned of Lucy Sells's *math filter.* She had noticed that Wesleyan's new coeds largely avoided math courses and majors that required calculus. These students were not "stupid," as some, including the students themselves, might claim, but anxious and apprehensive and often the victims of past intimidation and impatient teaching. Once identified, this condition could be faced and overcome, as shown by examples taken from a "math clinic" Tobias and others set up at Wesleyan. Her popular formulation took the concept out of the specialized realm of math educators and brought it to the attention of the broader adult reading public in *Ms.,* the *Atlantic Monthly,* the *New York Times,* and elsewhere. Anyone, even those in high places, could have math anxiety, and many men as well as women admitted that they did.[5]

Meanwhile, the new feminism was inspiring more sophisticated research on women and mathematics, supported by the small Women in Science Program at the NSF and the DHEW's new National Institute of Education, established in 1972. Among the studies revitalizing the field were those by Elizabeth Fenema and Julia Sherman, of the University of Wisconsin, who showed, among other things, that boys' perennially higher scores in high-school math achievement tests, which had previously been cited as evidence of their greater ability, were indeed largely and merely related to their having taken more high-school math courses than girls. They and others (Lynn Fox of Johns Hopkins, John Ernest of the University of California–Santa Barbara, and Edith Luchins at Rensselaer Polytechnic Institute, who studied 350 male and female members of the Association for Women in Mathematics, the largest population of women mathematicians ever available) generally documented sex differences (female deficiencies) in course-taking and test scores and attributed them to environmental factors, such as a lack of parental or teacher encouragement. These in turn became still more evidence that corrective intervention programs were needed. The researchers appeared in sessions together, for example, at the annual meeting of the AAAS in Boston in February 1976, which resulted in a published book and a subsequent report to the Ford Foundation on the current state and future needs of research on women and mathematics.[6]

At the same time, the leaders and members of the independent AWM, who had spent most of the 1970s raising consciousness and solidarity among women in and around university mathematics departments, provided some outreach and responded to key issues. In 1974 the AWM, whose central office was housed at Wellesley College, like Mills a women's college, began its Speakers' Bureau, which listed those members, mostly women, who were willing to give a lecture on a mathematical topic at a high school or college. This became one of the AWM's most influential activities. In 1982 the Alfred P. Sloan Foundation awarded the association twenty thousand dollars to support the bureau's activities, but its participants were so frugal in their expenses and so often donated any honoraria back to the fund that the AWM needed an extra year to spend it all. In 1985 some AWM members started local annual Sonia Kovalevsky Days (held preferably in January to celebrate the birthday of the pioneering nineteenth-century Russian mathematician), when they gave awards to female persisters and high achievers in area high schools and encouraged all girls to study more mathematics. For a time this was supported by funds from the National Security Agency.[7]

The AWM leaders also played a watchdog and spokesperson role. For example, in 1980 *Science* magazine published a report on the program for gifted students at Johns Hopkins run by Julian Stanley and Cecilia Benbow, who claimed to have shown that the main reason why so few of the 14-year-olds who were outstanding in mathematics, as measured by the Scholastic Aptitude Test in mathematics, were girls was simply genetic. Indignant, the past presidents Alice Schafer and Mary Gray fired off a scathing letter to the editor (which was printed as a signed editorial) in which they claimed, famously, that "anyone who thinks that seventh graders are free from environmental influences can hardly be living in the real world." Later, in 1992–93, AWM leaders also joined a media protest led by the AAUW when a new speaking Barbie doll said, among 270 messages, "Math class is tough."[8]

Meanwhile, a fourth group, members of the National Council of Teachers of Mathematics, largely high-school and elementary teachers and faculty at colleges of education, got interested and involved in the issue. Enthusiasm ran so high at a session on girls and mathematics at the NCTM's annual meeting in San Diego in April 1978 that attendees decided to set up a group of the their own, first called the Association for the Promotion of the Mathematics Education of Girls and Women, later shortened to Women and Mathematics Education (WME). Judith Jacobs, of George Mason University, organizer of the San Diego session, became its first chairperson. When one of its initial activities was a speakers' bureau, leaders of the AWM became alarmed at its duplication of their efforts. Some sort of merger or affiliation was discussed warily, but the WME went its own way, focusing primarily on the NCTM's members and regional and national meetings. Since its members were the nation's more motivated math teachers and their professors, increasing their awareness and providing them with tools to encourage more girls to study mathematics could have a substantial impact within the nation's classrooms.[9]

In fact, a 1981 "special report" to the Rockefeller Foundation about who would do science in the future by Sue E. Berryman, of the RAND Corporation (not to be confused with Susan V. Berresford, of the Ford Foundation) reported succinctly that "the key for women is pre–high school interests. These trigger an educational sequence that ultimately results in their underrepresentation among quantitative doctorates. . . . A strong preparation in mathematics in high school preserves the options of entering a college science major and a post-college quantitative career. Ironically, the high school tradition of offering more advanced mathematics as electives interacts with women's lesser interests . . . to foreclose these options to them. Removing choice during high school would preserve it after high school."[10] Even without Rockefeller support, all this (and other) work seemed to pay off, for the gaps between boys' and girls' achievement in high-school mathematics began to narrow. By 1989 the former discrepancy in achievement levels in mathematics between male and female eighth- and twelfth-grade students had been eliminated, and it was still insignificant in 2000.[11]

In fact in 1998, after twenty-plus years of operation, some criticisms of its practices, changes in its personnel, and a big increase in size (due to increased foundation support), the team representing the United States in the International Mathematics

Olympiad began to include a few girls. One in particular was Melanie Wood, of Indianapolis, who joined the team as a ninth-grader and began to win prizes, eventually including an international silver medal. So many of the other girls on American teams in this and other (lesser) international math competitions were Asian Americans or Romanian- or Bulgarian-born that their ethnicity began to attract notice. Several others were home-schooled or, like Melanie, children of mathematicians.[12]

Other Activities for Collegiate, High-School, and Younger Girls

Besides the high priority given to the subject of girls and mathematics in the 1970s and after, there were many other interventions and activities under way, some of which attracted foundation support and a certain media attention.[13] The media, themselves employing more women and moving toward lifestyle and human-interest stories, spread information about women and girls, particularly "firsts" and local activities. Parents and teachers learned of the new opportunities and encouraged girls' interest in science and engineering majors. Attracting national attention, especially in the *New York Times,* were those winning prizes in the annual Westinghouse Science Prize (renamed the Intel Science Talent Search in 1999), run by the Science Service. The annual publicity did not mention what one later study found, namely, that many of the women winners of the Westinghouse "talent search" did not persist in science. Five years after winning the 1983 prize, 41 percent of the females but only 12 percent of the males had left science. The Cornell anthropologist Rada Dyson-Hudson, a Westinghouse winner in 1947, was one who did persist.[14]

More females began to attend and graduate from the nation's technical high schools, such as the Bronx High School of Science, which first admitted girls in 1946 (the psychologist Naomi Weisstein graduated from there in 1957, as did the economist Claudia Goldin in 1963). Stuyvesant High School in Manhattan finally admitted women in 1969 (the physicist Lisa Randall graduated in 1980). In addition, several states and regions, especially in the South and the Midwest, started new, state-supported technical high schools or academies for gifted science students. The biochemist turned science educator Cecily Cannan Selby was among the founders of the first residential one (in North Carolina in 1980). Later the physicist Leon Lederman started one in northern Illinois. About ten other states and several regions followed suit.[15]

There was a lot of energy among those who sought to rid the curriculum of sexist (a new word then) materials. Several studies of math textbooks found that most math problems were about men and men's occupations, depicting the few females included stereotypically as nurses or dumb blonds. Most illustrations in science textbooks were of men or were of interest chiefly to them. The language was always about "he" and "him." In 1975 Lois Arnold, of Ossining, New York, documented the near total absence of any women in geology textbooks. In 1982–83 the women's committee of the American Association of Physics Teachers queried fifteen textbook publishers about their guidelines for depicting females; seven responded, and only five sent the guidelines. Others found similar situations in chemistry and astronomy books. Over time the situation seemed to improve somewhat as attention shifted to the more subtle area of classroom practices.[16] Teachers paid more attention to the boys, who were more disruptive in class and seemed eager to talk even when they did not know the answers. Thus when AAUW Educational Foundation started a new "initiative for educational equity" in the 1990s, it found plenty to deplore. Its first report, *How Schools Shortchange Girls*, prepared at Wellesley College's Center for Research on Women and published in 1992, had a chapter devoted to science and math.[17]

Some women scientists took an interest in the qualifications of high-school instructors. Margaret Law, of Harvard, chair of the Committee on the Status of Women in Physics (CSWP) of the APS in 1975, and others prepared a study of 349 female high-school physics teachers (members of the National Science Teachers Association who responded to a questionnaire) that showed that only about half (52%) had had any significant experience with physics before they had to teach it, and only 71 (20%) held any degree in physics. Male physics teachers had been shown to hold more degrees and to have chosen their vocation, but the women, who often also taught chemistry, biology, and general science, were taking additional courses in physics and were satisfied with their performance. When the editor of the *Physics Teacher* rejected Margaret Laws's article on the subject on the vague grounds that it was "unsuitable" for the journal's largely male readership, the physicist Eugenie Mielczarek, of George Mason University, who was active in the Chesapeake section of the organization, wrote a short summary of the article that was published.[18]

Other scientists and mathematicians wrote career booklets in response to the myriad inquiries that the newly established women's caucuses received from girls, teachers, and others. The level of sophistication and format—color, photographs, graphics—and the amount of information evolved rapidly. The mathematician Judy Roitman wrote an unillustrated, single-spaced text for women in mathematics in 1976 that proved quite popular, as it was cheap to produce and could fit into a letter-size envelope. Those running career days ordered it by the hundreds. Later others revised and updated it. Similarly, the physicists Laura Roth, of SUNY–Albany, and Nancy O'Fallon, of Argonne National Laboratory, prepared an informative career booklet on women and physics in 1977, but evidently it was less popular: guidance counselors did not bother to stock it, because so few girls asked about careers in physics. In 1981 Judy Franz, a former chair of the CSWP, obtained funds from the APS to produce a colorful packet with three booklets and a flyer for middle-school counselors and girls considering any sort of technical career, not just one in physics. It sold for three dollars.[19]

Starting in 1977 the NSF's Women in Science Program supported two programs for high-school girls, career workshops run by grantees all over the country and a program of visiting women scientists (VWS) run by the Research Triangle Institute of North Carolina. Each spring the institute recruited thirty women, to whom it gave a modest honorarium, to visit three to four high schools in their area for one day each, talking to a large group, to several smaller classes, and to individuals about opportunities and career prospects in their field. Hundreds of women signed up, and the program was deemed a success.[20] But it was terminated in 1979 by Congress, whereupon individuals and groups in some areas, such as several technical women at the 3M Corporation in St. Paul, Minnesota, who had been involved in the VWS program locally continued the activity on their own in an even larger way. By 1988–89 as many as 120–50 women from 3M were visiting local schools on company time and urging the students to persist in science courses. They also added a group of "Visiting Wizards," who went to elementary schools and did demonstrations, the most popular of which was cryogenic—freezing a banana and then using it as a hammer. Participants felt gratified by the favorable response and the publicity, which sometimes included their own children. One visit was described as "an eye-opener to some of our inner-city girls," and another by a pregnant woman scientist attracted many questions from the female students who wondered what the baby's birth might mean for the woman's job and career. The 3M effort also inspired women scientists at Honeywell, in nearby Minneapolis, to start a similar program there.[21]

In fact this sort of eager educational activism suggests that the rewards of such intergenerational outreach were mutual or reciprocal. The members probably sensed that increasing the numbers of future women scientists was in the long run in their own and their companies' best interests, and the specific activities not only helped expand the students' awareness but also provided the participants and their organization (such as 3M's Technical Forum) with a constructive activity that held them together year after year. Similarly, for example, when the Association of Women Geoscientists (AWG) was first organized in the late 1970s, one of its most successful activities was a career day run by its chapter in the San Francisco Bay Area. This gave employed women geoscientists, who were facing their own sets of intractable problems at the U. S. Geological Survey at Menlo Park, in local companies, and at Stanford University, UC–Berkeley, and other colleges in the area, a reason to get together, pool their talents, and work on an altruistic project to encourage young women to study geology and prepare for careers in the earth sciences, where for a time entry-level jobs abounded as a result of the energy crisis of the late 1970s. In the process, they could themselves feel some camaraderie, mutual support, and appreciation, and so with positive reinforcement from their peers in a sense broaden their own horizons.[22] Elsewhere the author of an article about Project WISE, an innovative program for high-school and college women at SUNY–Stony Brook that brought together women from Brookhaven National Laboratory, Cold Spring Harbor Laboratory, five nearby high schools, and the local AAUW chapter, talked explicitly and autobiographically of her pleasure in working with these like-minded women to put the program together.[23]

Other projects involved a series of presentations. One of these was TWIST (Training Women in Science & Technology), a joint project of the American Association of Physics Teachers and George Mason University and supported by two grants from the Fund for the Improvement of Post-Secondary Education (FIPSE) in 1982–84. Each Saturday morning for six weeks, thirty seventh-grade girls attended a workshop at George Mason, where they did what were coming to be called "hands on" activities and listened to a career presentation by a woman scientist from the area. On the seventh Saturday they toured a nearby corporate facility. The program exposed them to a variety of personalities and possibilities in computer science, mathematics, astronomy, engineering, physics (including health physics), and space science.[24] Along these same lines the educator Gail Whitney started her award-winning Saturday Academy in Portland, Oregon, in 1983 for 150 middle-schoolers, who visited local laboratories. When she retired in 2000 the academy was enrolling 2,000 students in grades 2–12 annually. On a smaller scale was the Auburndale, Massachusetts, Girls' Science Club, founded in 1991 by Maureen Castellana, a nurse and mother of two, who invited all her daughter's fourth-grade female classmates to start going to scientific activities together. They attracted much publicity in 1993, when as sixth graders, after preparing for several Saturdays, they entered a LEGO robotic contest at MIT and beat half the older entrants.[25]

Other programs were more intensive and might last two weeks or more and involve residence away from home, in itself a novel experience for many. Among these were relatively inexpensive on-campus residential and nonresidential summer programs. Some private schools (e.g., Miss Porter's in Connecticut), women's colleges (e.g., Douglass, Mount Holyoke, Hollins, Radcliffe, and the College of Notre Dame of Maryland), engineering schools (the Stevens Institute of Technology and Cornell University's college of engineering), and universities (Vanderbilt, the University of Michigan, and the University of Southern California, among others) ran summer programs for girls as young as eighth-graders, and others (the University of Michigan) offered internships in a woman scientist's laboratory for high-school girls.[26]

Some activities were so popular that in the new era of equal opportunity male students also applied. This could lead to an easily solved mini-crisis or to an official protest to the state government. For example, the independent Maria Mitchell Association, on Nantucket Island in Massachusetts, had had an annual summer program for college women for decades. The Yale astronomer Dorrit Hoffleit proudly reported in her autobiography that of the one hundred women who had been summer interns during her directorship of the program between 1957 and 1978, at least thirty-five, or almost two per year, had become professional astronomers. But when a male student applied in 1977, she was accused of "reverse discrimination" for accepting only women. After some reflection, she broke her longstanding tradition and accepted him, finding alternative housing for him nearby. The next year, her last, she accepted two young men.[27] Other, more expensive summer activities were camps in the great outdoors, where girls learned about computers or did ecology. One summer camp run by the Vermont Technical College, a state-supported institution, was so popular that in the 1990s a young man applied, was rejected, and then filed a complaint with a state agency claiming that he had been discriminated against. Eventually the Vermont Human Rights Commission upheld the college's right to run a women-only program as a way to attract women into traditionally male-dominated fields. By 2001 it had more than fifteen hundred alumnae.[28]

Among the many other projects termed "informal education" by the NSF (because they were not in classrooms or in camps) and largely funded by grants were several that tapped into traditional organizations for girls. Thus the Girl Scouts of America, which had started in the 1920s, developed several merit badges for mathematics and science (dubbed "math whiz," "science sleuth," and "computer fun"). In 1989 the American Association for the Advancement of Science began to teach troop leaders in three states (Minnesota, North Dakota, and South Dakota) enough science to implement the program there, and in 1993 the Franklin Institute of Philadelphia started a three-year, $1.5 million project to train 11,400 Girl Scout leaders in six big cities across the nation. Meanwhile, seven branches of the Girls Clubs of America ran Operation SMART (Science, Math and Relevant Technology), with informal scientific and computer activities after school for girls aged 6 to 18.[29]

Other traditional organizations also took part. Black churches ran programs for teenage girls. Science museums developed hands-on exhibits, and in 1985, with the help of an NSF grant, Chicago's Museum of Science and Industry prepared a traveling exhibit, "My Daughter, the Scientist," which included scattered bits of the history of women in science. Everywhere that it traveled, there were related lectures and workshops for teachers. Other museums, starting with the Center of Science and Industry, in Columbus, Ohio, in 1972 and continuing with the American Museum of Natural History, in New York City, which hosted 130 girls aged 9 to 12 in March 1989, held sleepovers in their planetaria or near the dinosaur exhibits. (Such events have since become regular revenue raisers.) More attention was also directed at girls in the science-for-children media. In the 1990s the bulk of the NSF budget for informal education went to the Children's Television Workshop, which had a program, the "Magic School Bus," for 2- to 5-year-olds with a female teacher who led adventurous hands-on activities and another, "3-2-1 Contact," for 8- to 12-year-olds. Thus many groups were working innovatively to integrate science education into the everyday educational, recreational, and entertainment experiences of female children and young adults.[30]

Feminism and Federal Support of Science Education: Roller-Coaster Resources

As may be evident from much of the above, a key factor in the level of activity in science education for girls was local inspiration and local funding by an array of corporations, colleges, and private groups. NSF support through WIS was also important early on (1975–81), but in 1981 President Ronald Reagan, who had campaigned to close down the newly created federal Department of Education, terminated the NSF's Science Education Directorate. Fortunately, this was only temporary, for at this low point, when all looked so bleak, Reagan's first secretary of education, Terrel Bell, adroitly reversed the situation by establishing his own National Commission on Excellence in Education, which in April 1983 issued the report *A Nation at Risk,* which urged the American people to take steps to reverse the widespread failings of American public education, including science education. As the result of a carefully arranged media campaign, the report was widely and favorably received, perhaps especially by those parents and civic groups (e.g., commissions on the status of women) that had become aware of and interested in issues relating to girls and math in the 1970s.[31]

Thus it is tantalizing to speculate that all the locally publicized women-in-science projects starting in the 1970s and partly funded by such federal groups as the NSF's Women in Science Project, FIPSE, and the WEEA, a brainchild of Congresswoman Patsy Mink (D-HI), but whose budgets were also often under attack, might have had something to do with the sudden revival of interest in and support for science education within the Congress, the White House, and the NSF in the mid-1980s. Yet there is little direct sign of any influence. Of the forty-one reports contracted for by Bell's commission, none had any ties to women-in-science personnel or issues. Of the many experts testifying at its six hearings around the nation, only one—Nancy Kreinberg, of the Lawrence Hall of Science at UC–Berkeley, EQUALS, and the Bay Area Math/Science Network—had any ties to this active sector.[32]

During Reagan's second term the National Science Board took up the challenge, appointing the physicist Homer Neal chair of a task committee on undergraduate science and engineering education, which issued its report in 1986. Of the forty-seven witnesses called to provide testimony, only the redoubtable Betty Vetter, of the Scientific Manpower Commission (as it was then called), and Carolyn Rozier, of Texas Woman's University, seem to have been at all connected to women-and-science issues. Nevertheless, the report called for a great expansion in funding for science education and outlined several areas, including "informal education," that needed support.[33] Congress agreed, as did subsequent presidents George H. W. Bush and Bill Clinton, and the corresponding appropriations for NSF's Directorate for Education and Human Resources soared in the 1990s, reaching the phenomenal $465 million in fiscal year 1992 (fig. 4.1).[34] In 1993 the NSF created a new Program for Women and Girls (PWG), with about $8.5 million in annual funding.

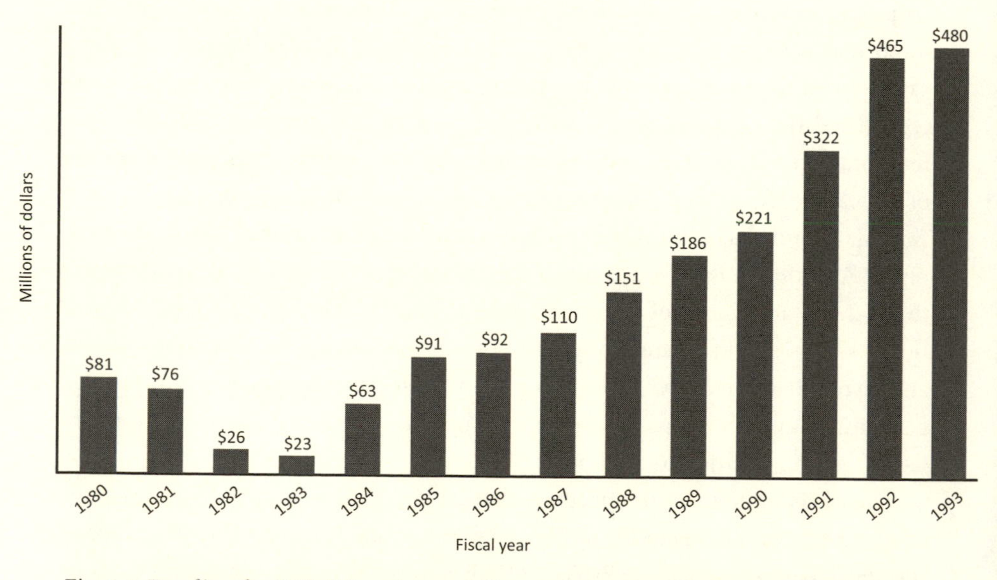

Fig. 4.1. Funding for NSF Education and Human Resources Directorate, fiscal years 1980–1993 (millions of dollars). Adapted from NSF, Directorate for Education and Human Resources, *Decades of Achievement: Educational Leadership in Mathematics, Science and Engineering* (Washington, DC, [1990]), 5.

(In 1998 this became the more politically acceptable Program for Gender Equity [PGE], to which men might submit proposals.) This big increase in funding benefited many of those already running local women-in-engineering or women-in-science programs, who could add a summer program or expand in some other direction. Although an NSF grant might not cover the entire cost of a new project such as a museum exhibit or a Girl Scout activity in a housing project, its reputation for high standards was a kind of endorsement that reassured other donors. Also, the NSF's peer-review processes generated and transmitted nationwide an ever higher level of expertise, which new projects were expected to attain.[35]

Conclusion

Starting in the early 1970s, thousands of women scientists began to reach out to girls and young women with the new message that it was now acceptable, practical, and even fun to study mathematics in high school and complete a college major in science—including mathematics, geology, engineering, computer science, astronomy, and physics. Such preparation could open the door to a rewarding and remunerative career. This unprecedented outpouring of energy was led by enthusiastic women who devised several innovative new organizations—Math EQUALS, the Bay Area Math/Science Network, Expanding Your Horizons, and Women and Mathematics, among others—to carry out this purpose. The organizations were usually based at traditional institutions, such as universities, women's colleges, and scientific societies, through which a few foundations could support them. The revival of the NSF's Directorate for Education and Human Resources in the late 1980s and 1990s greatly expanded the number and the size of many efforts, including those mentioned in chapter 3 at engineering colleges and coeducational institutions, which were also reaching out to high-school and even younger girls. Many of the women's colleges, to which we turn next, also began to see and use science education as a central justification for their existence.

CHAPTER FIVE

Using Science to Fight Back

Equal Opportunity at the Women's Colleges

The years after 1970 were difficult ones at many of the nation's women's colleges. Fewer than half of the roughly two hundred women's colleges in existence in the 1960s still held that status by the 1990s, for most had begun to admit men and thus had become coeducational liberal arts colleges. Yet many of the remaining ones (about ninety-three private colleges by the 1990s) determined to fight back and compete more vigorously than earlier for good students. This involved innovating in various ways—reaching out to reentry women; offering weekend, evening, and summer classes; and of special relevance here, featuring their science programs. When recruiting young women, many of the single-sex colleges began to publicize the accomplishments of their alumnae and to boast of their supportive environment. They also raised money to build new science facilities, pay competitive salaries, develop new majors (including a few in engineering), and award substantial financial aid. Several ran summer programs for women in science. As a result, although they enrolled fewer than 2 percent of the growing number of college women by the mid-1980s (but 2.5% by the mid-1990s because of the "Hillary factor"), they, like the historically black colleges and universities, retained a moderately important role for women in science. Many women scientists were alumnae, and some taught there or held administrative posts, which in the era of equal opportunity they could use as springboards to other jobs in the larger scientific and educational world.

These historic institutions faced at least two serious challenges starting in the late 1960s. First, they were stung by angry charges from their own alumnae, some of whom were the leaders of women's liberation, that they had not been feminist enough but instead had bowed to the back-to-the-home movement of the late 1940s and 1950s and trained women for their role as suburban housewives.[1] The second, greater economic and demographic challenge came from coeducation, when first Wesleyan University in 1968 and then a host of other previously all-male liberal arts colleges began to admit women, including, dramatically, fifty-seven who transferred from Mount Holyoke to Brown, Yale, and Princeton in 1972.[2] The Catholic colleges for women faced a third problem in that few young women were joining religious orders, and those that did preferred direct social action, such as running soup kitchens or shelters for the homeless, to college teaching.[3] All this left the leaders of the suddenly embattled women's colleges reeling. They hastily formed commissions to reconsider their role and identity. Many reaffirmed their mission as single-sex institutions for women and determined to fight back.

In 1969 Pauline Tompkins, president of Cedar Crest College in Allentown, Pennsylvania, held a meeting that was attended by representatives of about two hundred women's colleges. They quickly realized that although their institutions were competitors for a rapidly shrinking pool of students, they had a lot to gain by working together to hone their collective message and get it out effectively. In 1971–72, when the U.S. Congress was formulating the new Education Amendments Act, these women's colleges acted collectively to make sure that the legislation explicitly assured the continued existence of private, single- sex, undergraduate institutions, even if the public colleges that had historically admitted only women (e.g., Douglass, Mississippi University for Women, and Texas Woman's University) could not be justified. They also formed an interest group, the Women's College Coalition, in Washington, DC, under the aegis of the Association of American Colleges (AAC), to represent most of them. The coalition hired a part-time staff member to conduct studies and disseminate her findings to the press, which had recently begun to depict women's colleges as outmoded institutions. Yet in 1974, when the Department of Health, Education, and Welfare was preparing the guidelines for the enforcement of Title IX of the Education Amendments, the American Council on Education (ACE) turned to the economist Barbara Newell, president of Wellesley College, and not the fledgling Women's College Coalition, for a statement on how to implement equal opportunity at a women's college. She consulted many faculty and staff members at Wellesley and elsewhere and then contributed her statement: such colleges would admit women students only but would hire and promote without discrimination, pay equally, use unisex insurance tables, and otherwise be model employers. President Newell circulated her statement widely and was greatly praised for it.[4]

By the mid-1970s enrollments at the remaining women's colleges had started to rise again, fueled by a surge of premedical students,[5] prompting Bernice Segal, longtime chair of Barnard's chemistry department, to note in 1979 that while Yale had graduated just 4 women in chemistry, and Harvard and Princeton only 2 each, Wellesley had produced 27, her own Barnard, 21, and Mount Holyoke, 11.[6] When, however, two prominent women's colleges (Goucher and Wheaton) went coed in 1987 and 1988, respectively, pessimism returned. But the process was not inexorable, and in April 1990 the colleges' continued appeal was dramatically epitomized when undergraduates from Mills College, whose then president had recently advocated admitting men, not only held a two-week strike on campus but, even more effectively, cried on late-afternoon network talk television. That seemed to galvanize a certain recognition that indeed the students at these institutions cared deeply about them, which spurred their advocates to fight back, and the college reversed its decision.[7]

One aspect of the crisis of the late 1960s and early 1970s was that some alumnae and others, realizing how little they knew even about their own institutions, began to write histories of their colleges or, as Betty Friedan had done in the 1960s, to conduct statistical studies to find out what had happened to all the talented women of earlier decades.[8] Some scientists and social scientists went further and put their energies into defending and justifying these unique institutions. In particular, the physiologist M. Elizabeth Tidball, a Mount Holyoke graduate and alumnae trustee in the late 1960s, discovered about 1970 that the women's colleges had indeed in the past trained a disproportionate number of women who went on to earn doctorates and become scientists. She published several studies showing not only that Mount Holyoke had led in the number of women chemists that it had trained but that several other women's colleges had led in the number of women trained in nontraditional fields like physics and geology. Thus, contrary to what many persons thought (partly because many earlier studies of the backgrounds of American scientists had blithely omitted the women's colleges), many of these graduates had done rather well professionally.[9]

In 1976, together with her fellow Mount Holyoke alumna the physicist Vera Kistiakowsky, Tidball published an influential article in *Science* on the undergraduate origins (baccalaureate institutions) of women scientists.[10] This launched what became a mini-industry. The economist Mary Oates and the mathematician Susan Williamson extended these studies, which found a disproportionate number of graduates of the women's colleges in such indicators of achievement as listings in *Who's Who of American Women.* More recently, Norean Radke Sharpe and Carol Fuller have examined the baccalaureate origins of women scientists.[11] These studies offered persuasive evidence of the colleges' impact on talented women students in the past. Future graduates were expected to also become women of consequence, whether as a result of specific role models, mentors, or the college's advocacy of achievement. What better proof of the strength and effectiveness of their mission could there be than the actual achievements of the alumnae? Whether it would continue now that major men's colleges were open to women was uncertain, but most of the women's colleges embraced this rationale and featured their successful alumnae's achievements in their recruitment literature aimed at high-school girls and their parents, such as in airline magazines. It became their mantra that now that employment opportunities for women were expanding so noticeably, their documented success in training achieving women would be even more valuable in the years ahead. Their influence was greater than their enrollments might indicate, especially for women in science.

In particular as competitive pressures increased, some women's colleges used such alumnae fame and recognition to recruit young women. Chatham College, in Pittsburgh, for example, used its link to Rachel Carson, of the class of 1928 (who, ironically, had not been very happy there), to attract applicants interested in the environmental sciences. Mount Holyoke publicized its faculty's and alumnae's contributions in the sciences more directly, as in a booklet with historical as well as recent pictures of graduates and current students who were going on to graduate and medical school. Ambitious daughters of upwardly mobile immigrant parents, such as Elaine Chao, an economics major in Mount Holyoke's class of 1975 and later secretary of labor in President George W. Bush's administration, took note and came.[12] The *New York Times* helped in 1995 when it published an article on the many Wellesley-trained women economists in business and government. Students of Carolyn Shaw Bell and Marshall Goldman in economics, frequently the largest major at Wellesley, they formed an alumnae network in New York City and Washington, DC.[13] In 1999 the College of St. Catherine, the largest Catholic women's college in the nation, with campuses in both Minneapolis and St. Paul and more than four thousand students, made "women, science and technology" one of its four strategic areas.[14] Nancy Vickers, president of Bryn Mawr, boasted a bit provocatively at Yale's bicentennial in 2001 that routinely 35–40 percent of her seniors went on to do graduate work in the sciences, the highest percentage in the nation.[15] In March 2005 the Women's College Coalition had a full-page advertisement in the Sunday *New York Times* entitled "Women Can't Do Science?" and listing fifty-three women's colleges whose graduates had been "trailblazers in the sciences" over the last 170 years.

Besides forming a coalition, the leaders of the remaining women's colleges worked harder to make their own institutions more visible and distinctive. In particular they hired more women presidents (including scientists), mounted major fund drives, built new buildings, recruited more minority, part-time, and older students, and to a lesser extent started innovative curricula for women in science and even engineering. The result of all these realignments was that even though the total number of women's institutions continued to decline, the range of opportunities offered at the student and administrative levels was broadening at the remaining women's colleges.

In the 1970s the boards of trustees of the non-Catholic women's colleges began to hire more women presidents than ever before. More qualified female candidates were available, and governing boards, which themselves began to have more women members, were more willing to take a chance on them, especially when some financially imperiled institutions could not necessarily get the men they wanted for the post. In 1975 even Smith College, which had been headed by men for its entire first century, broke tradition and chose the historian Jill Ker Conway. By 1981 there were record numbers of women presidents—77 at the 118 women's colleges at the time—20 percent of whom were scientists. Although a third (26) were religious sisters, this number was declining, as laws in many states began in 1969 to stipulate that only colleges headed by laypersons could receive state aid.[16] Then as opportunities in academic administration at major research universities opened to women in the late 1980s and 1990s, there began to be movement into and out of the women's colleges. Some new presidents came from careers at coeducational universities, and several heads of prominent women's colleges, who might formerly have spent their entire career at that one institution, moved on to presidencies at larger, more prestigious, and more complex universities. Among them were the political scientists Donna Shalala, who moved from Hunter College to the University of Wisconsin (1988), and Nannerl Keohane, who moved to Duke University in 1993, after twelve years at Wellesley. A few new presidents of the 1980s and 1990s were African American women, leading to the new term *sister president.* Among them were Johnnetta Cole at Spelman in 1987, Carol Surles at Texas Woman's University (whose board of trustees admitted men to its undergraduate programs in 1994), and Ruth Simmons, formerly the provost at Spelman and the vice-provost at Princeton, at Smith in 1996. Simmons, a professor of French, was chosen as president of Brown University, the first African American chief executive officer in the Ivy League, in 2000.[17]

The presidency of Spelman, the Atlanta college for African American women, demonstrates another aspect of the changing racial ceiling in women's higher education. In 1976 Albert Manley, president since 1953, was retiring, and Marion Wright Edelman, distinguished alumna, prominent Washington attorney, and founder of the Children's Defense Fund, was chair of the board of trustees. After a lengthy search, the board voted, not for the finalist Jewel Plummer Cobb, a black woman biologist then at Connecticut College, but for Donald Stewart, an African American professor then at the University of Pennsylvania. Upon hearing this, the students blockaded the trustees for twenty-six hours. Although startled, the trustees nevertheless stuck by Stewart, who served for ten years and made many improvements. When he left for the presidency of the College Entrance Examination Board in 1987, the Spelman trustees, still headed by Edelman, decided that the time had finally come to hire a black woman and selected the anthropologist Johnnetta Cole, of Hunter College, a choice that was greeted with jubilation. After Cole stepped down in 1999, she was succeeded by two black women, first Audrey Forbes Manley,

MD, the wife of the former president, and then, in 2002, by the psychologist and acting president of Mount Holyoke College, Beverly Daniel Tatum.[18]

The case of the African American biologist and dean Jewel Plummer Cobb demonstrates one aspect of the interaction between the broadening of opportunity at mainline institutions, with, for example, more open searches for administrators, and the simultaneously shrinking number of top posts at single-sex institutions. Much of Cobb's career was built on a series of appointments as the first African American (professor, dean, president) at institutions that were ceasing to be women's colleges, a process somewhat akin to running up the down staircase, to borrow a phrase from a bestselling novel of 1965. Cobb had blossomed at single-sex Sarah Lawrence College in the 1960s, but after a divorce she had left there in 1967 (it admitted men in 1968) to become a dean of the college (i.e., of students) at Connecticut College for Women. This institution was also, however, in the process of admitting men (lots of them, to equalize the sex ratio quickly) and shortening its name. When before long its president retired, she was asked to apply for his position but was not appointed. Then, after being turned down for a deanship at the University of Michigan and the presidency of Spelman, in late 1976 Cobb accepted the tenuous position of dean at Douglass College, where her predecessor, the economist Margery Somers Foster, had resigned in protest over the ongoing evisceration of the college. Three years later after an "ugly" and eventually unsuccessful battle, as the *New York Times* vividly called it, for the deanship of the newly coeducational Hunter College, Cobb accepted the presidency of California State University at Fullerton in July 1981, the very month that the separate Douglass faculty was abolished and became part of Rutgers. Finally, fourteen years after leaving Sarah Lawrence, she had reached the top post at a stable and growing coeducational institution, though in the heart of conservative Orange County in southern California.[19]

Facing every women's college, its president, and its board of trustees was the subject of finances. With the energy crisis of the early 1970s and inflation later in the decade, even colleges that were not struggling needed more money. Trustees wanted their presidents to raise funds, and even quite wealthy colleges had major fund drives in the 1980s and 1990s. Since many of the alumnae had flourished financially and were increasingly willing to give to their own (rather than their husband's) alma mater, donations to the women's colleges increased dramatically. Fund drives accordingly had ever-higher goals. Smith launched a three-year drive for $125 million in 1987, raised its goal to $152 million, and then surpassed even that in 1990 with $163 million. In 1991 Barnard College set a goal of $100 million; ten years later it concluded with $163 million. In 1998 Chatham College announced a

Table 5.1. Women's colleges with largest endowments (in thousands of dollars), June 30, 2000

College	Endowment	College	Endowment
Wellesley College	$1,253,385	Barnard College	157,852
Smith College	906,942	Randolph-Macon Woman's College	147,144
Bryn Mawr	466,960	Sweet Briar College	111,212
Mount Holyoke	425,296	St. Mary's College (Indiana)	87,444
Agnes Scott College	408,141	Hollins College	86,308
Spelman College	219,754	Chatham College	64,467
Scripps College	206,994	Wells College	59,649
Mills College	175,911		
Simmons College	158,449		

Source: "Fact File: 569 College and University Endowments," *CHE,* Apr. 13, 2001, A40–A41.

goal of $21 million in three years, and by 2001 it had raised $31.7 million. In 2000 Wellesley College, which under president Nannerl Keohane in the 1980s had surpassed all previous fund-raising records, kicked off a capital campaign for $400 million, the largest to date by a liberal arts college; by 2008 her successor, Diana Chapman Walsh, had raised $470 million. Some Catholic colleges, such as St. Mary's of Indiana and the College of St. Catherine in Minnesota, also had record fund drives.[20]

Some women's colleges received spectacular individual gifts. Spelman College attracted national attention in 1988 when Bill and Camille Cosby gave it $20 million, the largest gift to a black college at the time, and told its new president, what must be rare at any women's college, to plan big. Then in 1992 the founder of *Reader's Digest* transferred $37 million in stock to the college. In 2000 the College of St. Catherine received an unrestricted gift of $20 million from its founding order, the Sisters of St. Joseph of Carondelet. Meanwhile, the value of the endowment of Agnes Scott College kept increasing as the price of its 4 million shares of Coca-Cola stock went up. In June 2000 a certain landmark was achieved when the endowment of Wellesley College, the largest of any remaining women's college, surpassed $1 billion (see table 5.1).[21]

All this new money led to many new and refurbished buildings, including especially "science centers," science laboratories, and even science dormitories. Among the new centers were one at Simmons College in Boston, completed in 1972, and that at Wellesley College, built in 1977 and expanded twice thereafter, which won an architectural award. Several others are listed in table 5.2. These "centers" offered certain efficiencies and opportunities by combining several (up to ten) formerly quite separate departments in one place, where they could share staff, facilities (e.g., a shop), equipment (including computers), and a library. Possibly there was some cross-listing of courses and cross-fertilization of ideas among the faculty in

Table 5.2. New science centers at women's colleges, 1972–2009

College	Date Completed	Size (sq. ft.)	Cost	Name of Building
Simmons	1972 + renovations	22,000	$6.7 million[a]	Park Science Center
Wellesley	1977[b]	130,000	$15 million	Wellesley Science Center
College of New Rochelle	1985 (started)	—	$10 million	Rogick Life Science Center and Science Hall
Chatham	2000	40,000[c]	$10.4 million	Science Complex
Spelman	2000	116,000	$34 million	Albro-Falconer-Manley Science Center
Hood	2002	68,000	$20 million	Hodson Science and Technology Center
Agnes Scott	2003	110,000	$36.5 million	Mary Brown Bullock Science Center
Mount Holyoke	2003	116,000	$36 million	Kendade Hall Science Center
Converse	2004	36,000	$10.6 million	Phifer Science Hall
Wesleyan (GA)	2007	42,000	$12.5 million	Munroe Science Center
Wilson	2009	76,500	$25 million	The Harry R. Brooks Complex for Science, Mathematics and Technology
Smith	2009	142,000	$73 million	Ford Hall

Sources: Clippings from *CHE;* college websites.
[a] Includes $2.9 million for renovations.
[b] Later additions, 1991.
[c] Renovations, 18,500 sq. ft.; new construction, 21,500 sq. ft.

their teaching or research as well. The new buildings also aided student recruitment. In 1977 Wellesley sent out flyers showing its new building to girls earning high math scores on the Scholastic Aptitude Test. However, the total number of science majors at Wellesley, which had risen sharply in the 1970s, did not increase markedly over the next two decades. The college did create a new department of computer science, introduce a major in psychobiology, and develop a joint course on evolution (geology, biology, and chemistry) and, in 1978, a forum on sociobiology.[22] A new building could also inspire the faculty to apply for research grants, which might include funds for new equipment. (Wellesley even created its Office of Sponsored Research in the early 1980s.) Starting in 1988 a few women's colleges—Barnard, Bryn Mawr, Mount Holyoke, Smith, Spelman, and Wellesley—also got substantial funds for undergraduate education from the prestigious Howard Hughes Medical Institute.[23]

One unusual new building was a dormitory specifically for female science students at Douglass College. The construction of the dormitory, named for two of its scientist-deans, Mary Bunting and Jewel Plummer Cobb, was part of a continuing campaign by women in New Jersey not to let the state-supported Douglass die. In 1970 Douglass was one of the nation's largest colleges for women, but it was struggling with the central administration of Rutgers University to maintain its separate existence and name in one form or another. This was a protracted process that involved many separate steps. The first step was taken in the 1970s, when Douglass's separate faculty was dismantled and reallocated to universitywide departments. (Interestingly, adding Douglass's four women chemistry professors to the Rutgers department, which previously had had no women, instantly propelled the combined department to the top of the American Chemical Society's list of doctoral departments with the most tenured women in 1971. It was still at the top—tied with UCLA at ten—in 2001.)[24]

Remarkably this drastic reorganization did not reduce the commitment of many to Douglass, for it retained its name, its deanship, its dormitories, and for a time its own admissions. After Cobb left in 1981, her successor as dean continued to introduce new programs for women, especially programs related to science and mathematics. In a sense the innovativeness in New Brunswick grew out of a determination to retain the name "Douglass" in one way or another. In particular, in 1986 Dean Mary Hartmann, a historian, started the Douglass Project for Women in Science and Mathematics, an umbrella for several programs headed by the historian Ellen Mappen and the biologist Carol Stuerks. One was the Mary Bunting Reentry Program, for women in mathematics and computers, which Bunting had initiated while president of the college in the late 1950s. Others were summer programs for high-school girls, internships at nearby industries, and other "co-curricular" programs on campus during the regular academic year.[25]

The most expensive innovation on campus grew out of Club Curie, a house inhabited by women science and engineering majors. Club Curie was so popular and oversubscribed that the college refurbished an older building as a separate women's dormitory for science majors and graduate students. (It may have been modeled on MIT's McCormick Hall, a dormitory built in the 1960s and open to women in all fields, which at MIT at that time would have meant mostly science and engineering students.) Since Douglass students majored in many fields, a special dormitory for science majors introduced a new kind of self-segregation. Officials obtained funding from the state of New Jersey in an unusual way. Among the college's many faithful alumnae throughout the state were several state assemblywomen, who introduced a bill that would provide money for women's studies and for special projects for women on campus. This bill, as well as support from the university, the New Jersey State Dormitory Authority, the NSF, and Merck & Company, of Rahway, NJ, provided funding for the project's special dormitory for women science majors. Opened in 1989, it housed one hundred women undergraduate and graduate students in science and provided in-house computer terminals, round-the-clock tutorial services, and a culture in which female science "wonks" might feel comfortable, make friends, and stay in science a while longer. It was successful to a degree, but its location far from the science laboratories and a reluctance among some students to live separately from nonscientists meant that, as with any dormitory, there was a certain turnover.[26]

Meanwhile the proportion of women on the faculty at many women's colleges was about 50 percent, lower than it had been earlier at some of them but still the highest anywhere in academia except at the remaining colleges of home economics (and even they were changing). (At coeducational institutions, women were usually just 10%–20% of the faculty.) This 50 percent ratio reflected the increase in the number of men hired and retained after 1945. Many men were hired in the 1970s and after, because, as Wellesley faculty explained it, women candidates were being be offered so many other opportunities at coeducational colleges and universities that the best applicants to the women's colleges were usually men.[27] But the distribution could be uneven, and by the late 1990s some departments, often departments of computer science and sometimes physics or chemistry, had no women, or at least no tenured women. For example, although three of the four tenured faculty members in Smith College's physics department were women, as were five of the thirteen in biology, there were no women among the six tenured faculty in its chemistry department. When a woman did come up for tenure in chemistry in 1996 and was turned down by the college, her subsequent lawsuit brought the college much negative publicity, such as an article in *Science* entitled "No Women Chemists at Women's College." This seemed to belie the college's own publicity about the great opportunities it was offering to women in science. The details were more complicated.[28] There were few women scientists of color at the women's colleges, except for Cobb, the psychologist Beverly Tatum at Mount Holyoke, and several at Spelman. The chemist Sheila Ewing Browne, at Mount Holyoke, was partly Native American.[29]

Many women's colleges created a department of computer science, expanded their geology staff, and added majors (but not departments) in biochemistry, molecular biology, or neuroscience in these years. Sister Mary Kenneth Keller, probably the first woman in the United States to receive a PhD in computer science (at the University of Wisconsin in 1965) founded the department of computer science at Clarke College, in Dubuque, Iowa, and chaired it for the next twenty years (the college admitted men in 1979). Lenore Blum started a department of mathematics and computer science at Mills College in 1974, and Wellesley started one in 1982 (although there had been a growing program there earlier). Wellesley's geology department, which dated from 1890, grew from one faculty member to five when enrollments skyrocketed in the 1970s, but then dropped to three when they subsided in the 1980s.[30]

Some other women's colleges began to move into the area of engineering in limited ways. Mills College started a five-year dual bachelor's-degree program with Stanford, Berkeley, Boston University, and later USC; Agnes Scott and Spelman both started a joint program with the Georgia Institute of Technology, also in Atlanta; Wellesley College started an exchange program with MIT; and Smith College started a minor in engineering in conjunction with the nearby University of Massachusetts at Amherst. The effort at Smith started because Smith's president, Jill Ker Conway, who served on the board of directors of IBM, thought more women should be trained for technical careers. When students found it too difficult to schedule classes on both campuses, Conway and the physicist Elizabeth Ivey made plans for Smith to start a program of its own. When in 2000 certain rules regarding the accreditation requirements for engineering education were loosened, Smith started its Picker Program in Engineering with considerable funding from the Ford Motor Company. It was not the usual four-year on-campus program but a two-year program of elementary courses that fed students into the more specialized upper-level ones at the University of Massachusetts. It attracted twenty students, none dropped out, and by 2004 Smith was awarding a bachelor of science degree in engineering.[31]

Although most women's colleges were rather slow to take up the emerging field of women's studies, Simmons College had a program by 1974. Barnard College housed the flagship journal *Signs* for a number of years, and its women's center hosted a series of important meetings under the title "The Scholar and the Feminist" in the 1970s and early 1980s. (At a meeting on biology in April 1977 Ruth Hubbard, of Harvard, met several historians of science, including Donna Haraway.) In 1981 Douglass alumnae in the New Jersey legislature got funding for a professorship of women's studies to be located at the college, and later Douglass created a center for women and politics and a women's leadership institute.[32]

Many on the faculty at the women's colleges were very interested in science education. Besides constantly improving and updating their courses and seeking funds for new and ever more expensive equipment, a few innovated in the science and especially the mathematics curriculum. Nationally known was Lenore Blum, of Mills College, a graduate of Simmons College in Boston, who, after earning a PhD in mathematics from MIT, had followed her husband to the Bay Area and held a lectureship at the University of California at Berkeley. When this was not renewed in the early 1970s, she accepted a position at the nearby, women-only Mills College. Finding very little interest among the students there in mathematics and computers, she started a set of courses that appealed even to "math avoiders," a new term at the time. Between 1974, when she arrived, and 1979 she doubled the enrollments in math offerings (from 331 to 698 out of a student body of 850), formed a department of mathematics and computer science, and obtained two grants from the Carnegie Corporation of New York to set up a resource center at Mills for the Bay Area Math/Science Network (see chapter 4).[33] It was a remarkable achievement.

Other departments also earned national distinction. In the mid-1990s the mathematics departments at Mount Holyoke and Spelman were cited by the Mathematical Association of America as having exceptionally effective undergraduate programs.[34] The Spelman mathematicians Etta Falconer and Shirley McBay also developed innovative mathematical programs in the 1970s. In the 1980s Sylvia Bozeman extended these to include a joint summer program for minority women with Bryn Mawr College, where Rhonda Hughes had joined the faculty.[35] The Mount Holyoke chemistry department continued to rank high among those departments training women who went on to earn a PhD. And Peter Beckmann, of the Bryn Mawr physics department, calculated that as of 1980 more than 5 percent of all the women members of the American Physical Society had earned at least one degree from that department.[36]

A few of the women faculty members at the women's colleges were elected or appointed to prominent positions in national scientific organizations and on federal advisory boards. Some of these appointments may have grown out of the first-ever meeting, in September 1979, of the heads of seventy women's colleges with DHEW Secretary Patricia Roberts Harris, arranged by the Women's College Coalition. Harris agreed to appoint a special assistant to deal with them and promised to appoint more of their women faculty to DHEW advisory committees.[37] Otherwise, the chemist Elizabeth Rock, at Wellesley, served on the Army Science Board for sixteen years after being nominated by a former student; the economist Carolyn Shaw Bell, also at Wellesley, testified before Congress, served on President Jimmy Carter's transition team, and had a monthly column in the *Boston Globe;* the Wellesley physicist Janet Guernsey served as president of the American Association of Physics Teachers; and the geneticist Evelyn Witkin, of Douglass College, was briefly the only faculty member at a women's college to be a member of the National Academy of Sciences (from her election in 1977 until 1979, when she became a professor at Rutgers).[38]

Into the 1980s the women's colleges had a certain niche and two superstars in national policy circles. Over the years, Jewel Plummer Cobb and Anna Jane Harrison held numerous positions in science policy, including terms on the National Science Board. In this regard they were following in the footsteps of earlier faculty women at women's colleges: Mary Bunting of Radcliffe, Mina Rees of Hunter, and Katharine McBride of Bryn Mawr all had served on the NSB. (See picture and Ch. 12.) Yet their service marked the end of an era, for after the mid-1970s the women on the board came from other institutions. Usually these professional and advisory activities took place off campus. Harrison's chemistry colleagues at Mount Holyoke carried her teaching load, and the administration even provided a part-time secretary, unusual at a small college. In the summer of 1983, while Harrison was president of the AAAS, she chaired an international meeting at Mount Holyoke on women and international technological development.[39]

The racial makeup of the student body at the women's colleges changed more dramatically than did that of the faculty in these years. Whereas formerly there had been very few black or minority faces in the yearbooks or alumnae magazines of these colleges, more were admitted in the later 1960s in response to the civil rights movement. Their proportion was higher at some of the women's colleges than at other institutions.[40] At Mount Holyoke the partly Native American chemist Sheila Ewing Browne even started two support groups: Native Spirit, for Native American students in the area, and Sistahs in Science, which hosted monthly meetings for women students of color.[41] The Mount Holyoke class of 1974 had forty women of color, or 7 percent of the class, while the Smith class of 1974, admitted in the fall of 1970, had sixty-nine, or about 10 percent of the class—both all-time highs. By the time the Smith class of 1974 returned for its twenty-fifth reunion, its black members could be proud of their individual personal success and eager to foster Smith's reputation. Perhaps their activism and support of the college were major factors behind its selection of an African American woman president, a first for the former Seven Sisters.[42]

When Cheryl Leggon sought to discover which undergraduate institutions had graduated black women who earned PhDs in the sciences between 1975 and 1992, she found that Spelman College had been by far the most "productive." Of 27 such doctoral bioscientists, 18 had their degrees from either Spelman or Bennett, the only two black women's colleges. Of the just 6 black women who earned PhDs in the physical sciences in these eighteen years, 4 were Spelman graduates and 2 had gone to Vassar. Of the 115 with doctorates in social sciences, 54 had graduated from Spelman or Bennett, 51 from the Seven Sisters (including Vassar), and 10 from Mills or Texas Woman's University.[43]

This relatively high proportion of women of color declined after 1975 because the price of tuition was rising and minority-group members often required a considerable amount of financial aid. There was also a hint in writings of the time that increased racial diversity was lowering the average test scores of college applicants. For example, in 1980 at Rutgers University male administrators bent on consolidating the Douglass faculty into a single unified Faculty of Arts and Sciences insinuated that the SAT scores of current students were not like those of past ones. Dean Jewel Plummer Cobb responded that the lower average reflected a nationwide trend; that the student body at Douglass had increased by 27 percent since

1965; that Princeton had admitted many top New Jersey women (as had the newly coeducational Rutgers itself); that the other women's colleges were recruiting more aggressively than Douglass (whose admissions were now handled—lukewarmly—by Rutgers); and that Douglass was a public state institution with programs in home economics, medical technology, and speech, most of whose students came from New Jersey. They and their scores should not be compared with those of students at elite, private, national women's colleges like Smith. The issue surfaced publicly in a 1985 article in the *Chronicle of Higher Education* that cited evidence that the average SAT scores of students at women's colleges had been dropping in recent years. (The phenomenon was also occurring at most liberal arts colleges across the nation.) In 1996 Smith College, with the support of a grant from the Howard Hughes Medical Institution, started its Peer Mentoring Program for Underrepresented Students in the Sciences. About twenty upperclassmen (white, of color, and international) receive training and read relevant materials and then meet monthly with their charges.[44]

Women's colleges also gave more than half of their honorary degrees to women. Mount Holyoke awarded several to scientists, including Vera Kistiakowsky, an alumna, and the biologist and NSF official Mary Clutter. From 1972 to 2000 Smith gave fourteen such doctorates to women scientists, but they were unevenly distributed over the twenty-nine years. Most years there were none, but in 1979 there were three, and in both 1980 and 1984 there were two, all during the presidency of Jill Ker Conway (1975–85), and then there were none after 1989. Among the recipients was Dixy Lee Ray, chair of the Atomic Energy Commission in the early 1970s, one of the few women to be a high government official at the time. Ray was a frequent recipient, being awarded at least nineteen honorary degrees, including eight from women's colleges. (The chemist Anna Jane Harrison, of Mount Holyoke, was another favorite, with a total of twenty, as was the university president and secretary of the Department of Health and Human Services Donna Shalala.)[45]

Several women's colleges had notable reentry programs. Women chemists at Rosemont College, outside Philadelphia, started a small reentry program for women chemists with NSF money in 1974. By 1991 it had forty-three graduates, including three PhDs and one MD, 80 percent of whom had been employed as chemists. Three nuns at Chestnut Hill College, in Philadelphia, and Edwin Weaver at Mount Holyoke also ran a few reentry programs in the 1970s. In 1975 Smith College started a different kind of transfer program: its Ada Comstock Scholars Program, for nontraditional women students. Many were low-income women, often with children, who transferred from community colleges. They lived in special dormitories, and Smith provided a staff to help them adjust to their new environment. More than 90 percent graduated, and some majored in the sciences. Mount Holyoke has a similar program for its Frances Perkins Scholars, and Wilson College has a Women with Children Program, which lets mothers house their families on campus.[46]

In other ways a few of the women's colleges offered support services to groups working to help the women's movement. The trustees of Wellesley College, for example, established a Center for Research on Women (CROW), a rarity at a women's college, in 1974. Directed by the political scientist Carolyn Elliott, it offered space to the chemist Lilli Hornig's Higher Education Research Service (HERS) and for a time to the Federation of Organizations of Professional Women. It also housed the *Women's Review of Books* and sponsored research on women's issues by social scientists, such as Paula Rayman and others, on childcare, displaced homemakers, science education, and what in the 1980s came to be known as "women's ways of knowing." In the early 1990s, researchers there prepared the AAUW's 1992 report on how public schools shortchange girls.[47]

This helpful support was notable in science and especially mathematics. President Barbara Newell of Wellesley offered the math professor Alice Schafer space and administrative assistance for the Association for Women in Mathematics office from 1973, when she became its second president, until 1993, well after Schafer's retirement from Wellesley. (She then taught at Simmons College and later at Marymount.) Both Newell and Schafer felt it was important for a women's college to support a women scientists' group. As mentioned in chapter 4, Mills College similarly provided a home for the Bay Area Math/Science Network, and several women's colleges offered summer programs, either remedial ones for the disadvantaged or enrichment ones for the already advanced. Bryn Mawr and Spelman cooperated on a summer program for college graduates planning to go on to graduate school in mathematics. In addition, Bryn Mawr co-hosted with the AWM a symposium in 1982 on the hundredth birthday of Emmy Noether, the eminent German mathematician, who had fled the Nazis in the 1930s and then taught at Bryn Mawr. Likewise, Radcliffe College and its Bunting Institute cosponsored a two-part symposium in 1985 honoring the nineteenth-century mathematician Sonia Kovalevsky.[48]

Going Coed After All

For the many other women's colleges that were eventually forced to merge, close, or admit men, the final decision came after a long and painful process. Just mentioning the word *coeducation* to the board of trustees felled many women's college presidents in the years after 1970, and turnover could be rapid. Usually the crisis arose because of faltering enrollment rather than a lack of money, though the two were related. (When asked in 2001 whether Hollins College was going coed, as the then president had recommended, one angry trustee remarked, "Not with a

$93 million endowment," and three students protested naked.)[49] Although the alumnae were often adept at raising money, their daughters tended to go to college elsewhere. Sometimes the current students were upset, as in 1990 at Mills, where they felt that they had something special and that admitting men would just make the college a smaller version of Stanford. Yet it was very hard to communicate this enthusiasm to potential applicants. Sister Janet Eisner, the president of Emmanuel College, in Boston, which finally admitted men in 2000, put it best: She said that the purpose of the college was to educate women, but since not enough would come unless it also admitted men, it would do this. But this was not a sell-out; after all, the college was still educating women.[50]

The impact of such a decision was usually relief to many and a great benefit for science. Low enrollments over a number of years often meant belt-tightening and even deprivation at the departmental level. When, for example, in the late 1970s the chemistry department at Emmanuel College lost its fourth faculty member and could no longer offer certain courses every year, the American Chemical Society placed its major on probation and shortly thereafter revoked its accreditation. This was quite a blow to Sister Dorothy Higgins, SND, longtime chair of the chemistry department, who had worked hard to build up the department and was very proud of its many alumnae. When a dean proposed dropping the separate chemistry, physics, and mathematics majors in favor of one in the "physical sciences," Higgins appealed to Anna Harrison for help. Harrison made a site visit, yet there was little that she could recommend. (Emmanuel awarded her an honorary degree anyway.) Thus it was perhaps almost merciful that twenty years later Sister Janet was able to broker a new arrangement that included a decision to admit men. Because of the college's prime location in Boston's Fenway, near Harvard Medical School and several hospitals, Merck leased a vacant lot for seventy-five years for $50 million and built a large new laboratory for 350 employees on it. This generated enough cash for the college to reacquire and refurbish a dormitory previously sold to the combined Beth Israel Deaconess Medical Center and make other improvements. But it was the shift to coeducation that turned a dormant campus into a vibrant one (or so the press coverage tells us), and numerically at least women students benefited at least as much as the men. At its low point in the late 1990s the college enrolled a certain number of freshwomen (say, $x = 200$). With the shift to coeducation, total applications quadrupled and first-year enrollments tripled (to $3x$). Seventy percent of the first-year students were women, or more than double the number of freshwomen the college had enrolled before, supporting Sister Janet's rationalization. Overall, the remarkable turnaround is reminiscent of Whoopi Goldberg and the other ingenious nuns in the recent film *Sister Act,* in which the poor but dedicated nuns take on the corporate, chemical, hospital, and real-estate interests and come out ahead. Perhaps it signifies future science-related possibilities for well-located women's colleges.[51]

Elsewhere, however, the move to coeducation and the need to bolster men's enrollments has led instead to the construction of a new football stadium, as at Seton Hall University in Pennsylvania, a former Catholic women's college. Its president, JoAnne Boyle, credited the addition of football with increasing the male enrollment from 19 percent to 41 percent in just four years. Other former women's colleges had attractive property and substantial endowments that their trustees sought to put to related uses. Newcomb created a new institute with a center for research on women, an alumnae office, and programs for students; Washington University at St. Louis established a graduate fellowship; and Radcliffe College, which had never had a faculty of its own but had been coeducational in practice since World War II, in 1999 transformed itself into an institute for advanced study that could take more advantage of Harvard's research opportunities. Many of its fellows were to be women, and some of its focus was to be on gender.[52]

Conclusion

The ever fewer women's colleges of the decades after 1970, energized by the increased competition for bright women students, fought back in various ways. Quantitative data assembled by alumnae scientists and social scientists showed that these colleges had produced a disproportionate share of past women achievers, and college officials took this up as a mantra, featuring it in advertisements. They formed a new organization, the Women's College Coalition, to publicize their strengths, including the successful careers of their alumnae; to recruit more vigorously; to strengthen their curricula, especially in the sciences; to accept more minority, older, and part-time women; and to wage aggressive fund drives. An integral part of their new message was the necessity for solid preparation for careers in science, mathematics, economics, and even engineering. Because many alumnae had flourished financially, several women's colleges completed record-breaking fund drives in the 1980s, the 1990s, and the first years of the twenty-first century, thus making possible new buildings, new programs, and more financial aid, which, one can hope, will lead to even more new opportunities in the future.

CHAPTER SIX

Surviving the "Minefields" in Graduate School

The coming of "women's liberation" in the late 1960s stimulated record numbers of female college graduates and reentry women to seek graduate degrees, especially in psychology and biology. The new promise of good jobs and perhaps even successful careers at equal pay made the necessary time and effort appear worthwhile to motivated women, especially as many of the traditional causes of attrition were also fading away—unequal financial aid had been outlawed, social isolation was dropping, and procedures were being devised for dealing with sexual harassment. By the year 2000 the number of women completing doctoral degrees in science and engineering at American graduate schools had more than quintupled, from 1,648 in 1970 to 9,396.[1]

This striking increase occurred in all fields of science and engineering at a time when the number of males earning doctoral degrees faltered until the mid-1980s and then rose again slowly. Thus the women's percentage of the whole first rose sharply and then continued upward at a slower rate. The most striking features, as shown in figure 6.1, were the continued dominance of psychology and biology, which continued to account for more than half of all doctorates awarded to women in all fields of science and engineering, and the persistence of wide differences among fields. In general, although the proportion of women earning doctorates rose in all fields, their rank order did not change much. Thus those disciplines that were the most feminized in 1970 (psychology, most social sciences, and some parts of biology), with about 15 percent or more women PhDs, remained so in 2000, though at the sharply higher level of 50 percent or more. At the other end of the spectrum, the least feminized fields in 1970 (engineering, computer science, and physics), at under 2.5 percent, remained the least feminized in 2000 but at the higher level of under 17 percent. By contrast, there were a few interesting and important changes in the middle fields. The earth, marine, and agricultural sciences, for example, which had largely excluded women in 1970, had a substantial number and proportion of them earning doctorates in 2000, moving ahead of astronomy, mathematics, and economics. Overall, then, a field's relative openness to women, when compared with that of other scientific disciplines, remained remarkably similar to what it had been years earlier. The range of percentages (and these are

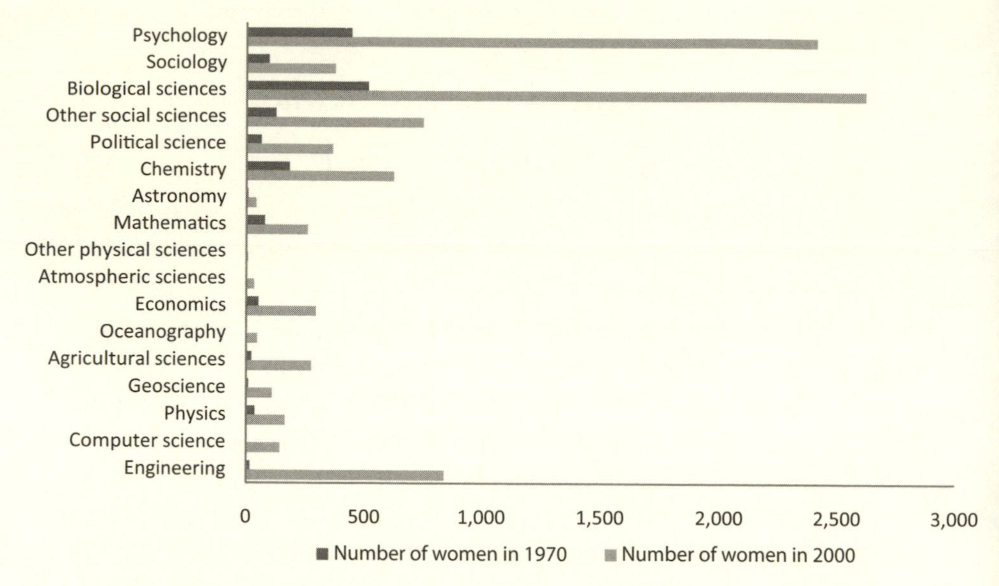

Fig. 6.1. Number of PhDs awarded to women in science and engineering, by field, 1970 and 2000. Adapted from Susan T. Hill, *Science and Engineering Degrees: 1966–2000,* NSF 02-327 (Arlington, VA: National Science Foundation, Division of Science Resources Statistics, 2002).

aggregates of many subfields) indicates that an individual in one specialty could have a very different graduate-school experience from that of an individual in another field at the very same time, a kind of "two cultures" phenomenon.

This remarkable increase in the number of women completing doctoral degrees was due not just to the passage of the Education Amendments Act of 1972, which banned discriminatory policies and practices, but also to a new level of motivation among female students and a new responsiveness at the graduate schools to students' complaints and difficulties. Individuals and groups within and outside the graduate schools—their first female deans, the students themselves, the women's caucuses, a few senior women scientists, and on occasion national organizations such as the Association for Women in Science (AWIS), the Association for Women in Mathematics (AWM), the Committee on the Status of Women in Physics of the American Physical Society, and the American Association of University Women (AAUW)—all took on expanded tasks, including projects with foundation support. This new willingness to speak out, identify problems, and press for change was especially evident after 1982, with the coming of writings on the "chilly climate," which reported on local efforts under way and encouraged others to take steps to improve their own hostile environments. Thereafter, surveys, reports, and interventions proliferated, including several by women scientists and engineers at MIT, some of whom were energized by membership on Lilli Hornig's committee at the National Research Council (briefly mentioned in chapter 1). As the graduate-school population became more feminized, each woman's chance of finding a friend or even a circle of acquaintances to talk to increased even more, and the attrition rate continued to drop.[2]

Most doctoral degrees in the United States were given, and still are, by just fifty large institutions, which, when compared with the variety of undergraduate colleges discussed in chapters 3 and 5, were organizationally quite homogeneous by the 1970s. All, for example, whether public or private, were by law coeducational. (This included the graduate school at Bryn Mawr College, which unlike its women-only undergraduate program, had been coeducational since the 1930s.)[3] Although each one is a collection of independent departmental units, in many respects like similar programs elsewhere, there were and are institutional differences of relevance to women students. For example the proportion of women earning doctorates at the University of California at San Francisco was high because the campus concentrates on the life sciences, in which women were and are relatively numerous. Similarly, the percentage of women at Texas A&M University jumped markedly in the late 1980s when it added a health-sciences campus to what had been an almost totally male engineering complex. By contrast, Princeton and the California Institute of Technology were for many years particularly difficult places for women graduate students, for few, if any, women had been admitted before the early 1970s. Breaking this tradition involved some unhappiness for some of the pioneers. Margaret Geller, for example, an honors graduate of the University of California at Berkeley and the holder of an NSF graduate fellowship, wanted to leave the physics department at Princeton University within a week of her arrival there in 1970. Hardly anyone spoke to her. Other women came but soon dropped out. Yet she stuck it out and became the third woman to earn a PhD in the department, following Glynnis Farrar in 1971 and Claire Max in 1972.[4]

Graduate education everywhere remained at its heart an intense one-on-one interaction with a major professor in one's chosen field. The nearly totally male faculty admitted and evaluated the students, allocated financial support, approved student progress, and awarded degrees. In this they were not accountable to anyone, although there were limitations based on hard realities such as laboratory space and facilities, teaching loads, and current and anticipated future financial support. Yet important as the relationship was, it could be fraught with uncertainties, for each professor had his own rules, expectations, and personality. Thus, individuals' experiences differed—by institution, by field, by laboratory, by sex, and by professor. Each department was a microcosm with its own culture of unstated assumptions and uncertain, even subjective expectations. All too often students who had been admitted because of their abilities and promise could flounder for reasons that were unclear. However arbitrary, biased, or unjustified faculty members' behavior might seem, their power differential was such that they could only be "encouraged" to be more humane and compassionate. Thus the faculty, for better or worse the centerpiece of any graduate school, remained unfettered. Accordingly, graduate education could offer a student the prospect of high points of discovery or breakthrough but also low points of disappointment or despair.

The bureaucratic entity known as the graduate school had developed around this central faculty core. Its staff handled such practical concerns as applications, recordkeeping, housing, immigration documentation, and financial aid. But in an era of increasing student (if not yet faculty) diversity these supportive functions could be expanded to provide timely advice and other relevant information to vulnerable students via orientation programs, newsletters, brown-bag lunches, speakers, and student groups. In the 1990s, when state legislatures sought more accountability from state-supported universities, the tremendous attrition rates of many departments came under scrutiny. This further strengthened the hand of the graduate school's staff, who pressed faculty members to explain and document their actions and propose remedies. No student was going to get a degree that she had not earned, but the faculty's high-handed practices of the past now might be questioned, scrutinized, and partly remedied via grievance procedures and appeals. The central administration was watching, but its enforcement powers or protections could be muted and even ineffectual, except in the emerging area of sexual harassment.

In the early 1970s graduate schools at many universities admitted a seemingly disproportionate number of men. This was the result of certain structural factors, as illustrated in a statistical study in the early 1970s of the University of California at Berkeley, one of the largest graduate schools in the nation. Accusations had been made that Berkeley's graduate admissions process was discriminatory, because overall 44 percent of the male applicants to its 101 departments for the fall of 1973 were accepted, compared with only 35 percent of the female applicants. The Berkeley statistician Peter Bickel and others found this conclusion to be misleading, because it was based on aggregated institutionwide totals rather than on departmental ones. Generally, women did not apply to the same departments that men did. They were most numerous among the applicants to several heavily subscribed humanities departments, for example, but were nonexistent in the applicant pools of sixteen other departments, mostly in engineering. Thus the authors concluded that most departments were behaving fairly. Although this conclusion rebutted the claim that Berkeley's admissions process was discriminatory overall, it highlighted the underlying imbalance that certain very large departments with much financial support (e.g., engineering and the physical sciences) attracted chiefly male applicants, while other, smaller and poorer departments (in the humanities and social sciences) attracted more women. Bickel noted the irony in the fact that this clustering of women's applications in the nonmathematical fields made admission in those fields more competitive than that in the highly mathematical ones.[5]

One person who in the early 1970s got very interested in assessing the whole graduate experience was Lucy Sells, a graduate student in sociology at Berkeley. In fact she wrote her dissertation on how male and female students fared at its graduate school. One of her first contributions, mentioned in chapter 4, was the identification and articulation of the "mathematical filter" in female course-taking in high school, which limited their applications for graduate work to relatively few fields. But the bulk of her dissertation had to do with attrition, or the many reasons why until recently so many women had not finished their Berkeley doctorates. She collected evidence from the Berkeley graduate division, data on recent Woodrow Wilson fellows, and questionnaires circulated by the Carnegie Commission on Higher Education, whose headquarters were in downtown Berkeley, to show that more than 60 percent of women graduate students dropped out, compared with only 45 percent of the men. This pattern varied a bit by field but was widespread. Sells recommended more student clubs so that the students could exchange information, build self-confidence, and thus in time attract the favorable interest of faculty members who were looking for capable protégés. By the spring of 1973 she had evidence that such measures were already proving effective in some fields. By then, consciousness-raising groups all over Berkeley were empowering women students to press on and complete their degrees. There at least, the longstanding gender differential in attrition was already dropping.[6]

At the same time, many other universities were undertaking status-of-women reports, which occasionally included the subject of graduate women. The University of Michigan, which, like Berkeley, had one of the largest graduate schools in the country, went further and commissioned a report entirely on graduate education for women, aptly entitled *The Higher, the Fewer.* Serving on this committee were several women faculty members, including the geographer Ann Larimore and the paleoecologist Margaret Bryan Davis. They recommended more flexibility at the graduate school, more options for reentry women and for part-time study, day care, greater ease of transferring between units, reexamining any policies that affected women students adversely, adding a staff person to the graduate school who would run programs and advise women and minorities, and urging departments to hire more women faculty in order to give the women students, whose numbers were increasing, more role models and understanding teachers.[7]

Graduate-School Deans and Initiatives

One remedy to the high dropout rate among female graduate students was the appointment of women faculty as graduate-school deans or associate deans, a step reminiscent of the appointment at many large coeducational institutions in the late nineteenth century of female deans of undergraduate women to perform a kind of "women's work" in the central administration.[8] Several of these new female graduate deans were scientists, starting with the psychologist Leona Tyler at the University of Oregon in 1966–73 and continuing with, among others, the anthropologist Helen Codere at Brandeis, the radiologist Alison Cassarett at Cornell, the psychologist Judith Rodin at Yale, the physicist Elizabeth Urey Baranger at the University of Pittsburgh, the geographer Risa Palm at the University of Colorado, the sociologist Gillian Lindt at Columbia, and the economist Anne Petersen at the University of Minnesota. When the linguistics professor Victoria Fromkin became dean of the graduate division at UCLA in 1979, she was the first such woman in the entire UC system. In 1988 she was also the first woman ever elected president of the Association of Graduate Deans of the prestigious American Association of Universities. By that time there were many others. In fact, in 1983, when 42 of 377 graduate deans in the nation (12.5%) were women, of whom almost 60 percent were scientists, especially life scientists, the Committee on Women of the Council of Graduate Schools conducted a survey to find out what differences there were between the men and the women deans. The survey revealed that the women were younger than the men, this being oftentimes their first administrative position, which many of them viewed as a possible stepping stone to even higher organizational posts. Indeed, the graduate deanship has turned out to be a key career move for some women scientists aspiring to top university and foundation positions. Some have become university presidents (e.g., the psychologist Judith Rodin at the University of Pennsylvania and the neurobiologist Susan Hockfield at MIT, both formerly at Yale) or chancellors (e.g., the planetary scientist Laurel Wilkening at UC–Irvine, formerly at the University of Arizona).[9]

But even where there was a male dean, the staff, knowing better than anyone else the high dropout rates for certain groups, began to expand their roles beyond the passive ones of data collection and recordkeeping to more active ones of offering an array of services and programs to inform graduate students of the risks of their plight and offer timely help in alleviating or preventing them. Starting in the 1970s, most graduate schools held orientation, brown-bag discussion groups, events, and workshops, circulated newsletters, and encouraged the creation of interest groups, such as one aimed at graduate women in science. Such organizations became a standard remedy, once there were enough women to form them. The hope was that such networks might enable women to form friendships and support groups outside of class or labs. Such clubs were formed at Stanford and MIT in 1974 and at Caltech also in the 1970s; their levels of activity waxed and waned, and little is known about their activities.[10] Group housing was another remedy. MIT renovated a former mansion on Memorial Drive for use as a dormitory for graduate women, which it named Ida F. Green Hall in 1983.[11] But this women-only housing was contrary to the usual pattern, according to which, as the marriage rate dropped and cohabitation and divorce increased, even former women-only on-campus dormitories, such as at Radcliffe/Harvard and Yale, were opened to both sexes.

Foreign and Minority Students

Much of the increase in the number of graduate degrees awarded in the period 1970–2000 is attributable to foreign students, especially in certain fields. This internationalization of the graduate school became so substantial that statisticians began to create new categories in the 1980s based on certain kinds of visas. Whether Indian, Iranian, Japanese, Chinese, Russian, or "other," the students came, filled laboratories and classrooms, and earned degrees. Overall, the percentage of women among these foreign students was lower than that of women among the American students, but in physics and economics, fields in which the percentage of U.S. women was very low, by 2003 the actual number of foreign women earning doctorates began to surpass that of American women doing so.[12] Although the international students were often better trained than American applicants and had fewer distractions to interfere with their completing their degrees, they may also have been even more isolated than their classmates, for professors, encountering language barriers and expecting the students to return to their home country, interacted with them even less than they did with their American students. Nor were the foreign students very active in social groups. But they were stoic and persevered, confident that the struggle was worthwhile, for people back home would be proud of them and their prestigious American doctorate. At times, however, political conditions at "home" deteriorated to the point where they could not return, at least not immediately. For example, there was sufficient unrest in Chile in the 1970s that the astrophysicist Maria Teresa Ruiz, who earned her doctorate at Princeton in 1975, did not return to a post at the Universidad de Chile until 1979.[13]

Over the years the number of minority women graduate students also rose, but not as dramatically as the number of women foreign students or as had been hoped or expected in 1970. One factor may have been that the definition of *minority* changed a bit over the decades. South Asian Indians, a rapidly growing group, ceased to be a minority in 1977, according to a decision by the Equal Employment Opportunity Commission, which designed and approved official definitions and so data collection. East Asians, in the 1980s a rapidly growing group with many women students, were a minority but not a disadvantaged one. The number of Hispanic women graduate students increased, but few were in science. Lydia Villa-Komaroff, a Mexican American, earned her PhD in cell biology at MIT in 1975, possibly the first Hispanic American woman to do so. The percentage of Native Americans earning doctorates in science has remained under 1 percent.[14]

In fact although many, and perhaps most, of the African American graduate students in the sciences were women, for many years more black men than black women completed the PhD. This may have been due in part to the presence and diligence of male African graduate students (and the scarcity of female Africans), for when the data are broken down by visa category as well as by race and sex, African American women received 50.8 percent of the PhDs awarded to black U.S. citizens.[15] Notable among the female African Americans who earned doctorates were Shirley Jackson, who completed a PhD in physics at MIT in 1973; Shirley Malcom, a PhD in animal ecology at Pennsylvania State University in 1975; and Jennie Patrick, a ScD in chemical engineering at MIT in 1979. In fact, MIT was for a time particularly active in minority issues and started a special program for minority physicists in the 1970s. Its physics department brought graduates of historically black colleges and universities (HBCUs), especially Morehouse and Spelman, to MIT for advanced training, but there were criticisms of how the program was handled, and its success was not great. For example, the Spelman College graduate Evelynn Hammonds, who had been inspired to attend MIT by Shirley Jackson, was discouraged upon her arrival in Cambridge. Rather than assessing her needs and giving her constructive advice, her faculty adviser strongly disparaged all her previous training. Distressed, she lost the will to persist. Later she earned a Harvard PhD in the history of science, joined first the MIT and then the Harvard faculty, where she still holds the Barbara Gutmann Rosenkrantz Professorship in the History of Science and Afro-American Studies. In 2006 she became a senior administrator at Harvard under the controversial president Lawrence Summers (see the epilogue). By contrast, later minority programs elsewhere were more successful. In 2000 the University of Maryland awarded doctorates in mathematics to three African American women, undoubtedly the first time this occurred anywhere.[16]

One program that may have helped some underrepresented graduate women to persevere was the Minority Women Scientists' Network, set up by the American Association for the Advancement of Science in the early 1980s, as described in chapter 1. Like other remedies, it offered connectedness as an antidote to isolation, on the theory that support and encouragement even from afar could help overcome frustration and despondency.[17] A successful initiative of the 1990s was a summer program for first-year graduate women in mathematics, especially minority women, supported by several foundations and government agencies. It grew out of undergraduate programs run by faculty at two women's colleges. In 1998, with the financial support of the NSF as well as grants from the Andrew W. Mellon Foundation and the National Security Agency, the mathematicians Rhonda Hughes, of Bryn Mawr, and Sylvia Bozeman, of Spelman College, started Enhancing Diversity in Graduate Education (EDGE). Aware of the high dropout rate of women, especially minority women, during or after their first year in graduate school, they set up a four-week summer program that taught the participants what to expect and kept in touch with them for a year or two thereafter.[18]

Financial Aid

Most financial aid was distributed by universities in the form of fellowships and research and teaching assistantships. By the mid-1970s men and women students (including those who were wives of male students) were receiving equal stipends, but in most fields the women and the minorities less often held the desired research assistantships that allowed students to work directly with the professor and possibly coauthor some publications.[19] Yet this was not the whole story, for the 1981 annual statistical report by the National Academy of Sciences on recent PhD recipients called attention to the fact that in recent years (since 1978) more women than men had had to support themselves during graduate study. (This was particularly striking in the field of medical sciences.)[20] In addition, older women who returned to graduate school for a master's degree or to prepare for a reentry position in industry or elsewhere did not get an equal share of financial support. (One remedy for this was, as Bernard Bulkin, of the Polytechnic Institute of Brooklyn, recommended, to get a job as soon as the course started and request that the employer pay the costs of their graduate training.) Thus, the several "career facilitation" programs supported by the NSF Women in Science Program between 1977 and 1981, which provided tuition support, were godsends to reentry women seeking mid-level training in computer science, engineering, and chemistry for related jobs.[21] Unfortunately, the federal government terminated the NSF program the next year.

For a variety of reasons, nationally competitive fellowships, awarded by outside panels, took longer to become equitable. Although many federal fellowship programs were cut or terminated during the later Nixon years (1970–74), the NSF continued its graduate fellowships, even increasing their number from 450 per year in the 1970s to as many as 950 per year in the 1990s. It also reallocated their distri-

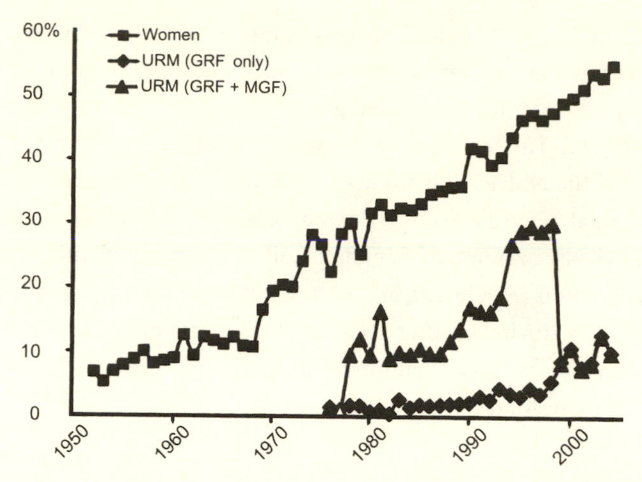

Fig. 6.2. Percentage of NSF graduate fellows who were women and/or minorities, 1950–2000. GRF = graduate research fellowship; MGF = minority graduate fellowship; URM = underrepresented minority. Adapted from NSF, Division of Graduate Education, Cumulative Index of the GRF Program and related data sets.

bution among the sciences away from physics, mathematics, and engineering and toward biology and the social sciences, all of which benefited women eventually. Initially only 10–20 percent went to women, even though they represented as many as 30 percent of the applicants. The situation improved slowly, as shown in figure 6.2, partly because there were more women applicants as the number of female undergraduate science majors grew and partly because more women scientists served on the selection panels. (Generally these were former NSF fellows.) The sparse data available indicate that women's success varied widely among fields. It was higher than men's in biology and biomedical sciences but much lower in mathematics, applied mathematics, computer science, and chemistry.[22] In 1990 the NSF added a graduate fellowship program specifically for women in engineering, with about seventy awards. In 1993, in response to the growing concern about the downturn in the number of women earning BS degrees in computer science, the NSF expanded this special fellowship program to include that field as well.[23]

Starting in 1976 the NSF broke its fellowship data down by race, which revealed that hardly any minorities were winning awards. Then in 1978 it started a Minority Graduate Fellowship Program, which grew and gradually became more equitable to women applicants. In 1985, for example, it awarded 21 (35.0%) of its 60 fellowships to women, who constituted 50.6 percent of the applicants. By 1992 the number of awards had doubled and the women's success rate was more nearly equal, with the NSF awarding 59 (49.2%) of the 120 fellowships to women, who made up

51.8 percent of the applicants (see fig. 6.2). Then in 1998, after a white male from Clemson University applied for a minority fellowship and was turned down as ineligible, he filed a lawsuit in federal court claiming that he had been discriminated against. After consulting with the Justice Department, NSF officials paid him $95,400 ($14,400 toward a fellowship and the rest to his lawyers) and then terminated the program, saying that they had been planning for some time to merge it with the regular NSF graduate fellowship program and to make awards to institutions, not to individuals. After this, the combined NSF graduate fellowship program granted more (about 10%) of its 1,000 awards to minority-group members, but this was down substantially from the 150 given annually by the minority program.[24]

Outside the federal government, several foundations, most notably the Ford Foundation and the Howard Hughes Medical Institute (HHMI), had large fellowship programs for scientists. That at Ford was aimed at minorities. In 1992, 588 (57.7%) of its 1,019 applicants were women, but only 57 awardees (49.1%) were female, with very wide differences among fields.[25] In 1987 the recently created HHMI developed a major graduate fellowship program for graduate students in the biological sciences (as well as others for medical students and postdoctoral fellows). The institute awarded fellowships to about sixty to seventy graduate students in the biological sciences annually, chosen from about one thousand to fourteen hundred applicants. Whereas men greatly outdid women in the first several years of the program, by 1992 50 percent of the applicants, as well as 50 percent of the winners, were women.[26] Besides these large programs, there were several other nationally competitive fellowships specifically for women. In 1974 individuals at Bell Laboratories, a branch of the large telephone monopoly American Telephone and Telegraph (AT&T), which had recently lost a class-action lawsuit for job discrimination, started its Graduate Research (Fellowship) Program for Women. Over the years this program attracted many outstanding women in the physical sciences and engineering, including Barbara Cooper, of Cornell (later its first female faculty member in physics), and several future Bell Labs scientists.[27] In addition, a few private philanthropists created new fellowships for women, such as the Cecil and Ida Green Fellowships in engineering and the physical sciences at MIT. (Cecil Green had been one of the founders of Texas Instruments.) At other universities, such as Washington University at St. Louis, where a former coordinate college for women was being discontinued, some of their endowment was converted into graduate fellowships for women.[28]

There continued to be several graduate fellowships supported by businesses and women's groups such as the Business and Professional Women and Zonta International.[29] Among the latter, the AAUW continued to be the largest provider of graduate fellowships to women in all fields, as it has been since the 1880s. In fact the energetic members of the AAUW not only continued but increased their prodigious fund-raising funds for graduate fellowships for women. In 1971 the AAUW Educational Foundation launched its Centennial Fund, with a goal of raising $10 million more for the endowment by 1982, the AAUW's hundredth birthday. By then it had increased the fellowships' stipends and added several new kinds of awards, and it was offering annually $2 million in more than one hundred American fellowships, of which 30 percent went to scholars in the life and physical sciences and 40 percent went to scholars in the social and behavioral sciences. In 1987, the year after the space shuttle *Challenger* exploded with the astronaut, engineer, and former AAUW fellow Judith Resnick aboard, the AAUW raised more than $320,000 for a special fellowship in her memory for a graduate woman engineering student. By 2002, despite a temporary downturn in the stock market, the AAUW Educational Foundation was awarding more than $4 million annually to fellows in engineering, business, law, and medicine, as well as the arts and sciences.[30]

A new fringe benefit in the area of parental leave emerged in 2006–7 at a few of the most prestigious graduate schools, and it was quickly copied elsewhere. Although the health benefits that were usually part of a graduate student's multiyear financial-aid package included prenatal care and the cost of a baby's delivery, there was no provision for any time off for the new mother, who lost her stipend temporarily if she went on leave. (In addition, if someone else had to fill in for her, as on a big team, funds for this had to come from somewhere else.) But beginning in 2006 with the chemistry department at Stanford, and followed by the graduate schools at Princeton, Yale, Dartmouth, and Stanford as a whole, several universities started providing these extra funds for up to twelve weeks of paid maternal leave.[31] This sudden interest in making graduate school more family-friendly seems to have grown out of efforts to retain more women students in the physical sciences and engineering after 2005 (see the epilogue).

Professors and Mentoring

As mentioned earlier, at the heart of the graduate-school experience was the intense interaction between faculty and students. Such men (and a few women, for there were not yet many female graduate faculty members) held almost all the power in graduate education and controlled directly or indirectly every aspect of it, including the power to allow students to continue their graduate work. Many had been trained, and even thrived, in highly competitive situations and sought to weed out weaklings. Many were looking for protégés or alternatively exploiting available hired hands for the continued staffing of their own projects. Foreign students whose continued stay in the United States depended on their visa status, which in turn depended on their continued enrollment, were especially vulnerable. Thus, getting along with and earning the praise of one's major professor was a graduate student's central concern.

If this relationship went well and the student lived up to or exceeded the faculty's expectations, the student might breeze through the requirements and complete his or her degree with minimal difficulty. Some success stories attest to this, such as those of the atmospheric chemist Susan Solomon, at the University of Colorado at Boulder, who was soon leading a team to Antarctica; the electrical engineer Denice Denton, at MIT, who quickly won a prestigious Presidential Young Investigator Award; and the molecular biologist Carol Greider, whose collaboration on telomerase with Elizabeth Blackburn at UC–Berkeley in 1984–85 later earned her a share of the 2009 Nobel Prize. They and others, especially the Sloan fellows, mentioned in the next chapter, lit up their fields and moved onto the fast track to major postdoctoral positions at top institutions.[32]

But too often there were problems, primarily communication or the lack thereof, inattention, low expectations, and even exploitation. One's relationship with one's professor remained an area of great anxiety, and occasionally male faculty admitted that they did not know how to handle their women students. The problem was pervasive in the field of mathematics, as revealed in a questionnaire that the mathematician Edith Luchins, of Rensselaer Polytechnic Institute, sent to 1,000 members of the AWM in 1975. The 350 responses from women and 52 from men showed that the main stumbling block facing many female graduate students (recently and in the past) had been their major professor's lack of encouragement. When asked who had discouraged their pursuit of a graduate degree in mathematics, three times as many women as men said that it had been their teachers and advisers, especially at the graduate level. Eighty percent of the total, including 100 percent of those under age 35, said that they had been treated differently because of their sex. Many recalled that their professors had questioned their seriousness of purpose and continued to hold negative attitudes and expectations and to take inappropriate actions.[33] But mathematicians were not the only ones to have difficulties.

In her autobiography, *Sea Legs*, the oceanographer Kathleen Crane tells of losing credit for her graduate work. When she got to Scripps Institute of Oceanography in 1973, her major professor was an ex-navy man who had never had a woman student before. He was absent a lot but turned out to be supportive, allowing her to go on some expeditions when she could find a female roommate. But another group leader was more of a problem: after initially laughing at her ideas, he stole them and published them under his own name in *Nature*. Crane moved to Woods Hole, Scripps's chief rival, soon after it started a doctoral program and eventually earned her degree there.[34] Similarly, in 1995 Antonia Demas discovered just weeks after earning her Cornell doctorate in nutrition education that David Levitsky, one of her committee members, was using her ideas in a grant proposal and in publications without giving her any credit. She protested and filed a lawsuit, but the judge sided with the university in finding nothing actionable, although many at Cornell admitted that Levitsky's behavior was not proper.[35]

The most notorious case of stealing credit from a female graduate student in these years was that of the bioscientist Candace Pert at Johns Hopkins University in the 1970s, which she too has recounted in a recent autobiography. Pert, a Bryn Mawr graduate who had studied with the biologist Jane Oppenheimer there, was excited to be working with Solomon Snyder and other leading figures in work on brain chemistry, a hot and well-funded area during the war on drug addiction. In 1973 Pert was the lead author of several key papers, including the one announcing a major breakthrough, the discovery of the opiate receptor. Word soon spread that the accomplishment was being considered worthy of a Lasker Award. When it received the award in 1978, she was not one of the three persons who shared the prize. Stunned, and not wanting to be another victimized Rosalind Franklin, she wrote a letter of protest to the donor, Mary Lasker. Snyder also asked the panel to reconsider its decision, but it refused. No one had taken Pert's participation seriously, despite the lead authorship, since she was a mere graduate student and the senior men chosen were felt to be long overdue for the serious recognition that a Lasker Award implied. (It has often preceded a Nobel Prize.) Years later, Pert retaliated by refusing to assist those nominating Snyder for the Nobel Prize. When he did not win, some blamed her lack of cooperation. In return, he was less than generous in aiding her subsequent career in AIDS research. A different person might have swallowed her anger, played along, and subsequently benefited in some way, but Pert, who had been denied a share of a major prize for the greatest discovery of her career, was not swayed. Why should she help the villains get even more recognition?[36]

One method of trying to engage the attention and advice of professors was mentoring, a practice that Bernice Sandler had discussed in the early 1980s and that became something of a fad in the early 1990s. In 1990 Stephanie Bird, a neuroscientist at MIT and the president of AWIS, received a three-year grant from the Alfred P. Sloan Foundation to support AWIS's Mentoring Project. Part of the project was to publish a guidebook (called a "paper mentor"), entitled *A Hand Up,* comprising fifty-eight short inspirational, practical, and often autobiographical essays, some of which had already appeared in the *AWIS Newsletter.* Only two were directed at graduate students, but most were useful to a variety of age groups. In fact the book was aimed at women scientists of any level of experience, from those beginners who needed pointers to those more senior women who by the 1990s might have good advice to pass on.[37]

Meanwhile, a separate study of mentoring in 1992 by three sociologists showed that often women students cited as mentors supportive individuals outside their field who had helped them emotionally. They also cited as particularly helpful conferences, especially those for women, that they had attended. Among these conferences were two held at Argonne National Laboratory in 1987 and 1989, led by the chemist Marion Thurnauer and the physicist Natasha Meshkov and others. Later there were others, such as the Baltimore Charter meeting at the Space Telescope Science Institute in 1992 (to be discussed in Chapter 10) and the eventually annual Grace Hopper Celebrations for computer scientists.[38]

A Hand Up may have been what inspired Joy Frestedt and other women graduate students at the University of Minnesota, who were fed up because they received so little attention from their professors, to hold a series of workshops on the issue in 1993–94. About 250 persons, mostly graduate students, attended the events, at which they aired their grievances and suggested improvements, including plans for a center for mentoring. Prospects for success looked good initially, as Frestedt had received support from several vice presidents, including three women. But before much more could happen, all three female vice presidents (including the economist Anne Petersen, who was dean of the graduate school) had left the university. Nevertheless, a much-needed Coalition of Women of Color Graduate Students was formed, and Frestedt moved on to solve the problem another way by becoming the national president of Sigma Delta Epsilon–Graduate Women in Science.[39]

Studies and Voices

In the early 1980s, as the concept of a "chilly climate" spread through academia, several studies of graduate students appeared. Finding in 1982 that hardly anything had been written on women graduate students in science and engineering, two researchers at Stanford University, strongly supported by the associate dean of graduate studies and research (the former physicist Jean Fetter), sent a questionnaire to Stanford's 1,172 male and female graduate students (627, or 54%, of whom responded) specifically to find out why the attrition rate for women was twice that for men. The women admitted that they were often under great stress; an alarming 23 percent (compared with only 9% of the men) said that they were often or always "on the edge of a nervous breakdown." The women were more often depressed than the men, and three times as many women as men (30% compared with 10%) had sought help for mental distress at the campus health center. But why this was so was not clear.[40] Meanwhile, at MIT the physicist/electrical engineer Mildred

Dresselhaus reported on a 1982 study of male and female graduate students in her very large department of electrical engineering and computer science (101 women and 541 men, with women constituting 15.7% of the total). Assisted by the graduate student Denice Denton (later a professor, dean, and university chancellor), and putting a positive spin on the situation there, Dresselhaus found that conditions had improved with the coming of more women students but that the percentage of women graduate students who left before completing their PhD was more than twice as high as that for men (27.5% compared with 11.9%).[41] Then in 1988 Sheila Widnall, a professor of astronautics at MIT and head that year of the American Association for the Advancement of Science (AAAS), devoted her presidential address, which was later published in *Science,* to a survey of the plight of graduate women in science. For a president of the AAAS to take the problems facing graduate women seriously enough to devote her most important public utterance to the topic was an acknowledgment that although improvements had been made, there was still a long way to go.[42]

In 1995 the Alfred P. Sloan Foundation made a multiyear grant to a group at the University of Michigan to discover what factors led to success among graduate women in science, especially those in mathematics, chemistry, and physics, where they were the most miserable. But this proved somewhat difficult, as the students feared reprisals and were reluctant to speak up in focus groups, even when they were offered anonymity. In fact, the more advanced they were, the more wary they were, for they had seen, as one graduate student reported, forty students start each year but only six finish. Although they had been admitted to the department and joined a lab, they found that once they had arrived, the faculty seemed uninterested in them. They felt powerless, bewildered, and a bit frightened; as one put it, it seemed as if they were in the midst of a "minefield."[43] Aided by other Sloan and NSF grants, such investigative projects proliferated, and by the early 1990s a literature on graduate women in science and engineering was developing. Several such studies were presented at a large meeting at the University of Minnesota in 1995, and many were later published in the new *Journal of Women and Minorities in Science and Technology,* which first appeared in 1994 with Carol Burgher, of Virginia Polytechnic Institute and State University, as its editor.[44]

Meanwhile, a few women graduate students, especially at MIT, had begun to find their own voices, write down their views, and on occasion publish them in underground magazines. Perhaps their increasing numbers and percentages emboldened a few of them to speak up. Or perhaps they were encouraged by the establishment on many campuses of graduate-student assemblies or senates, where delegates could voice their constituents' concerns. In any case, this new genre started among the many female graduate students in electrical engineering at MIT when in 1979 Candace Sidner, who had recently completed her degree, published an insightful critique, "Surviving as a Female Graduate Student," in *Sojourner,* a Boston underground newspaper for women. In it she acknowledged the help of several other graduate students.[45] Then in 1983 fourteen women graduate students in computer science at MIT, where female graduate enrollments in that area were about 15 percent, who had been complaining since at least 1976 about sexist remarks, invisibility, differential expectations, unequal treatment, and other issues bordering on sexual harassment, finally sat down to write a lengthy report outlining their complaints. Outraged at the persistent and widespread hostile atmosphere that they encountered daily, inspired by Roberta Hall and Bernice K. Sandler's recent work on the "chilly climate," and encouraged to put their feelings into words by MIT ombudsperson Mary Potter Rowe, they found their voice and detailed their experiences in a forty-page report, the first of its kind. They did not name individuals or try to embarrass anyone; they simply hoped to raise everyone's awareness of how offensive this behavior was: not only did it make them and other women already at MIT uncomfortable but it deterred untold others from applying or, if admitted, from enrolling. In 1983 MIT had no remedy for this widespread behavior, but in response to this report it did hold several faculty and departmental meetings, at which many of the men (both faculty and students) claimed to be unaware of how their comments, often meant to be humorous, were being received. They thanked the women for bringing the issue to their attention, saying that no one had intended any harm and promising greater sensitivity in the future. But the problem persisted. In fact, little seemed to come of the report, at least locally.[46]

Other women graduate students at MIT and elsewhere began to find their voice a little later. In 1986 Roberta Brawer wrote a lengthy open letter to the physics-department faculty at MIT recounting her difficulties and explaining her decision to withdraw. In 1988 Cornell University hosted a symposium for female science graduate students, inviting Sheila Widnall to repeat her AAAS address and encouraging graduate students to meet in groups and articulate their concerns. In 1992 Judith Fleischman, formerly an undergraduate physics major at MIT and by then a Columbia University graduate student in astronomy, wrote an article on the difficulties women faced for the newsletter of a Columbia-Barnard group she had formed.[47]

These student voices were heard, for at a 1990 meeting of chairs of physics departments one person asked whether something might be done to improve the chilly, if not frigid, "climate" in certain departments. In response, the women's committees of the American Physical Society (APS) and the American Association of Physics Teachers proposed a series of site visits to departments that asked for them. The APS funded an initial five visits in 1991 and 1992. On the basis of what was learned there, three physicists—the indefatigable Mildred Dresselhaus, Judy Franz, and Bunny Clark—obtained a subsequent NSF grant to pay for similar interventions at ten more physics departments that wanted to improve their climate toward women. Somewhat like a site visit from an outside funding agency, five women physicists would visit for a few days, during which time they would talk to everyone from the students and postdocs to faculty and administrators. Then they would send the department chair a report specifying the corrective measures the department might take, to which the chair was to respond within six months.

The key to the success of this venture was that the physicists making these visits be well-regarded fellow physicists—insiders known to departments and familiar with and appreciative of their good intentions, valid concerns, and traditional ways of doing things—rather than judgmental social scientists or other outside know-it-alls. By 1996 seventeen visits had been completed. Unlike the Michigan interviewers mentioned above, these female physicists found that the women graduate students were among the best sources of information about a department's atmosphere, as they were around a lot, observed various interactions, and were all too often its primary victims. (In fact they may have been among those who asked the chair to request such a site visit, as they were looking for ways to improve the situation.) Thus, when a preliminary questionnaire asked for their suggestions and remedies, they listed practical things, including more women speakers and greater nighttime security. But their chief suggestion was forceful and explicit: send the entire faculty out for sensitivity training, and if any refused, force them to take early retirement. (This was before retirement became voluntary in 1994.) Although this drastic step was never taken, the departments that had invited the site visitors later initiated some of their suggested reforms, such as having more women speakers, admitting more women students, and even hiring more women faculty. Another recommendation was to do what others had been doing all along: start a club for women graduate students, which would give them some autonomy. They could get together on or off campus, exchange information, critique practice talks, mutually empower each other, and perhaps suggest women speakers to the department chair, prod him to hire more women faculty, and get the university to do more about day care. Although it has not been formally evaluated, the program was so helpful that it obtained NSF support for five years and still exists.[48]

Another remedy that proved effective was simply writing down the department's rules and expectations, something that all too often had never been done before. The sociologist Mary Frank Fox has shown that what she calls the "department culture" is a large factor in the success of its women graduate students. This is somewhat related to the proportion of women graduate students in a given unit: the more women there were, the more comfortable the department was for all women. But even more important by the early 1990s was whether or not the department had written down and circulated its expectations for graduate students, for when they were not written down, misunderstandings were likely to occur, often affecting women disproportionately. These could lead to unexpected and painful withdrawals and dismissals, which better communication beforehand might have avoided. Often these written rules had followed an earlier unpleasant event, but sometimes it was related to the influx of new faculty members not familiar with a department's unwritten traditions. In fact, when Fox visited twenty-two departments of chemistry, computer science, electrical engineering, and physics at sixteen universities across the nation in 1993–95, she found that the best-run departments had learned from events of the two previous decades. They had chairs or heads who had dealt firmly with earlier complaints, including cases of sexual harassment, and had written down their guidelines. Departments whose faculty minimized or ignored previous episodes continued to have poor records of women applying, enrolling, and completing degrees.[49]

Off-Campus Groups

There also began to be alternative ways to learn how to cope with one's department and to find out about what was going on in one's field. One was to join the relevant women's caucus and read its usually quarterly newsletter. Although a woman might feel isolated, disillusioned, and even despondent in her current small group, the newsletter (and perhaps especially its photographs) might remind her that she was part of the larger national scene and exhort her to persevere. Convinced of the value of such sisterhood, Debbie Franzblau and Susan Landau, of the AWM, in 1982 got the names and addresses of hundreds of female graduate students from department chairs and had the AWM staff person send two thousand of them an application, a sample newsletter, and a letter of invitation that began as follows:

> Dear Graduate Student,
>
> Being a math grad student is hard. There are often reasons why being a woman makes it harder. It helps to know other women in math, and to work together. The Association for Women in Mathematics was organized a decade ago in response to that need. . . . Membership in AWM benefits you in several ways. The newsletter contains timely information. . . . AWM provides a network. . . . Many graduate programs are without a single female faculty member; knowing that there are women in mathematics, and that they are active, can be an important support through graduate school. Additionally, the contacts made through AWM can be extremely useful when seeking career advice, job

openings, sources of financial support, etc. Membership in AWM is $5.00 a year for students.[50]

About 150 joined, enough for the AWM to consider making this an annual effort.[51]

A worried or bewildered graduate student could also read the growing number of handbooks and guides offering practical advice on how to get ahead in graduate school and the generally Byzantine world of academia, where much of importance went unspoken. Reading these books could replace confusion and concern with enlightenment and lead to a liberating new sense of comprehension and even self-empowerment. Among these were the chemist Marjorie Farnsworth's *Young Woman's Guide to an Academic Career* (1974); the cell biologist Susan Goldhor's newsletter article on job-hunting (1976), which later became a booklet; the biologist Virginia Walbot's follow-up in 1979 on how to get a postdoc; a "Career Guide for Women in Science," distributed at the Cornell conference in 1988; C. M. Yentsch and C. J. Sinderman's *The Woman Scientist: Meeting the Challenge for a Successful Career* (1992); and special issues of *Science* on women in 1992 through 1995. In the late 1990s the Committee on Scientific, Engineering, and Public Policy (COSEPUP), of the National Research Council, put out two volumes—first one for graduate students, *Careers in Science and Engineering: A Student Planning Guide to Grad School and Beyond* (1996), and then one for faculty members, *Adviser, Teacher, Role Model, Friend: On Being a Mentor to Students in Science and Engineering* (1997).[52] Yet printed information had its limitations. To learn about the latest opportunities, to hear the gossip, to gauge one's personal prospects and perhaps get inspired, one needed to join the oral (and in the 1990s electronic) network in one's field. Many found that such connectedness had considerable advantages. One need no longer be totally dependent on one's professors for contacts and information. One could begin to build a network of her own and expand her options in ways advantageous to her own situation.

Off-campus get-togethers in particular offered a social and professional outlet that might even provide an illuminating perspective on what others elsewhere thought about one's own laboratory group. Fortunately, some of the women's caucuses, which existed at the national level, began to establish local chapters, often in college towns or metropolitan areas, that brought together a mix of graduate students, postdoctoral fellows, faculty, and visitors. These contacts could be valuable not only in remote areas but also, and perhaps even more so, in complex urban areas such as Cambridge/Boston, the San Francisco Bay Area, New York City, and Bethesda/Washington, DC, where there were many institutions with a large and ever-changing population. Such regional meetings offered a chance to "network," to use a new term for an old activity, or expand one's range of (generally female) friendships and thus get a broader idea of opinions and opportunities beyond one's own little unit. One might also hear of visitors in the area, possible future collaborations, or even forthcoming discoveries, job prospects, or funding possibilities. All this could be a boon to one's current mental health and future strategic planning. By the late 1970s the AWIS had branches in many places, as did the Association for Women Geoscientists by the early 1980s (especially in the Bay Area, Denver, and Houston). The AWM, which had started in the early 1970s as Mary Gray's newsletter, in the 1990s spawned regional groups in Berkeley, Madison, Cambridge/Boston, Connecticut, and Maryland, which they called "Noetherian rings," in honor of the eminent German mathematician Emmy Noether. The older women's organizations, such as Sigma Delta Epsilon and Iota Sigma Pi for chemists, which had long had chapters at major graduate schools, tried to be more responsive to young women in the 1970s. The former added the phrase "Graduate Women in Science" to its name in 1971 and then sometime after 1996 shortened and simplified it by dropping "Sigma Delta Epsilon" altogether.[53]

Sexual Harassment

Yet the usual remedies of student clubs, mentoring, and more women faculty could not solve certain continuing interpersonal problems. Before the 1980s the issue of gender or sexual harassment, which affected women more adversely than it did men, was shrugged off and not addressed.[54] Then various individuals and groups began to document and publicize the wide prevalence of sexual harassment, which surprised many, since it had hitherto been unreported. Affected students had generally tried to avoid the offender or quietly changed their fields. In documenting the phenomenon, psychologists led the way, and the American Psychological Association (APA) included it in a new statement of its ethical guidelines in 1981. Then a 1986 survey by the APA of 464 women clinical psychologists revealed that one female graduate student in six had been harassed in one way or another.[55] Five years later, in 1991, the astronomer Jill Price, of Bentley College, revealed that 40 percent of women (graduate and undergraduate) student respondents in that field had experienced some form of harassment, though far fewer had actually been physically assaulted.[56] (Five levels of activity were distinguished, from a hostile atmosphere, such as obscene posters or computer screens, on up to actual direct intimidation and physical assault.) The threat of sexual harassment at one level or another seemed to be a fact of life for graduate students.

But once the nation's courts began in the 1980s to consider hostile environments in cases of employment, universities began to take seriously complaints about it in the area of instruction. Circumstances changed in 1988, when five female undergraduates at Yale University filed suit in federal court for damages for retaliation when they refused faculty members' demands for sexual favors. Although the case went nowhere legally—four charges were dismissed for insufficient evidence, and the fifth was heard but denied—it brought the potential for culpability and so liability to the attention of college lawyers nationwide. Before long, various universities began to formulate internal enforcement policies that went through faculty committees, deans, and legal officers (rather than the dean of the graduate school) and that would keep as many cases as possible out of the courts.[57]

Occasionally a case was tried in the courts, such as the protracted and well-publicized one of the Cornell psychologist James Maas in 1994–99. (He fought his employer for depriving him of any female graduate students for some period of time.)[58] Others settled more quietly out of court. Unlike most aberrant faculty behavior, which was generally tolerated and shrugged off, this was one type of activity to which the university's central administration reacted forcefully, preferring self-enforcement to the threat of unfavorable publicity and the uncertainties of court proceedings.

Conclusion

Graduate school remained the central training experience for all scientists in the United States in the years 1970–2000. The major changes over these years were largely in the gender and nationality of the students enrolled, the broader range of fields the women studied, and the more supportive atmosphere. What had once been the domain of white American males now included women (50%), international students (possibly 25%), and minorities (10–15%). As a result, these decades were important in terms of records set by women earning higher degrees in science. This did not occur in a uniform way across all the sciences. As figure 6.1 shows, fields continued to differ, reflecting many existing social cultures and behaviors and perpetuating them into the next generation. Those in the biological and most social sciences already had, and would continue to have, one sort of career, while those in the physical sciences and engineering would have a rather different one. Even though there were gains, and women's numbers and percentages rose markedly in all fields after 1970, the disparity persisted.

Accompanying this change were efforts by persons within and outside the graduate schools to improve communications and clarify expectations in hopes of alleviating the brutal sink-or-swim atmosphere of yesteryear and thus reduce the graduate schools' traditionally high attrition rates. There is some evidence that by the 1990s, stipends, both federal and private, began to be more equitably distributed, as required by the 1972 Education Amendments for financial aid.

What was perhaps most important, watchdogs emerged over the years. Women graduate students had previously been pretty much ignored by all except possibly the AAUW, which on occasion wrote a report or held a meeting. In these years, individual women scientists and engineers, perhaps angry over their own treatment years earlier or because of concern for their daughters, though this was never stated, and various groups within the university—the dean and associate dean of the graduate school, women's commissions and advisory committees on the status of women, department chairs, student groups, and individual women themselves—began to take a sustained interest in the plight of women graduate students. They held workshops, conducted surveys, publicized problems, ran programs, provided counseling, and offered advice. A student in trouble began to be able to find staff members to whom she could turn and, as more women students enrolled, more friends responsive to her concerns. Even sexual harassment, a longtime bane of women graduate students, was admitted to exist, its prevalence documented, and appropriate procedures and penalties devised. Also, outside groups and individuals did studies, called attention to issues, developed a vocabulary (the "chilly climate"), got foundation support, even held site visits or other interventions, and raised fellowship funds. A few women graduate students even began to speak up and write their own critiques and reports, something they had never done before. In these various ways many individuals worked to make the graduate school a more humane place, more than something to be endured or survived. Yet significant problems remained, especially for minority women, of whom too few came and even fewer completed doctoral degrees.

The most frequently proffered remedy—the hiring of more women faculty—was occasionally acted upon, for many science departments did indeed hire their first woman faculty member in the 1970s and 1980s. In some cases their presence would, as many hoped, make a difference. But not necessarily, since, as we will see in chapter 11, they were fewest in the field where they were needed most, and many faced even bigger problems than the female graduate students did. Some might be confident, productive, happy individuals who transformed their departments and inspired their graduate students. But all too often they too were young, isolated, overworked, and frightened, hiding in their offices, unsure how to set up and run a laboratory, and resentful that on top of everything else they were expected to deal with the unhappy female graduate students—something they did not know how to do and were sure would get them into more trouble if they succeeded and perhaps even if they tried.

CHAPTER SEVEN

Postdoctoral Pathways

Preparation, Holding Pattern, or Jumping-Off Point?

Coincidentally, and fortunately for the growing numbers of women earning PhDs in science and engineering in the period after 1970, especially the biologists, the amount of "soft" money for research (i.e., short-term grants and fellowships) rose markedly in those years. Astonishingly, the annual budget of the National Institutes of Health (NIH) grew sixteenfold between 1970 and 1999—from $518 million to a whopping $8.19 billion—and then doubled again by 2003.[1] Because postdoctoral fellows from other laboratories brought new energy, ideas, and techniques (or "tacit knowledge") to a project at minimal cost, the easiest and best use of these increased annual appropriations was to fund a continuing flow of such short-term junior appointees.[2] American academia was able to handle this arrangement, since it was familiar with a tradition of prestigious programs of postdoctoral fellowships going back to the 1920s. In addition, since some of these awards were funded through national competitions and most others through peer-reviewed research grants to "principal investigators," primarily from the NIH, they added not only manpower but a certain element of prestige to a project or institution. Highly qualified personnel would work long hours for modest salaries, minimal benefits, and a short-term commitment because they thought they were training for a better and more secure future. At an hourly rate they were hard to beat. Funding agencies and grantees alike began to take advantage of (and even exploit) young scientists who wanted to stay in science a while longer even when the long-term job prospects were uncertain.

In fact this arrangement was so popular and cost-effective and served so many established interests that the numbers of postdoctoral fellows in the United States kept on rising (see fig. 7.1). Because the number of women with PhDs in the biological sciences, where most of the postdoctoral positions were, was increasing, their proportion of the total kept going up. For once the money was where the women were, rather than vice versa, but this presented a possibility of a new kind of "women's work" or even exploitation. Growing even faster were the numbers of foreign women postdocs, and by 2005 nearly half of the female postdocs were not U.S citizens (see fig. 7.2). Foreign women postdocs could be especially vulnerable

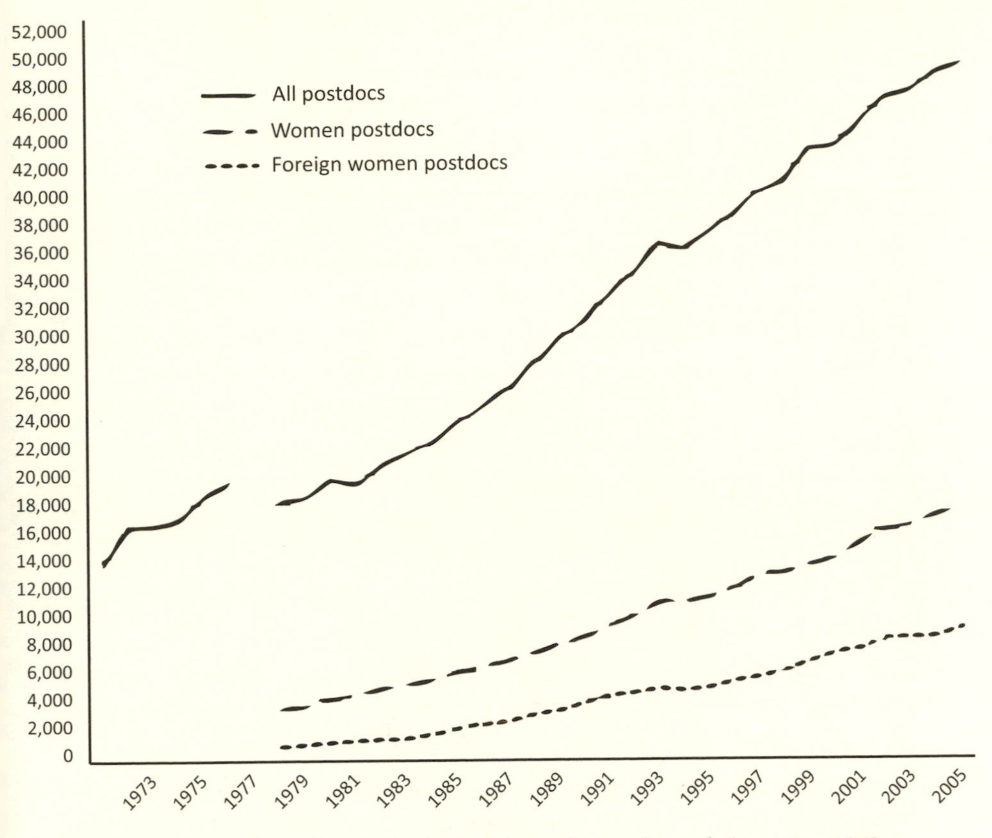

Fig. 7.1. Number of postdocs in the United States, by gender and visa status, 1972–2005. Adapted from National Science Foundation and National Institutes of Health annual Survey of Graduate Students and Postdoctorates in Science and Engineering, available online at https://webcaspar.nsf.gov/TableBuilder.

to exploitation or abuses, since their continued stay in the United States depended on visa renewals, which required current employment. Thus, the big expansion after 1970 became a mixed blessing, with some serious problems looming, perhaps especially for the women and foreign postdocs.

Expanding what had been a relatively rare experience for an exceptional protégé into a job category populated by more than fifty thousand postdocs by 2005 changed the position into just another step in a standard career. Over time some political and economic realities also demeaned and distorted the post's former elite role and status. First, the distribution of available positions reflected the interest of the U.S. Congress, so that even within the generally well supported biological sciences there were underfunded sectors, such as taxonomic work at museums, which

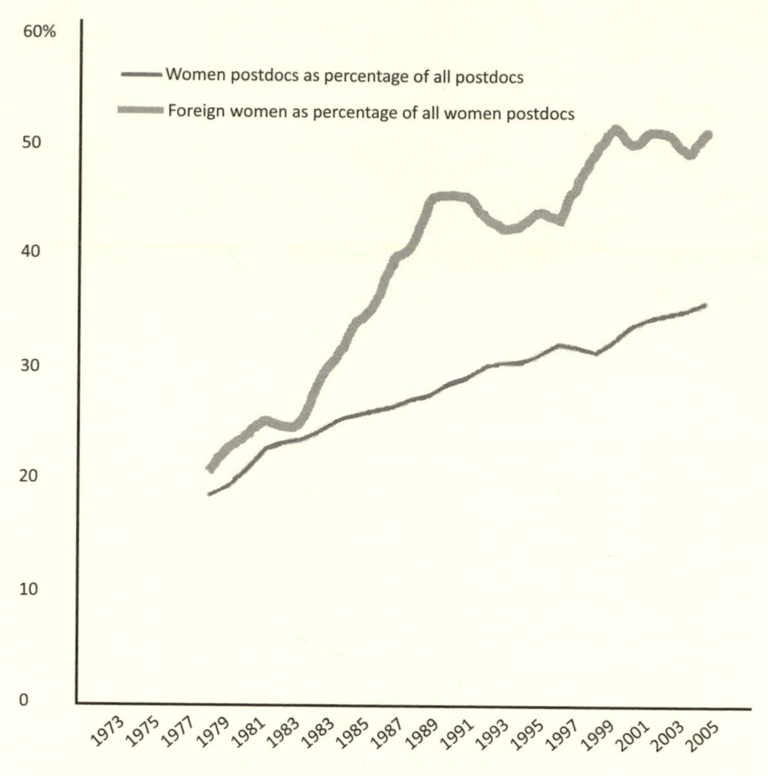

Fig. 7.2. Percentage of women postdocs in the United States, by visa status, 1973–2005. Adapted from National Science Foundation and National Institutes of Health annual Survey of Graduate Students and Postdoctorates in Science and Engineering, available online at https://webcaspar.nsf.gov/TableBuilder.

neither the federal government nor private philanthropy underwrote as well as they did molecular biology or cancer research. Second, there was the lingering burden of the "PhD glut" of the 1970s, when because of the near collapse of the academic job market, recent doctorates could not obtain faculty positions and somewhat unexpectedly stayed on in a series of temporary postdoctoral positions. The resulting "holding pattern" further increased the number of persons involuntarily stockpiled into what was becoming a kind of high-tech reserve labor force. Third, the very availability of so many postdoctoral fellowships and the expectation that all good job candidates would have held one or more increased the level of experience that could be required for a permanent academic post and so prolonged the time such aspirants spent in advanced training (much as the coming of "adolescence" had added a stage to human development earlier in the twentieth century). In biochemistry, for example, a second and even third postdoc became the norm. Thus, posts that were designed as temporary with no provisions for advancement or permanency were evolving into long-term apprenticeships and began to present many frustrations along with the learning opportunities. Positions that could be classified as "hired hands" on a professor's project were also open to certain abuses and came at a time in a woman's career (ages 28–37) when she might want to start a family and work part time. Worst of all, and something no one was really responsible for or able to deal with, was the underlying structural problem that the number of faculty positions at American academic institutions after 1970 was not rising. Unlike their predecessors of the expanding 1960s, the postdocs after 1970, when the women began to come in large numbers, were all too often left with an uncertain future. Thus, such a person could find herself working long hours and making many sacrifices in order to prepare for a future academic post that, in the ultimate betrayal, might well not exist after all.

Highly mobile and heavily international, they were a variegated, anxious, and increasingly discontented underclass. A few were on the fast track to top positions in academia; others would head for posts in the rapidly emerging biotechnology industry (discussed in chapter 8); and still others ended their careers in scientific research and, highly trained though they were, moved onto other, perhaps related activities. Most left quietly, as recommended in one of the pioneering guides to life as a postdoc, "On Choosing a Postdoc," by Virginia Walbot, published in the October 1979 newsletter *Women in Cell Biology.* It advocated quiet acceptance as the key to success: get along, be agreeable, work hard, do your best, do not complain, and move on.[3] Postdocs who were liked and contributed something to a project got the best letters of recommendation. Back then changing the system was not even considered; fellowships were something to be thankful for, not criticized and remodeled. But in the late 1990s the increasingly strident voices of the discontented began to be heard on the web, in *Science, Nature,* and the *Scientist,* and even at hearings before congressional hearings, as a few spoke out. In response, blue-ribbon committees (which had no postdocs as members) wrote reports on their economic plight, but their only proffered solutions were higher salaries, better health plans, and of course more federal money for even longer postdocs. Finally the postdocs began to organize and speak up for themselves. A national association of postdoctoral fellows was formed in 2003 with some support from the Alfred P. Sloan Foundation, but its agenda, as described on its webpage, was (deliberately) modest and helpful rather than strident and political. If there were any potential "troublemakers," this was where they would and perhaps should be. Few came forward, however, as most postdocs do not have the time, the skills, or the inclination to take on such a task. After, all they think they are postdocs only temporarily and need only look out for themselves.

Prestigious Postdoctoral Fellowships: Cutbacks and Criticisms

Overall, increasing women's proportion of prestigious fellowships was not the effective "entering wedge" to a more diverse academia that it might have been. The period started off on a gloomy note when in 1972, because of the rising costs of the war in Vietnam, President Richard Nixon terminated both the regular and the senior postdoctoral programs of the National Science Foundation. The former, in existence since 1952, had by 1972 awarded a trickle of about six fellowships per year to women (131 of 3,023, or 4.3%). These programs were never resumed, although for a time (1975–81) the NSF offered "energy-related postdoctoral fellowships" under the nonprestigious rubric of "national needs." Later, various parts of the NSF, such as the mathematics division and the ecology program, offered postdocs of their own, and in 1990 the Directorate for Biological Sciences and the Directorate for Social, Behavioral and Economic Sciences started a program for minority postdocs in the biological and social sciences.[4]

Another sign of the times was that in the early 1970s Bernice Sandler, of the Project on the Status and Education of Women at the Association of American Colleges, and then Judith Nies, of the Women's Equity Action League (WEAL), began to criticize the prestigious fellowships programs harshly. Their effort started in 1972, when a highly regarded law firm in Washington, DC, rescinded a job offer to a female lawyer on the WEAL staff because it wanted to hire a male applicant who it had discovered was a former Rhodes Scholar. Incensed, she, Sandler, and others at the AAUW and the American Council on Education got a grant from the Exxon Education Foundation, hired an intern to collect systematic data, and held a conference in November 1972 for officials of "the largest and best known fellowship programs." They also published and circulated widely a scathing report on the low numbers of women holding about fifty national fellowships, emphasizing in particular Rhodes Scholarships and White House Fellowships. When contacted, officials at both the John Simon Guggenheim Memorial Foundation and the Alfred P. Sloan Foundation reportedly were astonished that anyone would want data by sex from them, as no one had ever asked for it before and it had never occurred to them to break it down in this way. In 1974, 1976, and 1981, Judith Nies published three further reports on women's uneven progress in obtaining prestigious fellowships. She found that the adverse publicity helped temporarily, as the number of women awardees was higher in the late 1970s than earlier, but then it dropped off again.[5] Meanwhile, adding to the clamor, several women's caucuses, for example, the Committee on the Status of Women in Physics (CSWP) and the Women Chemists Committee (WCC), aided the cause when they began to report in their newsletters the names of any female Sloan fellowship winners in their field. Even more hard-hitting were Mary Gray's newsletters of the Association for Women in Mathematics (AWM), which not only mentioned approvingly any winners in mathematics but also commented acerbically on the lack of any in most years. Gray called on members of the AWM to apply or to urge others under age 35 to do so. Such attention, even if negative, may have encouraged increased numbers of eligible young women to apply.[6]

In fact, after a dismal start in the 1950s the Sloan fellowships became something of a bright spot for aspiring women scientists in the 1970s. Named for its founder, the head of the General Motors Company, the Alfred P. Sloan Foundation awarded about seventy-five postdoctoral fellowships per year in just three fields—physics, chemistry, and mathematics—later adding neuroscience in 1972, economics in 1980, computer science in 1983, and computational and evolutionary molecular biology in 2002. Very few of these went to women. In fact it was not until 1964 that the Sloan fellowship was first awarded to a woman, the Canadian applied mathematician Charlotte Froese.[7] Yet as the overall size of the program grew, and presumably as the number of female applicants increased, the women's numbers rose from about 2 awards (of roughly 70, or 2.9%) per year in the mid-1970s to about 10 (of about 90, or 11.1%) per year by 1989 to 17 (of about 115, or 14.8%) in 2003. Interestingly, in 1975–77, just when WEAL's and others' criticisms of the prestigious postdoc programs were at their peak, the number of Sloan awards to women jumped to 12, and the total awarded temporarily ballooned into the nineties, before reverting to a more measured growth rate (see table 7.1). The first woman to serve on the selection panels was Patricia S. Goldman-Rakic, a professor of neuroscience at Yale, in 1984. Many who won fellowships in the 1970s and 1980s have since become established and accomplished professors.[8]

The Damon Runyon Cancer Research Foundation, established in 1948, awarded nearly two thousand fellowships in the years 1970–2003. What is unusual about this program is that since the 1980s more than 30 percent of its awards have gone to women, which seems high until one recalls that the proportion of women earning MDs or PhDs in biology was more than 40 percent by then. Many Runyon fellows have since had distinguished academic careers.[9]

A third, more fully evaluated program that shows the limits of these prestigious programs was that named for Lucille P. Markey and established after her death in 1982. In her last will Mrs. Markey, heiress to the Calumet Baking Powder fortune, which she had invested in oil and natural-gas properties in west Texas, and owner of Calumet Farms, which bred racehorses for the Kentucky Derby (which they won eight times), established a "charitable trust" that was to spend all its funds (eventually more than $500 million) within fifteen years of her death. The trustees set up a postdoctoral program that was limited to the biomedical sciences. It provided up

Table 7.1. Female Sloan postdoctoral fellows, by year and field, 1964–1980

Year	Number Awarded	Chemistry	Mathematics	Physics	Neuroscience
1964	54		Charlotte Froese		
1970	76	Jean'ne Shreeve			
1972	79			Virginia Trimble	Melitta Schachner
1973	79	Alexandra Navrotsky			
1974	78	Martha A. Cotter	Karen K. Uhlenbeck Joan Birman	Helen R. Quinn	
1975	86	Wilma A. K. Olson	Nancy Kopell	Tanya Atwater Glennys R. Farrar Beatrice Tinsley	
1976	91		Linda P. Rothschild Jean E. Taylor	Kathryn Levin Roberta Humphreys	Jennifer H. Lavail Sarah F. Leibowitz
1977	95	Wray H. Huestis Geraldine A. Kenney-Wallace Kathlyn A. Parker	Marina Ratner Judith D. Sally	Sandra M. Faber Susan W. Kieffer	Lily Kung-Chung Jan Helen J. Neville Mary Lou Oster-Granite Ann-Judith Silverman Betty Zimmerberg
1978	79	Sally Chapman			Victoria Chan-Palay
1979	78				Anne C. Bekoff Susanna E. Blackshaw Carla J. Shatz
1980	78	Barbara J. Garrison Marye Anne Fox Maria C. Pellegrini			Eve E. Marder

Sources: Annual reports of the Alfred P. Sloan Foundation, online at www.sloan.org/fellowships/list.

to eight years of support, for three years after the PhD or MD and continuing five years into a tenure-track position at a major research institution. The selection committee, which included one senior woman, first Janet Rowley and then Shirley Tilghman, closely monitored the progress of the fellows through yearly reports and an annual conference. A subsequent evaluation found that all of those who had stayed in academia—some had gone into the biotechnology industry—had earned tenure quickly, mostly at prestigious places, a sign that they had been like Mrs. Markey's thoroughbreds, the cream of their crop, racing for the roses. But of the program's 113 fellows (about 16 per year from 1985 to 1991) only 22 (17.7%) were women, including Cornelia Bargmann, later at Rockefeller University, and Jennifer Doudna, at Yale and then Berkeley. Although this number was less than half of what might have been expected (women constituted 37.3% of the PhDs in the biological sciences in 1990 and an even greater percentage of the new MDs), evidently no one criticized. Part of the problem was that one had to be nominated by one's university, which involved a lot of infighting among the faculty sponsors pushing for their protégés and left women as only 22.3 percent of the applicants. Thus, this "entering wedge" was about half as wide as it might have been for talented women bioscientists in 1985–91, mostly because the selection processes was dominated by men, who largely identified candidates like themselves.[10]

Besides these programs aimed at immediate or recent PhDs, three other "postdoctoral" programs awarded fellowships that could be taken at any stage in one's career. Prominent among these were the John Simon Guggenheim Memorial Foundation fellowships, available to scholars and scientists of any age and in any field, which dated from the 1920s, when two bereft parents set up a foundation to memorialize their deceased son. Remarkably stable and sustained by infusions of fresh family capital, the foundation awarded 351 fellowships in 1971, but because of inflation thereafter, it had to increase its stipends substantially and then, in the late 1980s, cut drastically the number of its awards, which fell to 160 in 1991, as shown in figure 7.3. This reduction particularly affected its twenty-two science and social science fields, which were cut back from about half to about a quarter of the total fellowships awarded annually. Thus over the decades when the number of women scientists who might have applied for a Guggenheim (ten or more years after the PhD) was increasing, the number of awards was dropping significantly. Nevertheless, they did surprisingly well over the years 1971–2000, garnering 263 fellowships, or almost 9 per year, more than double their 113 in the previous thirty years. By far most of these were in the fields of anthropology, psychology, and sociology, followed by biochemistry/molecular biology and linguistics, with hardly any (5 or fewer in 30 years) in applied mathematics, astronomy, computer science, earth sciences, engineering, geography and environmental studies, and mathematics. In

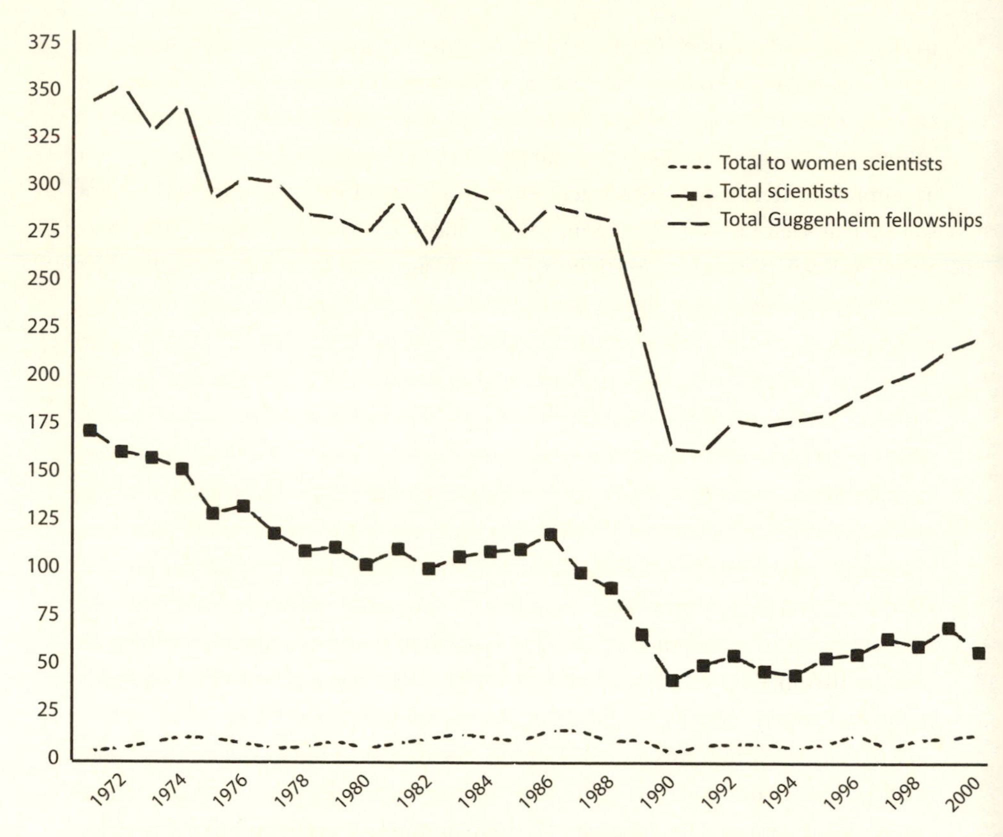

Fig. 7.3. Number of Guggenheim fellowships awarded to scientists, by sex, 1971–2000. Adapted from G. Thomas Tanselle, Peter F. Kardon, and Eunice R. Schwager, eds., *The John Simon Guggenheim Memorial Foundation, 1925–2000: A Seventy-Fifth Anniversary Record* (New York: John Simon Guggenheim Memorial Foundation, 2001).

statistics there were none. The published list of winners reveals a few female clusters, such as 4 psychologists in 1974, 5 sociologists in 1976, 5 anthropologists in 1996, and a stream of women in linguistics in the early 1980s. Several women on the move used the flexibility of their Guggenheims (which went to them personally and not to their employers) to facilitate a relocation. Thus the geneticist Ruth Sager, of Hunter College, used hers in 1972 to facilitate a move into cancer research and to the Harvard Medical School faculty; the political scientist Donna Shalala used one in 1975 to finish projects before going into a long career in administration; the demographer Judith Blake had one when she moved from UC–Berkeley to UCLA in 1976; the experimental psychologist Naomi Weisstein used hers in 1979 to leave SUNY–Buffalo and move back to Greenwich Village, where chronic fatigue symptoms overtook her; and the anthropologist Kay Warren had one in 1996, when she was moving to Harvard after two decades at Princeton.[11]

In 1980, when there were no women winners at all in the sciences and only five in the social sciences, Rada Dyson-Hudson, an anthropologist at Cornell, where the Cornell Eleven were starting their lawsuit against the university, wrote to Guggenheim president Gordon Ray not only to protest the low numbers of women winners but to point out also that by far most of the awards went to male full professors. She had thought younger scholars would have been favored. There is no record of his response, but the women's numbers increased in the next ten years, before falling back, showing once again that protest can make a difference for a while.[12]

Meanwhile, with considerably more fanfare the John D. and Catherine T. MacArthur Foundation, established in 1978, soon initiated its lucrative five-year, initially tax-free prize fellowships, soon dubbed by the press the "genius awards." One of its first fellows, in 1981, was the 79-year-old cytogeneticist Barbara McClintock, of the Carnegie Institution of Washington at Cold Spring Harbor, Long Island, who two years later won the Nobel Prize in Physiology or Medicine. The next women scientists to win a MacArthur prize were, in 1983, the mathematicians Julia Robinson, of the University of California at Berkeley, another older, highly regarded superstar, who had been elected to the National Academy in 1975; Karen Uhlenbeck, of the University of Texas; and the economist Alice Rivlin, of the Brookings Institution. The protein crystallographer Jane Richardson, of Duke, followed in 1985, a more interesting choice as she did not hold a PhD and worked with her husband at Duke University, where she was a research associate. Then in the mid-1980s very few women at all were appointed, for which the foundation incurred some criticism. Later the numbers of women fellows increased and included a series of anthropologists and archaeologists. In 1992 more than half of the fellows were women. It was as if the foundation, then headed by the historian Adele Simmons, and the fellows program, directed by the feminist scholar Catherine Stimpson, were making a concerted effort to identify and appoint women. The next year, the appointees included a cluster of six female biological and earth scientists. And starting in 1999 there were three to five women scientists per year from a more varied group of fields, including even engineering (see table 7.2).[13]

Besides these portable fellowships, which could be used anywhere, there were some institutes and centers that welcomed those with a fellowship and awarded some of their own, usually from block grants from foundations. Among these were the legendary Institute for Advanced Study at Princeton, New Jersey, with strengths in mathematics, physics, astrophysics, and social sciences. A year or two there early in one's career could be indispensable to later success in those fields, for in addition to the free time scholars had their to pursue their own work, they met daily at lunchtime and at teatime, interacting and forming friendships with others in their

Table 7.2. Women MacArthur Prize fellows in the sciences and social sciences, 1981–2005

Date	Total	Women	Scientists	Social Scientists
June 1981	21	3		Shelly Errington (anthropology)
December 1981	20	3	Barbara McClintock (genetics)	
August 1982	19	1		Francesca Rochberg (history of science)
February 1983	20	2	Julia Robinson (mathematics)	
August 1983	14	3	Karen Uhlenbeck (mathematics)	Alice Rivlin (economics)
March 1984	22	4		Sara Lawrence-Lightfoot (sociology)
				Heather Lechtman (archaeology)
				Judith Shklar (political science)
November 1984	25	3		Shirley Brice Heath (anthropology/linguistics)
July 1985	25	4	Jane Richardson (crystallography)	
August 1986	25	2		
July 1987	32	3		
August 1988	31	6	Helen Edwards (physics)	Ruth Behar (anthropology)
			Naomi Pierce (biology)	Anna Roosevelt (anthropology)
				Susan Rotroff (archaeology)
				Rita Wright (archaeology)
August 1989	29	8		Jennifer Moody (archaeology)
				Margaret Rossiter (history of science)
				Patricia Wright (primatology)
August 1990	36	12	Margaret Geller (astrophysics)	Sherry Ortner (anthropology)
			Mimi Koehl (biomechanics)	
			Nancy Kopell (mathematical biology)	
July 1981	31	9	Jacqueline Barton (chemistry)	Patricia Churchland (philosophy of science)
July 1992	33	17	Ingrid Daubechies (mathematics)	Evelyn Fox Keller (history/philosophy of science)
			Sharon Long (biology)	
July 1993	31	13	Maria Luisa Crawford (geology)	Nancy Cartwright (philosophy of science)
			Victoria Foe (biology)	Nora England (anthropology/linguistics)
			Jane Lubchenko (ecology)	
			Margie Profet (biology)	
			Ellen Silbergeld (biology)	
			Heather Williams (biology)	

July 1994	20	7		Faye Ginsburg (anthropology) Heidi Hartmann (economics)
July 1995	24	15	Sharon Emerson (biology) Susan Kieffer (geology) Pamela Matson (ecology)	Rosalind Petchesky (political science)
July 1996	20	7.5	Barbara Block (biology)	Joan Connelly (archaeology) Vonnie McLoyd (psychology)
July 1997	23	10	Eve Harris (biology) Nancy Moran (ecology)	Brackette Williams (anthropology)
July 1998	29	12	Leah Krubitzer (neuroscience) Rebecca Nelson (plant pathology)	Nancy Folbre (economics) Elinor Ochs (anthropology/linguistics)
July 1999	32	14	Jillian Banfield (geology) Carolyn Bertozzi (chemistry Laura Kiessling (biochemistry) Eva Silverstein (physics) Wilma Subra (chemistry)	Ofelia Zepeda (linguistics)
June 2000	25	12	Margaret Murnane (physics) Gina Turrigiano (neuroscience)	Susan Alcocok (archaeology)
October 2001	23	11	Lene Hau (physics) Cynthia Moss (animal behavior) Geraldine Seydoux (biology)	Kay Jamison (psychology)
September 2002	24	10	Bonnie Bassler (biology) Lee Ann Newsom (paleoethnobotany) Daniela Rus (robotics)	Janine Jagger (epidemiology)
September 2003	24	12	Deborah Jin (physics) Amy Rosenzweig (crystallography) Xiaowei Zhuang (biophysics)	
September 2004	23	13	Angela Belcher (chemistry) Daphne Koller (computer science) Amy Smith (engineering) Julie Theriot (cell biology)	Heather Hurst (archaeological illustration)
September 2005	25	10	Lu Chen (neuroscience) Claire Gmachi (engineering) Nicole King (biology)	Emily Thompson (history of technology)
Total	706	226.5	50	33

Source: MacArthur Fellows: The First 25 Years, 1981–2005 (Chicago: The John D. and Catherine T. MacArthur Foundation, 2005).

field from all over the world. Among the institute's women fellows in mathematics in the 1970s were Jean E. Taylor, a Princeton PhD with a faculty job at nearby Rutgers; Dusa McDuff, of England, who spent a year at the institute and then accepted a faculty post at SUNY–Stony Brook; and Judith Roitman, when she was between jobs at Wellesley College and the University of Kansas.[14] Similarly, the Center for Advanced Study in the Behavioral Sciences, adjoining the Stanford University campus in California, was a mecca for social scientists, broadly interpreted. Although there were far more women in the social sciences than in mathematics, Susan Ervin-Tripp was shocked to discover in the fall of 1974 that of the fifty fellows at the center that year, only three were women and only one was a minority. In the course of her time there she learned that one obstacle to women's appointment was the logistical one of a year away from one's family, but a second was that candidates for fellowships generally were already known to members of the selection panel before they applied. To expand this in-house list of future desired fellows, in the spring of 1975 Ervin-Tripp circulated a letter to many individuals soliciting names of suitable women future fellows. This may have been somewhat effective, for in 1978–79 one of the center's featured "special projects" was a study by a group of five female social scientists (the anthropologist Beatrice Whiting, the psychologists Janet Spence and E. Mavis Hetherington, the political scientist Nannerl Keohane, and the sociologist Jean Lipman-Blumen) on the topic "Sex Roles and Sex Differences." Among other women at the center over the years was the Berkeley sociologist and demographer Harriet Presser, later president of the Population Association of America, in the early 1990s.[15]

A third center with some potential for women scientists was the Bunting Institute of Radcliffe College, established in 1960, which had long had a fellowship or two for a woman scientist, generally funded by the Office of Naval Research. The marine biologist Llewellya Hillis-Colinveaux held it for two years in the mid-1980s, as did Mary Beth Ruskai, a chemist turned mathematician, shortly before Radcliffe College hosted the AWM's symposium on Sonia Kovalevsky in 1985. But laboratory science remained only a marginal pursuit there, because such a fellow would be off in another building most of the time. When as part of Radcliffe's 1999 amalgamation into Harvard University the Bunting became the larger, better funded Radcliffe Institute for Advanced Study, it began to appoint more scientists (men as well as women), including MIT and Harvard senior women faculty such as the protein biochemist Susan Lindquist and the physicist Lisa Randall, who needed a breather away from their labs but near to home. Perhaps more will happen there in the future, as its current director is Barbara Grosz, a renowned computer scientist.[16]

Because so few of these nationally competitive prestigious awards went to women, it was fortunate that about 14 percent of the many fellowships for women supported by the AAUW Educational Foundation went to postdoctoral scholars. By 1980 this was about fourteen per year in all fields of knowledge and rising. Among its scientist winners were the endocrinologist Patricia Brown, of Siena College, the chemist Mary K. Campbell, of Mount Holyoke, the physicist Cynthia Wyeth Peterson, of the University of Connecticut, and the botanist Nancy Slack, of Russell Sage College.[17]

Despite these and other programs, only a small number of these highly competitive national fellowships—about thirty per year—were going to women scientists in the 1970s, rising to possibly forty in the 1990s. The low numbers remind one of the 1997 study by two Swedish scientists who documented the presence of serious gender prejudice and cronyism on postdoctoral selection panels in that country and warned that when peer review is done in secret and is exempt from subsequent outside scrutiny, there is no assurance that it is done fairly.[18] In any case, these women winners represented something of an elite who were capable of obtaining a tenured post at a research university if they did not already have one. But they were just a small, visible fraction of the thousands of postdocs in science by the 1990s.

Regular Postdocs

Far more numerous and thus less prestigious were the thousands of postdoctoral fellowships awarded directly by the NIH for training on both its intramural projects in Bethesda, Maryland, and its extramural ones across the nation. Generally there was no lack of American and foreign candidates available. It was more a matter of mutual sorting through the many possibilities, with both sides identifying promising candidates, investigating them a bit, and then committing for a specified length of time before the competition did.

Ironically, the relative oversupply of good applicants came out of a political scheme for the ever elusive but frequently thwarted political goal of budget *reduction.* When President Richard Nixon faced budget pressures in 1974, he attempted to greatly reduce federal funds for training bioscientists. Congress, however, disagreed and arranged an ingenious way to seem to please both sides. It passed the National Research Service Award Act, which expressed a desire to limit the support for training by recommending that biennially a committee under the National Academy of Sciences (NAS) reassess the nation's manpower needs. But rather than cutting the level of support, as Nixon had hoped, this process opened it to continued expansion, for there were many reasons to assume that the nation's medical-research needs were constantly increasing: there were dramatic new therapies and scientific breakthroughs but also challenges like AIDS, resurgent age-old diseases like tuberculosis and diabetes, and an aging population facing Alzheimer's and Parkinson's diseases and macular degeneration. Once the NAS (hardly a disinterested party) claimed that an increase was justified, Congress, responsive to the public concern for improved health, voted ever-larger federal appropriations. Thus the funding of postdocs became the beneficiary of an open-ended, demand-driven budgetary process.[19]

Moreover, in the early 1980s this Committee on National Needs for Biomedical and Behavioral Research Personnel decided that because of the continuing "PhD glut," it should deflect its training funds less toward graduate students, of whom there were already too many and whose tuition had to be paid, and more toward postdoctoral fellows, which included those trained abroad, and over the next several years it shifted its emphasis accordingly.[20] There was little attention to, or concern for, the long-term academic prospects of these trainees, even though they might feel that this federal willingness to "train" them to ever higher levels implied a kind of commitment to supporting or employing them to continue using those skills later on. Meanwhile, the practice of supporting a growing number of postdocs was spreading to other federally and privately supported fields. Although two-thirds of postdocs were in the biological sciences and health fields in 1990, that too was changing. For example, by the mid-1990s two-thirds of physics doctorates were taking postdocs (compared with fewer than half in 1982), as were about half of those in astronomy. In other fields, however, such as mathematics, computer science, psychology, and the social sciences, postdoctoral fellowships remained rare.[21]

Because most federally funded research was concentrated in relatively few (under fifty) educational institutions and research institutes, the postdocs were centered on certain campuses and medical schools, often in big metropolitan areas where their spouses or partners might also find a post. One particularly large concentration was at the NIH itself, in Bethesda, Maryland, which awarded roughly twenty-two hundred individual fellowships annually. Principal among these were the Ruth L. Kirchstein Research Awards, named in 2002 for a distinguished, longtime bioscientist at the NIH. Some data on these F32 awards for 1984–93 indicate that in most years the success rate for men and women was almost identical. Generally, women received about 28 percent of the awards, but in one year, FY 1991, they won 39.6 percent. Even so, they were spread unevenly among seventeen institutes. At the most highly feminized end of the spectrum was the National Institute for Nursing Research, where 8 of the 11 postdocs (72.7%) were women; in the middle was the large National Cancer Institute, where 39 of the 141 postdocs (27.7%) were women; but at the National Center for Human Genome Research (NCHGR), only 1 of the 14 postdocs (7.1%) was a woman.[22] Thus, even at a large facility with thousands of postdocs they were distributed very unevenly.

Although many must have felt lonely and isolated in temporary positions in large laboratories at impersonal institutions far from home, there were initially few efforts to alleviate or overcome this situation. Women postdocs could and did join the local chapters of the national women's caucuses in their fields as they began to be formed around the nation, which enabled them to meet fellow specialists in the area. Then, starting in the late 1980s postdocs, in particular women postdocs, began to form their own groups. For example, in 1988 at MIT several such individuals established the MIT Association of Postdoctoral Women, or MITAPW, after the women's faculty group there invited them to a meeting, and Mildred Dresselhaus offered to pay for their monthly lunches out of the fund attached to her Abby Mauzé Rockefeller Professorship. Immediately the group set to work and filled a need. By 1990 members had written a guide for all postdocs at MIT, which covered topics from A to W, including such detailed entries as "authorship on publications," "childcare," "harassment," "medical care," and "women's organizations" (on and off campus), and had the name of a major woman scientist at the bottom of each page, constituting a kind of primer on the history of women in science. Members of the group had also negotiated a better health plan with the university, and a few had been appointed to relevant committees at MIT, such as on medical benefits and sexual harassment. They were no longer invisible hired hands or powerless migrant workers. They were settling in, becoming assimilated, and gaining a voice in MIT affairs.[23]

Similarly, postdocs at the Johns Hopkins Medical Center formed a group in the early 1990s that was officially recognized by the university in 1994, and the University of California at Berkeley started one in 1998.[24] Likewise, at the Salk Institute for Biological Research, in San Diego, California, established in the 1960s and a major consumer of NIH and March of Dimes funds, more than 280 postdoctoral fellows were in residence in 2005. With the constant coming and going of so many people, some felt the need to make an effort to "foster a sense of community" among the research fellows. Starting in November 1998, for example, the members of the Salk Society of Research set up an activities committee, which came to be composed mostly of women. This committee hosted a variety of events, such as monthly "SalkFESTs," a Friday-afternoon social time, "chalk talks," career days, an annual poster day, invited speakers, and other events. In 2003 the New York Academy of Sciences took advantage of its metropolitan location to host a monthly get-together for graduate students and postdocs from sixteen biomedical institutions in the area.[25]

Success Stories

This fluid international system of postdoctoral fellowships produced some strikingly successful women scientists, for these were the years in an individual's career that could make a big difference if one had chosen well and was in the right place at the right time. If they were well recommended, they joined strong groups, got onto interesting projects, and discovered something important or solved a significant problem, and by presenting papers and coauthoring articles they got at least partial credit for their work. Among these success stories was that of the physicist Gail Hanson, who just two years after earning her PhD at MIT joined the collaboration that was using the state-of-the-art colliders and detectors at the Stanford Linear Accelerator in California to discover the presence of hadron "jets," which confirmed the existence of quarks. Other success stories were those of the Australian cell biologist Elizabeth Blackburn, who in Joseph Gall's laboratory at Yale in 1975–77 identified telomeres, the short DNA sequence that forms the end of a chromosome, and their enzyme, telomerase; the molecular biologist Shirley Tilghman, who worked with Philip Leder in cloning the first mammalian gene at the NIH in the mid-1970s; the immunologist Linda Buck, whose work with Richard Axel at Columbia culminated in a 1991 paper on olfactory factors that won them both the 2004 Nobel Prize; and the neuroscientist Cornelia Bargmann, who used her postdoc with H. Robert Horvitz at MIT to move on to chemosensory behavior in *C. elegans,* the basis of her subsequent career.[26]

For the mathematician Sun-Yung Alice Chang, a series of short-term postdoctoral stints kept her alive professionally until she could land a suitable academic post at UCLA. The case of the unusually talented and determined astrophysicist Beatrice Tinsley illustrates the potential as well as the drawbacks of a series of short-term appointments. New Zealand-born and educated, she married an American, followed him to the United States, and earned her PhD at the University of Texas in 1967 with a pathbreaking dissertation on cosmological models. But the university would not hire her, and she then endured eight long years of postdocs and other temporary positions, a kind of walk in the wilderness, before finding a suitable academic post at Yale in 1975 and then receiving a Sloan fellowship in 1977. In a way it was fortunate that she had so much research time and published so many articles (thirty-six) early in her career before settling down at Yale, for her time there was cut tragically short by her death at age 40 in 1981. But by then she had trained several women graduate students and, as her obituary in *Physics Today* put it, "changed the course of cosmological studies."[27]

Drawbacks and Abuses

For a short period of time a postdoctoral fellowship might well be the exhilarating experience it was designed to be, but the situation could deteriorate. In 1977 the biologist Barbara Mandula, a former postdoc, wrote an article aimed at fellow women bioscientists that was tellingly entitled "Perpetual Postdocs." In it she said that the first two years were fine, but after that it was time to move on. Initially such posts looked attractive to recent doctorates because they offered the best of graduate school without the faculty pressures: they did not involve teaching or require grant proposals, and they allowed one to do research day and night in a well-equipped and, one hoped, safe laboratory. Yet over time this very lack of responsibility and status began to rankle, as a fellow remained dependent on the principal investigator (PI), who wrote (or at least signed) the proposals for the fellow's next project. Although fellows might begin to acquire a certain seniority in the laboratory hierarchy, they had no real authority in deciding how the work was to be done, which experiments to do next, or what and where to publish. They remained dependent and disposable despite having been trained to become independent. Although inertia and a desire to wait for the perfect job might tempt one to linger, it was better in the long run, according to Mandula, to move on quickly.[28]

Some postdocs complained of overwork. Their supervisors expected them to be in the lab all the time, including evenings and Saturdays. One extreme case occurred at the University of Idaho, where a British postdoc resigned after the chemistry professor Jean'ne Shreeve, who was also the vice provost for research and graduate studies, required him to sign a contract agreeing to work a fifty-five-hour week, a requirement that Shreeve felt was necessary since there had been misunderstandings with previous postdocs. Feeling exploited, the postdoc in this case refused and returned to England.[29] Despite the efforts of various individuals over the years, starting in the 1970s, there was usually no provision for part-time postdoctoral work. Even if it was feasible and humane for young parents, it was never enacted. Evidently it ran counter to the workaholic culture of upwardly mobile postdocs. As a result, when exhausted women did drop out, even temporarily, they were called "weaker sisters."[30]

Nor, until recently, was there any provision for pregnancy leave, a major concern for women in their twenties and thirties. Although some postdocs were considered staff members at some institutions and might become eligible for six weeks of paid leave, their supervisors' grants did not necessarily have the funds to pay both for the leave and for a substitute, even if a reasonable substitute could be found locally on short notice. Furthermore, the PI's research plans counted on the postdocs' being there to do the work toward the PI's next publication or presentation. Thus, to go on leave, even unpaid leave, would jeopardize the research project and could lead to a lukewarm letter of recommendation. In fact the expectation that a female postdoc would get pregnant and drop out long deterred the organic chemist (and Nobelist) Herbert Brown from accepting any women postdocs in his world-famous laboratory at Purdue. Finally in 1980 he accepted the persistent

Donna Nelson, later at the University of Oklahoma. When she did become pregnant a year later she disguised her state in extra-large lab coats, and since all went well, she was able to get back to work on a part-time basis within three days of her delivery. But for others the outcome was less pleasant. One set of deteriorating circumstances led a 38-year-old physics postdoc at SUNY–Stony Brook to file a lawsuit against the university in 2005 because her adviser had failed to provide her with a letter of recommendation and because of the subsequent derailment of her anticipated career in physics.[31]

Other studies revealed other problems. One 2005 survey found that half of the minority postdocs in one NIH-sponsored program had no mentor and were ignored.[32] Perhaps worse than the crushing workload and the future job uncertainties was the occasional loss of credit for one's work. The division of labor in most laboratories meant that though the junior personnel did most of the team's actual hands-on research, it was the PI who presented their results at national meetings and who was the lead author of most of the group's publications. Thus, the PI routinely took credit for work that he or, more rarely, she had not done personally. This was justified on the bureaucratic grounds that the PI had designed the project, survived the scrutiny of a selection panel, and obtained the grant that purchased special equipment, paid the added staff, and reimbursed the university for the cost of the lab space, utilities, security, library usage, and everything else included in the "overhead." This practice also helped boost his reputation and thus could lead to his getting tenure and future grants to continue the laboratory's work. But the postdocs also needed publications in order to build their reputations and launch their own careers elsewhere. Occasionally these practices slipped into exploitation and abuse, and once in a while they led to publicity in *Science* or *Nature* about denied recognition and even to lawsuits about the theft of ideas.

In 1991–93, for example, Sarvamangala Devi, a training fellow at NIH, developed a vaccine that was effective against a fungus that caused meningitis. Although she was the lead author, ahead of her two supervisors, on a paper published in the *Proceedings of the National Academy of Sciences,* the *Washington Post* reported the work as having been led by them. (The *Post* later printed her name as part of a clarification.) When subsequently she developed a vaccine for use against a fungus that caused infections in AIDS patients, the NIH listed her as "first or sole inventor" in its patent application. But then her fellowship ended, and she left for another postdoc at the Food and Drug Administration. When her successor at NIH took up her work, he claimed that he could not find her notebooks and took credit for the whole project. When Devi protested, he offered to add her name to his manuscript, but she complained that she had never seen it. Attempts by nonparticipants to mediate the dispute were largely unsatisfactory. The case attracted a great deal of publicity into 1995, largely because of the time lost in getting the vaccine into clinical trials.[33]

Another case of possible theft of ideas seems to have occurred to Reva Kay Williams, an African American astronomer with a PhD from Indiana University. In the course of identifying and applying for fellowships and postdocs in astrophysics in the early 1990s, she revealed some of her own ideas and criticized those of others. Then not only did she not get the fellowships but her ideas were subsequently presented without attribution in the papers of some of the people (and their students) on the selection committees. Her attempts at publishing these ideas were repeatedly rebuffed by major journals, whose peer reviewers in the small world of astrophysics were the very same people. Feeling used and abused, she managed to present her side of the story in a new online publication that did not involve peer review.[34]

Sexual harassment also could be a serious problem for postdoctoral fellows, and since any witnesses might be unwilling to testify, it could be hard to prove. Even if they did speak up, the harassers' buddies were not likely to penalize him. Some outside publicity might therefore be advantageous. Two cases at the NIH in the early 1990s received much newspaper attention, but in the end neither the psychiatry fellow Margaret Jensvold, of the National Institute of Mental Health, nor the neurologist M. Maureen Polsby, of the National Institute of Neurological and Communicative Disorders and Stroke, won her case. In 1998 the U.S. Court of Appeals for the Fourth Circuit refused to hear Polsby's last plea, commenting that "bad mentoring," as it was called, was not covered by existing laws.[35]

Whistle-blowing was another ethical dilemma for some postdoctoral fellows. When they witnessed (or thought that they did) others committing fraud in science, they might report this to the authorities, as did the bioscientist Margot O'Toole at MIT in 1986. But doing so could backfire on them. O'Toole's case was badly handled at the local level and led to the decade-long "Baltimore case," which quickly involved not only MIT but also the NIH and even the U.S. Congress in one of the most highly publicized causes célèbres in decades. From the start the Nobelist David Baltimore, coauthor of the paper in question, defended his colleague without looking at the evidence. After he stereotyped O'Toole "a disgruntled postdoctoral fellow," hardly anyone in the scientific community supported her (valid) claims. In the end the immunologist Teresa Imanishi-Kari, by then of Tufts University, was exonerated of deliberate fraud, but everyone lost something in this bizarre case, most of all Baltimore, who resigned the presidency of Rockefeller University, and O'Toole, who lost her job (a fourth postdoc, which was about to end) and had a hard time finding another.[36] With increasing numbers of vigilant and idealistic postdocs, chances for such accusations may rise, but the institutions' need to preserve their grantsmen's reputations and funding will cause them to cover up the evidence, scapegoat the accuser, and try to dispose of such claims quietly.

Protest and Some Reforms

By the mid-1990s postdocs—lingering too long, with no "real" jobs in sight, complaining about their lack of independence and status, wanting better health plans, and planning to settle down and raise families—found their voice, got organized, and protested on both the campus and, increasingly, the national level for better treatment. The coming of the Internet made it possible for postdocs to speak for themselves, organize on a national scale, and thus exert some political clout on their own behalf at minimal cost. First they organized the Young Scientists' Network, a web-based coalition, which published many articles on their plight in scientific journals. These began to attract sympathetic press coverage in *Science, Nature*, and the *Scientist,* which also started an annual survey of the best places to be a postdoc. The increasing press coverage also led to an occasional invitation to testify before Congress on scientific manpower issues alongside NSF and other experts on future manpower needs. The postdocs' personal complaints were fortified by statistical data on the continuing prolongation of the postdoctoral experience, such as the ominous fact that the average age at which a scientist got his or her first NIH research grant was by 2007 approaching 42, compared with 37 back in 1980. A few women were involved in some of these efforts, which generally centered in Washington, DC, though they must have been ambivalent, for in some ways such short-term posts served some of them personally at least as well, arguably better, than the tenure-track posts to which they presumably aspired.

In response to the newly vocal postdocs, the National Research Council (NRC) set up two committees that wrote reports on the postdoctoral experience in 1998 and 2000. Both favored some reforms but opposed any major changes that might cripple a good thing. Oddly, but symptomatic of the paternalistic way (or maternalistic, since two were chaired by senior women and involved the active participation of others) the biomedical establishment continued to treat the problem, the postdocs themselves, major stakeholders in the issue, had only marginal representation on these committees or none at all. One wonders what they would have said had they written the reports themselves. Also, in the tradition of earlier National Academy reports on postdocs, their gender was barely mentioned.

The 1998 committee, chaired by the molecular biologist Shirley Tilghman, of Princeton University, admitted that it was not in the faculty's interest (or that of their employers, which were often major universities highly dependent on the considerable overhead that came with large grants) to create much change. Nevertheless, there were problems, and so the committee recommended even more federal aid for senior postdocs to prepare them for impending independence. (This was later taken up by another NRC committee, and in 2006 the NIH introduced such a program, with a proposed 200 two-year awards of up to ninety thousand dollars each, called the NIH Pathway to Independence Awards.)[37] The 2000 report, chaired by the biochemist Maxine Singer, of the Carnegie Institution of Washington, and with the active involvement of MIT's Mildred Dresselhaus, reanalyzed the problem with better data. The authors highlighted the best practices at some places and deplored the chances for abuses. This report and the press response to it quickly induced the NIH, whose budget had doubled in the late 1990s, to give its postdocs the first raise in many years in 2001.[38]

Shortly thereafter, in January 2003, the National Postdoctoral Association was formed with a grant from the Alfred P. Sloan Foundation. The association has hired an executive director (Alyson Reed, formerly the executive director of the Maryland Commission for Women), is housed at the Association for the Advancement of Science headquarters in Washington, DC, has formed "collaborations" with other interested groups, claims it is not a union, and promises to work to improve conditions.[39] Meanwhile, many universities are forming their own offices of postdoctoral services within the graduate school. These offer some standardization of personnel procedures and certain useful services, such as grievance procedures, workmen's compensation, parking privileges, and placement assistance, which can alleviate the postdoc's former total dependence on the PI (who can often be a big part of the problem). The immense challenge of any meaningful reforms, however, may be evident from the experience of Jean Fort, the assistant vice chancellor for research at the University of California–San Diego, who spent six years on the Herculean task of creating one uniform health plan for the six thousand postdocs at the ten UC campuses, regardless of their funding sources.[40]

In 2005 Sigma Xi, a national association of scientists, released the results of its own survey of postdoctoral fellows, "Doctors Without Borders." Funded by the Alfred P. Sloan Foundation, the survey was administered through forty-five institutions, including twenty-seven universities and thirteen biomedical institutions. It got 7,537 responses, including 3,182 (42.2%) from women, of whom 2,488 (78.2%) were in the life sciences. Only 1,734 of the women (54.5%) were U.S. citizens or permanent residents, and only about 300 (4%) were African American. Eighty-one percent of those who responded were between the ages of 28 and 37 (but 9% were over 40). Sixty-six percent of the women (compared with 71% of the men) were married or partnered, but only 29 percent had children. Overall the respondents worked an average of fifty-one hours per week. Although most lived in expensive cities and spent a high proportion of their stipend on rent, they were less concerned about financial issues than about the quality and amount of their training. In fact the chief finding of the survey was that the postdocs wanted more direct, one-on-one training from their supervisors. How to get this was not clear. The NIH could mandate it, but that was unlikely.[41]

Chill Out

These successive reports on postdoctoral fellows offered surprisingly little information about what presumably preoccupied their thoughts, namely, what would happen to them next. Was all this expensive training being put to good use? What little we have on postdoctoral outcomes is scattered, almost happenstance, as if no government agency really wanted to know where they went. One study, for example—"The PhD—Ten Years Later," by Maresi Nerad and Joseph Cerny, of the University of California Graduate Division—collected data on the Berkeley PhDs of 1982–85 in six fields (biochemistry, computer science, electrical engineering, English, mathematics, and political science) who had held postdoctoral fellowships at a total of sixty-one universities. Ten years later most of those men and women who had completed postdocs in biochemistry were tenured, as were such men in mathematics. But the women in mathematics who had held postdocs were most likely to have left academia altogether to work for business, government, and nonprofit employers. The authors did not say why.[42]

Another study looked at faculty hiring at land-grant institutions. It showed that postdoctoral training was more common among women than among men hired for academic posts, but it was not clear why. In 1979 twice as high a percentage of the women (31.6%) as of the men (15.4%) hired by fifty-two such institutions in the agricultural sciences had held postdoctoral fellowships. By 1996 the rates for both sexes had increased markedly, but the gender gap persisted, with 41.2 percent of women and 27.0 percent of men hired in the agricultural sciences at land-grant institutions having held postdocs. What we can conclude from this is not totally clear. It would seem to indicate that postdoctoral training was worthwhile for women, but was it because the women needed the higher qualification of a postdoc to be hired? or that the women hired were in those laboratory fields in which postdocs were common, while the men may have been in those fields that did not require such advanced training, such as, perhaps, forestry or agricultural engineering?[43]

A different, almost inadvertent approach revealed an unsettling phenomenon that received little attention, as if appropriate job placement was now automatic. The sad fact was that in some fields the women postdocs' chances of getting a faculty post were only about half those of the men postdocs. For example, the physiologist Dean Smith, of the University of Wisconsin, noted in 1992 that women constituted 43 percent of the student members of the Society for Neuroscience and 36 percent of the PhDs in the field. They got 38 percent of the F32 postdoctoral fellowships awarded by the National Institute of Neurological Diseases and Stroke. Yet only 18 percent of the applicants for faculty positions, and just 12 percent of those hired, were women.[44]

From these ominous findings we can imagine that for many postdoctoral fellows what came next was not what they had desired initially. Instead of being swept smoothly into the next segment of the "pipeline," they faced a kind of jumping-off point into an alternative future. Maybe those with prestigious postdocs won in national competitions had a kind of fast track into and through the best labs on their way to tenured faculty posts at top-ranked universities. But for the rest there were a variety of options, including the fast-growing biotechnology industry.

Conclusion

Increasing numbers of female PhDs from the 1970s through the first decade of the twenty-first century, especially in the biological sciences, were holding postdoctoral fellowships. A few held prestigious ones earned in national competitions, chose among the top labs, had a positive experience, won other additional honors, and moved on to join the best departments. They were pioneering in realms where few women had gone before. But there were not very many of them.

Hordes of others, however, often supported by the NIH, constituted a highly skilled labor force that moved among laboratories in their field, looking for the big break that could put them on a path to a successful career. They faced a kind of unspoken weeding-out process with no time limit or clear cutoff point. One could linger as long as there was money, which made it even harder to give up or move on. All that education, from K–12 on up to a PhD and beyond had led to five years or so on someone else's big-science project, but now what? What they were to do for the rest of their lives or careers was up for grabs. There had always been a big gulf, especially for women, between education and employment. Now that so many of them were persevering to the end of an even longer educational stream, the hit-or-miss scrambling for opportunities seemed like an afterthought. Was it because so many of them were women that there was so little concern for their future? Maybe now that they are getting organized and heard in Washington, DC, they will begin to call for some structural changes to fit the realities of scientific careers in the twenty-first century.

Industrial and Self-Employment

Entering Wedges and Entrepreneurs

Corporate America took notice when, in January 1973, the federal Equal Employment Opportunity Commission settled its landmark lawsuit against the American Telephone and Telegraph Company, then the nation's largest private employer, demanding an epic $38 million penalty for pervasive sexual and racial discrimination. Before long, companies began to hire dramatically increased numbers of women, mostly young, in engineering, geology, and other technical fields in which previously they had routinely passed them over.[1] Thus in the mid- to late 1970s, when women in or near academia were disheartened about the slow enforcement of other equal-opportunity laws and the loss of the Sharon Johnson lawsuit at the University of Pittsburgh, those in industrial circles were quite optimistic about this unprecedented surge in the hiring of young women for entry-level jobs.

Analysts at the time were confident that this influx into the manpower "pipeline" would lead to continuous upward mobility and that lasting change was within sight. But as events would show, recruitment and initial hiring, even at an equal or better than equal salary, was no guarantee of sustained fair treatment. In fact, getting a first job was the easiest step, readily monitored by the federal government. It did not necessarily lead to a satisfying work history; instead, as this generation of female industrial scientists and especially engineers would find out, it was just the start of a career-long struggle with a succession of gender issues.[2] These women were in fact the "entering wedges" in an incremental modification of the corporate workforce that would continue for decades. Many would take an active role in changing their workplaces or in starting new ones, but only a trickle would emerge as corporate officers in the 1980s and 1990s. Among these "exceptions" who managed to reach new levels, generally that of vice president, or otherwise attract national attention were the engineers Betsy Ancker-Johnson and Esther Conwell, the computer scientist Anita Borg, the biotechnologists Mary-Dell Chilton, Gail Naughton, and Vicki Sato, the social scientists Marina von Neumann Whitman and Sandra Wood Scarr, and the chemists Esther Hopkins, Mary Good, Kathleen Taylor, and the greatest of them all, Gertrude Elion, who with only a master's degree actually won a Nobel Prize.

Overall, American private industry greatly increased its employment of scientists and engineers in the period 1973–99. The federal Census Bureau and the Bureau of Labor Statistics (BLS) calculated that the total number of full-time scientists and engineers working for manufacturing industries nearly doubled, from about 342,400 in January 1973 to 659,100 in January 1999, with the largest employers in the areas of electrical equipment, aircraft and missiles, machinery, and chemical and applied products. But growing even faster than these manufacturing industries was a collective category called "non-manufacturing," or service, industries. Here the expansion between 1973 and 1999 was from a paltry aggregate of 15,300 scientists and engineers to a substantial 338,700—more than a twenty-two-fold increase. This was largely, but not entirely, the result of the growth in computers and data-processing services, areas that were just beginning in 1973.[3]

This overall growth was, however, highly uneven, for America's high-tech industries were volatile and subject to boom-and-bust cycles. First the aerospace industry collapsed about 1970, after a sudden drop in federal spending. With the coming of the "energy crisis," the oil industry grew rapidly in the late 1970s but then downsized just as quickly in the mid-1980s, when many companies merged, consolidated, and reorganized, laying off many of the most recently hired. The computer industry, with IBM, Apple, Microsoft, and others, grew enormously despite layoffs in the mid-1980s, as production moved from mainframes to mini- and then microcomputers. By then a whole new industry of biotechnology, with its many small and financially insecure "start-up" companies, was emerging. Meanwhile, the already large pharmaceutical and chemical industries generally kept growing and diversifying despite successive recessions, layoffs, and retrenchment in the 1980s and 1990s. All this expansion and consolidation created not only opportunities but also difficulties for individuals planning careers in industry.

But just how many of these scientists and engineers were women is hard to determine, since, astonishingly, neither the Census Bureau nor the Bureau of Labor Statistics broke its published data down by sex. This means that to find women scientists and engineers, one must rely on various surveys and statistics obtained by other sources, such as the Engineering Manpower (later Workforce) Commission, the Institute of Electrical and Electronic Engineers (IEEE), the American Chemical Society (ACS), the American Association of Engineering Societies (AAES), and the American Geological Institute. These organizations sometimes collected data in collaboration with groups of largely industrially employed women, such as the Society of Women Engineers (SWE), the Women Chemists Committee of the ACS, or the Association for Women Geoscientists (AWG), established in 1977. Occasionally these groups did their own surveys, from which one can get a partial glimpse of the situation and issues facing women scientists and engineers in relevant disciplines. (Because of the availability of these data and the fact that many industries hired scientists and engineers in a variety of fields—for example, the chemical industry hired biologists, agricultural and medical scientists, and engineers as well as chemists—this chapter is largely organized by scientific fields rather than by industries.) The surveys are generally disappointing, however, in that they tended to document what is now obvious—that the numbers of women in industry were growing, that most of the women were young (under 30 or 35), many were unmarried, and few were mothers, and that starting salaries were rising but, as in Zeno's paradox, mid-career salaries never caught up with the men's. Only the chemists had enough long-term data to analyze the central issue of advancement, but in that field the older women, whose salary inequities were not corrected and even increased over time, tended to give up and retire early. This overall absence of data on the (lack of) progress of mid-level women, which is of greater interest and importance, is unfortunate, for their chief difficulties, especially at large companies, where many of the technical jobs were, centered on the "glass ceiling," a term coined in the *Wall Street Journal* in 1986.[4]

One remedy was to leave industry for self-employment, and after the mid-1980s women increasingly took this step, becoming consultants or entrepreneurs starting (or, rarely, inheriting) their own companies. Although women physicians and clinical psychologists had long had their own private practices, self-employment was rather rare for women scientists and engineers before the 1970s. In some fields, such as geology and geophysics, men had been doing this all along: after a decade at a big company or a consulting firm, they would start their own office, take on clients, and set their own hours and fees. As other women scientists and engineers grew disaffected with their treatment in the corporate or academic workplace, more of them learned how to set up a business of their own and took the plunge. In response to this nationwide trend, the Small Business Administration started a special initiative to help women entrepreneurs and even set aside special funds for woman-owned businesses. Whether these women wanted revenge or relief, they had to be willing to take risks. If one had a skill or a product that the marketplace might want, there were satisfactions in cutting out the bothersome bureaucracy and becoming one's own boss. As one article in the *AWIS Magazine* put it in 2002, "The flame of entrepreneurship is lit when you recognize you do something better than the others who are doing it."[5] This strategy was not going to change industry at large (as a victorious class-action lawsuit might), but it did open up new opportunities and possible satisfactions one company at a time.

In fact, women's elevated exit rate from industry, substantially higher than that of men, began to attract attention in the 1980s. So great was this retention problem that some large corporations that relied on government contracts started special programs in mentoring and management training for women, as well as offering flextime, parental leave, adequate day care, and other family-oriented benefits for everyone. In the 1990s the newly reestablished Committee on Women in Science and Engineering (CWSE) of the National Research Council called attention to these new programs in *Women in Industry: Why So Few?,* and Catalyst, Inc., a New York nonprofit organization with corporate support, publicized the continuing lack of women on corporate boards of directors and the near total lack of minority women at top levels.[6] In 1998 a woman editor at *Fortune* magazine took a more positive attention-getting tactic and started an annual special issue on the "50 most powerful women in business," thereby creating a new category of press coverage: the female business celebrity.[7] Thus by 2000 for most women in industry some doors had opened and starting salaries were just about equal, but a host of issues surrounding retention and advancement still needed to be addressed and solved.

Chemists

By far the largest group of women scientists in industry, especially pharmaceuticals, in the 1970s and after were the chemists, a group that had been hired (and underemployed) since the emergency of World War I.[8] When in 1975 the very large American Chemical Society—with more than eighty thousand members at the time—prepared its annual salary survey, it accompanied its report with its first separate supplement on women chemists, who then constituted 7.8 percent of the working members of the ACS (employed, under age 71, and not students), most often at the BS level and working in applied research. Traditionally they were concentrated more than men in the areas of writing, editing, library and literature work, and "other." They were rarely managers.[9] Later surveys in 1980, 1985, 1990, and 2000, in which the ACS's Women Chemists Committee, established in the 1920s, took an active part, also included questions about the women's perceptions of their situation. Were they satisfied with their salaries, their chances for advancement, and their work environment? if not, why not? Although chemistry, like engineering, is an immense field with many subcategories—analytical, organic, inorganic, and physical—each of which leads to different job prospects, the quantitative questions resulted in many tables and charts showing that overall the women were paid less than men at every level, in every specialty, by every type of employer, and for every job function, and the situation worsened the longer they were in the workforce. The qualitative remarks indicated that the oldest women, mostly at the BS level, were the most discouraged, or as Nina Matheny Roscher, the compiler of the 1990 survey, put it succinctly, "With experience there is a continuing disillusionment among women."[10]

Yet the exhaustive 1990 report found that over these years there had been some quantitative changes. The number of women chemists had increased to 15,247, their percentage of the ACS membership had risen to 17.3 percent, and more and more PhD women were forsaking academia and going directly into industry. The proportion employed in the pharmaceutical industry was higher than that in synthetic chemicals or polymers. By 1990 there was an occasional woman manager, but more than 40 percent knew at age 35 that they had no future in management. Some respondents raised issues of sexual harassment, and several volunteered that they were looking around for another company.[11] In 2000, when 24.8 percent of ACS members employed in the United States were women, their satisfaction rate with their chances for managerial (as opposed to professional) advancement had risen substantially, from a low of 49.3 percent in 1995 to 62.5 percent, perhaps because of changes they had worked to bring about in the workplace.[12]

Other evidence shows that by the mid-1980s women employees at several of the major pharmaceutical companies had begun to organize on the job and push for workplace reforms. This generally involved creating a company-approved women's group. One of the first and most successful of these was at Hoffman-La Roche, then headquartered in Nutley, New Jersey, where in 1971 a group of women began to meet off-site in the evenings. Perhaps fearful of where such discussions might lead, upper management encouraged them to become a company-approved employee organization and meet at the plant during lunch breaks. The company also provided a modest budget for occasional outside speakers. Named Concerned Women at Roche (CWR), the group had several early victories: permission for women to enter the plant after working hours; the choice of being listed as "Ms." in the company directory; the opportunity for returning employees to count their earlier time served toward seniority or pension benefits; and in 1974 the option of wearing pantsuits to work. A big accomplishment was the creation of one of the first on-site childcare facilities in the nation in 1977. It was assumed that improved benefits would aid recruitment and that increased retention would lead to longer, more productive careers. This evidently happened, for Hoffman-La Roche was able to increase the number of women chemists on its payroll from twenty in 1975 to two hundred by 1983.[13]

All these benefits earned Hoffman-La Roche a place in the top five on *Working Mother*'s 1988 list of the fifty best companies to work for. (The very existence of such a magazine is in itself a sign of the nation's changed workforce.) The other four were Merck, Apple, IBM, and an advertising agency.[14] Although one purpose of this kind of invidious ranking was to goad other employers to emulate the companies on the list, word had already spread at least within the industry, especially after the regional and local ACS sections made the institutional innovation of setting up women's committees of their own. No longer did one need to undertake the cost and disruption of attending a national meeting to overcome a sense of isolation and meet other women chemists. One could simply attend a local monthly event. Especially active were the Metro Women Chemists, run jointly by the New York City and North Jersey sections; FemChem, of Princeton, New Jersey; and the section in Columbus, Ohio.[15]

Unlike the experience at Hoffman-La Roche and perhaps more the norm in the industry, the women's committee formed at Miles Laboratories, in Elkhart, Indiana, worked on its own, without corporate support, and the company won no awards. In 1975 a group of salaried women began having lunch together once in a while. As their numbers grew and arrangements became more cumbersome, they drafted a constitution and became an official employee organization, supported by their own dues. They held Saturday workshops on career skills, nominated Miles women for the local YWCA's annual woman-of-the-year award, and generally provided "a forum for discussion, education, growth, and guidance, so that we can achieve personal and professional fulfillment within the corporate environment."[16]

Other technical women's groups and individuals undertook or joined activities outside the company. For example, the technical women at the 3M Corporation, in Minneapolis, some of whom had been active in the NSF's Visiting Women Scientists Program in 1977–79, were, as mentioned in chapter 4, stimulated by its discontinuation to give their own talks and demonstrations at the local public schools. Similarly, at Merck the African American chemist Jeannette Brown, a Hunter College graduate with a master's degree from the University of Minnesota who later recalled that although she had done good work, she had never reached the senior level she had aspired to, got involved in corporate outreach activities to black chemists. After participating in the NSF's Visiting Women Scientists Program in 1979, she became active in Merck's liaison group with the historically black colleges and universities, which for a time allowed her to visit their campuses, recruit summer interns, and donate money to their college laboratories. She was pleased to be the one to present Merck's check for its corporate donation to Project SEED (Summer Experiences for the Economically Disadvantaged), an ACS program that continues today.[17]

At the immense E. I. DuPont de Nemours & Company, which had long employed female organic chemists, there were several women scientists of distinction, including the analytical chemist Mary Kaiser, the mass spectrometrist Barbara Larsen, the physical chemist Sandy Issler, and the agricultural biotechnologist Barbara Mazur.[18] At the rival Dow Chemical Company in remote Midland, Michigan, more than twice as many women obtained patents between 1977 and 1988 as at DuPont (354 compared with 136). Polaroid's Vivian Walworth was in 1981–85 the first woman president of the Society of Photographic Scientists and Engineers.[19]

Beyond these major companies, a handful of female industrial chemists had careers of distinction in other companies. The analytical chemist Jeanette Grasselli (Brown) was the highest-ranking female at SOHIO (Standard Oil of Ohio) until the late 1980s, when BP (British Petroleum) bought it out and abolished her laboratory, and she retired.[20] Helen Free, of Miles Laboratories, famous for its main product, Alka-Seltzer, worked for decades with her husband pioneering the development of a new line of home products, including easy-to-use colored paper strips for urinalysis.[21] Meanwhile, Kathleen Taylor, with her 1968 PhD from Northwestern University on catalysis, rose to become head of the physical chemistry department of the research and environmental staff at General Motors.[22]

Other women chemists decided after a few years of bench work at a company that they should make a horizontal move within the company, such as to patent work and regulatory affairs, a rapidly growing area. Among these was Pauline Newman, who received her PhD in chemistry from Yale in 1952. She then earned a law degree from New York University in 1958 and worked as a patent attorney for FMC Corporation until 1984, when President Ronald Reagan appointed her to the U.S. Court of Appeals for the Federal Circuit.[23] Similarly, the African American chemist Esther Hopkins, after earning a Yale PhD in 1967, went to work for Polaroid, known for its minority training and employment programs. During her twenty years there she earned a law degree and turned to patent work, eventually leaving to take a high position in the Commonwealth of Massachusetts.[24] Others found opportunities in regulatory affairs, keeping abreast of forthcoming changes worldwide, writing reports, planning strategy, and overseeing employees, tasks presumably more amenable to the "mommy track" than were the duties of some other positions. Anyone aspiring to a post in top management could also expect to serve a stint in Europe or elsewhere.[25]

In 1997 a survey found that women still made up only 4 percent of the upper management of the nation's nineteen top chemical companies, especially at the level of vice-president.[26] The pioneering physical chemist Dorothy Martin Simon was a vice president and director of corporate research at AVCO Corporation from the 1960s until her retirement in the 1980s.[27] Another physical chemist, Joan Berkowitz, went to work for Arthur D. Little, Inc., in 1957, after completing her NSF postdoctoral fellowship at Yale; twenty-three years later, in 1980, she was made a vice president.[28] A year later the chemist Mary Good, at age 50, left the University of New Orleans (formerly Louisiana State University at New Orleans) to become the first woman vice president and director of research at UOP, Inc. (formerly Universal Oil Products), in Des Plaines, Illinois.[29] In 1987 Polaroid finally appointed its first woman vice president, Carole Uhrich, who had served successively as assistant scientist, principal engineer, and plant manager.[30] It was not until 1992 that

Dow Chemical appointed its first female vice president for core research and development. Trained as both a chemist and an engineer, Jacqueline Kelyman had been with the company in a series of posts since 1956.[31] Not to be outdone, DuPont had seven women vice presidents and one woman senior vice president out of a total of sixty-six officers by 1999. In 2009 Ellen Kullman, a mechanical engineer with a master's of business administration, became its first female president and, along with Stephanie Burns, PhD, since 2004 the chief executive officer of Dow Corning, a joint venture of Dow Chemical and Corning, Inc., in Midland, Michigan, one of the two highest-ranking women in the chemical industry.[32]

Promotion to a management position was not, however, the only path to success for talented women chemists, as illustrated by several careers that were marked by considerable outside recognition. Several women industrial chemists won the ACS's Garvan-Olin Medal for an outstanding woman chemist. With just a bachelor's degree (and nine children, including three older stepchildren), Helen Murray Free, of the Miles Laboratories, held every local and national office of the ACS, including the presidency in 1993.[33] The 3M chemist Patsy Sherman, who developed its popular Scotchgard, was elected to the National Inventors Hall of Fame, as was DuPont's polymer chemist Stephanie Kwolek, who in the 1960s had devised a new polyamide fiber used in Spandex and Kevlar, for which she was still being honored in 2001, when she was awarded the National Medal of Technology.[34] The research chemist Gertrude Elion was in a class by herself. After Burroughs-Wellcome hired her in 1944 following a lengthy job search, she spent her entire career there. Although she never entered upper management, she eventually headed a team that developed a whole new family of innovative and successful drugs for gout, herpes, rheumatoid arthritis, leukemia, malaria, immune disorders, bacterial infections, and AIDS. For these outstanding results she, her collaborator George Hitchings, and Sir James Black, of Britain, shared the ultimate accolade, the Nobel Prize in Physiology or Medicine in 1988.[35]

Engineers

For women engineers, who had made no significant inroads into industry before 1972, the sudden surge in recruitment starting in the mid-1970s changed their prospects dramatically. By the time they began to graduate from engineering schools in increased numbers, the economy had rebounded and even expanded, with the result that engineering firms pursued them in a kind of hiring frenzy at record starting salaries, equal and sometimes even more than equal to men's. Some women engineering graduates of the 1970s found their new jobs quite satisfying, as did the chemical engineer Leigh Anderson, hired in 1977 as a troubleshooter for Exxon

Research and Engineering Company in Florham Park, New Jersey. At that time Exxon had about six women chemical engineers in her area; two years later it had thirty.[36] Although such companies were hiring other young women like her, once on the job, the young engineer might feel lonely, bewildered, underutilized, and possibly sexually harassed. She might not be promoted (and could even be demoted) if she lacked assertiveness and confidence.[37]

A 1983 survey of the women members of the IEEE, 1,816 of whom responded, revealed that their most common employers were the aerospace and computer industries. They were young (64.3% were under 30) and still in their first job, and 75.6 percent were childless. Salaries started out almost equal to those of men, but the women fell farther and farther behind as they got older (or what was not quite the same thing, had more years of experience). Evidently employers were adapting somewhat to the needs of the most recently hired women but were not advancing those few women it had hired earlier.[38]

Almost a decade later, in 1992, the Society of Women Engineers cooperated with the American Association of Engineering Societies in a massive survey of the men and women members of twenty-two engineering organizations. The resulting report found that overall women made up 7.7 percent of the members, were substantially younger (the mean age for women was 33.9 years compared with 42.9 for men), were less often in management, less likely to be married (61% of women compared with 81% of men), and less likely to have children (31% of women compared with 49% of men). Although starting salaries were close to equal, and in some cases the women's were actually higher than the men's, by their thirties the men earned more, and for women over age 50 the discrepancy was substantial. Women engineers also tended to retire earlier than did men engineers. The women and men were very differently distributed among twenty-three fields of engineering. (Strikingly, women made up 32% of systems engineers, who were just 1.1% of all engineers.) Another interesting finding was that the women tended to work for large corporations rather than for smaller firms that specialized in engineering services. This was unfortunate for two reasons: the smaller firms constituted the fastest-growing sector, and the larger firms were more susceptible to the numerous organizational shake-ups of the time, as the electrical engineer Naomi McAfee found at Westinghouse in the mid-1970s.[39]

But none of these surveys spotted the most significant pattern of the mid-1980s, which was that many of these young women engineers were not only leaving their companies but leaving industry entirely. Attrition at, for example, the Corning Glass Works, in upstate New York, reached 15 percent in the mid-1980s. Too many of these newly recruited engineers were not fitting in and were moving on. For example, the chemical engineer Jennie Patrick, in 1979 the first black woman to earn a PhD in her field at MIT, worked at General Electric, in Schenectady, for just four years before moving to Phillip Morris, in Richmond, Virginia, for two and then to Rohm and Haas, outside Philadelphia, for three before settling down at the Tuskegee Institute in Alabama.[40] As a result of all this turnover, some of which is to be expected in a competitive market, several large engineering companies, like the chemical companies mentioned above, which might earlier have blamed the women as "bad risks," were sufficiently concerned about diversity (and the EEOC) to start special retention programs. Some of these identified candidates for management training, others focused on mentoring, and still others on family issues, as their personnel and benefits offices took on a "cafeteria" of lifestyle issues.

Thus when the NRC's newly revitalized CWSE held its first meeting, in November 1991, it solicited reports on "best practices" in industry as well as elsewhere, including programs at Hughes Aircraft, Corning, the Aerospace Corporation, and Xerox, and printed brief descriptions of practices at Bristol-Myers Squibb, General Motors, Monsanto, Hewlett Packard, S. C. Johnson Wax, Tenneco, Consolidated Edison, U.S. Sprint, and IBM, all of which employed scientists as well as engineers.[41] In fact the committee found so much to report on that it held a second meeting, devoted chiefly to industrial programs, in January 1993. Programs at Xerox, ALCOA, AT&T Bell Laboratories, Scios Nova, and the small Barrios Technology were described in upbeat terms.[42] Yet a report presented at the meeting by the economist Anne Preston indicated that these programs might not be enough. She gave chilling data on the high rate of exit for technical women from industry—double that for men and higher than that for women from government or nonprofit sectors. Nor were they departing just because of a sour economy or for motherhood or family responsibilities. This widespread defection was cause for alarm. Whatever they wanted—advancement or more flexibility—was not available.[43]

One result of these corporate initiatives was that by 1993, when President Bill Clinton signed the Family Medical and Leave Act, which applied to employers of more than fifty workers (and which President George H. W. Bush had vetoed earlier), most large industrial companies already offered such options.[44] More reports on these "best practices" were also presented by officers of Catalyst, Inc., of New York City, which had been working with twenty-eight companies since the early 1990s, at a large meeting on women in science and engineering sponsored by the New York Academy of Sciences in March 1998. The event was covered in a four-page article in *Chemical and Engineering News*,[45] a sign that by 1998 issues relating to women and science were considered of broad general interest to the chemical and engineering communities. Yet when this relative willingness by large companies to provide costly benefits to parents and working women is compared with its systematic lack of advancement for them, one begins to see upper management's trade-off: benefits were just money, one of the new costs of doing business, faced by all one's competitors in the United States; they did not threaten important things like control and power.

Geologists

Enrollments of women in geology were already rising when word went out in the early 1970s that petroleum companies, long the largest employers of male geologists and geophysicists, were now responding to the "energy crisis," exploring for more oil, and for the first time in a long while hiring women. This was a big change, since in the past the few employed women geologists had largely been relegated to elementary-school teaching, faculty positions at women's colleges, librarianships, or, in the 1920s, as sedimentary-core readers for oil-exploration companies.[46] There were exceptions, however. Doris M. Curtis, who earned a PhD from Columbia University in 1949 and was in 1991 the first woman president of the Geological Society of America, had worked as a staff geologist for Shell Oil in 1942–50 and 1959–79, in Houston and New Orleans.[47] Starting in the mid-1970s the "Big Seven" oil companies—Exxon, Amoco, Chevron, Gulf, Texaco, Shell, and BP—undertook glossy campaigns to recruit and hire women.[48] Among these new recruits was Theresa Flynn Schwartzer, with three degrees from Rensselaer Polytechnic Institute, who was hired by Exxon Exploration in Houston in 1972. She was a founder and active member of the women's committee of the large American Geological Institute and gave several papers based on its 1974 survey of its men and women members. In a presentation at the 1978 NYAS meeting on women in science she reported that women geologists' employment was rising and that 18 percent of the women who had responded to the 1974 questionnaire worked in industry, a dramatic increase since 1970, when a previous survey had found that only 120, or just 0.6 percent, of 21,000 geoscientists employed in the industry were women.[49] Among these young women was Susan Landon, who, with a master's degree from SUNY, was in 1974 the third woman hired by the Amoco Production Company in Denver. By the early 1980s she was in charge of all its exploration in Alaska, an unthinkable position for a woman just a decade earlier. But rather than moving on up at Amoco, she left in 1987 for two years with Exxon in Houston and then in 1990 returned to Denver as a consulting geologist. She remained active in professional affairs and in 1998 was elected president of the American Geological Institute, the second woman elected to the post.[50] Meanwhile, the math teacher Barbara Sue McBride joined Exxon USA in Midland, Texas, as a technician in the boom time of 1978 and earned a BS in geology at the University of Texas at the Permian Basin in 1980. She stayed with the company, and after holding a series of posts in Alaska and London, she became at the time of the Exxon-Mobil merger in 1999 its vice president in charge of exploration in South America.[51]

With the coming of environmental concerns to the oil industry in the 1970s, many women scientists, including the biologist K. June Lindstedt-Siva, of Atlantic-Richfield (ARCO), made careers out of oil spills, cleanups, and risk reduction. In 1988 the geologist Judith B. Moody, who had been denied tenure at the University of North Carolina and worked for Battelle Laboratories for a while, set herself up as president of J. B. Moody & Associates of Columbus, Ohio, experts in nuclear-waste disposal, but a suspicious auto accident put an early end to her third career. In 1989 the physicist and chemist Frankie Wood-Black moved into the area of "downstream technology" (or everything related to oil once it is out of the ground) at Phillips Petroleum in Oklahoma, the same company that made Barbara Price, an environmental engineer, who joined in the boom times of 1977, its first woman vice president in 1992.[52]

In 1983–84 the Association for Women Geoscientists, whose active Denver chapter had been formed in 1977 by fourteen women at Mobil Oil, conducted a survey of more than six hundred women geoscientists. It showed that by then 21.5 percent worked for the oil and gas industry, compared with about 31 percent of the eighty-plus men surveyed. But once the patterns of women geologists' employment began to parallel those of men more closely, large measurable differences showed up in the area of salaries, where there were substantial discrepancies. Although the women were mostly younger (75% were under age 33), neither age nor experience accounted fully for the disparity.[53] Then in 1985–87 there was a major retraction in the size of the petroleum industry, as several oil companies merged and reduced their technical staffs. Since these layoffs affected the recently hired women geoscientists unduly, a decade of equal opportunity was quickly undone; a few filed lawsuits.[54] Thus the "entering wedge" of women geologists had ridden the boom cycle to new highs and then been decimated by the subsequent bust. A few managed to stay on for better times in the 1990s, but most moved on to other things.

Computer Scientists and Engineers

Overall, the computer industry expanded tremendously after 1970, except during a short economic downturn in 1985, which resulted in excess capacity and layoffs. But totals are only part of the picture, for several levels of skills were involved in programming, operations analysis, and systems analysis, all of which were constantly changing. As the machinery, or "hardware," grew smaller and "smarter," with ever more sophisticated programming embedded within their chips, the jobs of computer personnel changed in two ways: those at the topmost levels, which entailed dealing with people, especially decision makers, grew more complex, while the lower-level, traditionally more feminized programming positions were eliminated. Thus computer scientists and engineers of both sexes were constantly retraining to deal with ever more sophisticated software, and those with people skills hoped to advance into management positions above the antisocial "nerds" of the field's celebrated "hacker" culture.[55] Some were freelance consultants or worked for temporary-manpower agencies. Usually these jobs lacked benefits, but they provided relatively high wages and flexible work hours, which appealed to single mothers, among others.[56]

Like the women geoscientists, the women in the computer industry, ubiquitous but isolated as they were, began in the late 1970s to form an organization, the Association of Women in Computing. Every woman in computing, it seemed, whether she currently had a job or not, was looking for the next position and so was interested in "networking." Local chapters in metropolitan areas seemed to be the obvious solution, but the organization had a hard time finding selfless women committed to running it—collecting dues, serving as officers, putting out a newsletter, running meetings, and coordinating activities. Hardly anyone could take the unpaid time to run the organization, and many of those who did resigned prematurely. Unlike the women engineers, who had SWE, or the geoscientists, who had the AWG, the computer professionals of the 1970s lacked the personal links or sense of community and commitment necessary for such organizational work.[57]

Since IBM long dominated the industry, many computer scientists have worked for it at one time or another as it developed and then moved beyond the mainframes of the 1980s toward specialized client "business services" of the 1990s and after. Two prominent early IBM women scientists were the mathematicians Jean Sammet and Frances Allen, who started in the days of FORTRAN and COBOL (late 1950s) and worked with compilers and other languages. Sammet developed ADA for the Department of Defense and was elected president of the Association for Computer Machinery (ACM) in 1974–76. Allen was one of several women (seven in 2008) to hold the status of fellow, signifying that she was one of the IBMs fifty top technical specialists worldwide. She was also among the nine IBMers elected to the Women in Technology International Hall of Fame.[58] Also of note is the applied mathematician Lilian S-Y. Wu, of IBM's Yorktown Heights, New York, research center, who has been active in women-and-science issues, serving as chair of the CWSE. Others, such as Jeanne Ferrante and Maria Klawe, spent several years at IBM in Almaden, California, before moving on to academic careers in computer science.[59] In 1988 the company expanded its already generous benefits package to include flextime; longer unpaid leaves (up to three years with benefits); "career service," or partial salary for retirees who volunteer for community service; and a test program of options for working at home.[60]

Finally, in the late 1980s, with (a bit ironically) the coming of electronic communication, an emerging "critical mass" of like-minded computer women in both industry and academe began to find one another and work together. As word began to spread of the plateau and then drop in the number and percentage of women earning bachelor's degrees in computer science (as discussed in chapter 3), there began to be a need for some sort of official or authoritative women's response and recommendations. Prominent in this coming together of mid-career women was Anita Borg, of the Digital Equipment Corporation's Western Laboratory, later at the Xerox Corporation, both in Palo Alto, California. Upon her return from a 1987 meeting of operations analysts, where she had met almost twenty other women, she started an electronic "list," then a new phenomenon, which she called "Systers." Although it was limited to women with degrees and jobs in the field, this easy means of communication and advice from the desktop quickly grew to include about 1,124 women in 1992, and by the time of her death in 2003, at age 54, it included 2,500 in thirty-eight countries. The women did not have to attend meetings, serve in offices, or pay dues to get the news and be up to date. By simply answering e-mail messages, they could be in touch with "systers" around the globe.[61]

Out of this new network evolved several projects, institutions, and traditions, as well as a cadre of women (governmental and academic as well as industrial) leaders in the field. As mentioned in chapter 3, the plateau in the number of undergraduate majors in computer science became an issue of concern to the NSF and led to some meetings, activities, and organizations that brought together several middle-aged women in the field. Building on these efforts, in 1994 Borg and Telle Whitney organized the first Grace Hopper Celebration of Women in Computing in Washington, DC, to honor the memory of a recently deceased pioneer in the field. It was an emotional, empowering, and even electrifying event for all. All the speakers were women, which in 1994 was a novelty for female computer scientists, who were used to marginal roles at their professional meetings. (The Grace Hopper Celebration has since become a tradition, now held annually and attracting considerable corporate and NSF support for student attendees.) Thereafter Amy Pearl, then at Sun Microsystems, in Silicon Valley, California, in 1995 organized a special issue of *Communications of the Association for Computing Machinery* on women and computers, which included a list of recently formed women's groups at ten major companies (the one at Microsoft was called "Hoppers"). In 1997 Borg herself started a nonprofit Institute for Women and Technology at Xerox, where she promoted studies of women and computers.[62] In 2002 one of these leading female computer experts, Maria Klawe, by then the first woman dean of engineering at Princeton University, was elected president of the ACM.[63]

By the 1990s a few other women were in the top management ranks of the computer industry: Barbara Willard served briefly as president of Computerland, which was founded by her father, before it was taken over by creditors; Kim Polese, founder of Java, the interactive web language, started her own company, Marimba, Inc., in Silicon Valley; and the nonengineer (a philosophy major with an MBA) Carly Fiorina, who had flourished in the executive-development program at AT&T after the 1973 consent decree opened it to women, was appointed chief executive officer of Hewlett-Packard in 1999. In six years there she faced a merger with Compaq Computers, a divisive stockholder revolt, and her own sudden ouster in 2005.[64] Thus by the start of the twenty-first century a central core of corporate and academic women computer scientists and engineers were working together and creating a place for themselves in the field. They were becoming more than an entering wedge: they had organized into a force primed for leverage and leadership within and beyond the profession.

Biotechnologists

Another young industry that took off in the 1970s, starting with the founding of Genentech in California in 1976, was biotechnology. Although what little information is available on the size of this field is not broken down by sex, many new companies hired chemists and bioscientists at the MS and PhD level at a feverish rate in the 1980s, when there was a temporary glut of biology doctorates.[65] Facing the challenges and vicissitudes of "start-up" companies everywhere, young biotech firms offered a variety of opportunities, especially to women with doctorates and academic experience.[66] They grabbed up several women who had started academic careers but had grown dissatisfied or had not been promoted. Offered high salaries, high ranks, including that of vice president, and potentially lucrative stock options, they were often glad of the chance for a second career. Accounts of work in biotechnology firms stress the chance to do basic research with the latest equipment, the lack of individual pressure to compete for scarce NIH grants, and the enormous flexibility in the workplace, but the researcher loses her independence, faces great business as well as scientific uncertainties, and works at a frenetic pace to develop and test a drug before impatient investors abandon the company.[67]

In 1983 Mary-Dell Chilton, a chemist and molecular biologist, left a tenured position at Washington University in St. Louis, where she had collaborated with Monsanto scientists, for Ciba-Geigy Seeds in North Carolina, of which she became a vice president in 1991. Meanwhile, even before the pathologist Sandra Panem was denied tenure at the University of Chicago in 1984, she had been interested in social issues. She was thus subsequently able to forge a second career as a venture capitalist, sitting knowledgeably on the corporate boards of several biotech companies for a series of financial powerhouses, including Solomon Brothers.[68] Likewise, Ellen Daniell, the first female assistant professor in Berkeley's molecular biology department, left in 1984, after being denied tenure, for the Cetus Corporation and later Roche Molecular Systems, both in the Bay Area, where she served, respectively, as head of human resources and director of licensing.[69]

Other budding biotechnologists learned about patenting, stock offerings, and meeting a payroll by starting or running companies themselves. In 1986, after a series of academic posts in New York City, the bioscientist Gail Naughton started Advanced Tissue Sciences to market her patent on a new kind of skin graft. After sixteen years at the company, where she found that her expensive product was more in demand as a cosmetic than as a treatment for burns, she moved on. She earned an MBA from UCLA in 2001 and a year later became dean of the business school at San Diego State University, the first biotechnologist to do so.[70] Similarly, in the mid-1980s Vicki Sato left the biology department at Harvard for a senior position at Biogen, later moving to Vertex Pharmaceuticals, both in Cambridge, Massachusetts, as chief scientific officer. In 2000 she became Vertex's second president, but she later moved back to Harvard as a professor of management practice at the business school.[71] In 1990 the AIDS researcher Flossie Wong-Staal, who had spent the 1980s working with the AIDS researcher Robert Gallo at the National Cancer Institute, became a professor at the University of California–San Diego, where she and others started Immusol, which she joined as its full-time chief scientific officer, in 2002.[72]

Some of these high-powered biotech executives participated in the activities of an organization called Women Entrepreneurs in Science and Technology, Inc. (WEST), which publishes a newsletter and holds career days to encourage other women to build their own companies.[73]

Others Fields

A few women scientists in other fields pioneered careers of distinction in industry.[74] The physicist Betsy Ancker-Johnson, for example, a Wellesley College graduate with a German PhD in plasma physics (and four children), worked for several companies—Sylvania, RCA, and Boeing Aerospace—before President Richard Nixon appointed her assistant secretary for science and technology at the Department of Commerce in 1973. Subsequently she served as vice president of General Motors, the first woman at that level in the automobile industry. The physicist

Mary Beth Stearns, who had surveyed women in industry for the original APS status-of-women report of 1971, spent most of her career at the rival Ford Motor Company. The solid-state physicist Esther Conwell also spent her distinguished career in industry, first at GTE and then at Xerox outside Rochester, New York.[75]

Worthy of note was the experience of Ann Branigar Hopkins, a Hollins College graduate who earned a master's degree in mathematics at Indiana University in 1967. After teaching for a year, she joined IBM as a programmer on a satellite project with the National Aeronautics and Space Administration. With the downturn in funding for space programs, she transferred her skills to accounting firms that were installing auditable (i.e., computerized) payroll and other financial systems in complex bureaucracies such as the United Mine Workers' health fund, the city of Washington, DC, and the Department of State. After joining Price Waterhouse (as it was then known) in 1978 and successfully completing several challenging assignments there, she was, however, refused a partnership five years later. Learning that although her work was appreciated, the men at the top did not think she looked like a partner, she filed suit. At the initial trial in district court in 1985 her lawyers had social scientists testify about the role of sexual stereotyping in personnel evaluations, thereby setting a precedent for others encountering such practices in the workplace. She won there, and then again at both the court of appeals and the U.S. Supreme Court, 6–3, thereby helping other women litigants nationwide.[76]

Among others in unusual businesses was the aquanaut Sylvia Earle, famous for her deep-sea dives, who in the 1980s started Deep Ocean Engineering with her third husband. Together he developed and she tested underwater rover apparatus for industry and the U.S. Navy.[77] Likewise, the mathematician Margaret Hamilton used the expertise she had gained in devising complex state-of-the-art computer systems for the Apollo and Skylab space programs to spin off her own software-development company, Hamilton Technologies, in Cambridge, Massachusetts.[78] Even more remarkable was the chemist Judith Osmer's new venture. After leaving the Aerospace Corporation in the 1980s (see chapter 10), she and the ex-physicist Virginia Carter, the founding president of LA NOW, set up and ran the J. O. Crystal Company, which manufactured and sold artificial rubies.[79] Another venturesome woman scientist who developed an avocation into a company was the Berkeley-trained biochemist Arlene Blum, who led a mountain-climbing expedition for women only to the summit of Mount Annapurna in 1978 and subsequently started a company that leads treks in Nepal.[80]

Among the few social scientists employed in private industry was Marina von Neumann Whitman, like Ancker-Johnson a Nixon presidential appointee, who later served as the chief economist and vice president of General Motors in the 1980s.[81]

By then the large banks and investment houses on Wall Street had also begun to hire women economists.[82] In 1995, in a seemingly different industry and after a full career in academia capped with many honors and several presidencies, the child psychologist (and mother of four) Sandra Wood Scarr accepted a corporate position as CEO of KinderCare Learning Centers, the largest childcare provider in the country, with fifteen hundred branches in nearly forty states. Her service in this corporate kind of "women's work" lasted only two years, however, as the company was purchased by New York investors who brought in new management.[83]

Conclusion

As federal enforcement agencies won highly publicized settlements against some of America's largest employers in the 1970s, private industry responded to the new demands for an equal-opportunity workplace. Although the technical industries proved to be a particularly volatile and unstable employment sector, they began to hire young women scientists and engineers in the 1970s through the 1990s. The jobs offered women more opportunities and higher salaries than ever before, but their on-the-job experiences varied. Often, rather than being welcomed, they faced what academics were beginning to call a "chilly climate," because they had been hired reluctantly and because government enforcement focused on initial hiring rather than advancement. A career in industry involved varied encounters and episodes, as older attitudes lingered in many places. In some corporations the "entering wedges" formed groups to press for changes; in others change came as a result of lawsuits about which few details are public. By the mid-1980s turnover was so high at some corporations that the personnel offices set up family-friendly policies and programs. As companies expanded and added new units, there were a variety of ladders to climb in the proliferating levels of management. Women started on these, but both they and their jobs could disappear with mergers, reorganizations, buyouts, layoffs, economic downturns, or technological obsolescence. By all accounts, until recently the ranks of upper management remained largely white and male, as the best-known minority women who started in industry moved out and on rather than up. Yet despite it all, by the first decade of the twenty-first century there were several female vice presidents of scientific and technical corporations and one Nobel laureate. The modest gains resulted in large part from pressures brought by the many new women's organizations formed at large companies, in most fields, and publicized by the NRC's CWSE. Women began to find solace or solutions to their job problems in "list servs," newsletters, meetings of professional organizations, or extravaganzas like a Hopper Celebration. If a woman could not make it to the top herself, as least she might try to make the middle a little more hospitable for the women coming along behind her. Or she could leave her employer, become an entrepreneur, and start her own company, which offered challenges and satisfactions as well as a bit of revenge.

CHAPTER NINE

Federal Employment

Lawsuits and Presidential Appointees

The years 1970–2010 were also tumultuous decades in the federal government, as women employees used President Richard Nixon's Equal Employment Opportunity Act of 1972 to force it to change, agency by agency. The new law allowed disaffected individuals to sue for themselves or for their "class" of fellow workers for systematically unequal treatment. If they won, the judge could order their one agency to correct its procedures and behavior. Thus, instead of a strong Civil Service Commission moving aggressively to implement Congress's vision and change the entire federal bureaucracy systematically, thoroughly, and relatively quickly, the legislation had placed the burden upon the workers themselves to get informed, hire lawyers, file lawsuits, collect evidence, wait years (especially as the district court in Washington, DC, was tied up in Watergate-related cases, another Nixon legacy), and then at best win a settlement with some pittance of back pay, a substantial amount for lawyers' fees, and instructions for that agency's current officials to finally take some long overdue steps to change its ways. The results were uneven, for not all officials complied. Often the women protagonists were laying siege to certain well-entrenched cultural patterns in which male medical or former military authorities were surrounded by devoted long-term female assistants who might also resist such change. At a few agencies where there was no lawsuit it was possible to adapt by broadening job titles and descriptions, a step that was perhaps justified as part of the agencies' otherwise changing mission, so as to open them to both sexes (e.g., "mission specialists" instead of "astronauts" or "natural resource managers" instead of "park rangers"). Meanwhile, other long-term women scientists persevered despite it all. Ironically, the most visible successes for government women in science in these years were not judicial victories or equitable employment practices but newsworthy political decisions: the acquiescence of the National Aeronautics and Space Administration (NASA) to pressure in the mid-1970s to start selecting women astronauts and President Bill Clinton's numerous appointees to top posts at several agencies in the 1990s. But even they were temporary.

Although the size of the federal workforce actually shrank by about one-quarter between 1970 and 2000, these were years of increasing opportunity for federal women scientists and engineers. As the bureaucracy took on new responsibilities in the areas of energy, education, equal opportunity, housing, transportation, and the environment, many agencies were reorganized and some new ones were created. Overall the government increased its numbers of technically trained personnel from 150,000 in 1974 to 190,000 in 1997. Remarkably, the numbers of women among them rose fivefold, from 6,600 (4.4%) in 1974, the first year the totals were broken down by field and sex, to 33,000 (19.4%) in 1997, the last year for which numbers were available in 2000. Women's numbers were up in seventy-four of seventy-eight technical fields, substantially so in agricultural, geological, and engineering sciences, where women's percentage of the total increased dramatically. Thus, the federal government's workforce was by 2000 leaner in size but more technically trained and more feminized than in 1974.[1] The result of it all was that by the mid-1980s in some agencies a well-trained young woman scientist such as, for example, the chemist Susan Solomon, at the National Oceanic and Atmospheric Administration (NOAA), or Pamela Matson, briefly at NASA, could be hired and achieve a distinguished career unscathed by the labor practices of the recent past.

By way of analysis here we can think of federal women scientists as being of three (not mutually exclusive) types. First, there were the activists, often longtime employees in dead-end positions, who had grievances, formed organizations, staged protests, testified before Congress, and, when all else failed, filed federal lawsuits. Second, there were the professional researchers, who had long careers of distinction and won prizes and medals. Perhaps they also had certain grievances, but they kept their distance from the activists. Third, there were the political, especially presidential, appointees, who came and went, presiding over whole agencies but all too often with remarkably little impact on or even interest in the equal-opportunity issues of their employees.

Activists: Lawsuits and Settlements

By the late 1960s there were already a great many mid-career women scientists in the federal government, especially in certain agencies.[2] In the then Department of Health, Education, and Welfare (DHEW), for example, women constituted for the time a fairly high percentage of total employees, but hardly any were in the top ranks. In others, such as the U.S. Forest Service (USFS), there were no women in the upper ranks and very few women even in the low ranks, and they were not paid equally. As such women scientists in government read eagerly about the coming changes in the *Federal Times* and the *Washington Post,* which had a regular column on federal employment issues, they became impatient to have longstanding abuses corrected. Some attended meetings with other malcontents and joined the new women's organizations that attempted to force bureaucrats to get beyond talking about change and start enforcing the new rules. President Nixon's Equal Employment Opportunity Act was just what they had been hoping for.

The story of women in the USFS, a part of the U.S. Department of Agriculture, is worth a volume in itself.[3] The saga started in June 1973, when the forester Gene Bernardi, one of the few women in Region 5 (California), filed a class-action lawsuit charging the USFS with sex discrimination in both hiring and promotion. The case never went to trial, but in 1981, after years of "discovery," a judge in San Francisco approved a consent decree that awarded Bernardi, who had by then left the service, $21,250 and authorized $1.5 million for compliance reports every six months for five years by two public-interest law firms in San Francisco. (One was Equal Rights Advocates, Bernardi's counsel, started a few years earlier by three women lawyers and funded for a while by the Ford Foundation.) The USFS did not admit any guilt but was instructed to come up with an affirmative-action plan that would remedy problems in the future.[4]

Meanwhile, in 1979, as a draft of the forthcoming consent decree was being circulated, Mary Albertson, of the USFS's Federal Women's Program, and others in Region 6 (Alaska and the Pacific Northwest) held a meeting at which participants brainstormed not only on how to help the increasingly vociferous women foresters but also, and especially, on how to train the male managers who now had to deal with "women's issues." One idea was to create a newsletter for and about women in forestry that could reach even the most remote locality. Edited for many years by Dixie Ehrenreich of the University of Idaho, *Women in Forestry* became a vigorous forum for issues confronted by women in nontraditional jobs at isolated sites. In 1987–88 it broadened its mission, as reflected in its new title, *Women in Natural Resources*. It tried to encourage and educate both women and men on the proper behavior for the new workforce. Over time, issues of family lifestyle, including maternity issues, began to be discussed, as some of these pioneering women stayed long enough to be promoted but by then were pregnant. It was difficult to combine the job with a lifestyle that was not traditional in the agency. Issues arose and were worked out on a case-by-case basis—for example, a woman in Alaska brought her newborn to the office—but the coming of liability insurance in the 1970s meant that there would be difficulties. The issues addressed in *Women in Natural Resources* also followed the progression of the young women up the management ladder.[5]

Nevertheless, there was considerable foot-dragging in the Forest Service. Once hired into entry-level positions, women foresters encountered a host of workplace conditions and employment practices, such as isolation, exclusion, and hostility, that left them ignored, marginalized, and even sexually harassed, a new term in the 1970s. So many left the government that turnover became a serious management problem. One 1986 article chillingly reminded women foresters that no one in charge had wanted to hire them; the recent legislation had forced them to.[6] In 1988 a federal judge found the secretary of agriculture in contempt for not having promoted enough women foresters in California and ordered him to spend another $1.5 million in the next three years to carry out his ruling. In 1991 still another federal judge in San Francisco repeated the decision. As a result, the percentage of women in professional positions at the USFS jumped from just 1 percent in 1976 to 26 percent in 1992, and by the 1990s a few women had become the first supervisors or deputy directors of their regions, sometimes transferring in from another federal agency. Thus, for example, in 1992 Elizabeth Estill, an ecologist and member of the Senior Executive Service, who had joined the USFS in 1988 after fourteen years at the Tennessee Valley Authority, became the first woman regional forester (for the Rocky Mountain Region).[7]

Also in the 1970s there was a hotbed of anger and activity at the National Institutes of Health (NIH) in Bethesda, Maryland, which had employed many women scientists since its founding in the 1930s. By 1970 many were clustered at the GS-12, GS-13, or GS-14 level—at the bottom of higher management. Since these women had many skills and much experience and outlasted the many political appointees above them, they performed many essential but unsung duties. (In the 1980s the term *glass ceiling* would be applied to their situation.) They felt that they were the mainstays of the institution and yet were taken for granted and even exploited. They collected tasks and did more and more overtime work, which was hardly noticed and rarely rewarded, as these added duties were not in their job descriptions. Someone had to do them, and they did, but they found that officially, such as at a personnel hearing, it counted for little. (For example, they might substitute over the Fourth of July weekend as an "acting" supervisor, or they might write the unit's annual report, but not get credit for this and later find the value of this responsibility discounted.) Often they had PhDs rather than MDs and so faced an additional conflict of cultures. In addition, this work relationship was structured such that the male MDs held the supervisory positions and titles, while the smart women were their "staff advisers," who in theory had no managerial authority but in practice could and often did actually run the place with a kind of vicarious shadow power. At the NIH the traditional roles of male MDs with very competent but silent and invisible female assistants even increased in the 1960s and 1970s, when reorganizations were frequent but few new positions were created. Thus some male MDs were running two or more units, often at different locations, and pushing as much work and responsibility as possible onto their overworked female assistants.[8]

Initially there was some progress at the DHEW and its subunit the NIH, under Secretary Elliott Richardson, who was late to believe in women's liberation but learned quickly. (In 1970 there were only 12 women at the super-grade levels GS-16 to GS-18, but two years later there were 24 [out of 370, or 6.5%].)[9] Meanwhile, the Civil Service Commission appointed various personnel officers for women and minorities to the NIH, but they were young and inexperienced, had minimal staff, left quickly, and made African American men rather than white (or any) women their priority. Already in 1970 some of the malcontents at the NIH began to meet at lunchtime, but their posters were taken down from bulletin boards. Angry, they formed an organization, the NIH Organization for Women, which eventually got some official recognition and was able to publicize and secure a room for its meetings. When participants got together, they discovered that they were interested in two issues: the lack of childcare and the lack of advancement. Before long the group split into two separate ones. The first got on-site day-care facilities fairly soon thereafter.

The second group, those interested in advancement issues, formed Self Help for Equal Rights, or SHER, in 1972 to focus on the various procedural and legal technicalities of securing equal opportunity. But they also sought a larger, more public role that included consciousness-raising or publicity for their claims. At a time in the early 1970s when, for example, few government women would discuss the issue in public or on the record for a journalist, SHER invited speakers—such as the feisty Estelle Ramey, of nearby Georgetown University, who gave witty, biting commentary on the (low) status of women at the NIH—whose talks were reported on in *Science* and the *Washington Post.* Members of SHER also collected statistical data (usually by GS level and institute), made presentations to various groups, talked to newspaper reporters, and even testified at government hearings, which infuriated their bosses.[10] The atmosphere at the NIH became embattled and intransigent. Some individuals, such as the anatomist Fann Harding (GS-14), fought their cases internally or through the EEOC. In 1976 she was finally promoted to the GS-15 pay grade but not to the position of chief, which she felt she deserved.[11] More significantly, the mathematician Rosalind Marimont, who won a promotion after complaining to the EEOC, also filed a class-action lawsuit in district court in November 1973 on behalf of seven thousand women at the NIH. The case was assigned to Judge John Sirica, of Watergate fame, and thus encountered numerous delays and was not decided until 1979. The eventual settlement gave Marimont back pay and attorneys' fees. Although it denied the suit class-action status, it required the NIH to develop "fair and definite" promotion procedures.[12] Some attributed the 1974 promotion of Ruth Kirschstein, after many years in lesser posts, to the directorship of the National Institute for General Medical Sciences—a female first—to pressure brought by the leaders of SHER.[13]

Nevertheless, the glass ceiling remained at the NIH, and SHER celebrated its twenty-fifth anniversary in 1997. SHER, which had considerable access to both the media and influential members of Congress, increasing numbers of whom were women, became a kind of permanent institutional gadfly and watchdog, which the NIH, still run by an entrenched male medical establishment even if it was headed by Bernadine Healy, MD, in 1991–93, continued to need. In fact the great disparity in grade levels (and so salaries and responsibilities) between men and women increased in the 1980s. More than half of all women at the NIH were at GS-13 or below, at which levels the number of jobs declined after 1976, while most of the men were at GS-13 and above, where the number of jobs greatly increased. The disparity continued into the 1990s.[14]

But the USFS and the NIH were not the only federal agencies with unhappy women scientists who, after repeated attempts within the agency, hired a lawyer, filed a lawsuit, and waited a long time only to find that even a court ruling in their favor and a new affirmative-action plan changed little. A third case occurred at the Department of Labor (DOL), where in 1974 the African American computer specialist Dolores Copeland (GS-13), of its data-automation unit, filed a class-action lawsuit against the department. Not only had she been denied further promotions but her superiors had withheld opportunities for the training that would have made her eligible for future advancement. Incensed, she hired the top-level law firm of Wilmer & Cutler, and sued. After many delays caused by the DOL's refusal to hand over key documents, the judge awarded Copeland a promotion to GS-14 and $6,200 in back pay and gave the Department of Labor a new affirmative-action plan based on Copeland's design. Eventually a "special master" awarded the other members of her "class" $33,000, while her lawyers complained that their $160,000 was not enough and in 1979 filed another lawsuit to get paid even more.[15]

A somewhat similar scenario took place in 1976–78 at the new Department of Energy (DOE), which traced its roots to World War II and the by then male-dominated field of atomic energy. Since many of the department's employees were former military personnel, the prevailing culture was one in which women were relegated to lower-level positions, underpaid, and routinely passed over for promotions. Although Richard Nixon appointed Dixy Lee Ray chair of the Atomic Energy Commission (AEC) in 1973, she evidently did little to improve the climate for women at the agency during her brief tenure. In early 1976 June Chewning, an internationally recognized nuclear manpower analyst and a founder of Federally Employed Women, an important group established in 1968 to keep an eye on the new Federal Women's Program, got so fed up with one particularly belittling supervisor that she filed a class-action lawsuit on behalf of 255 professional women against the Energy Research and Development Administration (ERDA), as the agency was then temporarily known. In July 1978, astonishingly, she won her case on statistics alone, because President Jimmy Carter's Department of Justice conceded that it could offer no defense of ERDA's entrenched "buddy system" of employment practices. Accompanying the consent decree was a new affirmative-action plan for the agency, and eventually 130 women shared a $2.2 million settlement, or about $17,700 each. Even so, it took years for the final details to be completed.[16]

In the meantime, in 1978 an unidentified person filed a complaint of sex discrimination against the U.S. Geological Survey (USGS), in the Department of the Interior, which had systematically underemployed women scientists ever since it first hired them early in the twentieth century. It had begun to hire more after 1973 and was starting to change some of its practices.[17] As part of the settlement, the chief geologist appointed a large committee (twenty-five USGS staff persons, including thirteen women, one of whom served as the chair) to consider the situations of all 263 women scientists employed before 1976 by the Geologic Division, the USGS's largest employer of technical personnel. After a substantial effort that included questionnaires and interviews, the committee reported that 17 women had indeed had serious problems and that 142 others also merited some remedies. The difficulties were numerous and included initial hiring at a lower GS level than men, underemployment as technicians rather than scientists, slower promotion rates, dead-end positions, a lack of opportunities for fieldwork, lower salaries, a lack of information and career counseling, limited opportunities for full-time work (after working part time), and no appointments above branch chief.[18]

One beneficiary of this new interest in personnel matters was Penelope Hanshaw, a 1953 graduate of Wellesley College, who had married, raised two children, helped her husband through seven years of graduate work, and held a part-time job as editor at the USGS. In 1981 she wrote the former Mount Holyoke geologist Christina Lochman-Balk: "I was still doing same [part-time editing] in 1975 when EEO hit in force and I was in the right place at the right time to move up into management and am now Deputy Chief Geologist [for Scientific Personnel]. It paid to hang in there all those years!"[19] One of her duties was to enforce the new personnel mandates and thus finally put the Geological Survey on the way to being the "showcase" that President Lyndon Baines Johnson had envisioned in the 1960s. For a while the USGS offered women scientists exciting opportunities. For example, Baerbel Lucchitta identified and mapped lunar, Mars, and Antarctic geological formations; the volcanologist Susan Kieffer, a Caltech PhD, was able to measure the lava flow at Mount Saint Helens, in Washington State, just after it erupted in 1980; and Mary Lou Zoback worked in Menlo Park, California, on tectonic stress and earthquake deformations.[20] The heyday for women geologists at the USGS was short-lived, however, for starting in 1995, budget pressures forced the agency to reduce its scientific staff substantially. By 2000 this had led to terminations for 525 of 1,850 scientists in the Geologic Division. When it was learned that this burden fell disproportionately on women (in the Denver office, 45% of the 44 women were laid off, compared with just 23% of the 215 men), two women filed a complaint with the EEOC. They persisted and eventually, six years later, after protracted and inexplicable delays, the Merit Systems Protection Board dismissed the claim in a 2–1 ruling.[21]

Two of the DOE's "national laboratories"—the Lawrence Berkeley Laboratory (LBL) and its sister, the Lawrence Livermore National Laboratory (LLNL), both in northern California—also had particularly troubled histories of sex discrimination. The chemist Mollie Gleiser had a harrowing career at the LBL, and when budgets were cut in 1971 she was laid off along with many others. Over the next several years she applied four times for various jobs in chemistry and editing but was never rehired. Then in 1978, after hearing that the LBL was hiring men less qualified than her, she filed a lawsuit. In 1979 the American Chemical Society extended her a loan of seven thousand dollars, but even so, she lost the case the next year, when the judge disallowed her statistical evidence.[22]

Even before 1980 women scientists at the nearby LLNL had been complaining to management about their lack of promotions and unequal salaries, which under California law were public information. A major obstacle to equal pay was the lab's method of evaluation, which its leaders considered sacrosanct. Instead of measuring a person's performance against that specified in his or her job description, managers ranked everyone based on their overall value to the unit and paid them accordingly. Women, even those with PhDs, tended to be in less "important" jobs and so received lower salaries and smaller raises each year. In the 1980s and 1990s some officials met with the women, made copies of their ever-better data (of which as computer experts they were quite proud), and promised to look into the matter, but nothing changed. Finally the women got fed up, and in December 1998 six past and current employees, led by the retired chemist Mary Singleton, filed a class-action lawsuit on behalf of more than three thousand women at the LLNL over unequal pay, nonpromotion, and the ranking system. (They chose an Oakland, California, lawyer who had recently won a million-dollar settlement from the UC–Berkeley for a woman in the architecture school.) Five years later they won big, with a $9.7 million settlement (plus $8.7 million for the lawyers) and the requirement of a new evaluation system.[23] Almost a year later the LLNL, which is run by the University of California, hired the physicist Cherry Murray, of AT&T Bell Laboratories, as its new deputy director for science and technology and Jane Long, a hydrogeologist at the University of Nevada and formerly of the LBL, as associate director for energy and the environment. As both replaced men who had been at the LLNL for decades, their appointments signaled two holes in the glass ceiling and a possibly significant change in the lab's traditional institutional culture. Unfortunately, Murray left for a Harvard deanship just two years later.[24]

When these generally successful lawsuits are compared with similar cases fought in academia and discussed in chapter 2, with their pervasive gloom and mixed results, several differences become apparent. First, the size of the "class" was usually much larger in the government cases, often numbering more than two hundred women, as job descriptions were more clear-cut and there was a larger number of eligible women clustered in the middle and so bumping up against the glass ceiling, which started at GS-13 or GS-14. Second, government cases may have attracted better attorneys, as a successful settlement would have been for a larger amount and so a higher fee, and the quality of lawyering available was much better in Washington, DC, even for a government case, than in many college towns, such as Ithaca or even Berkeley, although the quality in college towns did improve over time. Third, the judges had no qualms about intervening in the federal workforce or violating an agency's autonomy, as they often did with universities. Fourth, the level of proof was often lower. In the case against ERDA, the Justice Department was swayed by statistics alone, which in many academic cases, starting with Sharon Johnson's against the University of Pittsburgh, were accepted as merely showing the prevailing pattern of nonpromotion. For good or ill, the two groups of litigants seem not to have been very aware of each other. The academic women through their caucuses were well informed about, and sometimes urged their members to make donations to, the larger class-action cases against well-known universities and then cared enough to mourn when there was a loss, but they were unaware of the efforts by the government women, whose cases were on the whole more successful.[25]

Without Lawsuits

Other agencies managed to change entrenched practices and attitudes without lawsuits, although with some considerable publicity. The most controversy was at NASA, which in the mid-1970s not only changed its widely publicized astronaut program but also hired more women, including minority women, in its less-publicized earthbound technical positions. This institutional initiative, which involved Congress but not the courts, was unusually public and dramatic. The key events occurred in September and October 1973, or just weeks before the series of events surrounding Nixon's sudden firing of several high-ranking officials known as the "Saturday Night Massacre" (to which the NASA action was later linked). In September 1973 Ruth Bates Harris, the agency's African American deputy assistant administrator for equal opportunity, and two others reported to NASA Administrator James Fletcher, a white Mormon from Utah who had hired her two years earlier, that percentages of women and minority employees at NASA not only continued to lag substantially below those of other federal agencies but had risen only one percentage point, from 4.10 percent to 5.19 percent, since 1966. They recommended that more aggressive steps be taken, firm commitments made, and better results ensured. This call for action exasperated Fletcher, who summarily fired Harris and reassigned the other two. He said that Harris's direct style clashed with the style of the white men at NASA's headquarters and its southern installations, who were preoccupied with NASA's technical mission and remained unconcerned with personnel issues. When the news of Harris's firing broke, women's groups, including the Association for Women in Science, were outraged; they held meetings and wrote letters to congressmen, some of whom had recently formed the Congressional Black Caucus. The Senate Appropriations Committee and the Senate Aeronautical and Space Sciences Committee held hearings, which brought to light still more examples of NASA's noncompliance nationwide. They found that the Civil Service Commission was not doing much to monitor the situation or to force NASA to do better. NASA would have to change its ways fast, and astonishingly, unlike some other agencies, such as the DHEW, it did. Soon thereafter, Fletcher hired another experienced African American woman with a quieter style—Harriett Jenkins, a Fisk University graduate with a degree in mathematics and a doctorate in education, who had recently been the first black assistant superintendent of the Berkeley, California, school district. Although she lacked federal experience (for which she was criticized initially), she was less confrontational, more of a "gradualist," and in her more than two decades at NASA got many results.[26]

One issue that concerned many at the time of Harris's firing, including Vera Kistiakowsky, of MIT, a physicist and a member (later president) of AWIS, was whether NASA would choose any women astronauts anytime soon. During the 1960s, when the manned space program started, there had been no women astronauts despite the availability of several qualified candidates. Instead. the job was given to white men with, as the novelist Tom Wolfe put it, "the right stuff." There had been some criticism from the National Federation of Business and Professional Women and brief congressional hearings, but nothing had changed. Then in the late 1960s the northern California branch of the National Organization for Women took up the issue of women in space, and in the early 1970s articles appeared in aerospace and women's magazines, including a particularly timely one in *Ms.* in September 1973. By then the Lovelace Clinic, in New Mexico, was testing (presumably for NASA) twelve white female Air Force flight nurses for the effects of prolonged weightlessness. It may have been about this time that the decision was made to include women and minorities in the next set of astronauts to be chosen in 1978, although it was not announced until 1976. Thus the highly publicized Harris debacle may have contributed to NASA's finally resolving to make its long-postponed decision. (In fact a cynic might suggest that Fletcher deliberately fired Harris—it was not done in haste—in order to precipitate a brouhaha that would create an outburst of political support for the increasingly inevitable decision to recruit women astronauts, which was probably opposed by many within NASA.)[27]

After the successful completion of the spectacular moon landing in 1969, NASA began to plan new manned-flight projects. One involved a set of "space shuttles" that would take whole teams of astronauts into outer space to equip and then staff a space station and conduct experiments. The job description of these new "mission specialists" was broadened to include more scientific background, which some women might have, and less jet-pilot experience, which women did not have. In 1977 NASA started aggressive recruitment of women and minority applicants with a widespread advertising campaign and a nationwide recruiting trip by the black actress Nichelle Nichols, who had played Lt. Uhuru, the only woman on the starship *Enterprise,* on the popular television show *Star Trek.* NASA got 8,079 applications, including 1,544 from women. Eventually, in 1978, included among the thirty-five new astronauts were six women, including Sally Ride, a physics PhD from Stanford University. She was among those launched successfully into space in June 1983, an event that attracted much sexist publicity. Among her fellow astronauts were the geologist Kathryn Sullivan, who was the first American woman to walk outside a spacecraft in 1984; the electrical engineer Judy Resnick, who died in the *Challenger* disaster in 1986; and the durable biochemist Shannon Lucid, who long held the record for time spent in space by an American and continued to hold the women's record (188 days at the Soviet space station Mir in 1996). In 1987 the physician Mae Jemison was the first black woman chosen to be an astronaut, and in 1990 the electrical engineer Ellen Ochoa was the first Hispanic woman. Since 1978 there have been thirty-five women astronauts.[28] Thus, starting in 1973 significant change came to NASA without a lawsuit, but directly and dramatically with pressure from the public, the media, and especially Congress.

Another federal agency that employed hardly any women professionals before 1970 and managed to change without a major lawsuit was the National Park Service, in the Department of the Interior. The Park Service opened the ranks of its traditional military-inspired "park rangers," who wore a distinctive uniform and Stetson hats and sat erect on horses, to women in 1969. Shortly thereafter, there were battles, including a class-action complaint at the Independence National Historical Park in Philadelphia, not about advancement, but about the women's uniform. The big breakthrough for women scientists came in 1981, when the Park Service created a new job category, "natural resource specialist" or "natural resource manager." In the years 1982–93 forty-five women completed the training program for this new job designation and began to hold positions throughout the park system, which was not only introducing new techniques (for handling animal and fire control and archaeological remains) in its traditionally rugged outdoor national parks but also expanding into urban areas, historic buildings, and "heritage gardens," each of which has its own set of challenges. In the 1990s some of the women specialists moved into management.[29]

A third scientific agency that responded slowly and unevenly to the new era without a major lawsuit was the National Science Foundation, in which male science program officers and layers of male officials above them had had female assistants. In the early 1970s some women became assistant program directors and even temporary program directors, especially in biology and the social sciences, such as the ecologist and ornithologist Frances James, the ecologist Josephine Doherty, and the geneticist Rose Litman. Occasionally a woman was promoted to head of a section or, rarely, a division. The biologist Eloise E. Clark was the first to be a section head, a division head, and then a deputy assistant director, serving until 1976, when President Gerald Ford (a Republican) appointed her head of the new combined directorate for biological, behavioral, and social sciences. Over the years she hired other women, as did other directorates, such as those for science education, mathematics, astronomy, and in time even chemistry. Among these were the chemist M. Joan Callanan, the astronomer Laura Bautz, the mathematician Judith Sunley, and the biochemist Elvira Dornan, an African American, who was hired in 1978. Besides these career employees, the NSF had numerous temporary program directors, or "rotators," including Susan Bryant, of UC–Irvine, and Judith Willis, later at the University of Georgia.[30]

But this quiet, incremental change of the 1970s was abruptly challenged in the early 1980s with the coming of Reaganism, whereupon the science-education programs were terminated and Clark, who had been held over by President Carter, a Democrat, was forced to resign. (Vera Kistiakowsky, president of the AWIS, protested vehemently but to no avail.)[31] In 1983 Betty Vetter, of the Scientific Manpower Commission, pointed out that the NSF had no female program officers in chemistry or chemical engineering. As late as August 1983, of the fifty-eight NSF staffers with PhDs in chemistry only two were women, and their degrees were in geochemistry and biochemistry, leaving "not one female inorganic, organic, physical or analytical chemist at the National Science Foundation."[32] Eventually the situation improved, and as with the revival of the NSF's directorate in science education in the later 1980s, more women program officers were hired. Some were promoted, as was Sunley, in mathematics, to program director, deputy division director, and division director, perhaps partly in response to pressure from the Association for Women in Mathematics, whose Mary Gray had shared an office with Sunley at American University years earlier and continued to keep an eye on the NSF. Also in the late 1980s, the biologist Mary Clutter managed to rise, becoming the senior science adviser to director Erich Bloch, who needed coaching, as he came from industry and lacked a PhD. One of Clutter's claims to fame was that she refused to approve proposals for biological meetings that had no women speakers on the program.[33] Things continued to change, and by the late 1990s the electrochemist Janet Osteryoung, one of Caltech's first women PhDs, headed the NSF's chemical division, something nearly unthinkable just a few years earlier.[34]

One other agency at which several identifiable gender- and race-related trends were under way at the time was the Department of Agriculture (USDA), a traditional stronghold of southern and conservative white men with an enclave for women in home economics and nutrition. First there was the continuing erosion of the authority of the former Bureau of Home Economics, which had been created in 1923 and had a long tradition of strong female bureau chiefs going back to Louise Stanley, the first, and later Hazel Stiebeling, in the 1940s. In 1953, in the interest of economy and efficiency, this and several other old-time bureaus were amalgamated into the Agricultural Research Service. The nutritionist Ruth Leverton, who retired in 1974 after several lateral or even downward job moves, marked the end of this old tradition. Second, also in the 1970s, African American women nutritionists began to be hired and promoted, which led to a ruckus in 1986, when Edith Thomas, of the Expanded Food and Nutrition Education Program was fired for insubordination, or as she said, for paying too much attention to the needs of the historically black colleges. There was press coverage, and a report was written.[35] Third, there began to be female breakthroughs in the upper management of the USDA, such as in December 1973, when the fiber chemist Mary Carter became the first woman director of any of the USDA's twenty-nine research installations—the Southern Regional Research Center, in New Orleans, which led the world in cotton research. But then there was another reorganization, and in what may be a saga of the times, she moved up a notch but the chain of command got longer. (In 1978 the USDA combined four "services," including the whole Agricultural Research Service, each of which had formerly reported to an assistant secretary, into the Science and Education Administration, which reported to a director, who then reported to the assistant secretary.) In 1980 Carter became an associate administrator, and then in 1992 she became the area director for the South.[36] Finally, at the same time the USDA began to hire a trickle of junior women in the traditionally male-dominated field of soil science, where few women had worked for the federal government before the 1970s. By the 1980s some of these pioneers in the USDA's Soil Conservation Service were being promoted, and in 1988 Carol Wettstein became the first of several to serve as "state soil scientist," in 1989 Carolyn Olson became a "lead scientist" at the National Soil Survey Laboratory, in Lincoln, Nebraska, and in 2000 Maxine Levin became the national program manager of the Soil Survey Division.[37] Thus women scientists at the USDA, white and black, were moving mostly quietly but sometimes noisily beyond the segregated roles and fields of earlier years into the post-1972 pattern of "equal opportunity" with its omnipresent glass ceiling.

Personnel Reforms

These lawsuits and political dramas were just a small part of an even larger federal personnel crisis that crystallized during the presidential term of Jimmy Carter, when Congress finally took on this challenge despite the resistance of many entrenched groups. In 1978 it replaced the old Civil Service Commission, created with some fanfare in the 1880s, with the new Office of Personnel Management and an accompanying Merit Systems Protection Board, a quasi-judicial body that held hearings about complaints and terminations. At the same time, the long-standing "veterans' preference" of extra points on civil-service tests was dropped; a Senior Executive Service was created; a Federal Part-Time Employment Act passed; and some protections were accorded to whistle-blowers. Personnel responsibilities were decentralized to each agency, and the Federal Women's Program, created as part of the Civil Service Commission in 1967, was expanded to have at least one, preferably full-time officer in each agency to cover ethnic diversity and sexual harassment.[38]

However, it is hard to find evidence that any of these hard-fought and highly touted reforms mattered very much to women scientists. The Senior Executive Service offered the hope that top managers could earn more with various pay incentives and move among agencies, but very few women scientists seem to have taken advantage of this. Training programs were created to groom selected middle managers for this elite corps, but they were oversubscribed, and women were long underrepresented. The Merit Board was grossly underfunded at first, but in the 1980s it found its niche and issued reports on sexual harassment, whistle-blowing, and especially the glass ceiling, which was found to be ubiquitous in the federal government. The coming of the part-time option may have been a blessing to many, but it had been available to at least some women years earlier. In fact it had made one woman scientist's highly successful career possible. The physicist Johanna Levelt Sengers, for example, started to work part time in the heat division of the National Bureau of Sciences in 1964, after the birth of the first of her four children, and continued the arrangement for at least a decade. (She later won several awards, including election to the National Academy of Sciences.) Similarly, the geophysicist Mary Lou Zoback, an expert on earthquake deformations at the U.S. Geological Survey in Menlo Park, California, attributed her professional success and election to the National Academy of Sciences in 1995 to her part-time schedule for the previous thirteen years.[39]

One other change worth noting was the discontinuance after 1976 of the independently funded Federal Woman's Award, which dated from 1961, when women in government were presumed to be just secretaries. The prize had been created to honor outstanding longtime female employees who had defied the stereotype and made significant contributions. Over the years this award had featured scientists disproportionately, as about half of the six annual winners had been technical women, several of whom, most notably Rosalyn Yalow, unsung at the Veterans Hospital in the Bronx, had gone on to receive greater awards later, in Yalow's case the Nobel Prize in 1977 (see table 9.1). Although the prize had served its purpose at no cost to the government, its continuation in the era of equal opportunity as a separate compensatory award for women only, which had seemed so harmless and even chivalric before, began to raise concerns and embarrassment. One criticism was of the condescending press coverage of the awards banquet, which focused less on the winners than on the important men who accompanied them. More important was resistance from male nominators. At a time when there was pressure on personnel officers to treat women equally with men, to hire them in nontraditional

Table 9.1. Scientist winners of the Federal Woman's Award, by department or agency, 1971–1976

Department or Agency	Recipient
Department of Commerce	Joan Raup Rosenblatt, 1971
	Ruth M. Davis, 1972
	Marilyn Jacox, 1973[a]
	Evans Hayward, 1975[a]
Department of Agriculture	Ruth Mandeville Leverton, 1972
	Wilda Martinez, 1975
	Dorothy Fennell, 1976
Department of the Navy	Marguerite S. Chang, 1973
	Isabella Karle, 1973
	Marguerite M. Rogers, 1975[b]
NASA	Rita M. Rapp, 1971
	Marjorie R. Townsend, 1973
National Security Agency	Juanita Morris Moody, 1971
Library of Congress	Henriette Avram, 1974
Department of Labor	Beatrice Dvorak, 1975
National Institutes of Health	Marie U. Nylen, 1975
Office of Management and Budget	Joyce Walker, 1976

Source: Washington Post.
[a] National Bureau of Standards.
[b] Naval Weapons Center.

areas, and to include them in lists of persons to be promoted, the annual request for a list only of women who deserved a separate prize seemed odd and incongruous. This was especially so after the prestigious Arthur S. Flemming Awards for outstanding young men in public service had been opened to women. Thus, in 1972 Arthur Sampson, the acting administrator at the General Services Administration, refused to submit any names and suggested, as noted in the *Washington Post,* that perhaps the award should be modified to recognize the man or woman "who has made the most notable contribution to the cause of equal opportunity for women." Although the prize was funded by Woodward and Lothrop, a Washington, DC, department store, and got individuals and their agencies much free favorable publicity in Civil Service Commission publications and on the women's, or "society," pages of Washington area newspapers—much like the woman-of-the-year awards, or WOTYs, which also began to seem somewhat dated—no one spoke up to defend it, and it was allowed to fall silently by the wayside after the 1976 awards ceremony.[40] (For more on the whole issue of separate women's prizes, see chapter 12.)

Yet just a year later, in 1978, the Federal Women's Program of the new OPM approved a new single-sex organization, Women in Science and Engineering, or WISE, whose goals were to educate federal women scientists about their rights, enable them to meet one another across agency lines, and through their example popularize federal jobs in science to women students, much as the Federal Woman's Award had tried to do in its own way. Joan Humphries, of the NSF, headed the interagency group for a while, and Elvira Dornan, also of the NSF, participated in its science fairs, but it is debatable how attractive this new group was to women already busy in their own localities.[41]

Careers of Distinction

Despite all the controversy over the systemic mistreatment of women at various agencies, several fairly eminent women scientists did have long careers of distinction in the federal government, although in an era of increased mobility some left for greater opportunities elsewhere. Of course it is impossible to identify all of them, but some can be mentioned. There were a great many on the staff of the NIH, as listed both in table 9.2 and more extensively in a centennial volume of vitae published in 1988 to highlight its many women intramural employees. The highest-ranking woman career employee there was Ruth Kirschstein, who became, as mentioned earlier, a female first when she was appointed head of the Institute for General Medical Sciences in 1974. She served there until 1993, when she was promoted to deputy director of the whole NIH, a post she held until 1999. She later

Table 9.2. Selected women scientists employed at the National Institutes of Health, 1970–2000

Office of the Director	Bernadine Healy
	Ruth Kirchstein
Aging	Matilda White Riley
Allergy and Infectious Diseases	Janet Hartley
Cancer	Katherine Sanford
	Maxine Singer
	Elizabeth Weisburger
	Flossie Wong-Staal
	Anita Roberts
Child Health and Human Development	Florence Haseltine
	Antonia Novello
Dental Research	Marie Nylen
	Sharon Wahl
Diabetes and Digestive and Kidney Diseases	Judith Fradkin
	Elizabeth Neufeld
General Medical Sciences	Ruth Kirchstein
Human Genome Research	Elke Jordan
Division of Research Resources	Betty Pickett
Heart, Lung and Blood	Thressa Stadtman
	Martha Vaughan
Mental Health	Patricia Goldman-Rakic
	Candace Pert
Neurological Disorders and Stroke	Story Landis
	Ellen Silbergeld

Sources: NIH, *Women in Science* ([Bethesda, MD, 1988]); *AMWS.*

served as acting director of NIH and then as acting head of its new unit on alternative medicine in 2006 (when she was 80 years old).[42]

Most of the other distinguished women at the NIH did research and directed the work of others in their laboratories but rarely held higher administrative posts. In time several were elected to membership in the National Academy of Science (see chapter 12). What was different from earlier years was that several of these outstanding women did what had formerly been impossible, because academic medicine had been largely closed to women faculty: they left the NIH for an academic, nonprofit, or industrial post, sometimes in another part of the country. Thus Patricia Goldman-Rakic left for a professorship at Yale Medical School, one of the few places where her husband could work with primates, in 1979; Elizabeth Neufeld left to chair a department at UCLA in 1984; Maxine Singer left for the presidency of the Carnegie Institution of Washington in 1988; Candace Pert, of opiate-receptor fame, moved to a biotechnology startup, also in 1988; Flossie Wong-Staal, left for a professorship at the University of California at San Diego in 1990 and later formed her own biotechnology company; and Ellen Silbergeld left for a professorship at the University of Maryland in 1991.[43] This new mobility might also mean that those who stayed on would not be taken quite so much for granted.

At the same time, several women in the physical and biological sciences had careers of note at several of the DOE's national laboratories. In 1950 the Brookhaven National Laboratory had hired Gertrude Scharff Goldhaber as its first woman physicist when it employed her husband, something that rarely happened at universities in those days. She flourished there and was delighted to be elected to the National Academy of Sciences in 1972. After she retired, she was one of the founders of the Brookhaven Women in Science group in 1979.[44] Starting in 1970, the physicist Helen Edwards spent her career at Fermilab (as the Fermi National Accelerator Facility, in Batavia, Illinois, was called), where she designed and built its Tevatron, long the highest-energy particle accelerator in the world. She rose to become head of its Accelerator Division in 1987 and in 1989 seemed poised to be a major figure at the new Superconducting Supercollider in Texas, but it was never completed.[45] At Oak Ridge National Laboratory, in Tennessee, the biologists Dorothy Skinner and Liane Russell amassed long lists of publications in its famed Biology Division,[46] as did the bioscientist Mina Bissell at the LBL.[47] Meanwhile, as at the NIH, some senior DOE women left for academia in the 1980s. After starting her career at Oak Ridge in 1951, the nuclear chemist Darleane Hoffman followed her husband to the Los Alamos National Laboratory a year later. But her first moments there were unsettling, for officials at the personnel office told her that nothing was available for women other than clerical jobs. Taken aback, she nevertheless went to a welcoming cocktail party for her husband held by scientists, where one of those present quickly decided that her field fit exactly what his group needed and hired her on the spot. She stayed more than thirty years and was the leader of the isotope unit in the Nuclear Chemistry Division in 1984 when she left for the Lawrence Berkeley Laboratory and the University of California at Berkeley, where she succeeded the Nobelist Glenn Seaborg.[48]

The Argonne National Laboratory, in Illinois, which employed several accomplished women physicists, became over the years the model of the national labs in the area of women's issues and developed an active outreach program. Starting in the 1940s with Maria Goeppart Mayer, who shared the 1964 Nobel Prize in Physics, the line of women physicists and chemists at Argonne continued with Luise Meyer-Schützmeister, Natalia Meshkov, Caroline Herzenberg, Marion Thurnauer, and others, who formed an informal women-in-science group. In 1987 the group got funding from the laboratory's division of educational programs to sponsor a day-long program aimed at young women science majors in the area, an effort it repeated in 1989. That year the lab also started the Maria Goeppart-Mayer Distinguished Scholarships, which would bring two women scientists annually to the lab for up to twelve months. Then in 1991, when a DOE review of the national laboratories admitted that it did not have any programs designed to reduce women's turnover, Argonne's director, the African American physicist Walter Massey, started an official Women in Science Program to recruit and retain female employees, including reentry women. This activist tradition has continued, and in 2002 Argonne's associate director, Beverly Karplus Hartline, formerly associate director of the Thomas Jefferson National Accelerator Facility in Virginia, was one of the two convenors of the large International Congress on Women in Physics in Paris in 2002.[49]

The Department of Commerce employed several prominent women scientists in these years, as it contained both the National Bureau of Standards (NBS), which in 1988 changed its name to the National Institute of Science and Technology (NIST), with branches in Boulder, Colorado, and Gaithersburg, Maryland, and the new National Oceanographic and Atmospheric Administration, which contained the National Weather Service and several other units and employed meteorologists and oceanographers. The computer pioneer Frances "Betty" Holberton worked on UNIVAC at the NBS in the 1940s and on COBOL and FORTRAN later on until retiring in 1983. As the NBS was one of the few places that would hire a woman in physical chemistry before the 1970s, Marilyn Jacox worked on chemical data there from 1962 until her retirement in the 1990s. The physicist Jean Weil Gallagher worked on atomic-collision data first in Boulder and then in Gaithersburg, where she and Katherine Gebbie managed NIST's programs for standard physics reference. (A third physicist in Gaithersburg was Johanna Levelt Sengers, mentioned above.)[50]

Among the few women scientists at NOAA for a time was the pioneering meteorologist Joanne Simpson, who moved around a lot in her career and also worked at NASA's Goddard Space Flight Center, in Greenbelt, Maryland.[51] Also at NOAA was, and is, Susan Solomon, an atmospheric chemist who gained fame and acclaim in the 1980s for her work in Antarctica on the ozone layer. In fact, her extraordinary opportunities and success in the 1980s, when she was still in her thirties, mark her as someone of the younger generation who benefited from the battles fought and won before her, including the NSF's significant victory in 1970, when it got the U.S. Navy to allow women to do research in Antarctica, a small step forward for womankind.[52] Another NOAA employee of note was June Bacon-Bercey, an African American with an undergraduate degree in meteorology, who was a television "weather girl" in Buffalo, NY, before joining NOAA as chief of broadcast services. She then moved to California and became a forecaster with the National Weather Service. What makes her name known in meteorology circles is that in 1978, after winning $180,000 on a television game show, she endowed the June Bacon Bercey Scholarship, to be awarded by the American Geophysical Union to women pursuing careers in atmospheric sciences.[53]

Meanwhile, several women scientists and engineers had varied careers at NASA's various installations around the country. The astronomer Nancy Roman became the first woman scientist at the Goddard Space Flight Center when she moved to the newly formed NASA in the late 1950s from the Naval Research Laboratory (as did many others). She served as head of astronomy and planned future orbiting telescopes, which, because of their high cost, were not launched until 1990 with the Hubble Space Telescope. Roman did, however, manage to get Anne Underhill, who had earlier been a fellow graduate student at the University of Chicago, to head Goddard's laboratory for optical astronomy in 1970. But in 1977, perhaps as a cost-cutting move, it merged with NASA's laboratory for solar physics under the latter's chief. Underhill, who has been described as "unassuming in person," was thereupon pushed aside and relegated to the nonmanagerial role of "senior scientist" until she retired a few years later. Also at Goddard, Marjorie Townsend, the first woman to graduate from George Washington University with a degree in engineering, managed a project that launched small satellites into space to collect data. This was remarkable enough in 1973 to merit an article in *Science.* Out at NASA's Jet Propulsion Laboratory (JPL), in southern California, another lone woman engineer, Donna Shirley, worked on a variety of projects to Mars, Venus, Mercury, and Saturn, a few of which (most notably the Mars rover *Sojourner,* of 1997) came to fruition. Over the years the proportion of women engineers at the JPL increased to about 20 percent, but when the physicist Barbara Wilson came from Bell Labs in the late 1980s she was the first woman there in top management.[54] Reaching out beyond white women, NASA's new personnel director began to hire African American women scientists and engineers, including the several featured in occasional booklets such as *Women at Work in NASA* (1979). As a result of this publicity, these women, who may have been nearly invisible in the sea of white scientists and engineers at NASA, formed a large portion of those featured in a biographical directory of black women scientists and engineers, thus making NASA look like a model employer and showing how far it had come since the Harris affair of 1973, mentioned above. Among these minority women at NASA were the aerospace engineer Christine Mann Darden, who worked on sonic booms at the Langley Research Center, in Virginia, and the psychologist Patricia Cowings, at the Ames Research Center, in California, who patented the Ames Autogenic Feedback Training Mechanism, which astronauts use to minimize space sickness.[55]

There were a few women scientists of note in the Defense Department, especially the Army geodesist Irene K. Fischer, a world-renowned authority on the (ever so slightly changing) size and shape of the earth. The crystallographer Isabella Karle spent her entire career at the Naval Research Laboratory, outside Washington, DC, and won many awards, including such women's prizes as the Federal Woman's

Award and the Garvan Medal of the American Chemical Society, but she did not share her husband's Nobel Prize in 1985.[56] Also there was, more recently, the outspoken chemist and nanotechnologist Debra Rolison, discussed in the epilogue.

Elsewhere in the government, there were several significant women scientists at the Smithsonian Institution, where the astrophysicists Margaret Geller and Andrea Dupree and the planetary geoscientist Ursula Marvin worked at the Harvard-Smithsonian Astrophysical Observatory, in Cambridge, Massachusetts,[57] and at the Bureau of the Census, where the statistician Barbara Bailar had a productive career from 1958 to 1987, developing new methods of dealing with the perennial problems of non-sampling errors and undercounts.[58]

The National Science Board

Unlike other federal agencies and more like a university, the National Science Foundation has, in addition to its director, a governing group called the National Science Board (NSB), made up of twenty-four members appointed by the president. These political appointees were until the 1990s usually male university presidents or eminent scientists, although President Harry Truman appointed the Nobel laureate Gerty Cori, of his home state of Missouri, to the first board in 1952. Although women members were until recently quite few, several of them left a mark on this, the lowest science-policy post open to most scientists and more recently engineers (see table 9.3).

Starting in the 1960s and continuing into the 1970s, the few women "tokens" on the NSB tended to be scientists, often presidents of women's colleges, as mentioned in chapter 5. Continuing this tradition was Richard Nixon's appointment of Anna Harrison, a professor of chemistry at Mount Holyoke College, in 1972. She had brought herself to the attention of the powers that be in Washington, DC, when she testified before the House Appropriations Subcommittee on Housing and Urban Development, Space and Science in 1971 and 1972. Alarmed by cuts in the science-education budget proposed by the Nixon administration and by the recent resignation of the NSF's assistant director for science education in protest, she wrote her longtime congressman, Edward P. Boland (D-MA), a high-ranking member of the committee. He invited her to testify for fifteen minutes. But once under way, she was invited to stay on for two hours, as she held the congressmen spellbound with her discussions of the need for more federal funding of science education. Word of this remarkable performance reached the American Association of Colleges, whose staff had long wanted someone from a liberal arts college on the influential NSB. They nominated her, Nixon appointed her, and she served six frustrating years as the lone voice for science education on the board.[59]

Table 9.3. Women members of the National Science Board, by year appointed, 1950–2008

Year	Number	Names
1950	2	Sophie Aberle, Gerty Cori
1952	0	
1954	0	
1956	0	
1958	1	Jane Russell
1960	0	
1962	1	Katharine McBride
1964	1	Mina Rees
1965	1	Mary Bunting
1966	0	
1968	0	
1970	0	
1972	1	Anna Jane Harrison
1974	1	Jewel Plummer Cobb*
1976	1	Marian Koshland
1978	1	Ernestine Friedl
1980	2	Mary Good, Mary Jane Osborn
1982	0	
1984	3	Anneliese Anderson, Rita Colwell, K. June Lindstedt-Siva
1986	1	Mary Good†
1988	0	
1990	1	Marye Anne Fox
1992	0	
1993	1	Shirley Malcom*
1994	3	Eve Menger, Claudia Mitchell-Kernan,* Diana Natalicio*
1996	4	Mary K. Gaillard, M. R. C. Greenwood, Jane Lubchenco, Vera Rubin
1998	3	Pamela Ferguson, Anita Jones, Maxine Savitz
2000	2	Nina Federoff, Diana Natalicio*,†
2002	4	Delores Etter, Elizabeth Hoffman, Jane Lubchenco,† Jo Anne Vasquez*
2004	1	Kathryn Sullivan
2006	3	Camilla Benbow, Patricia Galloway, José-Marie Griffiths
2008	3	France Cordova,* Esin Gulari, Diane Souvaine

Source: NSB website, www.nsf.gov/nsb/members.
Note: The board has twenty-four members at a time; usually eight are appointed every other year to six-year terms, but not all complete their terms.
*Minority woman
†Second term

In 1974 Nixon also started what began to be a series of minority women on the board when he appointed the African American biologist Jewel Plummer Cobb, of Connecticut College, until recently for women only. Then after President Gerald Ford informed the board in 1975 that as part of the International Women's Year he wanted all agencies to do more to encourage women, Cobb persuaded the board to ask the NSF legal department whether the agency could support single-sex and minority programs. Eventually getting a green light, she also got the NSB to start a committee on women and minorities, which pushed and prodded the NSF's staff to develop and expand new programs for both underrepresented groups. By the time Cobb's term ended in 1980, the NSF had started several such programs, although its major program for women, run by M. Joan Callanan in the Directorate for Science and Engineering Education, was terminated about then, and its chief program for minorities, the Minority Institutions Science Improvement Program, or MISIP, headed by Shirley McBay, was sent to the new Department of Education.[60] But even this modest tradition of token women and minorities on the NSB barely survived the long chill under presidents Reagan and George H. W. Bush, although the capable chemist Mary Good, who had by then worked in both academia and industry rather than at a women's college, served two terms and in 1988–91 even chaired the NSB, another first for womankind.

Then several things happened in the early 1990s. In 1990 the National Research Council held a conference on recruiting and retaining federal scientists and engineers that recommended that "glass cutters" be found to break through the omnipresent glass ceiling.[61] And Anita Hill's gripping testimony at the nationally televised hearings on the nomination of Clarence Thomas for the Supreme Court in October 1991 had an important impact on the national elections a year later: record numbers of women were elected to both houses of Congress, and Bill Clinton was elected president. Overnight the situation for women scientists on the NSB changed. Several of the new congresswomen, including Anna Eshoo (D-CA), whose district included Silicon Valley and who had many technical women among her supporters, chose to be on the science subcommittee of the House Science, Space, and Technology Committee. During some congressional testimony, Eshoo asked an NSF official just how many women were on the NSB. Shocked to hear that at that point it was down to only one of twenty-four, she recommended an increase. Before long President Bill Clinton, who had campaigned on a promise of bringing more diversity to the federal government, complied and increased the number of women appointees to roughly three (of eight) every other year during his two terms. The cumulative result was that by 2002 there was a high point of nine women among the twenty-five members (twenty-four plus the director, Rita Colwell), or 36 percent.[62]

Other Presidential Appointees

Table 9.4 lists the more prominent women scientists, especially social scientists, appointed to high posts by the eight U.S. presidents since 1972—Nixon, Ford, Carter, Reagan, G. H. W. Bush, Clinton, G. W. Bush, and Obama. The numbers and level of positions provide a barometer of each administration's political responsiveness to the leverage exerted by the women's movement on the party in power at the time. (Another, weaker indicator is the awarding of the National Medal of

Table 9.4. Additional women scientist presidential appointees, 1972–2010

Richard Nixon, 1972–74		
Betsy Ancker-Johnson	Commerce	Assistant Secretary for Science and Technology
Dixy Lee Ray	Atomic Energy Commission	Chair
Marina von Neumann Whitman	Council of Economic Advisers	Member
Gerald Ford, 1974–77		
Eloise E. Clark	NSF	Assistant Director for Biological, Behavioral, and Social Sciences
Dixy Lee Ray	State	Assistant Secretary for Oceans, International Environment and Scientific Affairs
Jimmy Carter, 1977–81		
Eula Bingham	Labor	Assistant Secretary for Occupational Safety and Health
Ruth M. Davis	Energy	Assistant Secretary for Resource Applications
Alice Stone Ilchman	International Communication Agency	Associate Director for Educational and Cultural Affairs
Ruth L. Kirchstein	National Institute of General Medical Sciences	Director
Juanita Kreps	Commerce	Secretary
Barbara Newell	State	Ambassador to UNESCO
Janet Norwood	Bureau of Labor Statistics	Commissioner
Carolyn Payton*	Peace Corps	Director
Isabell Sawhill	National Commission for Manpower Policy	Director
Donna Shalala	HUD	Assistant Secretary for Policy Development and Research
Courtenay Slater	Commerce	Chief Economist
Mabel Murphy Smythe*	State	Ambassador to the Cameroons
Nancy Hays Teeters	Federal Reserve Board	Member
Ronald Reagan, 1981–89		
Bernadine Healy	Office of Science and Technology Policy, White House	Deputy Director
George H. W. Bush, 1989–93		
Barbara Bryant	Census	Director
Sylvia Earle	NOAA	Chief Scientist
Bernadine Healy	NIH	Director
Jane Henney	FDA	Commissioner
Bill Clinton, 1993–2001		
Mollie H. Beattie	Fish and Wildlife Service	Director
Rita Colwell	NSF	Director
France Cordova	NASA	Chief Scientist
Mildred Dresselhaus	Energy	Director, Office of Science
Christine Ervin	Energy	Energy Efficiency and Renewable Energy Chief

Table 9.4. Additional women scientist presidential appointees, 1972–2010 *(continued)*

Mary Good	Commerce	Undersecretary for Technology
Rhea L. Graham*	Bureau of Mines	Director
M. R. C. Greenwood	Office of Science and Technology Policy, White House	Associate Director for Science
Shirley Jackson*	Nuclear Regulatory Commission	Chair
Anita Jones	Defense	Director of Research and Engineering
Martha Krebs	Energy	Director, Office of Energy Research
Kathie Olsen	NASA	Chief Scientist
Anne Petersen	NSF	Deputy Director
Arati Prabhakar	NIST	Director
Alice Rivlin	Office of Management and Budget	Director
Donna Shalala	HHS	Secretary
Kathryn Sullivan	NOAA	Chief Scientist
Laura Tyson	Council of Economic Advisers	Chair
Sheila Widnall	Defense	Secretary of the Air Force
Janet Yellen	Council of Economic Advisers	Member
George W. Bush, 2001–9		
Wendy Chao	Labor	Secretary
Shannon Lucid	NASA	Chief Scientist
Kathie Olsen	NSF	Deputy Director
Barack Obama, 2009–		
Margaret Hamburg	FDA	Commissioner
Jane Lubchenco	NOAA	Administrator
Marcia McNutt	USGS	Director
Christina Romer	Council of Economic Advisers	Chair

*Black woman

Science and the National Medal of Technology, discussed in chapter 12.) The table shows that President Bill Clinton proved to be by far the most effective "glass cutter" of all time, but political appointees are usually quite temporary (some more so than others), and few have made equal-opportunity issues a high priority, as if just being there was enough.

There were a few such pioneering female appointees, perhaps best described as "tokens," under President Nixon at the subcabinet level—the physicist Betsy Ancker-Johnson as assistant secretary for science and technology in the Department of Commerce and the economist Marina von Neumann Whitman on the Council of Economic Advisers. They served quietly and competently far out of the limelight.[63] By contrast, Nixon's appointment of Dixy Lee Ray, of Washington State, to the AEC as a member in August 1972 and then chair in 1973 attracted much attention, for she was in many ways an unusual choice, as she was an outsider (and a threat) to the AEC's ingrown culture. (She was not the first biologist or woman commissioner of the AEC, as the microbiologist Mary Bunting had been both in the 1960s.) Ray lived in a mobile home and took her two large dogs to the office, which the press made much of—too much, in Ray's view—writing more about them and her than about energy policy. More importantly, Ray also claimed to be an environmentalist who cared greatly about Puget Sound and its environs, as she came from Seattle, the University of Washington, and the innovative Pacific Science Center, which she had headed for eight years. This stance upset some biologists, who did not think that any right-thinking life scientist could be in favor of nuclear power. One of the enigmas of her term at the AEC was the demise in November 1974 of Karen Silkwood, in Oklahoma, on her way to testify about the misuse of plutonium on the Hanford Reservation.

Because of demonstrations against nuclear power, the AEC was a rather controversial agency, which Nixon wanted Ray to shake up, a thankless task. At times that meant going around the entrenched old guard by commissioning fresh studies by outsiders. It also meant ruffling feathers by not reappointing people who had been there a long time and remaining calm when badgered by their friends on congressional committees or when embarrassed by their self-serving leaks to the press. After Ray accomplished the goal of splitting the AEC into two—the Nuclear Regulatory Commission and the Energy Resources Development Administration, soon to become the cabinet-level Department of Energy—Nixon's successor, Gerald Ford, appointed Ray assistant secretary for oceans, international environment and scientific affairs in the State Department. This was a less successful assignment. She complained publicly that her boss, Secretary of State Henry Kissinger, ignored her; he gave speeches on food and international science, she said, but the only time she had met him was when he came to her installation. She quit after only six months to return to Washington State, where she ran successfully for governor, and she was succeeded at the State Department by Patsy Mink, a former and later congresswoman from Hawaii, who also found the job difficult.[64] (In 1975 the economist Alice Rivlin became the first director of the Congressional Budget Office, an important post but not under the president.)

The strength of the women's movement within the Democratic Party in the mid-1970s is illustrated by the number of women social scientists, especially economists, appointed to high office by President Jimmy Carter. Among these were the economist Juanita Kreps, of Duke University, as secretary of commerce, just the fourth woman ever in the cabinet, and the first social scientist; the zoologist Eula Bingham to head the Occupational Safety and Health Administration (OSHA); the economist and Wellesley College president Barbara Newell as the ambassador to UNESCO; and the young political scientist Donna Shalala, from Hunter College, as an assistant secretary in the new Department of Housing and Urban Development.[65] Carter also appointed the economist and career employee Janet Norwood to head the Bureau of Labor Statistics, where she far outlasted his administration, not retiring until December 1991. He also reached out and appointed Carolyn Payton, an African American psychologist at Howard University, to head the Peace Corps.[66]

Then, however, for eight long years under Republican Ronald Reagan there were no presidential appointments of women scientists, which shows dramatically the weaknesses of presidential appointments as any remedy for lasting change. The Reaganites knew hardly any women scientists, and few of them moved in Republican circles. One exception was the photogenic cardiologist Bernadine Healy, MD, as deputy director of the White House Office of Science and Technology Policy in 1984–85.

After Reagan's eight years, President George H. W. Bush appointed the oceanographer Sylvia Earle as chief scientist at NOAA, but his best-known female appointee was the above-mentioned Dr. Healy to the hot seat of director of NIH in 1991. Educated at Vassar College and Harvard Medical School, she had served in the Reagan White House and had been head of the research division of the Cleveland Clinic from 1985 to 1991, during which time she also served as president of the American Heart Association in 1988–89. Her term started out well, but she served only two tumultuous years (1991–93), partly because Bush was not reelected but also because she was quickly beset by such old and new problems as scientific fraud (among them the long-running Baltimore case); the Gallo case, which also involved Congressman John Dingell (D-MI); her attempt to control overhead costs; her approval of a controversial teen-sex survey; her handling of the irascible head of the Human Genome Project, James D. Watson; and Republican opposition to fetal-tissue research. Although she wrote an editorial for *Science* expressing support for women in science and medicine, she loyally opposed Congress's proposed office for women's health as unnecessary. She also met with the malcontents at the NIH, appointed more equal-employment-opportunity officers, and issued new regulations on sexual harassment. Ironically, even she could do little to actually change practices within the NIH, for despite her title, most of the real power rested within the separate institutes.[67]

It was thus a dramatic change when President Bill Clinton, especially during his first term, appointed a flood of women scientists and social scientists, including many young and minority ones, to high office. These included the political scientist Donna Shalala, who was by 1993 chancellor of the University of Wisconsin, to the cabinet post of secretary of the immense Department of Health and Human Services (HHS). Shalala, like Kreps in Jimmy Carter's cabinet, served long and well without any major scandals. Clinton also appointed Sheila Widnall, a professor of aeronautical engineering at MIT, the first woman to serve as Secretary of the Air

Force, and the chemist Mary Good, who declined the directorship of the NSF, to the position of undersecretary for technology at the Department of Commerce, which she preferred, as it involved applied work.[68] In 1994 Clinton appointed the first woman deputy director of the NSF, the psychologist and statistician Anne Petersen, and then after her departure just two years later, he recommended the microbiologist Rita Colwell, of the Maryland Biotechnology Institute and a former member of the National Science Board, as her successor. But about that time the NSF director, Neal Lane, moved to a post at the White House, and Clinton nominated Colwell to direct the whole foundation, a significant first for womankind at the scientific agency whose budget was soaring, possibly even doubling. In her six years she did an exceptional job, and in 2003 the agency shared the Office of Personnel Management's Presidential Quality Award, for the best-run agency in the federal government, with the Environmental Protection Agency.[69]

Clinton's appointment of the 45-year-old Berkeley economist Laura Tyson to the coveted post of chair of the Council of Economic Advisers provoked considerable professional jealousy, condescension, and venomous backbiting from her fellow economists—she was too young, not theoretical enough, and not from the Ivy League. Bill Clinton did not care about such matters, but he liked the fact that he could understand her explanations of complicated phenomena better than others'. She stayed several years, and the economy boomed. He also appointed the physicist Martha Krebs, who had spent a decade at the LBL planning and overseeing new facilities, as the assistant secretary and director of the Office of Science at the Department of Energy.[70] Two other presidential appointees came from state government—the forester Mollie Beattie, the first woman to head the Fish and Wildlife Service, and the geologist Rhea Graham, the first woman and the first African American to lead the Bureau of Mines. Beattie had been commissioner of the Vermont Department of Forests, Parks and Recreation, and Graham had been the director of New Mexico's Mining and Minerals Division.[71] Other appointments that made many "firsts" were those of the Hispanic astronomer France Cordova as the chief scientist of NASA and the African American physicist Shirley Jackson to head the Nuclear Regulatory Commission.[72] Perhaps the youngest woman of all was the 34-year-old Arati Prabhakar, an Indian-born Caltech PhD and former program manager at the Defense Department's Defense Advanced Research Projects Agency, as NIST's first woman director. When interviewed, she said that she loved the job but felt the pressure. As a young, attractive, exotic female, she had to be perfect all the time; 99 percent perfect was not enough.[73] These women's role was not easy. Some had been promoted too fast and had not learned at lower levels lessons that would have helped them at the top. Others got less than the best jobs, such as Rhea Graham, at the Bureau of Mines, who learned that her agency was slated for extinction.

It is interesting in the light of all these female "firsts" in the Clinton administration to note the politics behind an appointment that did not go to a woman after all. In 1993 both President Clinton and his HHS secretary, Donna Shalala, wanted to appoint a female successor to Healy as director of the NIH, a female "second." But as word of this intention spread among female bioscientists, some of whom may have been contacted as possible nominees, opposition mounted, for many in the scientific community thought that Harold Varmus, an MD and Nobel laureate, would be the best choice, and he stated his willingness to serve. Before long, Shirley Tilghman, of Princeton, and Joan Steitz, of Yale, two prominent molecular biologists, but without MDs and so probably not candidates themselves, circulated a letter to forty-eight women bioscientists urging them to co-sign a letter to Shalala backing Varmus and disqualifying themselves.[74] It is not evident how many complied, but Varmus was appointed, and Ruth Kirschstein, who may have been among those passed over in the process, served as deputy director until his departure and then as acting director.

By contrast, President George W. Bush nominated hardly any women scientists to top posts during his eight years in office, even though by 2001 more of them than ever before were qualified for presidential appointments. (It was rumored for a time that the chemist Marye Anne Fox, who had known Bush in Texas, was in line for a top post, but it never materialized.) Rita Colwell finished out her six-year term in 2004, and Bush appointed the neuroscientist Kathie Olsen, formerly a staffer at the NSF and then the chief scientist at NASA under Clinton, as NSF's deputy director.[75] In the meantime, this generation of women scientists went back to academia and waited for the long G. W. Bush administration to end. They and many other bright and industrious women scientists were ready to serve when the Democrat Barack Obama was elected in November 2008.

Conclusion

A lot happened to both the federal government and its women scientists and engineers after 1972. Overall the federal government shrank in size, but its workforce grew more technical and more feminized as it struggled to face new challenges. The Equal Employment Opportunity Act of 1972 opened most personnel practices to legal redress. Several motivated women scientists filed class-action lawsuits against their agencies. A few were adjudicated speedily, but others took years. Often the settlements called not only for remedies such as back pay for the plaintiffs but also, because they insisted on it, sweeping changes at their agencies. The courts complied, but all too often the agencies did the minimum. It would take more than a successful lawsuit, even one overseen by a series of federal judges, to change entrenched cultures like those at the Forest Service and the National Institutes of Health. Congress could act, as it did in events at NASA, to make administrators accountable.

In some fields the supply of women scientists and engineers changed significantly in the 1970s and 1980s. Where before 1972 the federal government had employed very few women in agricultural, environmental, or engineering positions, by the 1990s many held degrees in these fields and had been hired by and were working for relevant agencies, and some were moving up into management after class-action lawsuits and settlements. Yet a lot remained the same, and perhaps even more so, as reorganizations and restructurings increased the numbers of middle levels of government. One could be promoted several times but still be far from the top. Women scientists, including those in the newly or barely feminized fields, began to cluster in the middle to upper-middle grade levels as a result of the same intractable problems with the glass ceiling that a few in other fields had faced and fought since at least the early 1970s. The period started with angry women facing promotion issues at the USFS and especially the NIH, one of the few agencies with enough women to make the problem systemic, and ended with the same pervasive problem at almost all agencies, as their first women engineers and scientists hired in nontraditional fields stayed for a decade or more of service and then expected to be promoted. What reforms did get implemented in some agencies, as at the USGS, were partly undone by the budget reductions of the 1990s.

Another tactic that seemed promising at the time, perhaps because so few women were rising to the top on their own, but had only a modest impact was the presidential appointment of female officials. Some administrations appointed many, others very few. Those who served were greeted with a lot of publicity, almost all positive and respectful, but they stayed only about three years each. (Ruth Kirchstein's twenty-seven years at the NIH, Janet Norwood's thirteen years as commissioner of labor statistics, and Donna Shalala's eight years at HHS were unusual.) Very few seem to have been able to do much to help the women career scientists at their agencies. For some, such as, perhaps, the minority women, just being there was an achievement in itself, a kind of demonstration effect. The idea that well-trained African American women would be running high-tech agencies in the 1990s—the physicist Shirley Jackson at the Nuclear Regulatory Commission or the geologist Rhea Graham at the Bureau of Mines—was unthinkable in 1972, even as legislation that would affect their careers was being signed. (But significantly neither of them had come up through the ranks.) Opportunities had broadened, workforces had diversified, and American science was the better for it, but problems remained. Bill Clinton's two terms were a high point, yet even then budgetary pressures forced "reductions in staff" of many of those very women who had fought so hard in previous years for employment, advancement, and equal pay.

The federal government was in some ways closer to being the "showcase" in 2000 that President Lyndon Johnson had envisioned in the 1960s. The years 1972 and 1978 had been landmark years. Progress had been substantial, and women scientists had persevered through a Republican-controlled Congress, an aging and increasingly conservative Supreme Court, and a Republican president hostile to science. The new Obama administration (2009–) raised hopes of bigger opportunities and achievements to come.

CHAPTER TEN

Nonprofit Alternatives

Speeding Up, Moving In, On, and Even Up

Overall the nonprofit sector, which had traditionally been a haven for talented women scientists, especially biologists, grew and changed after 1972, with important consequences for careers there. At first glance this sector was highly diverse, as it comprised several types of institutions and organizations—independent research institutes, including observatories, museums, zoos, botanical gardens, and marine biology stations; university-affiliated research institutes and medical centers; foundations; and other scientific and environmental organizations, many of which took on such science-related but invisible tasks as publishing journals, collecting and storing data, and running educational programs. For some of these institutions, new facilities and closer links to universities helped them overcome their often isolated locations and their traditionally seasonal activities. At its best the nonprofit sector offered highly motivated individuals a host of attractions—focus, intensity, flexibility, informality, risk taking, room for innovation, a lack of bureaucracy, and the chance to make a difference.[1]

Underlying this variety and vitality, however, was the common bond of budgetary pressure. As costs, especially salaries, rose in the inflationary 1970s, and as ambitions for new projects generally outstripped incomes even from large endowments, these traditionally self-supporting institutions became dependent on federal and other outside funds. (Even the once-affluent Carnegie Institution of Washington [CIW] had to drop its longtime aversion to applying for federal grants in the mid-1970s.)[2] At about the same time, the science budgets of several federal agencies began to increase enormously. Fortunately, most of the new funds—at the NSF, the new EPA, the reorganized Department of Energy, and especially the NIH and the National Cancer Institute with President Richard Nixon's lavishly funded "war on cancer"—were in the biomedical sciences, where many research institutes and women scientists were clustered. Table 10.1 shows the heavy dependency of many of these nonprofit institutions on a few federal agencies by 2004. In fact, in many ways these institutions were becoming quite similar to the DOE's "national laboratories" or the Federally Funded Research and Development Centers (FFRDCs), which were run directly for the federal government by nonprofit contractors such as the MITRE Corporation and Lincoln Laboratories.[3]

Table 10.1. Selected nonprofit institutions, by total federal support (in thousands of dollars) and federal agency providing the most support, FY 1999

Institution	Total Support	Most Supportive Agency (%)
1. Fred Hutchinson Cancer Research Center	$135,043	HHS (99.2)
2. Dana-Farber Cancer Institute	66,768	HHS (98.0)
3. National Academy of Sciences	65,839	NASA (25.9)
4. Memorial Sloan-Kettering Cancer Institute	63,812	HHS (96.6)
5. Whitehead Institute for Biomedicine	58,878	HHS (98.2)
6. Battelle Memorial Institute	55,150	DOD (43.0)
7. Stanford Research Institute (SRI International)	49,211	DOD (59.2)
8. Salk Institute for Biological Research	43,113	HHS (88.7)
9. Cleveland Clinic Foundation	42,358	HHS (82.3)
10. Joint Oceanographic Institutions*	36,177	NSF (99.9)
11. RAND Corporation	29,511	HHS (86.7)
12. Jackson Laboratory	27,907	HHS (97.3)
13. Cold Spring Harbor Laboratories	24,839	HHS (80.8)
14. The Institute for Genomic Research (TIGR)	18,845	NSF (50.5)
15. Wistar Institute	15,745	HHS (90.1)
16. Joslin Diabetes Foundation	13,674	HHS (100.0)
17. National Bureau of Economic Research	12,359	HHS (51.5)
18. Carnegie Institution of Washington	11,510	NSF (46.0)
19. Institute for Cancer Research (Fox Chase Cancer Center)	8,756	HHS (100.0)
20. American Museum of Natural History	7,252	NASA (79.1)
21. Nature Conservancy	6,789	Commerce (88.4)
22. Population Council	6,358	HHS (99.1)
23. National Opinion Research Center	5,573	HHS (72.6)
24. Marine Biological Laboratory	5,342	HHS (60.8)
25. Boyce Thompson Institute	2,301	USDA (31.7)
26. Lowell Observatory	2,157	NASA (57.2)
27. American Type Culture Collection	2,103	HHS (60.8)
28. Field Museum of Natural History	2,083	NSF (85.2)
29. Bernice P. Bishop Museum	2,000	NASA (100.0)
30. California Academy of Sciences	1,767	NSF (100.0)
31. National Biomedical Research Foundation	1,509	HHS (81.2)
32. American Association for the Advancement of Science	1,417	EPA (52.2)
33. Santa Fe Institute	1,325	NSF (75.8)
34. New York Botanical Garden	1,139	NSF (56.1)
35. American Geophysical Union	1,120	NSF (86.6)
36. Brookings Institution	658	NSF (49.1)
37. Center for Advanced Study in the Behavioral Sciences	93	NSF (100.0)
38. Institute for Women's Policy Research	25	Other (100.0)

Source: NSF, *Federal Science and Engineering Support to Universities, Colleges, and Nonprofit Institution, Fiscal Year 1999,* NSF-01-323 (Washington, DC, Apr. 2001), table B-27.

* Consortium of thirty-one oceanographic institutions, including the Woods Hole Oceanographic Institution, the Scripps Institute of Oceanography, and the Lamont-Doherty Geophysical Laboratory of Columbia University, organized in the 1970s to coordinate not only ocean drilling projects but ocean science research as well.

Some coastal nonprofit institutions, such as the Marine Biological Laboratory and Woods Hole Oceanographic Institution (WHOI) on Cape Cod, Cold Spring Harbor Laboratories on Long Island, and the Jackson Laboratory in Bar Harbor, Maine, which had a second disastrous fire in 1989, took advantage of this new money to increase their output by transforming their formerly seasonal operations into year-round, state-of-the-art facilities. Another quiet backwater, the Boyce Thompson Institute for Plant Research, established in the 1920s in Yonkers, New York, was initially devoted to long-term studies of seed germination and air pollution, but in 1978 it became a major player in the national competition for grants in the molecular biology of plants by moving to the Cornell University campus in Ithaca, New York, and constructing a new facility with specially equipped greenhouses. Even more dramatically, totally new nonprofit laboratories, such as the Salk Institute for Biological Research in San Diego, the Dana-Farber Cancer Institute in Boston, the Whitehead Institute for Biomedicine at MIT, and The Institute for Genomic Research (TIGR) in Rockville, Maryland, to name a few of the best known, were all designed from the outset to take advantage of all the new federal and private money for cancer research and the human genome project. (For example, the Whitehead Institute, established at MIT in 1981 with $120 million from a medical-instruments tycoon, had by 2004 an endowment of $300 million and an annual operating budget of $45 million, and 80 percent of its research support came from federal grants.) By contrast, older institutions such as the American Museum of Natural History, the Academy of Natural Sciences of Philadelphia, the New York Botanical Garden, the Missouri Botanic Garden, and the California Academy of Sciences, much of whose research was in the area of taxonomy, which was less well funded, struggled to keep up.[4]

The availability of and reliance on so much "soft" money from federal agencies meant that the tempo changed at certain institutions that had previously prided themselves on doing quiet, subdued, long-term research on unfashionable topics. They now became similar to, and more competitive with, top academic and even industrial organizations, as ambitions soared, plans for new facilities abounded, and new people, even whole teams, came and went.[5] Women shared the brunt of the increasing job insecurity that the new competitiveness entailed. The kind of splendid isolation or secure, grant-free lifetime refuge for talented women scientists of years past—such as those enjoyed by the geneticist Barbara McClintock at the CIW laboratory at Cold Spring Harbor, New York, and the anthropologist Margaret Mead at the American Museum of Natural History in New York City—was becoming a luxury that institutions could no longer afford, and long-term researchers with high salaries were expected to bring in grant funds for themselves and others. Yet this new necessity for outside support also benefited some highly productive women scientists, for the peer-review process that such support entailed brought their work under the scrutiny of their colleagues and potential employers at the universities. In time this end of isolation led to outside job offers that allowed a few mid-career and senior women to leave their nonprofit institutions for faculty positions at prestigious universities, something that had not been possible before the federal laws banning discrimination in academic hiring were signed in 1972.[6]

As discussed in chapter 7, some postdoctoral fellows supported by the federal soft money were able to spend a year or more at nonprofit institutions that would not have been able to support them themselves. Accordingly, there was a steady stream of recent doctorates, American and foreign, who came, learned, published, and moved on. They left their mark on the torrent of publications flowing from these institutions and took their experiences with them to their next location, which for a few might be another nonprofit laboratory.

In some ways the employment practices and attitudes at the nonprofits (except for the foundations in New York City, about which see below) were more flexible, accommodating, informal, and worker-friendly than those in place in the other sectors. First, there were no major lawsuits or scandals even in the tumultuous 1970s (except at the Aerospace Corporation, on which see below). Occasional difficulties and conflicts, such as with sexual harassment, attracted internal attention and outside publicity and led to sanctions. A few women's committees were formed and listened to, then faded quietly away. Second, in the 1980s, well before the major universities or even the federal government, some nonprofit organizations began to appoint women scientists and nonscientists, often with considerable experience elsewhere, to top posts, including that of chief executive officer. Even more remarkably, three of these top females were women of color. Like their predecessors, these women executive directors struggled to keep their institutions on course and to face new challenges. Yet seemingly few, perhaps only Maxine Singer, Shirley Malcom, and Heidi Hartmann, used this prominent pulpit to do much to help or encourage other women scientists.

Particular Places

Some of these institutions specializing in mathematics, atmospheric sciences, physics, and oceanography, which had traditionally had few, if any, women at their facilities, started to employ a trickle of token women on their staffs in the 1970s. Unfortunately, the Institute for Advanced Study, in Princeton, New Jersey, known for its high-powered mathematicians and astrophysicists, still had no women scientists on its prestigious "faculty" by 2010.[7] At the National Center for Atmospheric

Research, in Boulder, Colorado, established in 1960, the meteorologist Margaret "Peggy" LeMone, who arrived in 1972 and cofounded its first women's committee shortly thereafter, was still one of its very few women scientists when she was elected to the National Academy of Engineering in 1997. In 2000 it was the first nonprofit organization to invite what had come to be called the "climate for women in physics" site-visit committee of the American Physical Society to advise it on how to improve.[8] Meanwhile, the SLAC National Accelerator Laboratory in California had two women physicists, Helen Quinn and Gail Hanson, on its scientific staff in the 1970s; however, it would be decades before Persis Drell was hired as associate director of research in 2002 and promoted to director in 2007.[9] Little better was the Woods Hole Oceanographic Institution, whose director formed a women's committee in 1973, for as late as 1988 it still had just 7 women out of 100 on its scientific staff. This number rose sharply in the 1990s and reached 24 of its larger staff of 135 (17.8%) in 2003. Although WHOI's historian attributed this slow response to the limited number of women in oceanography, the presence of award-winning female visitors there in the 1980s indicates that some women were available earlier.[10]

Exemplifying the next stage of integrating more than a few token women scientists into their workforce were two institutions in the physical sciences, one old and one new. Long favored among physicists, mathematicians, and computer scientists was the nonprofit Bell Laboratories in Murray Hill, New Jersey, famed for its long string of famous discoveries and its six Nobel Laureates. Because part of the Equal Employment Opportunity Commission's historic 1973 consent decree with AT&T applied to Bell Labs, it immediately began to hire more women and minority "members of the technical staff" (MTS). Among its stellar recruits in the 1970s were Patricia Cladis, Evelyn Hu, Shirley Ann Jackson, Elsa Reichmanis, Cherry Murray, and Barbara Wilson.[11] Another sign of progress was the promotion in 1978 of the artificial-intelligence expert Erna Schneider Hoover to the post of manager, the first woman to head a technical department at Bell Labs. One unfortunate outcome of this influx of female staff members, however, was the recognition in 1981–82 that existing policies regarding sexual harassment were not sufficient. In response, several women scientists served on a committee that detailed the more effective steps that managers at each level were to take promptly.[12]

By the end of the 1980s, Bell's talented women were in demand elsewhere. In 1988 the physicist Barbara Wilson, by then a technical supervisor, left for the first of a series of management posts at NASA's Jet Propulsion Laboratory at Caltech, and Pam Surko, who had come from Princeton in 1980, moved with her husband to jobs in San Diego. Then in the 1990s, as had already begun to happen elsewhere, a few women scientists were promoted into Bell Labs' own upper management. In

1994 the chemist Elsa Reichmanis became the director of polymer and organic materials research; in 1997 the applied mathematician Margaret Wright, who had been hired in 1988, became head of the scientific computing department; and in 2000 the physicist Cherry Murray became vice president for physical-sciences research. But by then not all was well in Murray Hill, for a 1996 reorganization made Bell Labs part of the for-profit Lucent Technologies, which subsequently merged with the French telecommunications giant Alcatel. In 2004 Cherry Murray left for a brief stint as deputy director of the troubled Lawrence Livermore National Laboratory, while others, such as the chemist Maureen Chan, long active in the Women Chemists Committee of the ACS, retired early. Many feared that Bell Labs' glory days were over. Others were not so sure. In any case, some of its talented women were moving on to other nonprofits and national laboratories.[13]

Also moving on were too many women at the new Space Science Telescope Institute (SSTI), which was established in Baltimore in 1981 to handle the data coming back to Earth from the orbiting Hubble Space Telescope. By 1991 it had six female PhD astronomers on its staff of about fifty, a situation that, as at Bell Labs, led to issues of sexual harassment.[14] The institute was not the only institution where sexual harassment was a problem: two well-publicized surveys of astronomers in 1990 revealed that the problem was widespread in that field.[15] But it was staff women at the institute who decided that the problem was serious and pervasive enough to justify some public attention. They decided to address the issue boldly but obliquely by holding a national conference of mostly women astronomers at the institute in September 1992. (Expecting forty to fifty people to attend, they were astonished when about two hundred men and women showed up.) One notable feature of the meeting, which was organized by C. Megan Urry, who later left for a professorship and department chairmanship at Yale, was the enunciation and acceptance of the so-called Baltimore Charter, a kind of voluntary code of acceptable behavior later endorsed by the Council of the American Astronomical Society, the board of directors of the Association of Universities for Research in Astronomy (AURA), the NSF's Division of Astronomy, the Harvard-Smithsonian Center for Astrophysics, and other groups.[16] The charter put the leaders in the field on public record as ready to take action to enforce equitable working conditions at their institutions.

Yet the tactic's effectiveness at the SSTI itself was mixed, for after 1992 the number of women astronomers on the staff dropped drastically. By 2002 their turnover had risen to such an unacceptable level—five out of seven recently hired women had left, compared with one of nineteen men—that an independent review panel headed by Andrea Dupree, of the Harvard-Smithsonian Center for Astrophysics, was established to make recommendations. Later that year the board members of

AURA, the institute's governing body, which had signed the charter, showed their displeasure by setting guidelines for increasing the number of women on the institute's scientific staff and by renewing the director's next contract for just three years rather than the usual five. (He later left for a high administrative post in the University of California system.)[17]

As also can be seen from just these two cases, mid-career women scientists who might formerly have spent their entire career at one institute for lack of options elsewhere were beginning to move on. Even those without much teaching experience were able to get professorships or administrative posts at research universities. For example, in 1989 the crystallographer Helen Berman left the renowned Fox Chase Cancer Center, in suburban Philadelphia, which was the result of a merger in 1974 of a hospital and the Institute for Cancer Research, founded in 1927 as the first cancer research institute in the nation. It had a tradition of hiring women scientists (see below), and she had been a staff member for two decades, but she left for a professorship in the chemistry department at Rutgers University.[18] Also in 1989 the staff physicist Gail Hanson left SLAC, as it was called then, where she had been since 1973, for a professorship at Indiana University.[19]

Sometimes these moves were precipitated by a divorce, a remarriage, or a change in university nepotism rules. For example, in 1978 the recently divorced biochemist Judith Klinman left Fox Chase, where she had been a postdoc and then a staff member for several years, for a tenured professorship in the chemistry department at the University of California at Berkeley. (She later became its first woman chair.)[20] Similarly, the economist Irma Adelman left the World Bank in Washington, DC, and private employment in the early 1980s for faculty posts, first at the University of Maryland and then, after her marriage ended, in the agricultural economics department at UC–Berkeley.[21] And the geneticist Nina Fedoroff, who joined the staff at the Cold Spring Harbor outpost of the CIW as a postdoctoral fellow in 1976, left in 1995 for a professorship and directorship of a biotechnology institute at Pennsylvania State University.[22] In 1994 the entomologist Ann Hajek moved down the street from the nonprofit Boyce Thompson Institute for Plant Research, where she had been a staff scientist for a decade, to a faculty position in Cornell's entomology department, which had employed her husband for years, a move nearly unthinkable earlier.[23]

Other women scientists left one nonprofit for another, perhaps newer or richer one, as did the bioscientist Terry Orr-Weaver, of the Department of Embryology, in Baltimore, of the CIW when she moved to the new Whitehead Institute at MIT in 1986.[24] Or one could leave a nonprofit institution to start one's own business, as did the peripatetic geoscientist Judith Moody, mentioned in chapter 8, in 1988. A geologist by training, she had come to the Battelle Memorial Laboratories in

Columbus, Ohio, in 1980 after failing to get tenure at the University of North Carolina. But she left after a few years of work on a nuclear-waste recycling project there to start her own company. Unfortunately, but this third career was cut short by a suspicious fatal car accident in 1993.[25]

But some did stay on, rejecting outside offers and preferring to build up their field in familiar surroundings. Long-term careers were still possible, but rarely was there such a thing as lifetime tenure any more, for either men or women. More often there was a series of five-year contracts, each with an outside evaluation. As a scientist grew older and her salary increased, her value had to be continually re-justified. This required frequent retooling and access to the newest instrument or the latest technique for the next project. For those on the cutting edge of a fruitful line of work, this constant adaptation was a matter of course, but it was not necessarily cheap. For others it could mean the end of their job. In fact, James D. Watson, the longtime director of the Cold Spring Harbor Laboratory, advised that in running a competitive lab one should employ young scientists (under age 40), not middle-aged ones.[26]

Fortunately, this practice was not universal. For example, at the female-friendly Fox Chase Cancer Center the geneticist Beatrice Mintz, who had come in 1960 and was elected to the National Academy of Sciences in 1973, developed transgenic mice in the 1970s and incorporated a human gene into one in 1981.[27] The distinguished crystallographer Jenny Glusker, an English student of the Nobelist Dorothy Hodgkin, spent her whole career there. She came as a recent PhD to the laboratory of A. Lindo Patterson (famed for the "Patterson technique"), stayed on part time as a young mother of three, and succeeded Patterson as head of the unit upon his early death in 1966. Her research on carcinogenic compounds and metals in proteins, as well as her personal warmth and mentoring, made the lab a popular destination for crystallographers worldwide.[28] In the late 1970s the molecular biologist Shirley Tilghman came from the NIH and spent about a decade at Fox Chase before moving on to a professorship at Princeton, and in 1987 Anna Marie Skalka, head of a molecular biology lab at the Roche Institute of Molecular Biology, in Nutley, NJ, took up the post of scientific director at Fox Chase.[29] Even the biochemist Judith Klinman, who started her career there and, as mentioned above, left in the late 1970s, later said that the atmosphere at Fox Chase had been better for women than that at UC–Berkeley, as they succeeded more often there. With its ties to the University of Pennsylvania, it had the advantages of both worlds, as its scientific staff were often adjunct professors and it attracted many postdoctoral fellows and graduate students.[30]

On the staff of the historic, formerly wealthy, prestigious and far-flung CIW were several prominent women, including the astronomers Vera C. Rubin, a specialist in galactic motion, in the Department of Terrestrial Magnetism in Washington, DC, and Wendy Freedman, of the Carnegie Observatories in Pasadena, California, who arrived as a postdoc in 1984, later co-headed a team of twenty-seven scientists in calculating the Hubble constant to a new level of precision in the 1990s, and in 2003 became director of all the Carnegie observatories in the United States and Chile.[31] As Rubin later explained, one reason (besides its location in Washington, DC) that she had gone to Carnegie's Department of Terrestrial Magnetism in 1965 was its flexibility—it would allow her to leave each day at three in the afternoon so that she could be home with her four children (all of whom became scientists, one of them an astronomer). She later pointed out that such flexibility had advantages in both directions: although Carnegie considered her part time and paid her accordingly, she thought her job might have been better described as part *salary*, for she also "worked" at home in the evening. But Rubin's case was unusual even at the CIW.[32]

Other women scientists went to nonprofit institutions from universities, especially those that did not support their research adequately. The geneticist Ruth Sager, for example, left Hunter College, whose commitment to graduate work was waning in the early 1970s, for the expanding Dana-Farber Cancer Institute in Boston, where she spent more than twenty years; she was also a member of the faculty of the Harvard Medical School, where her new husband, Arthur Pardee, was also on the faculty.[33] Philippa Marrack left the University of Rochester in 1979 for the National Jewish Hospital and Research Center, in Denver, where her research on mouse T-cells flourished.[34]

Not all nonprofit institutions prospered in these years, however. In fact some that were not funded by the NIH or new foundations suffered in relative neglect. Among these were natural history museums and botanical gardens whose large collections of often rare specimens were the basis of descriptive work that was no longer adequately funded by the National Science Foundation. There taxonomists, whose careers were out of the limelight and who had never been very well paid, resisted retirement as long as possible, and then once pensioned, they continued to work as volunteers. For example, Janet Haig, at the Allan Hancock Foundation at the University of Southern California, and Jocelyn Crane, at the New York Zoological Society (Bronx Zoo), worked for long years on classificatory problems in crustaceans.[35] Also in New York City, the archaeologist Margaret Thompson, an expert on ancient Greek coins, spent her whole career (1949–79) on the curatorial staff of the American Numismatic Society.[36]

For the ecologist Ruth Patrick such determination and perseverance paid off big-time. The trustees at the Academy of Natural Sciences of Philadelphia, where she was a staff member, were unimpressed and even embarrassed by her research on diatoms, single-cell algae, which they considered unimportant. Yet they said that she could start a department of limnology, provided she raised the money for it. Undeterred, she pressed on and interested an industrialist who saw the work's value for pollution control, and began to make donations. Her fellow scientists were also impressed, and she began to win professional recognition. In 1970 she was elected to the National Academy of Sciences, its twelfth woman (see chapter 12), and in 1975 she won the John and Alice Tyler Prize of $150,000, one of the largest awards of the time, which she invested and then used in her research. Later she was elected chair of the board of trustees for the Philadelphia academy and raised the funds for a new wing to its building. Rarely does anyone initially so underappreciated get to turn the tables so completely.[37]

Even the immense American Museum of Natural History, in New York City, had continuing difficulties supporting research. The world-renowned Margaret Mead, who died in 1978, was frequently on leave or on the lecture circuit partly because her salary and benefits from the museum were so low. The curator Dorothy Bliss, an expert on shrimp, lobsters, and crabs, retired in 1980 after a long career there, and one of the museum's few young women curators was the ichthyologist Melanie Stiassny, who came from Harvard in 1987.[38]

Another woman scientist at the museum took advantage of its central location in New York City to start a kind of feminist collective of biologists, anthropologists, and psychologists, called the Genes and Gender Group. Led by the comparative psychologist and curator of animal behavior Ethel Tobach, its purpose was to contest wherever and however possible the recently published sociobiological views of E. O. Wilson and others and to defend whenever necessary the work of Margaret Mead from the posthumous criticisms of her rival Derek Freeman. They took on the role of "public intellectuals" by sponsoring or cosponsoring occasional symposia on relevant topics and subsequently editing and publishing the volumes. By the time the group disbanded in the mid-1980s, it had published numerous volumes on genetic determinism, male violence, rape, racism, sex differences, genetic screening, and other hot-button topics of the day. Among the group's contributors were many left-leaning biologists and social scientists from academia, including the Harvard biochemist Ruth Hubbard, the Boston University chemist Marian Lowe, the University of Wisconsin anatomist Ruth Bleier, and the City University of New York anthropologist Eleanor Leacock. Through the collective, Tobach and her associate Betty Rosoff were able to focus on producing the volumes, which found a ready market in colleges and universities, while avoiding the frictions and conflicts that a more hierarchical and long-term organization might have entailed.[39]

Advancement

When searching for a new president or founding director, the boards of trustees of some nonprofit organizations responded to the new times and, starting in the 1980s, occasionally chose a woman scientist or even nonscientist. This was a bit remarkable, for the job was growing along with the institutions and required kinds and levels of experience in business, personnel, fund-raising, and the law, as well as experience in research that only a few women scientists of the 1970s and 1980s would have had.

Possibly the first top female official of a major nonprofit institution was the immunogeneticist Barbara Sanford, who directed the Jackson Laboratory, in Bar Harbor, Maine, from 1981 to 1987, after wide experience at the Roswell Park Memorial Institute, the Massachusetts General Hospital, the Harvard Medical School (as research staff, faculty member, and director of research at the Sidney Farber Cancer Institute), the NIH, and the National Cancer Institute. With substantial support from the NIH, she was able to move the Jackson Laboratory into the area of mouse genomics and closer to its goal of becoming a year-round facility with state-of-the-art equipment and about eight hundred employees. Another was the applied mathematician Cathleen Morawetz, who headed the famed Courant Institute at New York University from 1984 until 1988.[40] In addition, two women scientists who would later be appointed to high government posts also began to move up in the mid- to late 1980s: the cardiologist Bernadine Healy, MD, who, as mentioned in chapter 9, headed the research division of the Cleveland Clinic from 1985 to 1991 before becoming director of the NIH; and the microbiologist Rita Colwell, director of the Maryland Biotechnology Institute from 1987 until 1998, when President Bill Clinton appointed her the first woman director of the NSF.[41]

A kind of turning point for the nonprofit sector (and perhaps for the nation as a whole) occurred in the late 1980s at the venerable CIW, where everyone knew that a successor to James Ebert needed to be found. The trustees may have been considering the MIT physicist Margaret L. A. MacVicar, who was becoming a national authority on science education, when they appointed her vice president in 1983. (More likely she was hoping for a higher-level post at MIT, for she held onto her job there and commuted to Washington, DC.)[42] Instead, in 1987 the board of trustees appointed the biochemist Maxine Singer, who had spent the previous three decades doing intramural research at the NIH. She was widely known and highly regarded throughout the scientific community, especially for being one of the organizers of the 1975 meeting in Asilomar, California, that examined ethical concerns about recombinant DNA. She had also a member of the board of trustees of Yale University since 1975 and the National Academy of Sciences since 1979.

During Singer's fourteen years at the helm, the CIW grew, had the major triumphs described in its annual *Yearbook,* and experienced no scandals. In addition, Singer won many awards and served on several boards and search committees, including chairing the influential Committee on Science, Engineering, and Public Policy (COSEPUP), of the National Research Council, which produced an important study of postdoctoral education. She even became something of a feminist late in her career, writing an occasional letter of protest, an editorial in *Science,* or a short piece in the newsletter of the American Society for Cell Biology. Early in her career, extremely busy with her job and four children, she had not been at all active, not even in Women in Cell Biology, but in an unusual trajectory her "consciousness" was raised later on. Perhaps she was sensitized by Vera Rubin, at the CIW; or by her children; or by what she was seeing in her lofty new position; or, most likely, by being passed over for the presidency of Yale in 1985. In any case, she "found her voice," supported several women's prizes, and protested corporate and recruitment practices that others, not moving in the same circles, did not see, shrugged off, or were unwilling to protest.[43]

In the 1990s and the first decade of the twenty-first century a few women were appointed to the top posts at other prominent nonprofit scientific institutions, including research institutes, museums, "corporations," collections, and even foundations, including some of the largest. Among these women was the microbiologist Claire Fraser, at TIGR, founded by her then husband, J. Craig Venter, in 1992. A major player in microbial genomics, it soon skyrocketed to fifth place among the largest federally funded nonprofit institutions. But when the mercurial Venter left in 1998 to start the rival Celera Genomics, his soon-to-be ex-wife, who had formerly worked at the NIH, became its chief operating officer, and she remained so after their split. In four years she doubled its budget, and by 2004 it had a faculty of 30 and a research staff of 245.[44] A second executive at a unique federally funded research facility was the microbiologist Yvonne Reid, who in 2000 was appointed director of the ever-growing American Type Culture Collection (ATTC), a research and storage facility for biological specimens that was highly dependent on funds from the Department of Agriculture. Located in Manassas, Virginia, its staff in 2000 included 32 percent minority-group members. An African American born in Jamaica, Reid had worked for the ATTC since 1980, even while earning a PhD from Howard University.[45] A third was Susan Lindquist, of the University of Chicago, who in 2001 was made director of MIT's Whitehead Institute, where in 2005 the biologist Lydia Villa-Komaroff, MIT's first Mexican American woman PhD and formerly on the faculty at the Harvard Medical School, served briefly as vice president for research and chief operating officer.[46]

An even more remarkable series of changes, perhaps approaching what the NSF would soon call "institutional transformation" (see the epilogue), occurred at the Aerospace Corporation, in El Segundo, California, a nonprofit corporation with about four thousand employees funded by the U.S. Air Force. In 1972 the computer systems engineer Shirley McCarty, one of its first women technical employees (hired in 1965), feared for her job when she dared to start a women's committee there. Later Judith Osmer, a chemist who worked with the ruby laser, filed a long-running class-action lawsuit for sex discrimination against Aerospace. But then things began to change. McCarty became head of human relations, and in 2008 Wanda Austin, an African American engineer who had risen through the ranks, was elected president. Among Aerospace's best practices by then were paid maternity leave, awards for women, equal benefits for secretarial and technical staff, and staff input on candidates for the board of directors.[47]

Natural history museums also experienced a resurgence, perhaps even a kind of golden age, in the years after 1972. This was less because of their research, for the support of taxonomic research was a tough sell to Congress and donors, than because the public, egged on by movies like *Jurassic Park* and public-television specials, became fascinated by dinosaurs, fossils, whales, dolphins, porpoises, oceans, black holes, and other subjects of scientific entertainment. Featuring special blockbuster exhibits and Omnimax theaters, museums attracted record levels of admissions, and in the 1990s the major ones fought over the best dinosaur specimens, including "Sue," the *Tyrannosaurus rex* from the Dakotas, who after a federal lawsuit and bidding war at Sotheby's in New York City ended up at the Field Museum in Chicago.[48] In 1993 the New York City lawyer and president of Barnard College Ellen Futter was appointed president of the Field Museum's main rival, the American Museum of Natural History, for whose scientific spectacles she raised much money.[49] More remarkably, a few new institutions were established, including some headed by women. Atlanta, for example, opened its new Fernbank Museum of Natural History in 1992. Originating decades before in a vision of a former Atlanta schoolteacher, the drive for the new museum was spearheaded by E. Kay Davis, herself a former math and science teacher, who raised $43 million for it in the 1980s. When the new building opened, she was its executive director, one of just ten women nationally to direct a major natural history museum.[50] After leaving NASA in 1993 and the NOAA in 1995, the astronaut Kathryn Sullivan served briefly as president and chief executive officer of the COSI Columbus, a new interactive science center in Columbus, Ohio.[51]

Aquaria became more numerous and spectacular as architects overcame technical problems and urban-renewal groups sprang up to remodel their downtown waterfronts with attention-getting edifices in Baltimore, Boston, Camden, Chicago, Cleveland, New Orleans, and elsewhere. The deep-sea diver Sylvia Earle was for a time associated with the Steinhardt Aquarium, of the California Academy of Sciences in San Francisco, and in 1997 the trustees of the new aquarium at Monterey Bay in California, supported by the David and Lucille Packard Foundation and considered by some to be the best of the lot, appointed Marcia McNutt, holder of an endowed chair in earth sciences at MIT, director of its research institute.[52]

Planetaria also enjoyed something of heyday after 1972, but two surveys in 1992 revealed women's continuing low status there. Noreen Grice, an "education coordinator" at the Charles Hayden Planetarium, at the Boston Museum of Science, found that only six of her sample of forty-eight planetaria currently employed women directors, seven had had one in the past, and thirty-five had never had one. Most employed women chiefly in clerical positions, but five did not even do that, and in all cases the women's salaries were far below those of men in comparable positions.[53] Related to planetaria was the venerable Maria Mitchell Observatory, established on Nantucket Island, off the coast of Massachusetts, in 1903, to honor Mitchell's memory and house her telescope. Open to the public in the summer and also running a small science museum, the observatory was known in astronomical circles for the modest researches of its part-time female directors and for the number of its women student interns who went on to higher degrees and sometimes even careers in astronomy. After the long-serving Dorrit Hoffleit retired in 1978, the trustees hired Lee Belserene and then Eileen Friel to run it, but in 1997 they broke tradition and employed their first man, Vladimir Strelnitski, a recent Russian émigré. Much of the observatory's support continues to come from NSF grants for undergraduate research.[54]

Of course not all such top appointments turned out well, and some led to frustration and early resignations. (In fact some of the jobs offered to women scientists had been turned down by men.) Some women may have felt that they were fighting against inexorable forces. One such case was that of the astronomer Sidney Wolff, head of the National Optical Astronomy Observatory (NOAO) in Arizona. An accomplished astronomer with considerable experience, including the directorship of the Kitt Peak National Observatory, she was appointed director of NOAO, which ran several telescopes, in 1987. But she resigned in 1999 when continuing financial stringencies became intolerable. Primarily dependent on NSF funds, the observatory faced obsolescence, a painful fact of life in the rapidly moving world of big astronomy. Wolff had to close down two of NOAO's five nighttime telescopes because of decreasing budget allocations from the NSF. To use an astronomical metaphor, they had been eclipsed by newer, bigger, and more expensive telescopes elsewhere.[55] A second case of mixed success was that of the psychologist Nancy Cole, who became vice president and then president of the Educational Testing Service (ETS), in Princeton, New Jersey, in the 1990s. As an expert on the biases of test scores, Cole, who had been a dean of education at the University of Illinois and was the past president of the American Education Research Association, was a logical choice. But her time in office was marred by continuing budgetary deficits and staff layoffs, as the ETS computerized its test-taking operations and many colleges ceased requiring an SAT score for admission. She retired in July 2000 at age 57.[56]

Science Policy in Washington, DC

Several nonprofit organizations at the intersection of science and public policy not only survived but even thrived in Washington, DC, on a series of soft-money projects that allowed them to persist and make contributions in areas not otherwise possible. A few women held major posts in these organizations, which had formerly been a domain of "old boys."

Most impressive of all was the cluster of activities at the American Association for the Advancement of Science headquarters. Starting in 1963, Betty Vetter ran its Scientific Manpower Commission, renamed Commission on Professionals in Science and Technology in 1986, publishing a monthly magazine, *Manpower Comments,* frequent reports, and a large compendium of educational and employment data on scientists, manpower statistics, thus establishing it (and herself) as a kind of clearinghouse in the complicated and rapidly changing world of technical manpower statistics. She did not conduct statistical surveys herself, but analyzed and highlighted for readers the many specialized and partial reports put out by other groups. Funded by subscribers and contracts, Vetter first developed a specialty on scientists and the draft; then, when that topic ended in the early 1970s, she started another on data on women and minority scientists, a growth industry thereafter. For two decades, until her death in 1994, Vetter was a frequent speaker at conferences on women in science and engineering, and she compiled a major reference work of the latest statistics in this increasingly data-driven area.[57]

Also at the AAAS was the Office of Opportunities in Science (OOS), started by Janet Welsh Brown in 1973 (see chapter 1). Over time it grew into the Education and Human Resources Directorate, headed by the African American biologist Shirley Malcom, with a sizeable staff and a whole set of programs having to do with education, women, minorities, handicapped scientists, and most recently postdoctoral fellows. Financed by a series of grants from various foundations, the NSF, and the NIH, these efforts generated substantial other programs in the areas of scientific outreach, science education, and scientific literacy. One program in the mid-1980s called Linkages "partnered" the OOS with local community groups (e.g., the Girls' Clubs of America, the National Association for the Advancement of Colored People, and the National Council of Negro Women) to provide technical assistance in projects aimed at inner-city youth, handicapped students, and disadvantaged women. Another initiative was international and built upon the World Conference on Science in Beijing in 1999 to form (with the AWIS, WEPAN, and the DOE) the Global Alliance for Diversifying the Science and Engineering Workforce. Malcom's unique contributions, unthinkable in previous decades, were recognized by her appointment to the National Science Board in 1993 and the awarding of the Medal for Public Welfare from the National Academy of Sciences, one of the very few awarded to women, in 2003.[58]

Also in the Washington area was the headquarters of the AWIS and its foundation, long headed by the biologist Barbara Filner, who started her career in the 1970s as a tenure-track assistant professor at Columbia University but then moved in 1979 to Washington, where she first served as a staff officer at the NAS Institute of Medicine and then, when the Howard Hughes Medical Institute expanded in the late 1980s, joined its staff as director of graduate fellowships. Similarly, for fourteen years (1990–2004) Catherine Didion served as the executive director of AWIS itself, running meetings, planning events, applying for grants, completing projects, testifying before Congress, and representing the members whenever necessary.[59] Bernice Sandler did much the same at her Project on the Status and Education of Women at the Association of American Colleges, in the higher-education cluster around Dupont Circle, until her funding from the Carnegie Corporation ran out in the early 1990s.[60]

Several women economists were also involved in nonprofit work in the nation's capital. Alice Rivlin, the first woman president of the American Economic Association, in 1986, worked at the Brookings Institution before and after her stint (1975–83) as head of the Congressional Budget Office, as did Isabel Sawhill, who had spent most of the 1970s at the Urban Institute.[61] Several other women social scientists, including Barbara Searle, Jayne Mansfield, and many Wellesley graduates worked at the World Bank, where Anne O. Krueger, later at the International Monetary Fund, was vice president for economics and research in the 1980s.[62] More entrepreneurial was the feminist economist Heidi Hartmann, who left Rutgers University in 1987 to start an independent nonprofit "think tank," the Institute for Women's Policy Research, whose staff conducted research on economic issues of concern to women, such as social security, welfare, and pay equity, and then testified on them before relevant congressional committees. One high point of her group's work at the intersection of policy and economics was the passage of the Family and Medical Leave Act in 1993, after six years of advocacy. Hartmann is one of those women who went to Washington and made a difference in other women's lives.[63]

Meanwhile, several scientific societies, largely headquartered in Washington, started to employ women scientists as their chief executive officers in such numbers that the job began to show signs of becoming an area of "women's work." In retrospect, this new receptivity to women in such posts seems a natural outgrowth of several factors, including the willingness of women scientists (perhaps in dual-career couples) to work in the Washington area and the waning of the old-boy network in certain quarters. By the 1990s, officials of scientific societies had learned to deal with and benefit from their women's caucuses, which were the source of much vision and energy (but also criticism) among the membership. Not least, at long last the members of the Cosmos Club, where much deal making reportedly took place, voted in June 1988 to admit women as members (see chapter 12).

The path to the top position varied. Some women spent their whole career on the staff, slowly rising through the ranks; the top level only became a possibility late in their career, if at all. For example, Kay Croker, long on the staff of the American Society for Pharmacology and Experimental Therapeutics and its executive officer from 1977 until 1997, later recalled the great reluctance of the society's officers to appointing her to that historically male post.[64] The political scientist Janet Welsh Brown moved out and up when she left the OOS, which she had led for six years, to head the Environmental Defense Fund, also headquartered in Washington. A few other women started their careers at scientific societies, but when they hit the glass ceiling, they left for a government job and later returned to a top job post. Madeleine Jacobs, for example, started at the ACS in 1969 as a journalist for its *Chemical and Engineering News (C&EN)*, then left in 1972 to work in public affairs and media relations at three federal agencies, returned to the *C&EN* in 1993 as its managing editor and later editor in chief, from which she rose to the executive directorship of the whole far-flung ACS in 2004.[65]

Other executive officers came after careers at universities or government agencies in Washington, DC, or elsewhere. In 1970 the geologist Susan "Sally" Newman left a research post at Caltech to serve as director of the Seismological Society of America, whose headquarters were in El Cerrito, California, across the bay from San Francisco. She was still there in 2006, achieving a sort of record for longevity.[66] At mid-career, Marcia Sward moved to the Mathematical Association of America in 1983 as associate director from an associate professorship at Trinity College, also in Washington, DC. She took on successively greater responsibilities, becoming the association's executive officer in 1989, and retired in 1999.[67] Barbara Bailar became the executive director of the American Statistical Association in 1988 (and one of the first women elected to the Cosmos Club) after a long career at the Bureau of the Census.[68] In 1994 the board of the American Physical Society persuaded Judy Franz, who had held several academic posts and had a long history of service to the profession, to become its executive officer at its headquarters in College Park, Maryland. She served three terms, until 2004, when she was succeeded by Kate Kirby, who had amassed a great deal of relevant experience as codirector of an institute at the Harvard-Smithsonian Center for Astrophysics in Cambridge, Massachusetts.[69]

Salaries varied considerably. For example, Madeleine Jacobs, chief executive officer of the very large ACS, reportedly earned an astounding $883,818, including incentives and bonuses, in 2005. She justified her earnings by citing her multiple roles in overseeing the society's finances (including more than $1 billion in three very large trust funds), running the Washington office with its three hundred employees, and supervising the operation of its subsidiary Chemical Abstracts, in Columbus, Ohio, with its staff of fifteen hundred.[70] But others protested their relatively low salaries, as did Catherine Didion, executive director of the AWIS, who reportedly earned forty-five thousand dollars in 1994, citing a survey of other executives' much higher salaries as part of a negotiation to raise her own.[71] In fact there is some evidence that some scientific organizations took financial advantage of their women managers. Particularly beloved among these society executives was the brilliant and energetic Turkish-born astronomer Janet Akyuz Mattei, who was the director for thirty years (1974–2004) of the American Association of Variable Star Observers, located in Cambridge, Massachusetts. After her death the society had a hard time finding someone equally qualified who would accept their proffered salary; officials finally admitted that it had probably taken financial advantage of Mattei's willingness to serve.[72]

Perhaps a sign of the future in the world of scientific-society executives is the career of Elizabeth Mark Marincola, who worked at the American Society for Cell Biology (ASCB) from 1991 to 2005, when she moved to head the venerable Science Service, publisher of *Science News* and organizer of the annual Intel (formerly Westinghouse) Science Talent Search. She did not hold any degrees in science, but she had other advantages. She grew up in Stanford, California, where her father was on the medical school faculty, and her first job was as an assistant to the wife of Stanford's president, the biologist Donald Kennedy. After earning an MBA at Stanford and marrying an NIH scientist, Marincola moved to Washington, DC, where she served as a deputy director of policy (i.e., budget) with the National Institute of Mental Health. When she saw that the ASCB was seeking to hire a new executive officer to replace Dorothea Wilson, who had been the mainstay on its staff for decades, she had the audacity to apply. Fortunately, some faculty members on the search committee, perhaps alerted to her candidacy, remembered "Lizzie" fondly from her youthful days at Stanford and thought of her as one of them even though she did not have a degree in cell biology. (Formerly a doctorate had been required.) By all accounts, her appointment was an inspired choice. She helped bring the society into the electronic age and into the Washington world of political lobbying. She worked with a succession of ASCB presidents (including Maxine Singer, Elizabeth

Blackburn, and Ursula Goodenough), tutoring them on Washington politics, as the society entered turbulent waters, withdrew from the Federation of American Societies of Experimental Biology, added a public-policy committee, and with several other scientific societies formed the Joint Steering Committee for Public Policy. In addition, the ASCB was active in the movements to double the NIH budget and to avoid government restrictions on stem-cell research. Her MBA also proved useful in negotiating society finances (or "revenue streams") in a world of open-access journals, which the society and the NIH favored. The ASCB, unlike the much larger ACS, for example, could afford to allow free online access to its acclaimed journal *Molecular Biology of the Cell* just two months after publication, because it could cover most of its expenses with earnings from its annual meeting, which was attended by ten thousand scientists from all over the world.[73]

Foundations

About 1970 most of the major American foundations were headquartered in New York City, where, according to a 1973 survey by the National Organization for Women, few employed any women above the clerical level or awarded more than 1 percent of their grants to them.[74] In fact the exceptional case of the Carnegie Corporation of New York, which focused on higher education, exemplified the situation. Although its titular head in the early 1970s was Alan Pifer, it was Florence Anderson, a Mount Holyoke graduate and former officer in the Marine Corps, who, with the deceptively modest title "corporate secretary," really ran the place from 1954 until 1975. It was Anderson who interacted with the applicants and grantees, often reminding them of their unmet obligations. She may also have been influential in the Carnegie Corporation's decisions to support continuing education projects for women since the early 1960s.[75] Building upon this precedent, in the early 1970s the Carnegie Corporation had prodded the higher-education establishment from within by supporting the psychologist Bernice Sandler's Project on the Status and Education of Women at the AAC; an internship program for future women administrators at the American Council of Education; a law textbook on women's cases; and Lenore Blum's Math/Science Network in the San Francisco Bay Area.[76] Meanwhile, at the mighty Ford Foundation the economist Mariam Chamberlain, long the only woman program officer in its national-affairs unit, funded some of the early work in women's studies, the growing number of "centers for research on women," and various projects of the OOS discussed in chapter 1, including a meeting on the thorny issue of setting up and maintaining women's rosters.[77]

But then in the 1980s several new foundations were created, including the John D.

and Catherine T. MacArthur Foundation, with headquarters in Chicago, the Gates Foundation in Seattle, the David and Lucille Packard Foundation in California, the Howard Hughes Medical Institute in Bethesda, Maryland, and others. Some of them began to do things differently. For example, the John D. and Catherine T. MacArthur Foundation broke new ground in 1989 with the appointment of the historian Adele Simmons as its new president.[78] In the 1990s, perhaps emboldened by President Bill Clinton's many top-level female appointees (see chapter 9) and the increased numbers of feminist men (e.g., W. T. Golden) on their boards of directors, some foundations began to appoint women to top positions, something that had been unthinkable and adamantly opposed just a few years earlier. Among these were the molecular biologist Enriqueta Bond at Burroughs Wellcome Fund in 1994; the health economist Karen Davis at the Commonwealth Fund since 1995; the urban-affairs specialist Susan Berresford at Ford in 1996; the psychologist Judith Rodin, who had recently stepped down after ten years as the first woman president of the University of Pennsylvania, at Rockefeller in 2004; and the biological anthropologist Leslie Aiello at the Wenner-Gren Viking Foundation in 2005.[79] What difference they might or did make is as yet unclear, but the potential was enormous.

Conclusion

Despite the high inflation of the 1970s, which wreaked havoc on the endowments of the nonprofit institutions, this diverse sector was able to remain a vital part of science, offering an attractive alternative to academic, industrial, and government settings. Some were old, going back to the early 1900s; several were new, the result of a major benefaction or a new endowment. Almost all strived to remain distinctive, selective, flexible, and innovative, even though most, whatever their origins, had joined the NIH grant race. As ambitions heightened and outstripped endowments, most of these institutions became custodians of federal funds, relying on overhead payments for current expenses, employing temporary personnel, and postponing new buildings and major repairs for capital campaigns. A lot was possible while the money lasted.

Some women liked these research and entrepreneurial positions at nonprofits, which seemingly freed them from the politics and turmoil of university departments, and many remarkable ones spent their entire career there doing outstanding work in their own style and more single-mindedly than would have been possible elsewhere. Others, in a new era of increased mobility, moved on to universities, other nonprofits, and small businesses. A few started their own institutes. Some, including a few women of color, moved up, even becoming the first women to hold an institution's top post, whence they guided the organization into the next decade. These new women executives were particularly evident at the scientific societies, some of which were taking on broader roles. A few encountered financial straits beyond their control and stepped down prematurely. But there were relatively few of the high-profile, forced resignations and bailouts (including a suicide) that beset top officeholders in higher education, to which we turn next.

The endocrinologist Neena Schwartz, one of the first copresidents of the Association for Women in Science, worked tirelessly for feminist causes for decades. Late in life she revealed in her autobiography that she had been a lesbian. (Photograph from Art Wise Photography)

The political scientist Janet Welsh Brown directed the new Office of Opportunities in Science at the American Association for the Advancement of Science in Washington, DC, from 1973 to 1978. On a string of small grants she and her small staff ran a variety of innovative programs and meetings and produced a series of useful publications: rosters, inventories, and reports. (Photograph by Mike Woodlon)

The MIT physicist Vera Kistiakowsky was prominent among those who stirred the American Physical Society to start a women's committee in 1971. She oversaw the preparation of its major report (with the help of a grant from the Alfred P. Sloan Foundation) and later was a watchdog on women's issues at MIT and in physics. (Photograph courtesy MIT Museum)

Upon passage of the Women in Science and Technology Equal Opportunity Act in 1980, Senator Edward M. Kennedy (D-MA) was surrounded by supporters, including, among others, *left to right,* Betty Vetter, Mary Gray *(behind Kennedy's right shoulder),* Shirley Malcom *(behind Kennedy),* Bernice "Bunny" Sandler *(in front next to Kennedy),* Marie Cassidy *(behind Sandler),* and Sheila Pfafflin *(right front).* (Reproduced by permission from *WEAL Washington Report* 9 [Dec.–Jan. 1981]: 4)

In February 1973 the biochemist Sharon L. Johnson pursued a pioneering lawsuit under new legislation against the University of Pittsburgh for its failure to award her tenure. The case seemed at the time to be a pivotal one and attracted much attention, but it dragged on for four years, whereupon the judge refused to rule in her favor. (Photograph from *Pittsburgh Press*)

The chemist Shyamala Rajender sued the University of Minnesota in 1973 for its refusal to interview or hire her for a tenure-track faculty position. In 1980 the case, by then broadened into a systemwide class-action suit, ended with a consent decree requiring "special masters" to carry out the terms of the settlement, which included future hires and all women mistreated since 1972. (Photograph from *Minneapolis Tribune*)

Starting in the 1970s, colleges of engineering, faced with declining enrollments and corporate demand for women and minority graduates, sought to recruit female students. This bright pink Georgia Institute of Technology poster attracted the disparaging graffiti at the bottom. (Reproduced with permission from Dora Skypek Papers, MARBL, Emory University)

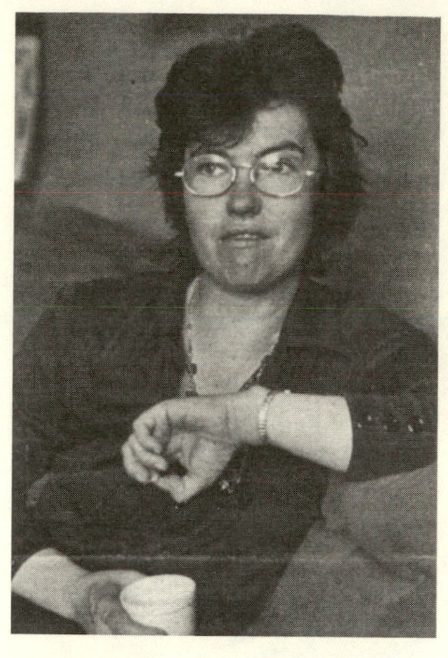

In the 1970s Lucy Sells, a graduate student in sociology at the University of California at Berkeley, identified the concept of "math filters," whereby female high-school students took so little mathematics in high school and college that they were not eligible for a great many fields at the graduate level. (Reprinted from *Daily Californian,* Oct. 11, 1973)

In the mid-1970s, with the help of others in the area, the mathematician Lenore Blum, of Mills College in California, started Saturday career days for women students in the Bay Area that blossomed into Expanding Your Horizon, now a worldwide venture. (Reproduced by permission from Teri Perl, *Women and Numbers: Lives of Women Mathematicians Plus Discovery Activities* [San Carlos, CA: Wide World /Tetra, 1993])

The United States has sent a team to the International Mathematics Olympiad since 1974. It was all male until the late 1990s, when Melanie Wood, of Indianapolis, became a member of the team and won two medals. (Photograph courtesy of the Mathematical Association of America 2010. All rights reserved)

The astronomer Dorrit Hoffleit, director until 1978 of the Maria Mitchell Observatory on Nantucket Island, off Cape Cod in Massachusetts, hired many female (and in the 1970s a few male) summer interns, many of whom later became astronomers. Shown here in 1975 are, *left to right,* Pattie Guida, Debby Carmichael, Valerie Mehlig (library assistant), Hoffleit, Mary Jane Taylor, Joan Lucas, and Melissa McGrath. (Photograph courtesy of the American Association of Variable Star Observers [AAVSO])

When she learned from Lucy Sells that girls intentionally took fewer courses in mathematics than boys, Nancy Kreinberg, a science educator at UC–Berkeley's Lawrence Hall of Science, started special classes for girls and formed Math Equals to work with area teachers. (Reproduced by permission from Teri Perl, *Women and Numbers: Lives of Women Mathematicians Plus Discovery Activities* [San Carlos, CA: Wide World/Tetra, 1993])

The New York Times
SUNDAY, MARCH 20, 2005

Women Can't Do Science?

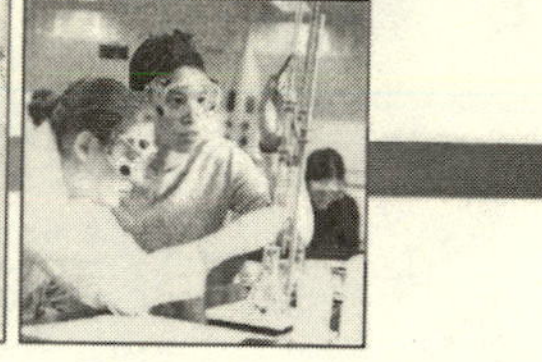

Graduates of women's colleges have a 170-year tradition as trailblazers in the sciences.

What accounts for our success?

It has everything to do with our first-rate faculties, curricula, laboratories, innovative research opportunities, and mentoring. But most of all, it has to do with our confidence in women's aptitude. At women's colleges, our students are expected to achieve in math and science...and they do.

In celebration of Women's History Month, the following members of the Women's College Coalition salute our many distinguished alumnae and faculty who have made major contributions in the sciences as well as those recent graduates and current students who will be the scientists and science educators the fiercely competitive global economy demands.

Agnes Scott College
Decatur, Georgia
www.agnesscott.edu

Alverno College
Milwaukee, Wisconsin
www.alverno.edu

Barnard College
New York, New York
www.barnard.edu.

Bay Path College
Longmeadow, Massachusetts
www.baypath.edu

Bennett College for Women
Greensboro, North Carolina
www.bennett.edu

Brenau University
Gainesville, Georgia
www.brenau.edu

Brescia University College
London, Ontario
www.brescia.uwo.ca/

Bryn Mawr College
Bryn Mawr, Pennsylvania
www.brynmawr.edu

Carlow University
Pittsburgh, Pennsylvania
www.carlow.edu

Chatham College
Pittsburgh, Pennsylvania
www.chatham.edu

The College of New Rochelle
New Rochelle, New York
www.cnr.edu

College of Notre Dame of Maryland
Baltimore, Maryland
www.ndm.edu

College of Saint Benedict
St. Joseph, Minnesota
www.csbsju.edu

College of Saint Catherine
St. Paul and Minneapolis, Minnesota
www.stkate.edu

College of Saint Elizabeth
Morristown, New Jersey
www.cse.edu

College of Saint Mary
Omaha, Nebraska
www.csm.edu

Columbia College
Columbia, South Carolina
www.columbiacollegesc.edu

Converse College
Spartanburg, South Carolina
www.converse.edu

Cottey College
Nevada, Missouri
www.cottey.edu

Douglass College of Rutgers University
New Brunswick, New Jersey
www.douglass.rutgers.edu

Georgian Court University
Lakewood, New Jersey
www.georgian.edu

Hollins University
Roanoke, Virginia
www.hollins.edu

Judson College
Marion, Alabama
www.judson.edu

Mary Baldwin College
Staunton, Virginia
www.mbc.edu

Meredith College
Raleigh, North Carolina
www.meredith.edu

Midway College
Midway, Kentucky
www.midway.edu

Mills College
Oakland, California
www.mills.edu

Mount Holyoke College
South Hadley, Massachusetts
www.mtholyoke.edu

Mount Mary College
Milwaukee, Wisconsin
www.mtmary.edu

Mount St. Mary's College
Los Angeles, California
www.msmc.la.edu

Peace College
Raleigh, North Carolina
www.peace.edu

Pine Manor College
Chestnut Hill, Massachusetts
www.pmc.edu

Randolph-Macon Woman's College
Lynchburg, Virginia
www.rmwc.edu

Regis College
Weston, Massachusetts
www.regiscollege.edu

Rosemont College
Rosemont, Pennsylvania
www.rosemont.edu

Russell Sage College of The Sage Colleges
Troy, New York
www.sage.edu

Saint Joseph College
West Hartford, Connecticut
www.sjc.edu

Saint Mary-of-the-Woods College
Saint Mary-of-the-Woods, Indiana
www.smwc.edu

Saint Mary's College
Notre Dame, Indiana
www.saintmarys.edu

Salem College
Winston-Salem, North Carolina
www.salem.edu

Scripps College
Claremont, California
www.scrippscollege.edu

Simmons College
Boston, Massachusetts
www.simmons.edu

Smith College
Northampton, Massachusetts
www.smith.edu

Spelman College
Atlanta, Georgia
www.spelman.edu

Stephens College
Columbia, Missouri
www.stephens.edu

Sweet Briar College
Sweet Briar, Virginia
www.sbc.edu

Trinity College
Washington, DC
www.trinitydc.edu

Ursuline College
Pepper Pike, Ohio
www.ursuline.edu

Wellesley College
Wellesley, Massachusetts
www.wellesley.edu

Wells College
Aurora, New York
www.wells.edu

Wesleyan College
Macon, Georgia
www.wesleyancollege.edu

Wilson College
Chambersburg, Pennsylvania
www.wilson.edu

The Women's College of the University of Denver
Denver, Colorado
www.womenscollege.du.edu

WOMEN'S COLLEGE COALITION • WASHINGTON, DC • WWW.WOMENSCOLLEGES.ORG

In 2005 the Women's College Coalition, formed in the early 1970s to promote a positive image of the remaining women's colleges, placed an advertisement in the *New York Times* featuring the past scientific accomplishments of their alumnae. (Advertisement from *NYT,* Mar. 20, 2005)

In 1987 the anthropologist Johnnetta Cole became the first African American woman to head Spelman College in Atlanta, founded in 1881 and still a college for black American women. Its earlier presidents had all been white women and African American men. She is shown here with the comedian Bill Cosby and his wife, Camille, generous benefactors of the college. (Photograph courtesy of Spelman College)

In the 1970s Wellesley College built its award-winning science center, which connected adjacent buildings and brought several formerly separate science departments together. (Photograph by Rick Mandelkorn)

After 1970 many women scientists held the position of dean of the graduate school, including the psychologist Judith Rodin at Yale University. She later served as provost at Yale, president of the University of Pennsylvania, and head of the Rockefeller Foundation. (Photograph by Michael Marsland)

While a graduate student at the Johns Hopkins University in the 1970s, the molecular biologist Candace Pert was the lead author on several pioneering papers on the opiate receptor. But the prestigious Lasker Award went to her professor instead, with no mention of Pert, who protested to no avail and later suffered professionally. (ADAMHA newsphoto, 1979)

In addition to her career as a solid-state physicist at MIT, Mildred Dresselhaus had a second career in scientific organizations and women-in-science activities. (Photograph courtesy of MIT Museum)

The demographer Judith Blake used her Guggenheim fellowship in the mid-1970s to leave Berkeley and start a new graduate program at UCLA. Unlike in the image here, she was said to be intense and intimidating, or in academic circumlocution, one who "did not suffer fools gladly." (Photograph for the *Chronicle of Higher Education* by Deborah Edelstein)

The whistle-blower biologist Margot O'Toole, a postdoctoral fellow at MIT in the 1980s, thought her professor's work was sloppy and her published claims inaccurate. When O'Toole reported this to authorities, they discredited her and defended the professor all the way through a highly publicized congressional hearing. It was a long time before she found another job. (Photograph by Douglas Goralski)

The chemist Gertrude Elion, a Hunter College graduate, worked for the Burroughs Wellcome pharmaceutical company for several decades, developing with a colleague a long list of important drugs. After winning the Nobel Prize in Physiology or Medicine in 1988, she spoke out about her own early difficulties. (Photograph from *Chemical Heritage*, Winter 2007–8, 40)

The chemist Mary Good left an endowed chair at the University of New Orleans after critics said it should have gone to her husband and flourished elsewhere as an executive in a chemical company, as a high official in the Department of Commerce, and as president of many organizations, including the American Chemical Society and the American Association for the Advancement of Science. She also chaired several boards, including the National Science Board, a first for womankind. (Photograph from *Science* 255 [1992]: 1372)

In the 1980s the computer scientist Anita Borg, of Xerox Palo Alto, who died young of a brain tumor, started a movement to bring the women in the field together with her electronic list of "Systers." In the 1990s she and others initiated the Grace Hopper Celebrations for women in the field. (Heinz Awards photo / Jim Harrison photographer)

In June 1983 the physicist Sally Ride became the first American woman astronaut in space. The cartoonist Jim Borgmann, of the *Cincinnati Enquirer,* took a poke at the sexist press coverage. (Reprinted by permission from Jim Borgmann, *The Great Communicator* [Cincinnati, OH: Colloquial Books, 1985], 115)

By the time the atmospheric chemist Susan Solomon started graduate work at the University of Colorado in the 1970s and continued her career at the National Oceanic and Atmospheric Administration, many doors had opened. In the 1980s she led a team of American scientists to Antarctica, where they discovered the "ozone hole." She won many honors at a young age. (Photograph courtesy of NOAA)

Shown here with Director H. Guyford Stever in 1975 are five then current and former members of the National Science Board. *Left to right,* Jewel Plummer Cobb, Mina Rees, Mary Bunting-Smith, Stever, Sophie Aberle, and Anna J. Harrison. (Photographs and Prints Division, Schomburg Center for Research in Black Culture, The New York Public Library, Astor, Lenox and Tilden Foundations)

The physicist Betsy Ancker-Johnson, a Wellesley graduate and vice president of General Motors, served as assistant secretary for science and technology in the Department of Commerce under President Richard Nixon. (Association of Physics Teachers [AAPT], courtesy AIP Emilio Segré Visual Archives)

The cardiologist Bernadine Healy, of the famed Cleveland Clinic, was one of the few females appointed to a high government post by President George H. W. Bush. She directed the National Institutes of Health for two tempestuous years in the early 1990s. (Photograph from Judy Sarasohn, *Science on Trial: The Whistle Blower, the Accused and the Nobel Laureate* [New York: St. Martin's, 1993])

In 1987 Donna Shalala, a political scientist and president of Hunter College, became the first female chancellor of the University of Wisconsin at Madison. She later served two full terms as secretary of health and human services in the Clinton administration. ("Donna E. Shalala," *Social Security Online*)

When in 1992 a group of young women scientists at the Space Science Telescope Institute, in Baltimore, planned for a meeting of women astronomers, they expected 40–50 participants. They drafted the "Baltimore Charter," a kind of code of ethics for professional behavior, which has been widely endorsed. (NASA / courtesy of nasaimages.org)

The biochemist Maxine Singer spent most of her career at the National Institutes of Health, in Bethesda, MD, but she was also a longtime trustee of Yale University and the first woman president of the Carnegie Institution of Washington, 1988–2002. (Photograph courtesy of the National Institutes of Health)

These seven women were the nucleus of a small group of Bay Area women scientists who met every other Thursday for more than a decade to advise and console one another about challenges in their personal and professional lives. Shown here on the statue of Albert Einstein outside the National Academy of Sciences building in Washington, DC, are, *left to right,* Mimi Koehl, Christine Guthrie, Suzanne McKee, Carol Gross *(top),* Ellen Daniell, Judith Klinman, and Helen Wittmer, a retired department administrator. (Photograph by Steven P. Gross)

In 1986 the molecular biologist Shirley Tilghman left the Institute for Cancer Research, outside Philadelphia, for a professorship at Princeton University. Fifteen years later she became its first female president. (Photograph by Michael Pirrocco, *Princeton Weekly Bulletin*, Sept. 25, 1989)

Starting in the late 1980s a few women were appointed chancellors of campuses within the University of California system. Shown here are the first two: the psychologist Barbara Uehling, of UC–Santa Barbara, and the biophysicist Rosemarie Schraer, of UC–Riverside. (By permission of the University of California Press)

The physicist Chien-Shiung Wu, a professor at Columbia University, won many prizes and was in 1975 the first woman elected president of the American Physical Society. As such, she worked with President Gerald Ford. (American Association of Physics Teachers [AAPT], courtesy AIP Emilio Segré Visual Archives)

After being refused more lab space for her work with zebrafish, the molecular biologist Nancy Hopkins grew dissatisfied with her treatment at MIT's School of Science and started to document the inequities. When the MIT Report was published in March 1999, it caused a sensation. (Reproduced with permission from "Biography of Nancy Hopkins," by Christen Brownlee, *PNAS* 101 [Aug. 31, 2004], copyright 2004 National Academy of Sciences, USA)

Academia after Rajender

Programs, Publicity, and Pressures

By 1980, after a decade of modest tokenism, intense legal confrontations with mixed results, and an increasing feminization of the undergraduate and graduate student populations, the proportion of women on college and university science faculties still had not changed very much; in fact the percentage had declined at some places. Nor, since full-time faculties were highly tenured and few retirements were expected, would there be much change anytime soon. The coming of Reaganism to the federal government in 1981 promised to slow and perhaps terminate even the minimal enforcement of the equal-opportunity legislation of the late Carter years. Slow accommodation and quiet out-of-court settlements were more likely, if anything happened at all, since universities' legal offices were less confident of winning in court, labor specialists were exploring and advocating alternatives to litigation, and even feminist lawyers were recommending lawsuits only as a last resort.[1] Yet every report on improving the experience of female undergraduate and graduate students stressed the importance of hiring more women faculty. Thus, there was an impasse that was also a kind of opportunity, for many expected the numbers of women on science and even engineering faculties to rise substantially. In the 1980s and 1990s this lack of numerical progress became the story, as attention focused on those institutions and fields that were lagging. Articles in the *New York Times*, the *Chronicle of Higher Education*, *Physics Today*, the *Chemical and Engineering News*, *Science*, the *Scientist*, the *Boston Globe*, and elsewhere presented recent data accompanied by the recurrent query, Why still so few?[2] At a time when women were moving into and up in medicine, the law, and even business, why was academia, especially science and engineering, not changing too? Why were some institutions and some fields even worse than others? When PEER, a branch of the National Organization for Women, presented Cornell University with its Silver Snail Award, for the least progress in the Ivy League, in 1979, it set the tone for much of the media coverage of the next two decades.

In the 1980s two new programs were established that provided resources and incentives to facilitate hiring some pioneering women science faculty and to reassure skeptics that the presence of these tokens would not mean the end of higher education as they knew it. The combative climate of the 1970s settled in the 1980s into a calmer period that was ripe for philanthropy and for an innovative government program. Fiscally stable and responsible organizations such as universities began to accept financial inducements to hire temporary faculty as a kind of on-site demonstration that might, it was hoped, have widespread ramifications throughout academia. Over the years these programs made modest inroads at several universities, and some left a groundwork useful for later NSF programs.

Also starting slowly in the 1970s and 1980s and increasing in the 1990s, search committees and boards of trustees, whose own memberships were changing, began to appoint women scientists, especially biologists and psychologists, including an occasional one of color, to top-level administrative posts, such as provost or president, at major academic institutions, as those at some nonprofit institutions were already doing and as Bill Clinton was doing with great gusto in the federal government. Thus, by the end of the 1990s, although parts of academia (the faculty) were still lagging noticeably behind other sectors, fields, and institutions, large changes from below in the student body and from above in the higher administration might be harbingers of coming changes. In addition, some individuals were getting angry, speaking out, and even on occasion starting new projects. What was missing was some sort of trigger event or outrage to unleash the anger and energy necessary to restart a movement or at least reenergize those in a position to do something. That 1999 event is discussed in the epilogue.

Two New Faculty Programs

In the 1980s two innovative programs to increase the hiring of women faculty in the sciences appeared. The earlier one was the program of visiting professorships for women scientists (VPW), created at the NSF by the passage of the Women in Science Bill of 1980 (see chapter 1), which the mathematician and activist Mary Gray had criticized at the time as woefully inadequate. Among its provisions was a program of about $1 million to $2.5 million annually for awards to from twenty-five to thirty women for from six to twenty-four months' support at an educational institution of their choice. It ingeniously utilized the traditional (i.e., nonthreatening) academic practice of financing a short-term visiting faculty position at a host university, which in fact could receive the overhead from the grant, a positive incentive. It was hoped that this experience would prove beneficial both to the woman and to the institution, where attitudes might begin to change and thus affect subsequent hiring practices. Perhaps it did, but in retrospect, the program proved to be an effective "entering wedge" both at the universities and within the NSF itself, for it was to be the precursor of a series of other, larger programs for women scientists and engineers.

In addition to diversifying a department for a short time by collaborating with its faculty or by doing her own research (and thus enhancing her own future career), the visiting woman professor, or VPW, was to spend 30 percent of her time teaching and encouraging the women students and scientists at the host institution in some way. By thus working to increase the number of women persisting in scientific careers, she might become a kind of catalyst for reform in the local culture. There was, however, some criticism and even opposition to this notion of using researchers as agents of change, especially from women mathematicians, who bristled at being expected to host brown-bag lunches and perform outreach. They complained that such visible activities called more (adverse) attention to their gender. Just being on the faculty of a prestigious (and hostile) mathematics department and doing solid research while there was, they felt, enough of a demonstration effect—their simple presence showed that despite widespread skepticism, an occasional rare woman could in fact do advanced mathematical research.[3]

This criticism was somewhat ungrateful, as mathematicians were somewhat overrepresented among the VPWs. Though the Women in Science Bill had mentioned "national needs," the fields in which the professorships were awarded largely reflected the applicant pool, except that mathematics was favored and the biological sciences were underrepresented.[4] A 1991 evaluation—itself an ominous sign, as one is often undertaken to forestall criticism—reported that for some VPWs the year had been a turning point in their career. In particular, most of the nearly three hundred women who had won these awards to date had used them to relocate or to get into a new line of research. Although it was not feasible to document systematically the short- or long-term success of any of their informal outreach efforts, relatively few had either sought to or been able to change a department or an institution in any significant way.[5] Nevertheless, a total of roughly four hundred women, spread over many fields and institutions, might make a difference later, for over the fifteen years of the program's existence many of them got tenure, moved to better positions within academe, and would be there to assist in later NSF initiatives. In 1996 the program was terminated, evidently the victim of several forces—a reorganization of the by then several programs at NSF for women and girls, a turn against affirmative action in the Republican Congress, and a preference at the NSF for concentrating funds on junior women faculty fighting for tenure.[6]

The second new program of the 1980s came from an entirely unexpected and improbable source. In October 1987 Clare Boothe Luce, a former Republican congresswoman, U.S. ambassador to Italy in the 1950s, writer, and the wealthy widow of Henry Luce, founder and publisher of Time Inc., bequeathed a substantial sum to the Henry Luce Foundation to create the Clare Boothe Luce Program, an innovative addition to the tradition of "creative philanthropy." Clare Boothe Luce, who was not a feminist or a college graduate and had shown no previous interest in science, set up a fund of $70 million from which to endow assistant professorships in the physical sciences and engineering at private, especially Catholic universities, many of which (Boston College, Georgetown, the College of the Holy Cross, and others) had admitted their first women students (outside of nursing) in the 1970s. Fourteen colleges and universities, including those that had given her an honorary degree, were each given the interest on $3 million in perpetuity. The rest of the fund was to be divided among other institutions that would be invited to apply.[7]

Over the years, as the endowment grew, the program was expanded to include the biological sciences, five-year professorships, and graduate and undergraduate fellowships and generally made it easier for these often financially strapped colleges and universities to diversify their science faculties and student bodies more than they might otherwise have been able to do. As the program expanded, it led to many firsts not only at Jesuit and other Catholic institutions but also at women's colleges and coeducational ones that lacked female faculty in the sciences, such as Spelman, Caltech, Stanford, and the University of Washington. By 2007 there had been 137 Clare Boothe Luce Professors and more than 800 undergraduates and 390 graduate students supported by Luce. With such external funding, a great many institutions that had never had women faculty in a field or at all found themselves able to broaden their faculties, and the young Luce Professors especially liked the flexible research fund attached to their professorships. The program still exists, with Jane Z. Daniels, a founder and former director of WEPAN, as its director since 2001.[8]

These two programs opened the door to more extensive programs at the NSF and elsewhere in the foundation world. What inroads lawsuits had accomplished with a stick (and continued to do with the Jenny Harrison case of 1986–93, mentioned below), philanthropy and government incentives were quietly furthering with a carrot.

Publicizing the Lagging Fields

As more departments began to reach the stage of tokenism, and a few women in those departments found their voices, the intermittent (but rarely answered) lament Why so few? spread to even more fields. Sometimes alone, sometimes with others, and sometimes as part of a committee, scientists collected and published data that showed that women were there in certain numbers but always underemployed and underpaid. This was true in astronomy,[9] which had many links to physics and shared certain similarities with it; economics, which was in some ways like mathematics and where a marginal group of "feminist economists," who sought to change the basic conceptions of the field, by their own account had made only a modest impact;[10] the earth sciences;[11] oceanography;[12] and meteorology.[13] Numbers and percentages had their uses, as in showing a pay gap of eighteen thousand dollars a year between men and women professors of meteorology. But action was more likely to result when the working conditions or the humiliations got personal.

The worst numbers (and barely improving, if at all) were in the departments of mathematics. Since its founding in 1971 the Association for Women in Mathematics (AWM) had kept a sharp eye on these numbers, but its leaders had never bothered to put together a list of zeroes, as the chemists had, perhaps because the nearly total absence of women from the top departments was so well known that it would be superfluous and depressing to remind everyone visually of this central fact of their professional existence. Besides, all the data were publicly available each year in the September issue of the American Mathematics Society's journal, *Mathematical Notices.* Yet as the years passed and top departments in other fields began to hire and promote women, all-male math departments began to seem anomalous. This pattern finally came to public attention in 1992, when *Science* magazine, which had been following Jenny Harrison's eventually successful tenure case at Berkeley (see below), spotlighted mathematics in a special issue on gender. As part of its story it printed (and later corrected) a chart showing the number and percentage of women, tenured and on the tenure track, in the top ten mathematics departments. Both sets of numbers were appallingly low. Of the 288 (or 303) such positions, only 3 (or 5) full professorships were held by women. This 1 percent, however counted, was well below the high and rising proportion of women earning PhDs in mathematics (from 7% in the early 1970s to about 30% by 2000), mostly awarded by these same departments.[14]

Yet there did not seem to be much that could be done. Occasionally a foreign woman accepted a post but then left after a few years, and some departments claimed to have standing offers out to a few American women who never accepted. Mathematics had no journal like the weekly *Chemical and Engineering News* or even the monthly *Physics Today* that could publicly criticize the laggards, and the field remained at the level of zeroes. The only thing to do was to train more women to the postdoctoral stage in hopes that sooner or later one would receive tenure at a top institution and actually stay. Meanwhile, a stream of intimidated VPWs visited the top departments and then left, female assistant professors hid in their offices to avoid unpleasant encounters, and a rising proportion of female PhDs got jobs and even tenure elsewhere, especially at liberal arts institutions, where they could have a satisfying career and possibly train future women mathematicians.

Even this rare publicity on mathematics departments grew out of the long-running but eventually victorious Jenny Harrison case at Berkeley, 1986–93. Berkeley had hired Harrison in 1978 as its third woman, after the Russian-Israeli Marina Ratner in 1975 and the award-winning faculty wife Julia Robinson in 1976 (Robinson died in 1985). When Harrison, a geometric topologist, came up for tenure in 1986, the department was badly split, voting 19 for and 12 against, with 7 abstentions, which was not enough for tenure. In 1988 her Berkeley job ended, whereupon she got a one-year visiting professorship at Yale University from NSF's VPW, and in 1989 she filed suit, demanding to see the evidence against her. Fortunately, in 1990 the U.S. Supreme Court ruled unanimously in another case, that of Rosalie Tung versus the Wharton School of Business at the University of Pennsylvania, that universities had to turn over any "confidential" personnel files to the plaintiff in a discrimination suit. At this point the administrators of the University of California, who had hitherto upheld their mathematics department and the sacred confidentiality of its personnel files, became more flexible and began to think of finding a way to settle out of court. Fortunately for all, by the time another outside tenure panel evaluated Harrison she had come up with some new "significant results," as mathematicians say, that were important enough to justify not only tenure but a full professorship.[15] Interestingly, and sadly, the one tenured woman in the Berkeley department at the time, Marina Ratner, herself a controversial hire partly because she had not gotten tenure in Israel and did not yet speak English, strongly and widely opposed Harrison, circulating letters saying so to the *AWM Newsletter* and the San Francisco newspapers. Meanwhile, the leaders of the AWM, which in the 1970s had taken a strong interest in the Berkeley math department, to which many of them had ties, was divided on the Harrison case and did not take a stand or provide financial support, although the AAUW Legal Advocacy Fund did.[16]

In the 1990s, after this highly publicized incident, the number of women tenured at the top ten mathematics departments increased to 9 (of 288), thus approaching 3 percent, as Princeton hired 2 (Ingrid Daubieches from Rutgers in 1993 and Sun-Yung Alice Chang from UCLA in 1998) and the University of Michigan hired 3 more, for a total of 4. (One analysis of hiring at the University of Michigan claimed that although the proportion of women scientists hired in the 1990s was better than earlier, it was stuck at the level of women awarded PhDs back in 1979.)[17] Meanwhile, other, lower-ranked departments had hired a few women, among them the reentry mathematician Judith Sally at Northwestern in 1972; Judy Roitman at the University of Kansas in 1978; Karen Uhlenbeck at the University of Texas in 1988; and the Frenchwoman Michele Vergne, who visited the MIT department intermittently. Other departments of note were at Northeastern University, with Nancy Kopell (later at Boston University) and Chuu-Lian Terng; Rutgers, with Jean Taylor and Barbara Osofsky; and the University of Illinois at Chicago Circle, with Louise Hay, Karen Uhlenbeck (early on), Vera Pless, and Bhama Srinivasan. Thus, by 2000 the picture was only slightly better than in 1980 at the middle and even the highest levels.[18]

Self-Interventions

Also in the 1990s, as described in chapter 6, the physicists began in a modest but innovative way to do something themselves about this continuing problem in their field. The American Physical Society had set up a team of site visitors, which by 2000 had visited twenty-four departments that had invited them. Perhaps the reports from this innovative voluntary intervention, supported by the NSF for several years, gave officials and advisory committees at the NSF ideas for strengthening their own evolving women's programs. Helping those in a department who wanted to change its climate from within, which would not have been possible in the 1980s, was now seen as a prudent use of funds. Thus, this initiative in the early 1990s may have been a precedent for the NSF's own enlarged women's programs later on. But the visits may also have helped the departments themselves, for later evidence indicates that there had been some progress in the top physics departments. When in 2001 the chemist Donna Nelson and her team of undergraduates at the University of Oklahoma (see the epilogue) surveyed fifty top physics departments, she found 131 (6.6%) women among 1,988 faculty, up considerably from data for 1982 (see chapter 2). By rank, 74 (5.2%) were full professors, 27 (9.4%) were associate professors, and 30 (11.2%) were assistant professors, finally showing some progress among the recent hires. (Even more appalling were her racial data: 23 were Asian, but only 7 were Hispanic, and there were no African Americans or Native Americans.)[19]

Meanwhile the chemists, led by Sister Agnes Ann Green, IHM, of Loyola Marymount College and the Women Chemists Committee of the American Chemical Society (ACS), continued to depict and deplore the glacially slow changes in chemistry departments in her biennial reports. Occasionally the ACS's weekly *Chemical and Engineering News* and its monthly *Journal of Chemical Education* publicized Green's findings. Based on the directories published by the ACS annually of faculty in chemistry departments granting PhDs (and biennially of faculty in those granting BS degrees), the list had been nearly all zeroes in 1970–71, when men held 98.5 percent of the faculty positions in chemistry (hence the nickname "list of zeroes"). More than a decade later, in 1982–83, men's proportion had dropped only slightly, to 95.9 percent. The field was in a steady state. Overall the number of departments was not increasing, nor was the size of department faculties; in fact some were shrinking. Yet bit by bit, department by department, there began to be a modest

Table 11.1. Women on the largest chemistry faculties, assistant professor and above, 1978 and 2000

	1978			2000		
University	Total Faculty	Tenured Women	Untenured Women	Total Faculty	Tenured Women	Untenured Women
Berkeley	56	1	0	51	4	1
Indiana	52	0	2	30	0	2
Ohio State	49	0	1	44	2	2
Texas–Austin	48	1	1	49	1	2
Purdue	46	1	1	46	4	2
UCLA	42	0	0	50	7	2
UCSD	42	2	1	44	4	0
Minnesota	42	0	0	41	3	0
Wisconsin	41	0	1	40	2	1
Illinois–Urbana	40	0	1	39	3	1
Michigan	39	1	0	38	3	1
MIT	38	0	2	29	3	1
Northwestern	38	0	0	24.5	0	2
Michigan State	38	0	0	—	—	—
Penn State	38	0	0	30	3	1
Florida State	37	1	1	36	3	3
Washington	34	0	0	42	2	2
Colorado	34	0	2	38	5	2
Florida	34	0	1	46	3	1
Iowa State	30	0	0	—	—	—
Caltech	29	0	1	27	2	1
Chicago	29	0	0	26	1	2
Cornell	29	0	0	32	2	0
Case Western Reserve	28	0	0	—	—	—
Oregon	27	0	1	—	—	—
Harvard	25	0	1	20	1	0
Notre Dame	24	0	0	28	0	2
Yale	24	0	0	23	1	1
Carnegie-Mellon	24	0	0	—	—	—
Brown	24	1	1	—	—	—
Princeton	22	0	0	25	1	0
Stanford	21	0	1	25	1	0
Brandeis	20	0	1	—	—	—
Columbia	19	0	0	21	1	1
Johns Hopkins	18	0	1	16	0	1
Total	1,182	8	21	950.5	62	34

Sources: "Only Eight Women Have Tenure on Major Chemistry Faculties," *C&EN* 56 (Sept. 11, 1978): 29; Janice R. Long, "Women Chemists Still Rare in Academia," ibid. 78 (Oct. 1, 2001): 56–57.

change, and by the year 2001 several of the zeroes had become ones, twos, or even threes, as shown in table 11.1. Some women chemists even got tenure, including two at the University of Minnesota, where the consent decree in the Rajender case required that two of its next five faculty positions in chemistry go to women. In 1989 Cynthia Friend got tenure at Harvard, its first woman and for a long time the only one. Thus, overall there was modest progress, for by 1992, after two decades of rising percentages of women earning doctorates in chemistry, only three departments had no women faculty—at Oregon State, the University of South Carolina, and the University of Virginia. The top departments had moved onto the next stage of tokenism (ones and twos), and the "list of zeroes" could be retired.[20]

But as the numbers grew, so too, almost hydralike, did related issues that were even more difficult. Tales began to circulate of the unequal and insulting work practices encountered by some women chemists. When Kristin Bowman-James became department chair at the University of Kansas in the late 1990s and doubled its number of tenured women to six, this so upset the older male faculty that they voted not to give her a second term.[21] Signs began to emerge in the 1980s and later that many female graduate students in chemistry were so disgusted by what they saw of the lifestyles and behavior of both their male and their female faculty that many of them were deciding not to apply for jobs in academia but to go into industry instead.[22]

In the 1990s a group of about twenty mid-career women chemists emerged determined to do more than document and deplore the dismal situation. The triggering event seems to have been an article on chemistry in the same special issue of *Science,* in March 1992, that included the abovementioned article on women mathematicians. Included were comments by Mary Good, a former president of the ACS, and Jacqueline Barton, a MacArthur Prize fellow at Caltech, the latter quoted as saying a bit too cavalierly that "there are no obstacles if you work hard." This outright denial might have been tolerable earlier, but now it prompted fourteen chemists (including four men) to write five sharply critical letters to the editor of *Science* expressing the wish that this were so. They were dissatisfied with the conditions they saw, and with no prospect of change in the foreseeable future, they were getting angry and speaking up.[23]

Nothing changed immediately, but finally, in 1998, Geraldine Richmond, a professor of chemistry at the University of Oregon, and Jeanne Pemberton, a chemist at the University of Arizona, two signers of the 1992 letters to the editor, got together with several other women professors of chemistry and brainstormed about what needed to be done. They knew firsthand all the stories behind the statistics. Women chemists, even previously and seemingly successful senior ones, were facing difficulties in their careers: their lab space was taken away, they got little or no time off after childbirth, they were paid lower salaries, they had heavier teaching loads, they often lost credit for their ideas, and they had a higher suicide rate. Not only did all of this affect them and their careers but the younger women around them noticed these things and grew discouraged about continuing in academia. It was time for someone to start doing something, but who and what? All these women chemists needed "coaching" by professional counselors.

Finally the group took matters into their own hands. They bypassed the ACS and its women's committee and formed an independent organization, the Committee for the Advancement of Women in Chemistry (COACh), to collect the women's stories and hire professional counselors to run workshops on assertiveness training, negotiation tactics, management skills, and other survival tactics that were not taught in graduate school but would help individuals defend themselves and fight back. These sessions could be held on-site the day before the national meetings, which everyone would be attending. Fortunately, COACh was able to obtain seed money from the long-established Camille and Henry Dreyfus Foundation, dedicated to the "betterment of the chemical sciences," which had recently begun to support projects relating to the needs of women and minority chemists. By 2006 COACh had secured more than $1 million in grants from the DOE, the NSF, and the NIH, had helped more than three thousand women chemists cope with the difficult people and situations around them, and could report several success stories of careers salvaged. With a little NSF and other outside support, women chemists, like the women physicists, could help others to help themselves.[24]

Senior Women: Contented or Discontented?

Although little was written about the success of academic women in any of the sciences, a few senior women, especially bioscientists, emerged in the 1970s and after as major laboratory directors and grantswomen. (A few of these were involved with or spoke out upon women-in-science issues.) Among this group were Charlotte Friend, Ruth Sager, Margaret Bryan Davis, Neena Schwartz, Shirley Tilghman, and Joan Steitz, as well as the sociologist Judith Blake, in demography. There were also a few in the physical sciences, such as Mildred Dresselhaus and Judith Klinman. In the years after 1972 they were chairing departments, running big NIH- and NSF-supported research and training grants with numerous graduate students and postdocs, establishing centers, serving on review panels, and attending and occasionally running Gordon conferences (on which see chapter 12). Aside from the stresses of building a research unit based on temporary grants, they enjoyed the experience enormously, or as Ruth Sager wrote to her mentor when she returned to her laboratory after an absence, "It seems to be the same old rat race but I must admit I love every moment of it."[25]

Of course they liked it even more when they did not have to apply for grants. Some female bioscientists benefited from the decision in 1985 by the newly reorganized Howard Hughes Medical Institute (HHMI) to award about one hundred (later increased to about three hundred) multiyear investigatorships to outstanding individuals at major biomedical institutions and thus free them from writing grant proposals for five years. These highly coveted super-professorships were initially limited to four (later five) sub-areas: cell biology and regulation, genetics, immunology, neuroscience, and structural biology. Although these were relatively feminized specialties, there seem to have been very few women among the initial investigators. But turnover was significant, as renewal was dependent on a rigorous review by the HHMI's scientific review panels. By 2007, when there were about three hundred HHMI investigators at universities worldwide, about sixty, or 20 percent, were probably women, to judge from their first names, a percentage that could have been higher. Among the best-known Americans women were Shirley Tilghman, of Princeton; Susan Lindquist, at the University of Chicago; Joan Steitz, at Yale; Elaine Fuchs, at the Rockefeller University; and Susan Taylor, at the University of California at San Diego.[26]

These top women offered few public hints until the 1990s that not all was well at the top of the academic ladder. When Berkeley's first tenured woman chemist, Judith Klinman, a very busy single parent, arrived in the late 1970s, she tried at first to go it alone, but in 1981 she joined a select off-campus group of Bay Area women science faculty (and one department administrator), saw what communal wisdom she had been missing, and became a faithful participant thereafter. At biweekly evening meetings in the members' homes, seven women exchange experiences and sought and shared advice on how to cope with the difficulties and stresses of the unknown personal and professional terrain they were encountering. Started in 1977, with the help of some "radical psychiatrists" after one of the founders had attempted suicide, the group went public with a panel at a session at the annual meeting of the American Society for Cell Biology in San Francisco in 1994. Their need for such collective mentoring had not lessened over the years.[27]

In the 1980s another new area was opening to legal remedy—that of sexual harassment, or the existence of a hostile environment on the job. This had not been covered by the civil rights legislation of the 1960s and 1970s, but by the late 1970s it had emerged as a legal issue and by 1990 some women were winning lawsuits. One was Jenny Jew, an associate professor of anatomy at the University of Iowa, who sued the university in 1985, after being denied promotion to full professor because one professor had told the others that she had only gotten tenure in the first place because she had had a sexual relationship with the former department chair. Despite her repeated protests to the proper authorities, the university had done nothing to stop this defamation, which she, as a Chinese American, found particularly humiliating and offensive. In fact the university's lawyers had fought her all the way in court. In 1990 she finally won and was awarded promotion to full professor, a guarantee of a nonhostile work environment, and $1.1 million—$176,000 for damages, back pay and benefits since 1985, and $895,000 for her lawyers.[28]

Then in 1991 Frances Conley, MD, a professor of neurosurgery at the Stanford Medical School, took the shrewd and unusual step of not filing a lawsuit (and thus incurring all the limitations that that entailed) and instead doing what had never been done before—she simply and bravely resigned noisily over the unprofessional behavior of one of her colleagues. The incident was highly publicized, partly because it followed closely upon the riveting Anita Hill testimony on sexual harassment at the Senate hearings on the appointment of Clarence Thomas to the Supreme Court. Conley was widely interviewed on television talk shows and spoke on many campuses. Then in another remarkable twist, after a university committee made some amends, she returned to the faculty and wrote a riveting book about her experience, *Walking Out on the Boys.* She insisted in the book that the conditions she had faced were not full-blown "sexual harassment"—and that journalists had linked her case to Hill's and blown everything out of proportion—but something else that was gender-related, demeaning, and dismissive, though she had no term for it. How widespread the problem was in academia and what ought to be done about it were not addressed. Although Conley's case may have been unique, Nancy Hopkins, at MIT, would show that the problem was widespread.[29]

Perhaps emboldened by Conley's case, the Princeton biologist Shirley Tilghman, another single parent, was remarkable for her two articles on the Op-Ed page of the *New York Times* in January 1993, in which she suggested abolishing the tenure track in order to provide flexibility for junior women faculty in their thirties who wanted to start families.[30] Then the sociologist Henry Etzkowitz, who visited thirty departments in the early 1990s, found that when there was more than one woman on a department's faculty, they were often factionalized. Rather than working together to make significant reforms once they had reached a "critical mass," as sociologists had predicted would happen, they worked against each other. In fact all too often the senior women scientists presented something of a problem for the younger ones, who were trying to make a place for themselves. The older "superwomen" who had "made it" when times were tough generally wanted the young women to be as overprepared as they were, a strategy most of the latter thought unfair. This incompatibility made it hard to work together to bring about lasting changes. Thus, even though the number of women faculty in a department or a university might be rising, and some senior women might be visibly successful, it was not clear in 1994 that working conditions or practices were going to change significantly.[31]

Moving into Academic Administration

On another front, a few women scientists began to move into academic administration, where few had trod before 1972. The turning point was either the appointment of the political scientist Donna Shalala to the post of chancellor at the University of Wisconsin–Madison in 1987 or the meteoric rise of the psychologist Judith Rodin from department chair at Yale in 1986 to its provostship in 1991 (after a series of men resigned) and the presidency of the University of Pennsylvania in 1993. Before 1987 there were a few, often embattled appointments to such high posts, and after 1993 there was a steady stream of less publicized female provosts and even presidents.

In 1973 the American Council on Education established an Office of Women in Higher Education (OWHE), which began to hold workshops for women aspiring to become administrators. Many skeptics were surprised when these programs were vastly oversubscribed. In 1977, with the support of the Carnegie Corporation of New York, the OWHE, long headed by Emily Taylor, a former dean of women at the University of Kansas, and then by Donna Shavlik, started a "national identification program," which sought to "address the belief that few women are prepared for or interested in administrative positions" and to locate, encourage, and train likely candidates for this challenging work. Similarly, the former chemist Lilli Hornig started in 1972 and ran for many years another group, Higher Education Resource Services (HERS), long centered at Wellesley College and later at the University of Denver. Starting in 1977 it ran a summer program at Bryn Mawr College that (for a substantial fee) trained women in the skills of educational administration. This special training of "baby deans" began to pay off in the 1980s and especially in the 1990s, but it also became something of a stigma, as men seemed not to need such extra credentials.[32]

Before 1972 hardly any women scientists held even the lowest-level administrative post, that of department chairman, as it was then called, but thereafter a few women began to hold this appointed position, which increasingly was referred to by the new, gender-free term *chair.* They seized the opportunity and rose to the task. Among these pioneers were Elizabeth Scott, chair of the statistics department at the University of California–Berkeley from 1968 until 1973, and her friend Florence Nightingale David, at UC–Riverside from 1970 to 1977, who once complained that predatory men there from other academic units fought to take over her fledgling statistics department. Notable others were Neena Schwartz, who served as chair of the biology department at Northwestern from 1974 to 1977, and the legendary Elizabeth Hay, who had built her career on the new electron microscopes of the 1950s and after and who served from 1975 to 1993 as chair of the anatomy department at the Harvard Medical School. (As such she was quite indignant with the feminists running the irreverent *Women in Cell Biology Newsletter,* who portrayed all department chairs as reactionary male chauvinists.) After a three-year stint as a full professor of biology at Yale in the 1970s, Margaret Bryan Davis set up and chaired a separate, successful department of ecology at the University of Minnesota. By comparison, Judith Rodin was appointed chair of the psychology department at Yale in 1986 only in response to an offer of the presidency of Radcliffe College. But even this first step into administration was still slower in coming in less-feminized fields, such as engineering, where as late as 2000 a female department chair was still unusual. At MIT it was 2003 when the earth scientist Maria Zuber became the first woman to chair a department in the School of Science, and then in response to unusual circumstances (see the epilogue).[33]

At the next administrative level, that of deanships, the pattern was mixed. As mentioned in chapter 6, several women scientists became deans of their graduate schools in the 1970s and after and often used the post as a springboard to higher posts. Less welcoming were the deanships of colleges of arts and sciences and such former male bastions as the colleges of agriculture, engineering, medicine, and veterinary medicine. In fact, even the rumor of the possible appointment of the black biologist Jewel Plummer Cobb to the deanship of letters and science at the University of Michigan in 1975 was met with such virulent press coverage and strong faculty resistance that it was withdrawn. Yet by 2000 a few women had held these posts, including the geneticist Madeleine Goodman in arts and science at Vanderbilt and June Osborn as dean of public health at the University of Michigan. Eleanor Baum was perhaps the nation's first woman dean of engineering when she was appointed at the Pratt Institute in New York City in 1984, but Denice Denton claimed that her appointment at the University of Washington in 1996 made her the first woman dean of a major engineering school.[34]

Women provosts and vice presidents for academic affairs were a novelty when the physics professor and dean of the graduate school Kathryn McCarthy, at Tufts University, was appointed provost in 1973. The anthropologist Nancie L. Gonzalez served as vice president for academic affairs at the University of Maryland–College Park from 1977 until 1981, and in 1979 the biologist W. Ann Reynolds was appointed provost at Ohio State University, the start of a long and generally successful but also troubled administrative career (see below). In 1981, after chairing Berkeley's Department of Nutritional Sciences twice, Doris Howes Calloway became the first woman provost of its professional schools and colleges, and in 1983 the NSF's Eloise E. Clark became provost at Bowling Green State University in Ohio.[35] Then there was a long drought before the door finally opened wide with the appointment of the chemist Marjorie Caserio as vice chancellor for academic affairs at UC–San Diego in 1990 and that of Judith Rodin, by then dean of the Yale Graduate School, to the provostship at Yale in 1991. The wide and favorable publicity surrounding Rodin's appointment signaled a certain acceptance of women in such a powerful post, and in the following years several more women, including some scientists, were appointed to these positions, such as the biochemist Mary Sue Coleman at the University of New Mexico in 1993, the anthropologist Alison Richard at Yale in 1994, the psychologist Nancy Cantor at Illinois in 1997, and the biologist Karen Holbrook at the University of Georgia in 1998. Almost all were immediately considered prime candidates for presidencies at other major universities.[36]

Several of the first women presidents and chancellors of "flagship" campuses had a tumultuous time. Evidently the first was the biochemist Lorene Rogers, who taught nutrition in the department of home economics at the University of Texas at Austin (UTA) before serving as president from 1974 to 1979. Termed a "red-headed dynamo" by the physics Nobelist John Wheeler, she was elected interim president of UTA in 1974 and as such was "apparently the first woman ever named chief executive officer of a major coeducational U.S. university," according to the *Chemical and Engineering News.* When UTA's board of regents, which included several women, then elected her president, there was opposition from students, who held a protest rally, and from some faculty, who refused to attend any meeting she chaired. Nevertheless, she served a full term, recruited several important physicists to the Austin campus, and retired in 1979 at the then regular retirement age of 65.[37]

Other women campus presidents of the 1980s had varied experiences. The economist Barbara Newell, formerly president of Wellesley College, served as chancellor of the Florida State University system for four years in the early 1980s, and the biologist Jewel Plummer Cobb became chancellor of California State University at Fullerton in 1981, the first and only African American woman chief executive officer in the nineteen-campus system. She served to mixed reviews until 1990, when she reached age 66 and was forced to retire.[38] Meanwhile, the biologist Evelyn Handler, formerly dean of science at Hunter College and briefly president of the University of New Hampshire, served as head of Brandeis University from 1982 to 1990, finally resigning after a difficult time, as the university was reducing its overtly Jewish identity in an attempt to attract more Asian American students.[39] Overshadowing all of these was the dramatic appointment in 1987 of the political scientist Donna Shalala, president of Hunter College, to head the Madison campus of the University of Wisconsin. This was a breakthrough at a major research university, and Shalala, characterized in 1990 by the *New York Times* as a "master politician," served successfully until 1993, when President Bill Clinton appointed her to the even more demanding post of secretary of health and human services (see chapter 9). After eight years at that post, in 2001 she became president of the private University of Miami, known for its football team and its medical school, two big and continuing challenges for any president.[40]

Table 11.2. Women scientist chancellors in the University of California system, 1987–2010

Years	Campus	Name	Field
1987–92	Riverside	Rosemarie Schraer	Biophysics
1987–94	Santa Barbara	Barbara Uehling	Psychology
1993–98	Irvine	Laurel Wilkening	Planetary science
1995–96	San Diego	Marjorie Caserio*	Chemistry
1996–2004	Santa Cruz	M. R. C. Greenwood	Nutrition
1999–2006	Merced	Carol Tomlinson-Keasey	Psychology
2002–7	Riverside	France Cordova	Astrophysics
2004–	San Diego	Marye Anne Fox	Chemistry
2005–6	Santa Cruz	Denice Denton	Electrical engineering
2009–	San Francisco	Susan Desmond-Hellman	Cancer/drugs
2009	Davis	Linda Katehi	Electrical engineering

Sources: CHE; AMWS.
*Interim

Perhaps inspired by Shalala's appointment, a 1988 report by a committee of the National Association of State and Land-Grant Colleges, which surveyed women (and for the first time minorities) at the 113 land-grant and other institutions, showed that of their 112 CEOs only 3 (2.6%) were women. Meanwhile, officials at the more comprehensive OWHE, which covered all of higher education, including many small women's colleges, counted 350 female presidents at various institutions nationwide (11% of the total). In 1990 the OWHE hosted a national gathering of all the women college presidents in the country. About 165 current and former chief executives came, met one another, and networked. Before long, members of this growing pool of talent would begin to move out from presidencies of current and former women's colleges to the Big Ten and the Ivy League in a kind of secondary migration.[41]

Also in the late 1980s, the University of California, with its nine (later ten) campuses, began to appoint a few women scientists, including eventually a few physical scientists and engineers, to some of its nationally visible top posts. It started cautiously by appointing the Penn State biophysicist Rosemarie Schraer as vice chancellor of its Riverside campus in 1986 (and then chancellor a year later). She was followed by the psychologist Barbara Uehling at the Santa Barbara campus in 1987 and the planetary scientist Laurel Wilkening at Irvine in 1993 (others are shown in table 11.2). For various reasons turnover was high with the average length of service about five years.

The Ivy League was slower in making top-level firsts. Yale could have appointed the historian Hanna Holborn Gray as its president in the 1970s or the biochemist Maxine Singer in 1985, but it held back, evidently preferring to let other institutions go first. It was not until 1993, when the trustees of Yale passed over the award-winning psychologist and provost Judith Rodin, by then age 50, for its presidency that the more enlightened trustees of the University of Pennsylvania, which had the most women faculty in the Ivy League (about eighty), took the historic step and elected her its (and the Ivy League's) first woman president. Many considered this an inspired choice, as Rodin had grown up in Philadelphia, graduated from the Philadelphia High School for Girls, and was a Penn alumna. She served successfully for ten difficult years, as the university's medical school and hospitals responded to the coming of managed care. The trustees responded financially, and for a time she was the best-paid woman university president in the country, receiving just under seven hundred thousand dollars a year.[42]

Also in 1993, Duke University appointed the political scientist Nannerl Keohane, the president of Wellesley College since 1981, to its top post. She served with distinction for twelve years. In 1995 the biochemist Mary Sue Coleman was appointed the first woman president of the University of Iowa, where she served until 2002, when she left to head the University of Michigan at Ann Arbor. In 1999 Shirley Jackson left her post as head of the Nuclear Regulatory Commission to become the first woman, and the first person of color, to serve as president of Rensselaer Polytechnic Institute in Troy, NY, whose ambitious trustees wanted her to build ties to industry and raise money.[43]

But not all women presidents were so fortunate or so popular and successful. Some faced the public humiliation of a forced resignation, usually because of forces beyond their control. In 1996 the planetary scientist Laurel Wilkening, chancellor of the University of California at Irvine, "retired" at the age of 52, after criminal behavior by several faculty physicians at the university's fertility clinic besmirched the notable achievements of her administration. In 1999 the anthropologist Yolanda Moses, who had stepped in as president of the City College of New York in 1993, when it had a crisis with its ethnic studies department, was asked to step down after several years of budget cuts. Others have bounced back. The physiologist Judith Ramaley, for example, has had the resilience to head three universities—Portland State, the University of Vermont, and Winona State University in Minnesota.[44]

Graceful exits may be especially rare for women heads of big public systems of higher education. These are tough jobs, beset with a constant barrage of budgetary, political, and personnel issues. An early and notable example of such a system administrator was the biologist W. Ann Reynolds, who, as mentioned above, became provost at Ohio State in 1979, and then she became chancellor of the whole California State University system (nineteen campuses) in the 1980s. Yet Reynolds encountered political difficulties of her own at California State, the catalyst for her dismissal being claims that she had misspent state funds in restoring the chancellor's residence in ritzy Bel Air. In 1990 she moved across the country to head the troubled City University of New York system in New York City, and in 1997 she went to head the University of Alabama at Birmingham, which she left in 2002 and later sued for age discrimination.[45]

A notable flameout was that of the psychologist Judith Albino, formerly dean of the graduate school at SUNY–Buffalo, who after just one year there as a vice chancellor accepted the chancellorship of the four campuses of the University of Colorado system in 1991. She probably should have been dissuaded by the withdrawal of the five male candidates ahead of her or by the salary, which, unbeknownst to the Colorado trustees, was lower than her salary at Buffalo. It was an embattled situation, with the trustees themselves badly divided. Given the circumstances, it is remarkable that she lasted five years before resigning in 1995. Even more dramatic was the suicide, for unknown reasons, of Denice Denton, the recently installed chancellor of the University of California at Santa Cruz in 2006. The sudden death of such a promising and highly accomplished electrical engineer shocked everyone.[46] By contrast, the mild-mannered economist Katharine Lyall served as president of the twenty-six-campus University of Wisconsin system for twelve years (1992–2004), after ten years there in lesser posts. She worked hard, was polite to everyone, and remained outside the political frays that so often derailed others.[47]

By the late 1990s it remained to be seen what these women provosts and presidents would accomplish, for if achieving the top job was the culmination of a long chain of events going back decades, it was just the beginning of another. With the right circumstances and resources, the posts offered the chance to transform traditional institutions substantially.

More Federal Initiatives

In 1995 NSF officials and their advisers were once again pondering what to do about their women's programs, and Congress was getting ready to double the NSF budget. In December 1995 the NSF hosted a large open conference, "Women & Science: Celebrating Achievements, Charting Challenges," which was attended by about seven hundred individuals, to assess what programs had worked best in the past and to collect ideas on what to do next. About a hundred staff members attended, as the goal was to include all the research directorates as well as that of education and human resources. At the meeting, Luther Williams said that the NSF's goal was to support the existing VPW program, but by the time the report on the conference came out in 1997, NSF officials had decided to terminate it and combine its several other programs for women and girls into one large program, Professional Opportunities for Women in Research and Education (POWRE), with a budget of $8 million to $12 million. This new program would provide $75,000 grants for two years primarily to junior women scientists trying to get tenure. Although the program was criticized by some for restricting its applications to women, it was very selective, with an average applicant success rate of merely 26 percent, and just 20 percent in its first year. (Critics were probably not aware of how competitive it was.) It made six hundred awards in its four years of existence, a sign that it had found its niche. Its administration was less visible than that for the VPW program, for its applications went through the research directorates rather than through the one for human resources.[48]

Yet before long POWRE too came under criticism, as had other programs before it, not so much for what it did but for what it did not do. In the spring of 1998, when the NSF director (Rita Colwell) started a task force to review POWRE, it was becoming clear that it alone was not going to change universities or science departments appreciably. Even if all of its hundreds of grantees got tenure and they all started to work on gender-bias issues, they were dispersed over numerous departments at fifty to one hundred or more universities and thus would not be likely to change their institutions significantly in the near future. Advisers and officials were frustrated that everywhere in academia women scientists and engineers were battling essentially the same entrenched and pervasive practices, attitudes, and behaviors. The problems they faced were indeed systemic, or "industrywide." Perhaps a new approach was needed, but no decision was made, and POWRE was allowed to continue to exist.[49]

Meanwhile, in October 1998 Congress finally established a new congressional commission on women in science and engineering, a longstanding goal of Congresswoman Constance Morella (R-MD), of Montgomery Country, just outside Washington, DC, who served from 1987 to 2003. In 1992 Morella, who had chaired the Maryland Commission on the Status of Women in the 1970s, had submitted a bill to create such a temporary commission. It was passed by the House, where the Berkeley mathematician Jenny Harrison was among those testifying in its favor. But the bill failed in the Senate, where it did not have a sponsor (not even Orrin Hatch or Edward Kennedy, both of whom had championed the 1980 bill on equal opportunities for women in science).[50] Finally, in 1998, when the tenacious Morella reintroduced the bill, she added "and minorities" to its name, President Clinton spoke approvingly of it, and this time both the House and the Senate, where Olympia Snowe (R-ME) sponsored it, passed it. Most of the eleven members appointed to the Commission on the Advancement of Women and Minorities in Science, Engineering, and Technology Development (CAWMSET) were new faces from technical businesses, and several were persons of color. Its staff, largely NSF employees from the director's office, held hearings in Arlington, Virginia, Bethesda, Maryland, and Seattle and issued its final report, *Land of Plenty: Diversity as America's*

Competitive Edge in Science, Engineering and Technology, in September 2000. Interestingly, the report said nothing very new and did not recommend major changes (e.g., "institutional transformation") in current NSF programs, even though such discussions were under way there at the time. Mostly it exhorted others to make greater efforts in hopes of keeping women-in-science issues alive at the NSF and in Congress during the upcoming transition to the next presidential administration—which limited purpose may have been Morella's goal.[51]

Conclusion

In the twenty years after the historic Rajender decision at the University of Minnesota in 1980 not much changed at the top research universities in most of the sciences. Some departments hired their first and possibly their second woman scientist. Sometimes she got tenure. As the number and especially the percentage of doctorates going to women skyrocketed in many fields, numerous reports by individuals and groups revealed continuing disparities in rank and salary at major institutions. They created a recurring drumbeat, Why so few? In the biological and social sciences a few senior women responded to opportunities to become grantswomen, run major laboratories, and hold HHMI investigatorships. Similarly, a few women bioscientists and social scientists were beginning to be provosts and presidents of campuses and even systems. Some flourished and then moved on to higher, more prestigious posts. But others were hired with too little experience for impossible jobs, got caught up in the conflicts and scandals that swirled around them, and then were forced to resign as sacrificial lambs for the university's image.

But no one seemed to have a plan as to what might be done differently, let alone any leverage or resources. The only tool seemed to be publicity, such as an article in the *Chemical and Engineering News* or an Op-Ed piece in the *New York Times.* Yet all too often their authors posed the rhetorical questions that left no one accountable. Only the women physicists and chemists seemed willing to innovate and start new programs that might make a difference, such as COACh. The federal government was doing little in the way of enforcement; lawsuits were settled out of court with gag orders; and aside from occasional finger-pointing, the issue of women's advancement was moving to the back burner. One positive voice was that of Maxine Singer, who in 1992 recommended to women in biomedicine that they not leave their continuing situation to others but take matters into their own hands, chiefly by doing such excellent research in the expanding area of women's health that they would control future resources in that area.[52]

Somewhat promising was the increasing number of women, including scientists, appointed to high office starting in the late 1980s by trustees, who were more interested in equity issues than the faculty were. This was similar to what was happening near the top of the nonprofit institutions, especially after 1987 with the appointment of Maxine Singer as president of the Carnegie Institution of Washington, and in the federal government after 1993, when the Clinton administration appointed record numbers of women, including many scientists, to top posts.

Also in the mid-1990s, discontent began to surface among the seemingly privileged women scientists at the top, those few full professors at prestigious institutions like MIT. This reframing shifted the focus to issues other than numbers and the targets to university administrators, who could be held accountable, rather than whole fields, which could not. Starting about 1997, new voices in and around academe, new organizations, and a new set of events began to emerge that, while not a unified movement, cumulatively began to attract considerable attention, recruit fresh leaders, and embark on new strategies for the now possibly attainable goal of "institutional transformation."

Taking the Scientific Societies beyond Recognition

Largely but not totally outside the domain of legislation was the world of scientific recognition run by independent and autonomous groups of specialists and led by meritorious elders usually elected by the members. The actual size of this realm is difficult to gauge, as it includes international groups as well as American ones, but more than two hundred such scientific organizations were affiliated with the American Association for the Advancement of Science in 2000.[1] Yet even without legislation, activities in this area were significantly changed by the new attitudes toward women after 1972. Formerly the chief role of scientific societies had been to run meetings, publish journals, and present awards for contributions to science, all of which was done quietly, behind the scenes, by long-serving staff and a changing cast of older white male officers. These tasks continued, but beginning in the late 1960s many societies, whose members were buffeted by economic, social, and political issues, also began to take stands on issues outside of science, including the Equal Rights Amendment to the U.S. Constitution, and to undertake occasional outreach projects, especially in the areas of science education and the public understanding of science. Thus, over time many forces, including their women's committees and the availability of foundation grants and contracts, influenced these traditionally apolitical societies to diversify their officers and staffs and to become more open to, and even take public stands on, a broader range of issues affecting the present condition and future standing of their fields.

Women had long been members of most (but not all) scientific and technical associations, although before 1970 very few women were involved in society functions in any major way.[2] This changed drastically during the 1970s largely because of the efforts of newly formed women's committees and caucuses, as almost every scientific society and even a few of those in engineering elected women officers, including their first female president. Although often mere tokenism as well as a historic first, this step was a symbolic acknowledgment of the presence and participation of women in their field. Then once their brief terms in office were over, these women past presidents could join the ranks of distinguished elders in the field. In the decades after 1970 the several American honorary academies, which until then had rarely elected women members, increased their numbers of women substantially while also increasing their overall size. In addition, by 2009 more American women had won other awards, including nine Nobel Prizes, than ever before.

The changes occurred partly owing to the increased consciousness across society that women's contributions should be valued more visibly and partly because women scientists themselves became energized, formed caucuses and committees, and collectively began to demand a larger role in societies' politics and affairs. As the scientific and technical workforce changed in the years after 1972, its leadership, as reflected in these societies, changed too, but to a lesser extent and after a struggle. One could also claim that women's greater prominence also invigorated the societies and their disciplines in various ways, for it broadened their base, engaged members more actively, and expanded their activities into more areas of usefulness.

Admission and Levels of Membership

For several organizations that could be termed private clubs, those that often had their own premises or clubhouses, the admission of their first woman was a long-avoided step. In the early 1970s, however, some began to admit their first women voluntarily, if reluctantly. Among these were the Chemists' Club of New York City, established in 1921, which admitted women in 1971. The Biosystematists of the Bay Area (San Francisco, Berkeley, and Stanford), which met at the Men's Faculty Club at Berkeley, voted to admit its first (five) women in 1971, and the Boston-based Nuttall Ornithological Club, founded in 1873, admitted women in the mid-1970s.[3]

At several other, more prominent clubs the issue was still being thrashed out a decade and more later. The all-male Explorers Club of New York City, founded in 1905, finally voted in April 1981, "after months of emotional, if civilized, debate," including some threats of resignation, to admit women members, and the next fall it elected fifteen, including the aquanaut Sylvia Earle and the astronaut Kathryn Sullivan.[4] Many other private men's clubs held out until the bitter end, that is, until June 1988, when the Supreme Court unanimously upheld a New York City human-rights law banning sex bias in membership in those private clubs in which business was transacted and in which employers paid members' fees or 20 percent or more of the members sought tax deductions for their dues. Two days before that decision was handed down, the die-hard members of the Cosmos Club, of Washington, DC, long the gathering place of male scientists in the area, finally voted to admit women. This came after more than a decade of criticism by women and their supporters, three previous votes among club members in favor of women's continued exclusion, a demonstration by NOW activists, a threatened (but often postponed) hearing by the Washington, DC, Human Rights Commission, and a final highly publicized resignation, this time of Supreme Court Justice Harry Blackmun, who joined a long string of other officials who had previously publicly terminated their memberships. Among the first women to be admitted were Barbara Bailar, of the American Statistical Association; the economist Janet Norwood, the U.S. commissioner of labor statistics; Rita Colwell, then director of the Maryland Biotechnology Institute; Elaine Sarkin Jaffe, deputy chief of the pathology laboratory at the National Cancer Institute; and Nina Roscher, a professor of chemistry at the American University and an active member of the American Chemical Society's Women Chemists Committee (WCC).[5]

Quantitative data on women's membership in scientific societies are sporadic and hard to find. (Fewer than half of the 112 affiliates of the AAAS that responded to a questionnaire in 1976 collected such data, and only one-quarter considered the data valuable.) Some societies traditionally had a higher proportion of women members than others, often but not always reflecting the varying percentages of women in different fields. Even within organizations there could be large differences. In the American Psychological Association, for example, some divisions were more highly feminized than others: the Psychology of Women division was 96 percent female in 1990; Developmental Psychology, 51 percent; and Psychoanalysis, 50 percent, whereas the least feminized were Theoretical and Philosophical Psychology (15% women), Applied Experimental and Engineering Psychology (12%), and Military Psychology (11%). In almost all societies in the physical and field sciences the proportion of women rose after the early 1970s, when the proportion of PhDs awarded to women in those fields began to increase substantially. Accordingly, when data on women members are disaggregated by age, the proportion of women in the younger groups was significantly higher (10–20 percentage points) than in the older age groups. In physics, for example, in 1990 women constituted 14 percent of the members of the American Physical Society (APS) under 30 but just 3 percent of those aged 41–45. In astronomy this discrepancy was even more striking a decade later, when women constituted 42 percent of the members of the American Astronomical Society aged 23 to 27 but only 15 percent of those aged 43 to 47.[6]

Some societies had long had the practice of electing certain members to the level of "fellows." In a period of rapid feminization of the general membership, as this was, such an elective honorary status could serve as a hedge against the women's too rapidly assuming influence. For example, in 1984 women constituted 32.8 percent of the total membership of the American Psychological Association but only 16.2 percent of its fellows. Four years later, when 5,700 more women had joined (compared with 3,000 men), their percentage of the membership was up to 37.1 percent, but their percentage of the fellows had just inched up to 17.7 percent.[7] The Entomological Society of America had elected several women fellows in 1940–52 but then did not elect another until 1993, when Dorothy Feir was honored. Since then there have been several. Similarly, in 1994 women made up 15 percent of the 29,400 members of the American Geophysical Union, a sharp increase in recent years, but they constituted only 3.1 percent of its fellows (18 of 576), where their distribution among subfields was quite uneven. Also notable was that half of the women fellows (9 of 18) were under age 50, which was true of only 16 percent of the men, and 6 of the younger women had even won the prestigious Macelwane Award in the previous ten years. Young women were being favored, and in the American Geophysical Union, as all too often occurred elsewhere, accomplished older women, once passed over, never caught up.[8]

This pattern of female underrepresentation at the fellows' level took its ultimate form in one scientific organization, the deliberately small and selective Society of Experimental Psychologists, in which all the members were fellows. Founded in 1904 to nurture bonds among young and old academic men, it did not admit its first women until 1929. In 1970 it elected only its fourth woman (Dorothea Jameson Hurvich), but after this there were more each decade, including the mother-daughter pair of Frances Graham (in 1974) and Norma Graham (1983) and the husband-second-wife dyad of Russell De Valois (1968) and Karen De Valois (1992). In the 1990s up to three women were among the nine new persons elected annually, and by 2004 the society's membership was about 15 percent female (about 30 out of 210), a high for the society but far below the proportion of women earning doctorates in the subfield of "experimental, physiological, and comparative" psychology (21% in 1971, 33% in 1981, and 46% in 1991).[9]

Rethinking the Women's Prizes

In the early to mid-1970s the purpose and role of three longstanding separate prizes for women in science came under examination. Created in the 1930s, when women were not considered for the main prizes offered by scientific organizations, these separate prizes, for which the women members had had a hand in arranging, had long offered a kind of "compensatory recognition," for which the women were grateful.[10] But in the age of equal opportunity the role of these prizes seemed anomalous to some, and if their endowments were to be increased to meet new economic realities in an age of inflation, their purposes needed to be rethought and reevaluated.

This need first occurred unexpectedly in 1971, when the astronomer E. Margaret Burbridge created a well-publicized stir when she refused to accept the Annie Jump Cannon Prize of the American Astronomical Society (AAS). Everyone was surprised, because it had been given triennially since the 1930s to a distinguished woman astronomer, and Burbridge was an obvious choice. In fact she was the only woman whom the society had awarded even half a prize (shared with her husband). It had just been a matter of time before she was chosen. She complained that the award was in fact not very selective, since there were so few women in the society that sooner or later all of them would have won it. The immediate result of this refusal was that the society created a committee to study the status of women in the field more generally. Among its several recommendations in 1972 was one that acknowledged that while a separate prize designated for a senior woman had been a necessity in the days when women had not won the other prizes awarded by the society, this practice was no longer suitable in the era of equal opportunity. Yet Annie Jump Cannon had given the money to the society to help women astronomers, some of whom still needed financial support. The committee shrewdly recommended that a financial award would be valued by, and helpful to, a promising young woman if it were presented as a prize for her future promise rather than, as was currently the case, dismissed as a stigma by an older woman passed over for other prizes. Thus the Annie Jump Cannon Prize was reconfigured as a fellowship for a junior woman (under age 35; later changed to within five years of having received a PhD), and its endowment was transferred to the AAUW Educational Foundation, which had long awarded many other fellowships to women.[11]

Since 1974 there has been a long list of winners, the first being Beatrice Tinsley, who used the money for postdoctoral work in cosmology. The winner in 1976 was Catherine Garmany, who was especially grateful, as it led to a postdoctoral fellowship at the Joint Institute for Laboratory Astrophysics at the University of Colorado and later a faculty position with the astronomy department there. As she wrote to the AAUW in 1977, "If future recipients can be helped as much as I have, I believe this Award will be one of the most valuable ones in the field of Astronomy." At some point the stipend was increased substantially, from one thousand dollars to five thousand. All in all, this was a clever and felicitous reformulation of a well-intended remedy for a continuing situation. But the main problem, the underrecognition of women astronomers by the society, remains: since 1972, despite a substantial rise in the percentage of women in astronomy, they have won fewer than 20 percent of the society's other major prizes.[12]

A somewhat similar situation presented itself to the ACS, which had been awarding the Francis P. Garvan Medal to an outstanding American woman chemist since 1937. By the mid-1970s there had been a string of distinguished winners (all white and many Jewish), but the principal of the endowment was shrinking. Also the question was being raised whether a separate prize for a woman was still appropriate, now that "equal opportunity" was the law of the land. It might be time to discontinue the prize or open it to men, or, as the astronomers did, reconfigure it in some way. The topic was discussed inconclusively at several luncheon meetings of the WCC, which finally decided to air the issue more broadly in the committee's newsletter. Two past winners prepared statements pro and con, respectively, which appeared in the March 1976 issue. Marjorie Caserio, the Garvan Medalist for 1975, a former research fellow at Caltech who had moved in 1965 to the new University of California at Irvine, made several key points in favor of keeping the award. First, if they wanted to improve the status of women, as presumably they all did, they should not abolish this award, which was the only one designated for a woman. This prize was needed because in practice it was the only one that women were winning; the other twenty-one prizes awarded by the ACS rarely, if ever, went to women. The Garvan Medal was also open to women in all specialties, while most other ACS awards were restricted to particular specialties. (A later winner claimed that since the Garvan Medalist was chosen from the more than seventeen thousand women members of the ACS, it was a more selective prize than the other ACS awards, for each of which an average of only seven thousand chemists were eligible.) Second, prizes played an important function: they honored not only the recipient but also her institution and those around her. Caserio did not say, but it was evident in the WCC newsletter, that many past winners of the Garvan Medal had gone on to win other, greater awards, as if the prize's publicity had suddenly cast the limelight upon them (which might be termed a *spotlight effect*).[13]

Jean'ne Shreeve, of the University of Idaho, the 1972 Garvan Medalist, countered with the "con" statement. She was proud of her medal but claimed that "women have always had to make sacrifices in order to make progress." If women chemists wanted to be treated equally and considered for all the other ACS prizes, they had to give up this prize. In fact its very existence was a major reason why no woman had ever won any of the other ACS prizes, for it relieved other selection committees from considering them and deterred nominators from even submitting the full documentation that the bureaucratized ACS required. Also, it was entirely proper that women won few other prizes, for they made up only 8 percent of the ACS membership, and largely in those areas and jobs—biochemistry, editing, and teaching—for which there were few ACS awards.[14]

In time the WCC adopted the "pro" view and recommended retaining the prize to the ACS Council, which concurred and authorized a fund drive. In 1979 the W. R. Grace Company agreed to support the award for five years, and the Olin Corporation has done so since 1984, when the award was renamed the Garvan-Olin Prize. Along the way the stipend was increased to five thousand dollars plus expenses. It is still in existence, and women still do not win very many of the other ACS prizes.[15]

By contrast, two other women's prizes floundered and were discontinued. As mentioned in chapter 9, the Federal Woman's Award, created in the early 1960s to honor women in the federal government and long supported by the Woodward & Lothrop department store of Washington, DC, dropped out of sight after 1976. A few years earlier an official of the General Services Administration had refused to nominate anyone for the prize once the Arthur Flemming Award, formerly restricted to outstanding men in public administration, had been opened to women. Evidently no one—not former winners, Woodward & Lothrop, or members of the Civil Service Commission—spoke up to support its continuation, and it was allowed to lapse.[16]

Among other women's prizes of interest to women scientists for many years was the AAUW Achievement Award, presented annually since 1943 to an outstanding intellectual or academic woman. At first it honored the truly legendary women of the time, including many scientists, when they were still rather young, such as Barbara McClintock in 1947, Rachel Carson in 1956 (well before *Silent Spring*), the astronomer Cecilia Payne-Gaposchkin in 1957, and C. S. Wu in 1959, but then in the late 1960s it seemed to become more eclectic and no longer identified the very top women. It included Dixy Lee Ray in 1975, Jessie Bernard in 1976, Margaret Mead in 1978 (the year of her death), and Grace Murray Hopper in 1983, but it was also becoming redundant, for these women had already won many other prizes and honorary degrees by the time the AAUW got around to honoring them. To its credit, starting in 1987 and for its last twenty years about half of the winners, six of thirteen, were women of color (including Johnnetta Cole, the first black woman president of Spelman College, in 1991, and Mae Jemison, the first black woman astronaut, in 2007), a rarity in the world of recognition. In 1991 the award began to be given every other year, and then it was discontinued in 2008 as part of a restructuring of programs at the AAUW Educational Foundation. Once a coveted prize, it lost its unique spot in the firmament.[17]

Yet at the same time other women's groups enthusiastically created several new prizes, as if these doubts or this debate had never occurred. Several of the women's caucuses created prizes in honor or in memory of some of their early leaders or other heroines, as did the Association for Women in Mathematics for Louise Hay, Ruth Satter (sister of Joan Birman), Alice Schafer, and Emmy Noether; the Committee on the Status of Women in Physics for Maria Goeppart-Mayer (funded by General Electric) and Luise Meyer-Schützmeister; and the American Women in Science Educational Foundation for Judith Pool, Julia Apter, Frieda Salzman, and recently Barbara Filner.[18]

Other, newer groups and an individual also created new women's prizes. In 1994 the American Committee for the Weizmann Institute of Science, led by its longtime member the biochemist Maxine Singer, created the Weizmann Women & Science Award, of twenty-five thousand dollars, to be given every other year in any field of science and technology. Its winners have included Joan Steitz, Vera Rubin, Jacqueline Barton, and Carla Schatz; it was discontinued in 2008.[19] In 1997 the Maria Mitchell Association, on Nantucket Island, off the Cape Cod coast, created an annual Women in Science Award to honor men and women who have encouraged the advancement of girls and women in science or technology. This award has gone to those who have established and run successful educational programs for American women and girls, such as Jane Zimmer Daniels, Cinda-Sue Davis, Suzanne Gage Brainard, Susan Staffin Metz, and Denice Denton, among others.[20] In 1998 L'Oréal's Fondation d'Entreprise, representing a multinational hair-color corporation, created the international L'Oréal-UNESCO Awards for Women in Science, which presents five awards per year to accomplished women in science worldwide.[21] Finally, in 2004 a male Nobelist at the Rockefeller University created the Pearl Meister Greengard Prize of fifty thousand dollars annually in memory of his mother (who had died giving birth to him) for an outstanding female in biomedical research. Its first award went to the French embryologist Nicole Le Douarin.[22] These old and new awards seem to fill a varied but continuing, perhaps even growing need. Although their perpetuation requires a certain amount of organization and funding, they host pleasurable meaningful events and garner much favorable publicity for both the donors and the awardees. For a time they generate a kind of altruistic, heartwarming feeling all around, as the woman-of-the-year awards were originally designed to do.

Criticism of the WOTYs

Journalists started woman-of-the-year awards (or WOTYs, as some called them) in the 1940s as a way to counter the more usual man-of-the-year awards. They were usually run by the "society page" editor of big city newspapers as a salute to local women achievers and volunteers who otherwise would have gone unrecognized. Sometimes one woman might be honored, but other times there might be a whole array, such as an athlete, an entertainer, an educator, a political figure, and, if one could be found, a physician or scientist.[23] Sometimes the award presentation ceremony was tied to a fund-raising event for a local charity. One such ceremony in Los Angeles in June 1974 led to a bizarre twist. That year a group of society matrons who each year raised one hundred thousand dollars for Caltech by hosting a lavish black-tie dinner and benefit ball in honor of a prominent "man of science in southern California" gave their award instead to Dixy Lee Ray, the new chairman of the Atomic Energy Commission. The burly Ray graciously donned evening attire and acknowledged the one-thousand-dollar scholarship that was donated in her name to Caltech. But the headline in the *Los Angeles Times* the next morning called her the 1974 "Man of Science," though the title of the award was quickly changed to "Woman of Science" for the evening edition.[24]

Some of the national media took up WOTYs in the 1970s. In January 1976 *Time* magazine included two physicists (C. S. Wu and Vera Kistiakowsky) among its one hundred women of the year, thus proving to skeptics that there were a few scientists among the female high achievers.[25] But the *Ladies' Home Journal* took this genre to even greater heights in the mid-1970s, making it into a major television extravaganza. Starting in 1973 the *Journal* honored eight to ten women annually from a variety of fields. Because the votes cast by readers of the *Journal* were a part of the selection process, name recognition was an important factor and assured that celebrities predominated. Over the years the honorees included such well-known female personalities of the time as Barbara Walters, Billie Jean King, Lady Bird Johnson, Barbara Jordan, Betty Ford, Beverly Sills, Maya Angelou, Rosalynn Carter, and Katharine Hepburn. Among the women scientists and physicians so honored were Virginia Apgar in 1973, Dixy Lee Ray in 1974, Margaret Mead in 1976, Elisabeth Kübler-Ross in 1977, and Rosalyn Yalow in 1978. In 1974 CBS televised the award ceremony live from Lincoln Center, and Clairol sponsored it. One male television commentator wrote that although the show was a step up from a beauty pageant, overall it was still condescending to women, especially because the advertisements trivialized women's problems by implying that all that stood between them and success was "icky hair."[26]

This well-intentioned celebration of women of achievement came under more serious attack, however, in June 1978, when the recent Nobelist Rosalyn Yalow used her moment of fame to criticize publicly such separate prizes for women. That same month (six months after she won the Nobel Prize in Physiology or Medicine), Princeton University was also awarding her an honorary degree. In the commencement procession, some of the Princeton professors commented negatively about the *Journal*'s award. Yalow decided to decline it and did so in a letter that she sent not only to the *Journal* but also to the *New York Times,* which published it on its Op-Ed page under the heading "Thank You, but No Thank You." In the letter, Yalow claimed that separate prizes for women in science were inherently discriminatory. Earlier, in 1961, when she had been awarded one of the first Federal Woman's Awards, she had had some doubts but had accepted it and chosen not to protest. However, now that she was a Nobel laureate and was being honored by Princeton University, she felt differently, and she was glad that the prize had been discontinued. Lenore Hershey, editor of the *Ladies' Home Journal,* responded that such separate awards were still very much needed and refrained from pointing out that perhaps Yalow's earlier acclaim might have brought her some of the notice that led to her more recent awards.[27]

Since then the *Ladies' Home Journal* has continued its women's awards but under slightly different rubrics and configurations. Instead of an annual event for ten women in its November issue, it offered lists of the one hundred "most important" American women then living in all fields in 1988 and 1999, of the ten in just medicine in 1995, and of the twentieth century's one hundred best known or most admired (it was not specified) in the world, living or dead, in 1999. (This article, entitled simply "thelhj100," was also published in enlarged form as a separate issue of the magazine and as a book.) The *Journal* also carried a list of the "most fascinating women" of 1997, which included the astronaut Shannon Lucid; and the winners were featured on another CBS-TV special sponsored by Clairol. All the lists included at least one scientist, and there is no evidence that anyone besides Yalow ever refused to be included. Most were probably flattered by the attention and pleased by the publicity, which was intended as a compliment.[28]

Then the concept of WOTYs moved from a women's magazine to the world of science journalism, which previously had not shown much interest in female scientists. In 2002, after three years of effort, the associate editor of *Discover* unveiled her list, with photographs, of the fifty "most important" women in science. Her criteria were not spelled out, and there were some interesting omissions (e.g., Linda Buck, soon to be a Nobelist), but her selections highlighted a cohort of female rising stars, some still in their thirties, and included MIT professors, HHMI investigators, AIDS researchers, astronomers, oceanographers, and others. Rather than feeling adversely singled out, as Yalow had, some of the winners mentioned the fact of their selection on their websites, as if it were a regular scientific honor. By then science journalists were copying both the WOTYs and the celebrity culture of *People* magazine and the movie fanzines, which had long featured female stars, to create a new cultural entity, the female science celebrity.[29]

Scientific Meetings

Women scientists began to complain in the 1970s about the programs and arrangements of many scientific and mathematical meetings. Even in biology, women were rarely presenting papers, even less often chairing sessions, and almost never giving invited plenary addresses. By the 1980s the situation had become so intolerable that the ecologists Karen Blumer and Jessica Gurevitch collected data on organizers and presenters at two annual meetings (1983 and 1987) of the Ecological Society of America. They found that when women organized sessions, the number of women making presentations correlated well with their membership in the society and their rate of publication in the society's journal. But when men issued the invitations, the number of women speaking was drastically lower, all too often zero. Gurevitch published their results, which the society's immediate past president Margaret Bryan Davis, of the University of Minnesota, then read. She too was aware of the problem, since she often read the programs of upcoming events and had noticed the lack of women presenters. Since some of these events were sponsored by funds from the National Science Foundation, she wrote a forceful letter to its director, Erich Bloch, who had recently spoken to members of the National Academy of Sciences of his desire to help women and minorities in science, enclosing a copy of Gurevitch's article. He apparently delegated the problem to Mary Clutter, the senior biologist on his staff, who gained some notice by thereupon making it a matter of policy to refuse to fund any meeting that did not have women on the program.[30]

Long before then it had became clear that the gender of the person organizing the meeting made a big difference, for the organizer tended to invite those in his or her network of colleagues. The Gordon Conferences, first organized in the 1930s, had long been dominated by men, which was not surprising since they were supported by the chemical, petrochemical, and pharmaceutical industries and focused on topics of interest to them. The conferences were held in the summer at boys' prep schools in New Hampshire, where attendees lived in the dormitories and shared bathroom facilities, so the very few women who did attend were housed separately in the (air-conditioned) infirmary. But over time, as the NIH and the NSF supported Gordon Conferences on more biological topics, a few more women were invited and attended. Some later recalled their terror at giving their first talk in front of the established luminaries in the field, all seated in the front row ready to pounce. Occasionally one could trace a later job offer to a contact initially made at a Gordon Conference.

In the early 1970s the situation began to change. Gordon Conference officials invited the biochemist Maxine Singer to cochair the 1973 conference on nucleic acids, which led to her subsequent lengthy involvement with public policy on recombinant DNA. They also invited Charlotte Friend to organize the 1973 conference on virology, which turned out to be unusual for the time, because 23 percent of the participants were women, who were all housed in one of the student dormitories. Several later commented on how different these meetings were from others they had attended: more of the women spoke up, more chaired sessions, some marveled at the deft and unpretentious style of the organizer, and most felt that they had gotten a lot out of the meeting and looked forward to participating in another one. But in the 1990s female participation, never very high in some other fields, such as analytical chemistry, fluctuated wildly, perhaps because the conferences moved outside the United States to locations in Europe and Asia.[31]

In this changing context an all-female meeting of primatologists held at the University of California–Santa Cruz in 1990 attracted attention. When challenged, both female organizers claimed that their choice of only women participants had not been deliberate; they had listed all those whose work would be of interest and only later realized that none were men. By the time they noticed this, they decided not to add any token men. They felt that women, especially junior ones, would speak up more if only other women were present. According to one account of the meeting, there was much less "male posturing and filibustering" at this event than at the usual ones, and that much more progress was made in just a few days than had been expected. When interviewed by *Science* magazine, several uninvited men were incredulous that a meeting could have succeeded without them.[32]

Nevertheless, because childcare remained nonexistent at scientific meetings (except those of the American Anthropological Association), it was difficult for mothers of young children to participate or even attend. Court rulings upholding liability claims and the skyrocketing cost of such insurance made hotels and organizations reluctant to take any risks. Nonetheless, societies encouraged women to attend, helped create lists of local childcare providers, and took steps to ensure the participation of more women on the program.[33]

Equal Opportunity

Some scientific societies took modest steps to further women's professional and educational opportunities by financially assisting those members facing job difficulties. For example, the American Chemical Society had long had a fund that provided loans to help chemists suing their employers. When, however, their employers were also members of the ACS, this could prove divisive. For example, when Emerson Venable, a member of the ACS board of directors, wanted the society to help Sharon Johnson with her pioneering lawsuit against the University of Pittsburgh in the early 1970s, he was criticized by her colleagues at the university, who were also ACS members. After her successful suit for an injunction, but before the long drawn-out trial over discrimination, the WCC invited her to talk at one of its luncheons, which she did. Later, as her trial was protracted and she could have used some support, she was not aided. Nor was Shyalama Rajender, who had served on the WCC in the early 1970s, preparing one of its first lists of zeroes. But the ACS board's Committee on Professional Relations did help other women fighting employment battles, including Molly Gleiser, formerly of the Lawrence Berkeley Laboratory, and Elise Brown of the NIH.[34]

In the 1970s some scientific societies raised outside funds to support projects in the area of equal opportunity, an innovation at the time. The APS administered a grant for ten thousand dollars from the Alfred P. Sloan Foundation in 1971 to support the work of the new temporary committee on women in physics in preparing its roster and report on women in the field. Then in the early 1970s some of those women referred to in chapter 4 as working in the area of "math mobilization" moved another scientific society into the still-new realm of foundation-sponsored outreach. There the longstanding but poorly funded committee on visiting lecturers of the Mathematical Association of America (MAA) broke new ground in the early 1970s when the iconoclastic Mary Gray wrote a grant proposal that induced IBM (and later others persuaded other foundations) to support a very active Women and Mathematics program for nearly two decades. Although supporting such an active committee was at first an extra burden for the association's already busy staff, in time the project's budgets included an overhead fee to the society. Encouraged by this welcome new revenue stream, the MAA's leaders later sought other grants, while broadening the usefulness of other committees' activities (see chapter 4).

The big event of the period from 1977 to 1982 was, as mentioned in chapter 1, the joining by the AAAS and 150 other scientific and professional societies in the effort to get three more states to ratify the Equal Rights Amendment (ERA). They did this by boycotting convention sites in six key states. The council members of scientific societies were not comfortable voting on such social issues, and their votes were often narrow and divided. In the end their actions had no consequences, for the hotels took in other groups, and the state legislatures continued to vote against the ERA. The members of the scientific societies must have felt relieved and chastened to discover that their actions had such little political impact.[35]

Then in the early 1990s, as mentioned elsewhere, the APS supported (as it still does) perhaps the most innovative intervention by a scientific society, a program of site visits to departments that had asked for advice. This program proved so successful with the first five departments visited that the APS then applied for and got a multiyear grant from the NSF to continue and expand the program, and when that expired, the APS resumed support itself.[36] In fact the experience of working with the energetic feminists, at first a challenge for the longtime staff of many organizations, in time began to change and diversify them. Some staff welcomed the feminists' interest in APS affairs and helped with their efforts to change the society's traditional ways, while others, perhaps annoyed, delegated dealing with them to junior staff members. Over time, as the society had more women members, committee chairs, and other officers, those who worked well with them tended to be promoted or move on to other scientific societies. Bit by bit, as explained in chapter 10, more of these experienced women began to hold the top job of executive officer themselves.

Women Officers

Some of the status-of-women reports of the early 1970s, such as those of the APS and the AAS, had pointed out that women had never or rarely held office in mainstream scientific organizations. This situation began to change in the 1970s, when some were added as tokens, and it entered new territory in later decades as women began to head important committees, including the nominations committee or the committee on committees, whose appointments might change the society from within. One clue to the change under way appeared in an article on women in the American Society of Mammalogists, which noted that sometime in the 1980s and 1990s women began to move from the society's service committees to those offering possibilities of leadership.[37]

Some societies have local or regional "sections" that elect officers every year. Increasingly there were women among them, such as the engineer Vickie Carr, who in 1976 chaired the large New York City section of the immense Institute of Electrical and Electronics Engineers. Similarly, the ACS, for example, has more than one hundred local and regional sections, a few of which women have chaired ever since 1930, when Icie Macy Hoobler led the Detroit section. More women started doing so in the 1970s, perhaps because the male professors who had led sections in the past had less secretarial help than before. In 1986, when all the officers of the St. Louis section of the ACS were women, the fact was noted with a picture in the *Chemical and Engineering News*. A year later it was noted that women officers often brought a fresh vitality to the local groups, for of the six ACS sections that won awards in 1987 for "outstanding performance," an unusually high percentage had women officers—four had female chairs, and two had slates of predominantly women officers.[38]

Most of the largest scientific organizations also have specialty or topical divisions. Some, such as the ACS's Division of Chemical Information, as it is now known, and its Division of the History of Chemistry, are rather feminized, while others, such as the Division of Analytical Chemistry, are much less so. Over the years a few women have chaired divisions of the ACS, including Anna Harrison, an award-winning teacher at Mount Holyoke College, who chaired its Division of Chemical Education in 1971.[39] Starting in the 1970s a few women made inroads into even higher echelons of some of the largest chemical organizations in the country. In 1971 Mary Good, then of Louisiana State University, became the first woman ever to serve on the board of directors of the ACS. The situation changed rapidly after that, for by 1977 there were four women members (of fifteen), three of whom were on the board's executive committee. In 1978 Good became the board's first woman chair, which may have facilitated her rare career move out of academia and into industry (as vice president and director of research at Universal Oil Products) in 1980.[40] Meanwhile, in 1977 Margaret Skillern, of Texas, was the first woman ever to serve on the twelve-member board of directors of the American Institute of Chemical Engineers, which had forty thousand members. Over the years many women have been elected to four-year terms on the eight-member board of directors of the AAAS. They constituted a third (nineteen of fifty-seven) of those serving from 1964 to 2000. But the pattern there and elsewhere was not simply one of steady feminization; it was uneven, and sometimes there were significant downturns. For example, the twelve-member Board of the American Psychological Association, which had no women in 1975 had four in 1988 but fell back to just two in 1991.[41]

As women held more and more of these lesser posts in the 1970s and 1980s, a few became qualified for the top positions. Some nominating committees could identify a cluster of them and shock the members, who were used to tokenism. When in 1994 the nominating committee of the American Sociological Association presented four women for its top two offices, there was some reaction, but that was rebuked by a letter to the association newsletter supported by twenty-one members, including five men. The four women were all highly qualified, having made major contributions to the sociology of gender, which had become an important part of the field in the previous two decades.[42]

Although it is hard to document and thus measure what the women did once in office or what impact they had, one remarkable contributor suggests the possibilities of the most dedicated. After a very busy year chairing the initial committee on women of the APS in 1971–72, an experience that showed her how remote the society's governance was from many of its members, Vera Kistiakowsky was appointed chair of its committee on committees. There she added more women members to serve in the society's myriad posts. Then in the mid-1970s she chaired a committee that was charged with recasting the society's constitution and later its bylaws to allow both more input by members and more sustained discussion of possible society action on the many social and political issues concerning physics or physicists. One result of this was the creation of the Panel on Political Affairs (POPA), which would meet twice a year to consider pertinent issues and to recommend appropriate action to the society's larger council, whose members met only once a year and could not be expected to investigate all the pros and cons of complex issues themselves. These topics ranged from the safety of nuclear reactors to energy issues, the plight of the "disappearing" scientists in Argentina, and, as it turned out, the ERA. In the later 1970s Kistiakowsky herself served an eventful term on the panel, and in 1978 she was one of three physicists nominated for the presidency of the APS. When she lost to Marvin Goldberger, an eminent researcher less involved in society affairs who was the new president of Caltech, she philosophized that although society work was useful and necessary, when voting the members put more value on eminence in research.[43] Her own example suggests that often it is members working on the infrastructure, rather than the figurehead presidents, who are in fact the agents of change, for the modified APS governance allowed the society to get more involved in social issues, including energy, defense spending, and the ERA.

Women Presidents

In almost all scientific societies, electing a woman president was an attainable goal in the 1970s. Most had among their senior members or fellows a few women of substantial accomplishment in research laboratories, at the women's colleges, and increasingly at prestigious universities. Often a women's caucus having been formed and having trained itself in the society's politics and procedures would get her name on the ballot, although she herself might be embarrassed at being selected in this way. One example of this was the nomination and election of Bodil Schmidt-Nielsen as the first woman council member of the American Physiological Society in 1971 and then the society's first woman president in 1975.[44]

A long but inevitably incomplete list of women presidents of scientific organizations from 1971 to 2000 is given in table 12.1. Only three were known to be women of color—physicist C. S. Wu and microbiologist Alice Huang, both Chinese, and African American physicist Shirley Jackson. Over time some societies had more than one woman president. During the thirty-year period 1971–2000 the very large AAAS had eleven, despite a membership that was only 19 percent female in 1992 (the one year for which information on the gender of members is available). Several other societies had three or more. Sigma Xi elected its first in 1984 and then had three more in 1989–99.[45] A few women were president of more than one society, and several were the first women to be head of a society. The heyday for "firsts" was the late 1970s, followed by a few laggards into the 1990s. Frances James was the first woman to head the American Ornithological Union in 1984, and Janet Spence was the first president (of either sex) of the new Association for Psychological Science in 1988. Some such elections broke particularly long traditions, as when the Connecticut Academy of Arts and Sciences, established as early as 1799, admitted its first woman member in 1911 and finally elected its first woman president, the mineralogist (and Mount Holyoke alumna) H. Catherine W. Skinner, in 1985.[46] In 1991 Doris Malkin Curtis, who headed the Society of Economic Paleontologists and Mineralogists in 1978 and the American Geological Institute in 1980, became the first woman to head the Geological Society of America in 1991, after

Table 12.1. Women presidents of selected scientific organizations, 1971–2000

1971	Anne Anastasi	American Psychological Association
	Marcia Guttentag	Society for the Psychological Study of Social Issues
	Mira Komarovsky	American Sociological Association
	Mina Rees*	American Association for the Advancement of Science
1972	Virginia Evans	Tissue Culture Association
	Grace Goldsmith	American Society for Clinical Nutrition
	Leona Tyler	American Psychological Association
1973	Elizabeth Hay	American Society for Developmental Biology
	Mirra Komarovsky	American Sociological Association
1974	Nancie Gonzalez	Society for Applied Anthropology
	Jane Oppenheimer	American Society of Zoologists
1975	Ernestine Friedl	American Anthropological Association
	Charlotte Friend	American Association for Cancer Research
	Frances Graham	Society for Research in Child Development
	Janet Guernsey	American Association of Physics Teachers
	Margaret Mead	American Association for the Advancement of Science
	Bodil Schmidt-Nielsen*	American Physiological Society
	Annemarie Weber	Society of General Physiologists
1976	E. Margaret Burbridge*	American Astronomical Society
	Elizabeth Hay	American Society of Zoologists
	Isabella Karle	American Crystallographic Association
	Barbara Palser	Botanical Society of America
	Elizabeth Russell	Genetics Society of America
	Helen Whiteley	American Society for Microbiology
	C. S. Wu*	American Physical Society
1977	Cynthia Irwin-Williams	American Society for Archaeology
	Evelyn M. Kitagawa	Population Association of America
	Mildred Mathias	American Association for the Advancement of Science—Pacific Division
	Neena Schwartz*	Society for the Study of Reproduction
	Beatrice Sweeney	American Society of Plant Physiologists
1978	Dorothy Bliss	American Society of Zoologists
	Mildred Cohn*	American Society of Biological Chemistry
	Doris Curtis	Society of Economic Paleontologists and Mineralogists
	Margaret Bryan Davis	American Quarternary Association
	Charlotte Friend*	New York Academy of Sciences
	Anna Jane Harrison*	American Chemical Society
	Elizabeth Hay	American Society for Cell Biology
	Eileen B. King	American Society of Cytology
	Elizabeth L. Scott	Institute of Mathematical Statistics
	Jane Setlow	Biophysical Society
	Betty Twarog	Society of General Physiologists
	Rosalyn Yalow*	Endocrine Society
1979	Dorothy Bernstein*	Mathematical Association of America
	Cynthia Deutsch	Society for the Psychological Study of Social Issues
	Elisabeth Gantt	Phycological Society of America
	Jenny Glusker	American Crystallographic Association
	Berta Scharrer	American Association of Anatomists

Table 12.1. Women presidents of selected scientific organizations, 1971–2000 *(continued)*

1980	Doris Curtis	American Geological Institute
	Margaret Dayhoff	Biophysical Society
	Florence Denmark	American Psychological Association
	Beatrice Sweeney*	American Institute of Biological Sciences
1981	M. Margaret Clark	American Anthropological Association
	Elizabeth Hay	American Association of Anatomists
	Patricia Holmgren	Botanical Society of America
	Mary Jane Osborn	American Society of Biological Chemists
	Helen A. Padykula	American Society for Cell Biology
	Alice Rossi	American Sociological Association
	Vivian Walworth	Society of Photographic Scientists and Engineers
1982	Margaret Barr Bigelow	Mycological Society of America
	Marilyn Gist Farquhar	American Society for Cell Biology
	Marian Koshland	American Association of Immunologists
	Mary Lou Pardue	Genetics Society of America
	Julia Robinson*	American Mathematical Society
	Neena Schwartz	Endocrine Society
	J. Mary Taylor*	American Society of Mammalogists
1983	E. Margaret Burbridge	American Association for the Advancement of Science
	Lucille Hurley	American Institute of Nutrition
	Janet Spence	American Psychological Association
	Barbara Webster	Botanical Society of America
1984	Mildred Dresselhaus	American Physical Society
	M. Patricia Faber*	Sigma Xi
	Anna Jane Harrison	American Association for the Advancement of Science
	Marjorie Horning	American Society for Pharmacology and Experimental Therapeutics
	Frances James*	American Ornithological Union
	Mildred Mathias	Botanical Society of America
	Beryl Simpson	Society for the Study of Evolution
1985	Victoria Fromkin	Linguistic Society of America
	E. Patricia Morse	American Society of Zoologists
	Matilda White Riley	American Sociological Association
	Alice Rivlin*	American Economic Associations
	Elizabeth L. Scott	Bernoulli Society
1986	Jean Bennett	Optical Society of America
	Jean Brenchley	American Society for Microbiology
	Nancy Deloy Fitzroy	American Society of Mechanical Engineers
	Mary Ellen Jones	American Society for Biochemistry and Molecular Biology
	Mary Lou Pardue	American Society for Cell Biology
	Bonnie Strickland	American Psychological Association
	Anne Marie Vidaver	American Phytopathological Society
1987	Martha Christensen	Mycological Society of America
	Margaret Bryan Davis	Ecological Society of America
	Mary Good	American Chemical Society
	Mary Jo Nye	History of Science Society

(continued)

Table 12.1. Women presidents of selected scientific organizations, 1971–2000 *(continued)*

	Shirley Tucker	Botanical Society of America
	Sheila Widnall	American Association for the Advancement of Science
1988	Helen M. Berman	American Crystallographic Association
	Shirley Frye	National Council of Teachers of Mathematics
	Elisabeth Gantt	American Society of Plant Physiologists
	Alice Huang	American Society for Microbiology
	Janet Spence*	Association for Psychological Science
	Jeanette Thorbecke	American Association of Immunologists
1989	Lida Barrett	Mathematical Association of America
	Elizabeth Gantt	American Society of Plant Physiologists
	Patricia Goldman-Rakic	Society for Neuroscience
	Penelope M. Hanshaw*	Geological Society of Washington
	Eva King Klein	American Society for Pharmacology and Experimental Therapeutics
	E. Patricia Morse	Sigma Xi
	Harriet Presser	Population Association of America
	Judith Shklar*	American Political Science Association
1990	Melinda F. Denton	American Society of Plant Taxonomists
	Judy Franz	American Association of Physics Teachers
	Anita Payne	Society for the Study of Reproduction
1991	Patricia Calarco	Electron Microscopy Society of America
	Rita Colwell	Sigma Xi
	Doris Curtis*	Geological Society of America
	Lynn Riddiford	American Society of Zoologists
	Beryl Simpson	Botanical Society of America
	Annette Weiner	American Anthropological Association
1992	Brina Kessel	American Ornithological Union
	Margaret G. Kidwell	American Genetic Association
	Sally Gregory Kohlstedt	History of Science Society
	Jane Lubchenko	Ecological Society of America
	Gale Michener	Animal Behavior Society
	Alexandra Navrotsky	Mineralogical Society of America
	Elizabeth F. Neufeld	American Society for Biochemistry and Molecular Biology
	Maria New	The Endocrine Society
	Janet D. Rowley	American Society of Human Genetics
	Sidney Wolff	American Astronomical Society
1993	Eloise E. Clark	American Association for the Advancement of Science
	Susan Gerbi	American Society for Cell Biology
	Barbara Grosz	American Association for Artificial Intelligence
	Beryl Simpson	American Society of Plant Taxonomists
	Martha Sloan	Institute of Electrical and Electronics Engineers
1994		
1995	Rita Colwell	American Association for the Advancement of Science
	Ursula Goodenough	American Society for Cell Biology
	Cathleen Morawetz	American Mathematical Society
	Yolanda Moses	American Anthropological Association
	Barbara Schaal	Botanical Society of America
	Susan S. Taylor	American Society for Biochemistry and Molecular Biology

Table 12.1. Women presidents of selected scientific organizations, 1971–2000 *(continued)*

1996	Lynne Billard	American Statistical Association
	Andrea Dupree	American Astronomical Society
	Anne Krueger	American Economic Association
	Jane Lubchenco	American Association for the Advancement of Science
	Elinor Ostrom	American Political Science Association
1997	Mina Bissell	American Society for Cell Biology
	Nancy Dengler	Botanical Society of America
	Mildred Dresselhaus	American Association for the Advancement of Science
	Sue Piper Duckles	American Society for Pharmacology and Experimental Therapeutics
1998	Carol Baskin	Botanical Society of America
	Elizabeth Blackburn	American Society for Cell Biology
	Doris Carver	IEEE Computer Society
	M. R. C. Greenwood	American Association for the Advancement of Science
	Judith Klinman	American Society for Biochemistry and Molecular Biology
1999	Peggie Hollingsworth	Sigma Xi
	Diana Wall	Ecological Society of America
2000	Patricia Gensel	Botanical Society of America
	Mary Good	American Association for the Advancement of Science
	Mary Hendrix	Federation of American Societies for Experimental Biology
	Marcia McNutt	American Geophysical Union
	Anneila Sargent	American Astronomical Society

Sources: AMWS and web pages.
*First woman president of the organization

102 men. But even for her there were still certain limits: although she was a petroleum geologist, she never did head the American Association of Petroleum Geologists, a male bastion.[47]

Some of these honored women were not aware of or appreciative of the political actions that had gone into their nomination and election to the presidency. Rosalyn Yalow, trained as a physicist, who did important work on radioimmunoassay, was the first woman president of the Endocrine Society in 1978–79, a year after winning the Nobel Prize. She evidently thought she had been elected to the high post solely on her own, for in the opening remarks of her presidential address she said that her selection showed that it was time for the society's women's caucus to cease to exist. This was taken as an ungracious slap by those women in the society—which had had only male presidents since its founding in 1916—who had worked long and hard to have her elected.[48]

By contrast, the much-honored geologist Doris Curtis, mentioned above, was more appreciative of and reflective on what her professional societies had done for her. In 1978, as the first woman president of the Society of Economic Paleontologists and Mineralogists, she took advantage of her position to write a guest column entitled "Rewards of Professional Participation" for the *Bulletin of the Houston Geological Society.* Looking back on her long career, largely with Shell Oil, and on her many years of activity in several professional organizations, she wrote of the satisfactions of keeping up with the field, of understanding the inner workings of scientific societies and the profession, and especially, despite the early snubs, of knowing and being recognized by others in her line of work, not just those at Shell. This was such an unusual statement of a widespread but rarely articulated sentiment that it was often reprinted in other geological journals. Another sign of her unusual devotion to geological societies came years later, in 1991, when she died before completing her term as the first woman president of the immense Geological Society of America (GSA). As she requested, her ashes were buried beneath a tree planted in her memory at the society's headquarters in Colorado. Another marker was the creation in her memory of the Subaru Outstanding Woman in Science Award, a postdoctoral scholarship first given by the GSA in 2001.[49]

Since most presidents held office for just one year, it is hard to know what such women accomplished or what difference their being in that position meant to them or other women in the field. But a few examples have come to light. In 1975 the members of the APS elected C. S. Wu, of Columbia University, as their first woman president. She had come to the United States as a student in the 1930s, and in 1956 she had designed and run the historic experiment that showed that parity was not conserved, an achievement that won others the Nobel Prize. During her one-year term she managed to make some notable contributions in the area of scientific advice to the president of the United States, a topic of particular concern to physicists, who have long been advisers on missiles and other aspects of national defense. A few years earlier, President Richard Nixon had in frustration abolished the Office of Science and Technology, but his successor, Gerald Ford, was amenable to restoring it. At the same time, a new group, the Committee on Scientific Society Presidents, was formed to give the major scientific associations more clout with regard to national policy. (Previously the presidents of the chemical, physical, and other scientific societies had testified individually before government bodies, but it was felt that one joint committee might have more impact.) Wu was one of the few women in this group by the mid-1970s, and she proved particularly effective. At the end of her term (and his, as it turned out) Gerald Ford sent her a note of thanks for her help in getting Congress to pass the National Science Policy and Organization Act, which restored science policy to the Executive Office of the President.[50]

Similarly, Anna Harrison was no mere figurehead in her year as president of the ACS. Frustrated on the National Science Board in the early 1970s, when science education remained a very low priority, she made her first run for the ACS presidency in 1974 but lost to the Nobelist Glenn Seaborg. The next year she ran again but lost very narrowly to Henry Hill, an African American industrial chemist. Encouraged by her supporters, who wrote hundreds of letters of support, and colleagues at Mount Holyoke, who contacted its many alumnae who had become chemists, she ran a third time and won in 1976, the culmination of more than forty active years in the ACS. She served in 1978, a year during which significant issues came to the fore, and she represented a quiet and somewhat feminized constituency within the ACS—those who taught chemistry, especially at colleges. While president, Harrison traveled a lot, especially to college campuses, and spoke often of the need to interest nonmajors in chemistry. Something of a bridge builder, she also got the ACS, most of whose members were industrially oriented, to work with the Environmental Protection Agency to set standards for dealing with toxic substances. Although herself ambivalent on many women's issues, Harrison's very presence broadened the acceptance of women and women's issues within the powerful ACS. While president she received letters of complaint from women members about the sexist cartoons that still appeared in the association's official publications. She challenged the editors, who expressed bewilderment but eventually changed their ways. Perhaps her presence on the ACS governing boards was a factor in persuading the ACS's substantial political-affairs staff to support Senator Edward Kennedy's Women in Science Bills of 1978 and 1979 (though the staff was even more impressed that the Berkeley Nobelist Melvin Calvin had signed a petition favoring such new programs). Yet when the issue of the ERA boycott came up at the ACS in late 1978, Harrison, who might have played a forceful role in urging the ACS to move its 1979 meeting place away from Florida, did nothing, and not only did the society meet in Miami in 1979 but its committee on meetings and exhibitions voted that political considerations should not play a role in site selection.[51]

Past presidents retained an elevated status and sometimes were called upon to play a significant role in society affairs. Helen Whitely served as president of the large American Society for Microbiology (ASM) in 1977, but it was her subsequent decade as chairman of the society's Publications Board from 1980 to 1990, while fighting a prolonged illness, that her obituary discussed at length. She initiated two new journals, expanded others, started a book series that included textbooks, and prodded the ASM into electronic publishing. A statement in the *Journal of Bacteriology* in January 1991 read: "Dr. Whiteley's guiding hand was felt in all aspects of the ASM publications program, and her presence will be sorely missed."[52]

Women's Organizations

As the mainstream scientific societies made more space for women and included them in their activities, awards, and leadership positions, there might have seemed to be less need for the older women's organizations, such as Iota Sigma Pi, Sigma Delta Epsilon, and the Society of Women Geographers, which had filled an important niche before 1972. There is some indication that indeed there was less activity among those whose organization and structure dated from an earlier era of honor societies or scientific fraternities focused on local sociability rather than on the status of women in the field. By the 1970s they had older members and a decentralized structure with chapters in campus towns or major metropolitan areas. Organizationally they tended to rely on volunteer help, not to have a headquarters in Washington, DC, not to engage with the major disciplinary societies, and not to take on grant-supported outreach projects. Some struggled to keep their newsletter going and to put out an occasional membership roster and hoped that someone would rise to the task of updating the organization's history. To a certain extent they had been supplanted by the newer groups of younger women with a paid staff at national headquarters. But they did have endowments and numerous awards, and Sigma Delta Epsilon, as an affiliate of the AAAS, had a vote on its council. What will happen to these organizations is unclear. They may dissolve, merge with other groups, or become revitalized by young new members, as has been the case with the Society of Women Geographers, whose member Kathryn Sullivan, an astronaut, took the society's coveted flag with her into space, and whose incoming president in 2005 was Kimberly Crews, an energetic African American demographer at the U.S. Census.[53]

Joining the Club: The National Academies

In the early 1970s women scientists and engineers also began to play a larger role in the giving and receiving of the many honors conferred by the national academies and other prestigious organizations, including the Nobelstiftung in Stockholm. Starting from almost zero in the early 1970s, the numbers of women elected to the three national academies, considered by many to be the highest honor an American scientist or engineer or medical person can hope to achieve, began to rise ever so slowly. The most women and the highest percentage were in the Institute of Medicine (IOM), and the least women and the lowest percentage were in the National Academy of Engineering (NAE), both fairly new organizations. The IOM was established in 1970 with forty charter members and a goal of about four hundred by 1980. Included were a wide range of fields, the social sciences and nursing education, as well as basic biomedical science and medical administration. It also took note of the African American medical, dental, nursing, and social-work communities. Accordingly, about a quarter of the early members were women, and several were black. Rather than stopping at four hundred members, however, officials modified their charter to allow for further expansion. As the number of persons elected rose to sixty-five per year by 2002, the number of women rose, but their proportion of the whole did not. In fact it dropped below 20 percent in the 1980s, as fewer women were elected just when one might have thought that even more were qualified for selection. Also, because the IOM's domain partially overlapped with that of the National Academy of Sciences (NAS), some of the women bioscientists elected to the IOM (Jewel Plummer Cobb, Susan Taylor, and Nancy Hopkins) really belonged in the National Academy of Sciences, a larger, older, and more prestigious organization that was bursting at the seams.[54]

Meanwhile, the NAE, established in 1964, had one woman member (Lillian Gilbreth, the famed industrial psychologist and mother of twelve) in the 1960s. After her death the academy elected the computer pioneer Grace Murray Hopper in 1973 and the MIT electrical engineer Mildred Dresselhaus in 1974, followed by a trickle of others. In the 1990s it developed ways of rewarding those sections that elected women with extra slots for even more men. (This may be why women in neighboring fields such as physics and mathematics were elected.) In 1997 the number of living women members reached forty-two, and by 2008 the NAE's website could list one hundred women ever elected.[55]

By then, however, the NAE's goal was broader than just electing more women members. In 1997 it received a grant of two hundred thousand dollars from AT&T to create a website and hold a series of "life-transforming" conferences for young women interested in careers in engineering. In April 2003, with funding from several public and private agencies, the NAE hosted an Engineering Societies Diversity Summit with representatives from twenty-eight societies to increase the impact of their organizations' efforts to attract more women and minorities to careers in engineering. Thus the movement to bring more women into engineering, which had started at the engineering schools in the early 1970s, had by the first decade of the twentieth century reached the pinnacle of the engineering establishment. In 2004 the NAE hired Catherine Didion, the longtime executive director of the Association for Women in Science, and in 2007, when the three academies merged their women's committees into one—the Committee on Women in Science, Engineering, and Medicine (CWSEM)—she became its program officer.[56]

By far most of the women elected to the NAS between 1970 and 2000 were in the biological sciences (see table 12.2), a phenomenon that led to even more being elected, since only members can nominate and elect other members. The virologist Charlotte Friend is a good example. Upon her election in 1976 she received many congratulatory notes, including one from the medical physicist Rosalyn Yalow, elected the year before and a fellow graduate of Hunter College, who wrote, "I am sure that you, like I, will enjoy being a member of 'The Club.'"[57] A year later

Table 12.2. American women elected to the National Academy of Sciences, by field, 1970–2001

Year	Biological Sciences (84)	Physical Sciences (26)	Social Sciences (25)	Math/Statistics (10)
1970	Rebecca Lancefield (microbiology)			
	Ruth Patrick (ecology)			
1971	Mildred Cohn (biochemistry)		Eleanor Jack Gibson (psychology)	
1972	Elizabeth S. Russell (genetics)	Gertrude Goldhaber (physics)		
1973	Beatrice Mintz (genetics)			
	Helen M. Ranney (internal medicine)			
	Helen B. Taussig (cardiology)			
1974	Estella Leopold (botany)			
	Sarah Ratner (biochemistry)			
1975	Dorothy Horstmann (virology)	Rosalind Yalow (medical physics)	Frederica deLaguna (anthropology)	
			Dorothea Jameson (psychology)	
			Margaret Mead (anthropology)	
1976	Charlotte Friend (microbiology)			Julia Robinson (mathematics)
1977	Elizabeth Neufeld (medical sciences)		Elizabeth Colson (anthropology)	
	Ruth Sager (genetics)			
	Evelyn Witkin (genetics)			
1978	Elizabeth Miller (oncology)	E. Margaret Burbridge (astrophysics)	Mary Haas (linguistics)	
	Mary Jane Osborn (microbiology)	Isabella Karle (crystallography)		
1979	Salome Gluecksohn-Waelsch (genetics)			
	Gertrude Henle (virology)			
	Maxine Singer (biochemistry)			
1980	Eloise R. Giblett (hematology)			
1981	Marian E. Koshland (biochemistry)	Vera C. Rubin (astronomy)		
	Thressa Stadtman (bioenergetics)			
1982	Margaret B. Davis (ecology)			

1983	Lynn Margulis (biology) Mary Lou Pardue (cell biology) Joan Steitz (biophysics) Jean Wilson (internal medicine)			
1984	Marilyn Farquhar (cell biology) Elizabeth Hay (anatomy) Mary Ellen Jones (biochemistry) Janet Rowley (oncology)			
1985	Mary Dell Chilton (biotechnology) Martha Vaughn (metabolism)	Mildred Dresselhaus (solid state physics) Sandra Faber (astronomy)		
1986	Liane B. Russell (genetics)	Susan W. Kieffer (geophysics)		
1987			Jane E. Buikstra (anthropology)	Karen Uhlenbeck (mathematics)
1988	Ann Graybiel (neuroanatomy) May West-Eberhard (entomology)		Frances K. Graham (psychology) Patty Jo Watson (anthropology)	
1989	Phillippa Marrack (immunology) Ora M. Rosen (oncology)		Jane Menken (demography) Barbara Partee (linguistics)	
1990	Gertrude Elion (biochemistry) Nina Fedoroff (molecular genetics) Patricia Goldman-Rakic (neurobiology)	Esther Conwell (solid state physics)	Sarah Hrdy (anthropology)	Cathleen Morawetz (applied mathematics)
1991	Mary Edmonds (biochemistry) Susan Leeman (physiology) Jane S. Richardson (biochemistry)	Mary K. Gaillard (physics)	Victoria Bricker (anthropology)	
1992	Carol Gross (bacteriology)	Margaret Geller (geophysics) Susan Solomon (astronomy) JoAnne Stubbe (chemistry)	Olga Linares (anthropology)	
1993	Christine Guthrie (molecular biology) Nancy Kleckner (genetics) Sharon Long (developmental biology)	Alexandra Navrotsky (geochemistry)	Eleanor Maccoby (psychology)	Marina Ratner (mathematics)

(continued)

Table 12.2. American women elected to the National Academy of Sciences, by field, 1970–2001 *(continued)*

Year	Biological Sciences (84)	Physical Sciences (26)	Social Sciences (25)	Math/Statistics (10)
1994	Mary Avery (pediatrics) May Berenbaum (entomology) Lucille Shapiro (developmental biology) Ellen S. Vitetta (microbiology)	Marye A. Fox (chemistry) Judith Klinman (chemistry) Pamela Matson (soil science)	Matilda Riley (sociology)	
1995	Clara Franzini-Armstrong (cell biology) Judith Kimble (biochemistry) Carla Schatz (neurobiology)	Mary Lou Zoback (geophysics)	Anne Krueger (economics)	
1996	Carolyn Cohen (biophysics) Elaine Fuchs (molecular genetics) Elisabeth Gantt (botany) Margaret Kidwell (genetics) Jane Lubchenco (ecology) Maria New (pediatric endocrinology) Susan Taylor (biochemistry)	Johanna Levelt-Sengers (thermophysics)	Cynthia Beall (anthropology) Victoria Fromkin (linguistics)	Nancy Kopell (biomathematics)
1997	Susan Lindquist (molecular genetics) Lois K. Miller (entomology) Linda Randall (biochemistry)	Tanya Atwater (geophysics) Neta Bahcall (astrophysics) Sylvia Ceyer (chemistry)	Joyce Marcus (anthropology)	

1998	Elizabeth Craig (biochemistry)		Norma Graham (psychology)	Ingrid Daubechies (mathematics)
	Susan Gottesman (genetics)			
	Eva Neer (medicine)			
	Joan Ruderman (biology)			
	Audrey Stevens (biology)			
	Susan Wessler (genetics)			
1999	Marlene Belfort (genetics)	Margaret Kivelson (geophysics)	Elizabeth Spelke (psychology)	Dusa McDuff (mathematics)
	Joanne Chory (biology)	Cherry Murray (physics)		
	Patricia Donohoe (surgery)			
	Janice Miller (microbiology)			
	Barbara Schaal (genetics)			
2000	Rita Colwell (microbiology)	Martha Haynes (astronomy)	Lila Gleitman (linguistics)	Marsha Berger (computer science)
	Joan Massague (medical sciences)		Susan Hanson (geography)	Grace Wahba (statistics)
	Barbara Meyer (genetics)			
2001	Pamela Bjorkman (biology)	Inez Fung (atmospheric sciences)	Elinor Ostrom (political science)	
	Joan Brugge (biology)			
	Mimi Koehl (biology)			
	Lynn Landmesser (neuroscience)			
	Patricia Zambryski (microbiology)			

Sources: *CHE, NYT.*
Note: The table includes American citizens only. "Foreign associates," such as the Canadian citizen Shirley Tilghman, of Princeton University, who was elected in 1996, are not included.

the chairman of the women's committee of the American Society for Microbiology asked Friend if she would nominate other women in the field that the committee might recommend. Friend (and the also recently elected Dorothy Horstmann, of Yale Medical School) agreed, and their efforts were rewarded by the election in 1979 of two women previously passed over, Salome Gluecksohn Waelsch, aged 72, and Gertrude (Brigitte) Henle, whose husband and lifetime collaborator had been elected in 1975 without her, an omission that greatly distressed them both.[58] Another chain starter was the astronomer E. Margaret Burbridge, the first woman astronomer elected, in 1978, who not long after that nominated Vera Rubin, who as soon as she was elected in 1981 worked on nominating other women astronomers, including Sandra Faber and Margaret Geller.[59]

But even their energies waned over time, and many others did not spend time nominating or pushing for others unless asked by special friends (as Maxine Singer did for Marianne Grunberg-Manago in 1979 and for Jacqueline Barton in 1998).[60] Thus, only ten women mathematicians and statisticians were elected between 1970 and 2000, and among the chemists (and biochemists) Gertrude Elion was elected in 1990, at age 72, two years *after* winning the Nobel Prize in Physiology or Medicine. By contrast, the atmospheric chemist Susan Solomon, famed for her work on the ozone hole in the 1980s, was only 36 years old when she was elected in 1992. Aside from electing other members, Charlotte Friend found that the major advantage of being an academician was that she was entitled to submit up to six manuscripts per year to the *Proceedings of the National Academy of Sciences*, a real boon that she shared with her many international colleagues, who did not have access to it directly.

As the number of women in the NAS inched upward in the 1970s, there were certain related developments. In 1984 the academy's council finally voted to change all the male pronouns in its constitution to gender-neutral ones.[61] In 1990 Vera Rubin and Maxine Singer invited all the other women members to write to the NAS president, Frank Press, urging him to take some special initiative to increase the number of women members and participants on National Research Council boards and committees.[62] Then in 1992–93 some of the younger newly elected women, such as the geophysicist Susan Kieffer and the mathematician Karen Uhlenbeck, disillusioned with what they had seen of the academy—its selection procedures and its programs that bypassed topics of interest to women—discussed with the physicists Cécile DeWitt-Morette and Vera Kistiakowsky a possible national academy of women scientists or consultant bureau for women in science, whose members, which might include men and non-academicians, would take up topics of interest to women scientists. Unfortunately, nothing came of these provocative ideas.[63]

Nobel Laureates

Out of a total of nearly two hundred Nobel Prizes awarded worldwide between 1970 and 2000, four went to American women (including one Italian American with dual citizenship): Rosalyn Yalow in 1977, Barbara McClintock in 1983, Rita Levi-Montalcini in 1986, and Gertrude Elion in 1988.[64]

Trained as a physicist Rosalyn Yalow, a Hunter College graduate aged 56, was initially popular with the New York press, who considered her a hometown girl made good. The *New York Times Magazine* accordingly dubbed her "The Madame Curie of the Bronx." She got half of the prize for her work on radioimunnoassay, the detection of minute amounts of radioactive hormones in biological processes, in the 1950s and 1960s with a since-deceased colleague at the Veterans Administration Hospital in the Bronx. An outspoken New Yorker who held strong views on a number of topics and was willing to be interviewed and quoted, she at first seemed to be something of a feminist. Her remarks at a formal dinner for laureates in Stockholm were described by the *New York Times* as "a ringing appeal for sexual equality and for greater opportunity and effort by women." In a subsequent interview on Swedish television she said that there had never been any women on the Nobel selection committee for medicine, something no one else had ever commented upon. But before long, as in the WOTY episode above, she demurred from some aspects of the women's movement. Thus she might be considered a conservative feminist; while she acknowledged that women had it harder than men, she was not in favor of special treatment or programs for women. They should do, as she had done—work harder and strategize better than men.[65]

The geneticist Barbara McClintock was 81 when Stockholm finally awarded her the whole 1983 prize—she was only the third woman ever to win the prize on her own, after Marie Curie in 1911 and Dorothy Hodgkin in 1964, both in chemistry—for work she had done in the 1940s on "jumping genes," or the movement of genes within chromosomes. Somewhat frail, she managed to travel to Stockholm to deliver her address and receive her prize. Like Yalow a New Yorker, McClintock, who lived nearby at Cold Spring Harbor on Long Island, was called by the press a "brilliant loner" and left alone. But the inadvertent publication, also in the fall of 1983, of Evelyn Fox Keller's biography *A Feeling for the Organism,* which was based on a series of interviews with McClintock in 1978, heightened interest in her early career difficulties and suggested to some the elusive concept that maybe she (and other women) did science differently than men. This reportedly infuriated her, but she had little to say publicly before her death in 1992.[66]

In 1986 Rita Levi-Montalcini, of Rome, Italy, and formerly of Washington University in St. Louis, shared the Nobel Prize in Physiology or Medicine with Stanley

Cohen, a longtime collaborator, for their work in the 1950s on nerve growth factor. In the press she was chiefly known for helping young Italian scientists find placements in the United States and for dressing in the latest Italian fashions. In the late 1980s she published a rather elliptical autobiography, *In Praise of Imperfection,* about her twin sister, an artist, and herself.[67]

By contrast, the forthright Gertrude Elion used her fame as a Nobelist to remind everyone on numerous occasions how difficult it had been for her to get a job in the chemical industry in the 1930s. A third New Yorker, with just a bachelor's degree from Hunter College and a master's degree from New York University, she had spent four decades developing at Burroughs-Wellcome, first in the New York area and then at Research Triangle, North Carolina, a long list of safe, effective, and highly profitable new drugs for a variety of diseases (see chapter 8). This had been possible only because George Hitchings, her longtime collaborator, had hired her in 1944, when finally wartime pharmaceutical laboratories began to take on women chemists. At her memorial service in 1999 two colleagues summed up her approach to experimental analysis as follows: don't tell me your conclusions, let me see the data and I'll decide for myself.[68]

Near Nobels

There were also several near misses, women who might have or even should have won the Nobel Prize. When Milton Friedman won the Nobel Prize in Economic Science in 1976, Anna Jacobson Schwartz, with whom he had collaborated for more than thirty years, did not share the prize. No one commented on this publicly, though a conference in her honor was held in New York City a decade later. By contrast, when in 1985 the crystallographer Jerome Karle shared the Nobel Prize in Physics with two other men but not his wife, Isabella Karle, who had spent her entire career—more than forty years—working with him, many people were shocked. Their daughter, Louise Hansen, a scientist at Brookhaven National Laboratory, complained in the *New York Times* of her mother's omission, and later Karle was awarded the National Medal of Science. In 1997 the atmospheric chemist Susan Solomon narrowly missed sharing a Nobel with some colleagues, but she professed in an oral history not to be upset, as she had been consulted beforehand and the decision was a correct one. (In 1999, perhaps in gratitude for her cheerful cooperation and self-abnegation, she too was awarded the National Medal of Science, one of the youngest recipients ever.)[69]

Another woman scientist who was given what might be called the "Nobel shove" was Louise Chow, whose web page states that "the 1993 Nobel Prize in Physiology or Medicine was awarded to her collaborator" for work done in 1977 at Cold Spring Harbor Laboratory. In fact, she had had a key role in interpreting certain key electron micrographs that showed that genes of higher organisms contain long segments of "nonsense" DNA called "introns." There was room to include Chow if the Nobel committee had wished, since the prize was shared by just two others. But while her collaborator's backer (James D. Watson) had urged him to prepare a long memo outlining his contributions, no one had asked Chow, born in Taiwan and a 1973 PhD from Caltech, to do the same, though her American husband, who had also taken part in the work, sharply criticized the collaborator's version. Later, after the (mis)award, her Caltech adviser admitted to the press, "She's a woman, an Asian woman who's a little quiet. . . . Sometimes they get ignored."[70]

Solomon's ready acceptance of a major professional slight and her later significant reward casts light on the further adventures of the bioscientist Candace Pert, who, as described in chapter 6, had been omitted from the group awarded the Lasker Prize in 1978 for the discovery of the opiate receptor, though she had played an integral role in its discovery and had even been the lead author of one of the key papers. Although she had protested publicly after being omitted from the award, and various rationalizations were given, nothing had changed. In fact, as she has described in her autobiography, circumstances grew worse, as she was labeled "controversial" and she refused to help Solomon Snyder win the Nobel Prize. The resulting ostracism was one factor that induced Pert to leave the NIH in 1988 and embark on a short-lived (as it turned out) career in the biotechnology industry. Years later, in her autobiography, she wondered what might have happened if she, like Susan Solomon, had cooperated with the "old boys" and helped them get "her" Nobel Prize for Solomon Snyder.[71]

National Medals of Science and Technology

Although the National Medal of Science (not to be confused with the National Medal of Freedom) was first awarded in 1962, there were no women winners until Barbara McClintock in 1970. Then, as shown in table 12.3, there were several during the 1970s and 1980s, although in some years no medals were awarded. The best years for women were 1983–2002, perhaps a kind of golden age, during the presidential administrations of George H. W. Bush and Bill Clinton, for there was at least one (but never more than two) in each of the years the medals were given. In the years 2002–7, during the presidency of George W. Bush, there were only four women medalists. (The three awarded in 2009 "for the year 2008" were given by President Barack Obama.) Among the thirty-six women winners to 2008 were four

Table 12.3. Women winners of the National Medal of Science, 1970–2008

Year	Total Number Awarded	Number Awarded to Women	Women Winners
1962	1	0	
1963	5	0	
1964	11	0	
1965	11	0	
1966	11	0	
1967	12	0	
1968	12	0	
1969	6	0	
1970	9	1	Barbara McClintock*
1971	0	—	
1972	0	—	
1973	11	0	
1974	13	0	
1975	16	1	C. S. Wu
1976	15	0	
1977	0	—	
1978	0	—	
1979	20	1	Elizabeth Crosby
1980	0	—	
1981	1	0	
1982	12	1	Mildred Cohn
1983	19	2	E. Margaret Burbridge, Berta Scharrer
1984	0	—	
1985	0	—	
1986	20	1	Joan Steitz
1987	20	2	Anna Anastasi, Rita Levi-Montalcini*
1988	20	1	Rosalyn Yalow*
1989	19	1	Katherine Esau
1990	20	1	Mildred Dresselhaus
1991	20	2	Mary Ellen Avery, Gertrude Elion*
1992	8	2	Eleanor Jack Gibson, Maxine Singer
1993	8	2	Vera Rubin, Salome Waelsch
1994	8	1	Elizabeth Neufeld
1995	8	1	Isabella Karle
1996	8	1	Ruth Patrick
1997	9	1	Darleane Hoffman
1998	9	2	Cathleen Morawetz, Janet Rowley
1999	12	2	Lynn Margulis, Susan Solomon
2000	12	2	Nancy Andreasen, Karen Uhlenbeck
2001	15	1	Ann Graybiel
2002	8	1	Evelyn Witkin
2003	8	0	
2004	8	0	
2005	8	0	
2006	8	2	Rita Colwell, Nina Fedoroff
2007	8	1	Fay Ajzenberg-Selove
2008	9	3	Joanna Fowler, Elaine Fuchs, JoAnne Stubbe
Total	474	36	

Sources: The National Medal of Science, 2002 (Washington, DC: NSF, 2001); www.nationalmedals.org.
*Nobel laureate

Nobelists plus the near Nobels Wu, Karle, and Solomon. All had previously won many other awards.[72]

Here too C. S. Wu had an interesting experience. In 1976, as the incoming president of the APS, she received a letter addressed to a "Mr. C. S. Wu," president of the APS, asking for nominations for the National Medal of Science. Enclosed was a list of previous winners, and she would have noted that to date McClintock was the only woman. Despite this awkward start, she was one of the awardees "for 1975," although the award ceremony and dinner with Gerald and Betty Ford presiding was not held until October 1976. Three years later Wu was invited to the January 1980 dinner for the 1979 winners, which included the next woman medalist, the anatomist Elizabeth Crosby. Wu was also on the selection committee for the 1983 winners, which included two women, the astronomer E. Margaret Burbridge and the anatomist Berta Scharrer.[73]

The National Medal of Technology was established in 1980 to honor great engineers and industrialists. Of its many winners since 1985, the only women have been the physicist Helen Edwards in 1989, the computer scientist Grace Murray Hopper in 1991, the DuPont chemist Stephanie Kwolek in 1996, and the engineer Esther Takeuchi in 2008. When interviewed in 2001, Kwolek commented that perhaps she was being honored so much because she was a woman. After all, she had discovered the acrylic fiber Kevlar, which is used in bulletproof vests, in the 1960s. She felt that her receiving the award some three decades later must have meant that there had not been many other qualified women since then.[74]

Conclusion

The nation's many scientific and technical associations were transformed after 1972 from generally quiet establishment-oriented, male-dominated groups of like-minded individuals to larger, more diverse, and more active and open forums for action on scientific and technical issues at many levels. They still award prizes and publish journals, but buffeted by social, economic, and political forces, they have taken on additional activity on the political issues of science, including research funding, science education, and education and careers for girls and women. Via their websites many are trying to reach an audience beyond their members. The active participation of women, initially unwelcome in some quarters and at times strongly resisted, has had a lot to do with this broadening and resurgence. Many forceful women members expected more from their scientific societies and worked hard to open them up and to make them work for the betterment of all the members and even nonmembers. The societies, which were generally open to all qualified scientists, offered women opportunities to band together, to focus on certain targets (e.g., nominations and elections), and to achieve significant goals (e.g., supporting the ERA or electing the first woman president). Over three decades their increasing presence and demands have gotten some of these societies, including even the national academies, to commit themselves to somewhat greater equity, diversity, and opportunity, and thus to reflect feminist values more broadly.

A New Era of Institutional Contrition and "Transformation"

As the century was ending and the aforementioned events were unfolding, two bombshells—the first at MIT in March 1999 and the second at Harvard in January 2005—exploded and made the whole subject of female scientists, especially the continuing difficulties of senior women at major research universities, front-page news across the nation. Before long these two incidents had catapulted the whole topic into what might be termed a new era of institutional contrition and voluntary change. There would be more female presidents of universities, more women members of the National Academy of Sciences, and the biggest, most ambitious new program yet at the National Science Foundation, which aimed at nothing less than the "institutional transformation" of research universities. These two episodes and the subsequent reaction raised the priority and thus the urgency and funding of the issue, shifting it away from the relatively lackluster impersonal and quantitative appeal (Why so few?), though that continued to exist, to the qualitative or personal appeal of the mistreatment of senior women, the survivors at the end of that extraordinarily long "pipeline" (Why so lonely? or Why so isolated?). This reframing of the issue reverberated with the experiences of women faculty across the nation, creating an atmosphere of impatience and putting the establishment, temporarily at least, on the defensive.

The initial publicity soon had consequences for plans and projects already under discussion elsewhere. In the two years after the 1999 MIT report, a period still dominated by the presidency of Bill Clinton, several significant events occurred at the national level. This time, unlike in the 1970s, there would not be lawsuits, which would take years, cost millions, and end in some kind of consent decrees. Instead there would be promises of (eventual) change, some new female top officials, leaks to the press, and the increased involvement of various arms of the federal government. They did not constitute an organized "movement" but were a series of actions and tactics by old and new persons and organizations. Various groups (women chemists, boards of trustees, members of the National Academy of Sciences) and individuals took advantage of the new atmosphere to press for goals they had never imagined before—record numbers of women presidents at elite universities, double the number of women in the National Academy, harsher criticisms of laggards in the science media, talk of stronger enforcement of Title IX, and more help from the NSF. Then in January 2005, when the MIT brouhaha had quieted down, a major lapse by the top official at Harvard, followed by profuse apologies, promises to improve, his eventual resignation, and the appointment of a woman president (its first), seemed to open the door to further change even at Harvard, a place less enlightened than MIT had once been. But the George W. Bush administration (2001–9) did not capitalize on the Harvard spectacle to press for greater action, and by 2008 the economy had collapsed and university endowments were falling catastrophically. It was a busy decade, and the results are not yet in.

The MIT Report

In March 1999 a report was posted on the MIT website that revealed that even its top women, the seventeen faculty members who had earned tenure in its School of Science, did not enjoy the status and power that should come with their professorships. Having been treated unequally for years, they had accepted the situation and only now were realizing and admitting that they were marginalized and miserable. At the peak of their careers they were isolated, powerless, and almost invisible. Their male colleagues worked around them, as if they were not there. They had been so thrilled to get tenure at MIT that they had tried not to notice that they were not part of the inner circle that chaired departments and search committees, set salaries, allocated lab space, and met outside offers. The result of all this avoidance or denial was that they received demonstrably lower salaries, less workspace, fewer resources, and hardly any awards. When they did get job offers from elsewhere, their department chairs allowed them to leave. Moreover, none of the six departments in the School of Science had ever been chaired by a woman. The situation could no longer be ignored. In fact, one of the several odd things about the episode was that the major events had occurred five years earlier, starting in 1994, and had by the time of the posting of the report on the MIT website in March 1999 been largely resolved amicably. (A recent, unexplained rise in the number of junior women on the faculty of the School of Science was one sign of this new vigilance.) Astonishingly, MIT's dean and president, faced with what might have been grounds for a high-profile class-action lawsuit, had readily admitted that the findings were true but said that they had come about unintentionally and claimed that the administration was working to rectify conditions.

The incident presented surprises at several levels. First, how could this be happening at MIT, an institution that in the 1970s, under President Jerome Wiesner, had seemed to strive so hard to become a model of equal employment opportunity? Evidently this success had led to a certain complacency in the 1980s, which had gotten worse in the early 1990s. If this could happen at MIT, it might well be happening at all the other research universities across the nation. Second, the incident demonstrated the strength of the women faculty's tactics—quantitative data and firm solidarity. The data, which were largely sensitive and confidential, took years to collect, but the near-total solidarity—sixteen of seventeen eligible faculty members signed the key letter to the president—was spontaneous. Led by the molecular biologist Nancy Hopkins, they were in this together and stayed unified for the duration. Finally, there would be no lawsuit. The complainants did not want confrontation; they were senior women who wanted officials to improve conditions and practices so that they could stay where they were and keep on with their work. Having persuaded the dean of the School of Science and the MIT president that their unequal treatment, though unintentional, was a fact, and having obtained their promises to improve conditions, they did not object to a public acknowledgment of conditions for senior women in this one school at MIT. No one there had meant to treat women faculty differently or unequally, it was claimed, and yet the daily practice had evidently resulted in significant career differences. (Of course, this was reminiscent of Mary Rowe's phrase for the phenomenon—"Saturn's rings." It was almost invisible but omnipresent and thus had a large cumulative impact.)[1]

Reaction across America was swift. President Bill Clinton and his wife, Hillary, invited Hopkins to the White House in April 1999 to help celebrate Equal Pay Day, a visit that may have influenced President Clinton's subsequent interest in and support of women-in-science issues.[2] Also in April the members of the National Academy of Sciences discussed women's careers in science for the first time in a lively session chaired by member (and at the time chancellor of North Carolina State University) Marye Anne Fox, a prelude to some significant change at the NAS over the next few years (see below). Meanwhile, Nancy Hopkins went on the road, speaking at about one hundred campuses across the nation, where women scientists and others flocked to see and hear her. They were electrified by what she had to say, as it resonated well with their own experiences. Most (except for one critic in Alaska) felt with a sense of triumph and relief that a breakthrough had been made—that after decades of denial and disbelief, two initially skeptical male officials had finally acknowledged the existence of a pervasive situation and taken steps to change it.[3]

Mixed Results

Yet the euphoria at MIT did not last, and little changed on the MIT faculty in the next several years, aside from the geophysicist Maria Zuber's becoming chair of the earth sciences department. When in 2006 Nancy Hopkins, who was becoming a feminist in spite of herself, reported that aside from certain spurts in hiring women faculty at MIT, under Jerome Wiesner in the 1970s and under Dean Robert Birgenau in the late 1990s, little was happening. It took publicity, pressure, and the personal commitment of a top administrator for women to be hired, and maybe not many even then. When that pressure lessened, as it had under presidents Paul Gray and Charles Vest, erosion and backsliding set in. Hopkins found this disheartening, since scientists, like most on the MIT faculty, were not used to such regression. They expected that once "solved," problems disappeared and thus required no further attention.[4] But the problems kept surfacing at MIT. In 2006 a young female biologist turned down a job offer there after a Nobel laureate on the faculty refused to work with her, and in 2007 the *Boston Globe* reported that only one woman (compared with twenty-four men) received tenure at MIT that spring.[5]

Meanwhile, the MIT report energized the faculty at Caltech, where the biologist Alice Huang was just completing one report on its women, to establish a committee to write another with "concrete recommendations" for action. By December 2001 this task force, chaired by the astronomer Anneila Sargent, had interviewed all twenty-nine women on the faculty and an equal number of similarly situated men. It found that there were no women in the administration, that few women held endowed chairs, that women's salaries were lower than men's, that mentoring was poor to nonexistent, and that the women were significantly more dissatisfied than the men. The task force recommended, among other things, hiring more women, raising salaries, increasing the women's visibility, monitoring their progress, and starting a special fund drive to meet women's needs if necessary. In September 2002 President David Baltimore issued a statement "reaffirming and extending Caltech's commitment to attract and retain a diverse faculty" and appointed its first female vice president (for student affairs).[6]

The highly publicized Hopkins episode at MIT was also a factor in precipitating a series of events at the National Academy of Sciences in Washington, DC, where in April 1999, just a month after the brouhaha started at MIT, the topic of women and science appeared for the first time on the program of the annual spring meeting of the NAS under the heading "Who Will Do the Science of the Future? A Symposium on Careers of Women in Science." Moderated by the chemist Marye Anne Fox, the day-long event covered the ground from K–12 on up to scientific "leadership." That year the academy's governing boards voted to increase the number of new members elected annually from 60 to 72, perhaps as a way not only to find space for more women but also to reduce the median age of newly elected members, a more pressing concern to many. But neither outcome occurred. When the increase went into effect in 2001, the number of women chosen was even lower—just 7 of 72, down from the 8 of the previous year's 60—and the median age still went up. Increasingly the press (the *New York Times, U.S. News and World Report,* the *Chronicle of Higher Education,* and Daniel Greenberg in his *Science and Government Report*) criticized and even mocked the continuing old age of the men and the low numbers of women being elected.[7]

Finally in June 2001 the academy's council set up a new Committee on Nominations and Elections in the 21st Century to consider fresh ways to identify, nominate, and elect more women. Significantly, among the ten members of this committee were six women: Vera Rubin, Maxine Singer, Mildred Dresselhaus, Sharon Long, Pamela Matson, and chair Susan Solomon. After discussing a possible goal of 25 percent women members within ten years, they dropped it as unattainable, but the committee was able to implement procedural changes (e.g., more open nominations) and trade-offs among sections that brought significant results in just eight years. In 2002 ten women were elected, in 2003 and 2004 seventeen each (including Nancy Hopkins), and then in 2005 nineteen, the highest ever. But in 2006 there were only twelve, and in 2007 just nine, almost back to where the number had been in 2001. Nevertheless, by 2007 the number of women members had doubled in just the eight years since 1999, and cumulatively they constituted 10 percent (189) of the academy's 1,899 active members.[8]

The MIT episode may also have played a role in accelerating the recent trend in appointing women presidents of major universities that had never had one before. Although the actual work of search committees is usually confidential, they too were becoming more diverse and impatient. By the late 1990s the pool of qualified women candidates was growing, as more were coming up through the academic ranks and as the Clinton administration, with its many women in top posts, was coming to a close. For example, in 2001 the molecular biologist Shirley Tilghman, who had served on many NRC committees but had never been a dean or even a department chair, was appointed president of Princeton University, an incredible turn of events. (Her candidacy had been so improbable that she had been appointed a member of the search committee.) Her appointment offered particular hope to young women scientists, for in the early 1990s she had published two Op-Ed pieces in the *New York Times* that proposed abolishing the tenure system as the only way to help aspiring mothers advance in academia. Although once in office Tilghman did not do this, she did appoint a woman provost (the political scientist Amy Gutmann) and several women deans (including Maria Klawe in engineering).This provoked some criticism but also raised the hope that the era of tokenism and glass ceilings had passed. But it proved difficult to hold onto such an array of talent, and within a few years, both Gutmann and Klawe had moved onto presidencies of their own.[9]

The trend continued into the first decade of the twenty-first century, as described in chapter 11, in the Big Ten and the University of California system. But it was still a surprise in 2004 when MIT's board of trustees, known as the Corporation, appointed the neuroscientist Susan Hockfield, Yale provost and former dean of its graduate school, as MIT's first woman president. Almost immediately thereafter the University of Pennsylvania topped this and made history again by appointing Princeton's provost, the political scientist Amy Gutmann, as its second woman president, to succeed the psychologist Judith Rodin, who was stepping down but later headed the Rockefeller Foundation. There seemed to be a message in these appointments. Prestigious institutions, with their particular ways of doing things, were not admitting any previous wrongdoing, but their trustees were willing to break with age-old traditions and perhaps begin to do things differently. It will be interesting to see what these pioneering women will be able to accomplish and who will follow them as presidents.[10] Maybe the atmosphere had changed and a new era was beginning, but many old problems remained as well. In fact, as the biochemist Phoebe LeBoy, of Penn, said, it might be better for women in science if the new female appointees were department chairs rather than presidents, for it was the chairs who handled the faculty personnel issues.[11]

One other outcome of the Hopkins episode was a meeting at MIT almost two years later, in January 2001, supported by the Ford Foundation, of the heads of nine similar large research institutions. There David Baltimore of Caltech, Lawrence Summers of Harvard, and Richard Levin of Yale, among others, promised that they too would examine the treatment of their female senior faculty and take steps to correct any inequities in salaries, size of laboratories, teaching loads, numbers of graduate students and postdocs, chairmanships, involvement in decision making, and meeting of outside offers. This group, subsequently termed the Nine Universities Group, met once again in 2006, after the Lawrence Summers affair at Harvard (see below). By then four of the nine university presidents were women, and three of the group (two women and one man) cosigned a statement disapproving of Summer's utterance and published it in the *New York Times.* Yet little seems to have come from this group. It met after crises, responding to events in the spotlight, promised reforms, and then faded away.[12] One exception was a report prepared by a task force appointed by Tilghman at Princeton to advise her on what to do to improve the status of women faculty there. Chaired by the molecular biologist Virginia Zakian, the task force found that the situation was mixed: some departments were nearly gender-balanced, and others were not. It recommended the appointment of an administrator to work on the issues and the creation of a $10 million fund for faculty recruitment and retention.[13]

Nelson's Numbers

The MIT report soon began to have a ripple effect, energizing individuals elsewhere to undertake new projects. In 2000 the chemist Donna Nelson, of the University of Oklahoma, clipped an article from the *Chemical and Engineering News* on the low numbers of women in top chemistry departments and posted it on her office bulletin board. The undergraduates there read it and asked about collecting similar data on minorities in chemistry. At first Nelson, who is one-quarter Native American, thought that the preparation of such a list of zeroes (and maybe occasional ones and twos) would be a waste of time, rather like documenting the obvious, but the students were interested, and no one else was doing such a thing at the time. By writing to the chairs of the nation's top fifty chemistry departments, they were able before long to compile a list of about forty-five minorities (African Americans, Hispanics, and very few Native Americans) nationwide. Then, with the aid of websites, which were new at the time, Nelson was able to communicate with these individuals and to introduce them to one another and get some publicity. At that point various interest groups began to get nervous about the embarrassment this information, which no one had bothered to collect before, might cause. Meanwhile, Nelson and her students expanded their data collection to include the sex and race of faculty in the top fifty departments, as ranked by NSF funding, not only in chemistry but in thirteen (eventually fifteen) other fields as well. (They have since expanded their coverage to the top 100 departments.) In a sense they were replacing and expanding and improving upon the work done earlier by the ACS's women's committee (especially the deceased Sister Agnes Ann Green, Nina Roscher, and Betty Vetter) and published the data in a variety of articles as well as on the easily accessible website of the National Organization for Women.[14]

The numbers showed not only how male but also how pale (white) most scientific fields still were, as Nelson devised a way to show both the sex and the race of each rank in each department (e.g., 25.01 meant one woman among a total of twenty-five full professors) with a different column for each race (e.g., 2.01 in the column marked "Hispanic" would mean one Latino and one Latina in that department). In addition to such totals and readily calculated percentages, Nelson devised a "utilization factor," which compared the proportion of women at the assistant-professor level in a field with the proportion of doctorates awarded to women in that field in the previous nine years. This would indicate, as shown in table E.1, that only about two-thirds of the many recent female PhDs in psychology or the biological sciences were going on to top academic posts, whereas in most fields of engineering, where doctorates were less common, almost all of the few women

Table E.1. Female PhD attainment, FY 1993–2002, and assistant professors, FY 2002

Field	% PhDs	% Assistant Professors	Utilization Factor
Psychology	66.1%	45.4%	0.69
Sociology	58.9	52.3	0.89
Biological sciences	44.7	30.2	0.68
Political science	36.6	36.5	1.00
Chemistry[a]	31.3	21.5	0.69
Economics	29.3	19.0	0.66
Mathematics	27.2	19.6	0.72
Chemical engineering	22.3	21.4	0.96
Astronomy[b]	20.6	22.0	1.07
Computer science	20.5	10.8	0.53
Civil engineering	18.7	22.3	1.19
Physics	13.3	11.2	0.84
Electrical engineering	11.5	10.9	0.95
Mechanical engineering	10.4	15.7	1.51

Source: Nelson and Rogers, *A National Analysis of Diversity in Science and Engineering Faculties at Research Universities,* 3, available online at http://now.org/issues/diverse/diversity%5Freport.pdf.
[a] FY 2003.
[b] FY 2004.

who received a PhD took an academic post. In a few fields with a utilization factor over 1.0 departments were hiring women with degrees from other fields or from foreign universities. This information added specificity and substance to the discussion of where (and where not) the record numbers of women PhDs were being employed. But it raised new questions as well. Were they being discriminated against by top departments that should have been hiring even more of them? Or was this drop-off voluntary, as might be the case if they were not applying for academic jobs in the numbers that might have expected? And if it was voluntary, was this because of the attractions of industry or because conditions in academia were so intolerable that they sought jobs elsewhere instead? If the latter, then perhaps something ought to be done about changing academia. Nelson was daring to document and publicize the unwelcome data that the more cautious NSF and NRC were deliberately avoiding.[15]

Title IX

In March 2000 another chemist, Debra Rolison, of the Naval Research Laboratory, published a provocative editorial in the *C&EN* that attracted a lot of attention and, in 2002, some congressional action. (Another sign of the changing times was the increasingly critical press coverage of women-in-science issues, partly because the formerly staid *C&EN* had a new editor in chief, Madeleine Jacobs, who published Rolison's editorial.)[16] Rolison argued that as the situation for women chemistry faculty was not improving very quickly, the federal government should do more to enforce equal opportunity in higher education. It could and should, in fact, use the authority granted by Title IX of the Education Amendments Act of 1972 to require equal treatment at academic institutions receiving any federal funds for scientific research. Enforcement should not just be left to the Office of Civil Rights in the Department of Education, which even in the Clinton administration had done little, but should be taken up by all the science funding agencies—the NSF, the NIH, NASA, and the DOE.[17]

Although Title IX permits civil rights agencies to cut off federal funds to universities whose programs discriminate against women in any way, intentionally or inadvertently, it has primarily been used to bring gender equity to the world of college athletics. This has led to a blossoming of women's sports in recent decades, especially after a long lawsuit against Brown University *(Cohen v. Brown University)* was settled in 1995. Thus it was something of a novel idea to suggest extending this "hammer," as Rolison called it, to the world of academic employment in science and engineering, where, as everyone knew, and as Donna Nelson was documenting and publicizing, all those unsatisfactory numbers were barely budging. A few well-chosen federal-compliance reviews carrying the threat of a cutoff of federal funds might hasten changes at the major research universities, for as Rolison noted, experience had shown that when investigators found violations of federal rules in research on human subjects, the matter was settled within forty-eight hours. Rolison later expanded on her ideas in a hilarious talk at a meeting sponsored by the Chemical Sciences Roundtable, a branch of the NRC, which later published it.[18]

Some considered Rolison's suggestion so radical that its implementation might cripple major universities, and the incoming Bush administration did not respond. But the idea caught the attention of Senator Ron Wyden (D-OR), who was chair of a subcommittee on science of the Senate Committee on Commerce, Science, and Transportation. Wyden held two hearings on Title IX enforcement, in July and October 2002. At the latter meeting, his constituent chemist, Geraldine Richmond, from the University of Oregon and, as mentioned in chapter 11, a founder of COACh, testified, and Wyden chastised the representative of the Bush administration's Office of Civil Rights for inaction.[19] One result of the October hearing was that Congress asked the Government Accountability Office to report on how many science funding agencies were conducting Title IX compliance reviews. The GAO's July 2004 report indicated that none were doing anything and that in general the authority was "underutilized."[20] A second result of the 2002 hearings was that Congress requested that the NSF fund a report on women on science and engineering faculties by the NRC's Committee on Women in Science and Engineering.[21] After this the NSF, the DOE, and NASA quietly increased the number of compliance reviews at major universities. Before long, complaints about the one at Columbia University surfaced, and in 2008 the commentator John Tierney published an essay on the Op-Ed page of the *New York Times* calling attention to and ridiculing the compliance reviews. Whose fault, he asked, was it that women were going mostly into biology and not into computer science? That was their own choice, and neither Title IX nor the federal government could make everything equal. But the situation may change, as the Obama administration has promised to increase the number of compliance reviews.[22]

More Philanthropy

In 2000 two anonymous donor, perhaps stimulated by the MIT report of 1999, gave substantial endowments (about $20 million) to Stanford University and to the University of Southern California to provide support for women scientists. Stanford used its $20 million to set up three programs—one for students, one for graduate students, and the Gabilan Provost's Discretionary Fund for faculty.[23] The few but energetic women scientists at USC used their aged alumna's gift as a springboard for a host of efforts, for spending her money also brought them together. They oversaw an office created to support an array of projects and to get matching funds from other parts of the university wherever possible. They also created a network of faculty women for mutual support and encouragement and participated in numerous searches, helping to markedly increase the number of women faculty there in seven fields—biology, chemistry, math, physics, astronomy, earth science, and all of engineering—from 15 in 2002 to 32 in 2007 and 38 in 2009 out of about 300 permanent posts. It was a model of what some outside money, committed self-help, and willing administrators could accomplish on one campus.[24]

The NSF's ADVANCE

Meanwhile, the 1999 brouhaha over the Nancy Hopkins incident at MIT, as well as a push from President Clinton, may have contributed to the NSF's bold decision in the spring of 2000 to do something dramatically different with its recently consolidated women's programs. This was unusual for the NSF, hitherto so hesitant to get involved in policing the behavior of its grantees (the very departments that were so unwilling to hire women PhDs) and recently so intimidated by fears of lawsuits from persons not eligible for its minority (but not yet women's) programs. In fact the newly vocal constituency aroused by the publicity over the MIT report and the support by President Clinton provided some political cover for NSF officials wishing to expand its program for women in science. The decision to start ADVANCE in the spring and summer of 2000 was one of the last initiatives of Rita Colwell's directorship.

The creation of ADVANCE is a classic example of what has been called "mission creep" in the NSF's series of innovative programs to help women faculty at research universities. In 1980 the Visiting Professorships for Women program (VPW) had been, as described in chapter 11, an imaginative and useful combination of self-help, federal aid, and institutional strategy. A token program (initially funded at less than $1 million), it provided motivated individuals a way to improve their own situation, diversify university departments even if only briefly, and extend some help or encouragement to other women at the institution to persevere and succeed. It had been terminated in 1997 and replaced by the much larger ($9–12 million) Professional Opportunities for Women in Research and Education (POWRE), as also explained in chapter 11. Yet before long it too came under criticism, as had other programs before it, not so much for what it did as for what it did not do. In the spring of 1998, NSF Director Rita Colwell started a task force to review POWRE, but no decision was made.[25]

Then in March 1999 the Nancy Hopkins saga dominated public attention, and many in Washington, DC, took notice. NSF officials, whose budget was about to double, began to think that the need to do something big and bold to change the academic climate was pressing. The institutions funded by the NSF needed substantial fixing before the next generation of women scientists encountered the same old practices. Also, there were advantages to making a few large grants rather than hundreds of small ones, and awards to officials of institutions avoided the political problems associated with grants for which only female principal investigators could apply. Thus NSF officials, perhaps responding to some remarks by Bill Clinton in a Rose Garden ceremony on Equal Pay Day in May 2000, embarked on designing a new, more ambitious program called ADVANCE, one of whose components was to bring about what sociologists called "institutional transformation." The new program would provide, among other awards, hefty five-year, $3–5 million grants to institutions whose top officials agreed to set up programs to help women at all levels and to review existing practices to make them more gender-equal. Such grants thus mobilized and empowered concerned individuals at institutions to work to transform them from within. Then perhaps all fields, including engineering and computer science, would begin to have a certain minimum number of women faculty members, such as at least one and maybe even two, per department. The time was ripe for this type of program, since some such women (including some former VPWs) were by 2000 highly placed, as department chairs, deans, or even provosts or presidents. They might be willing and able to use such grants to change their institutions significantly. Perhaps ADVANCE would help the NSF attain the newly articulated (but ever elusive) goal of making educational institutions hospitable to women in science, even and especially in the "chilliest" fields, engineering and computer science. Thus the NSF, with its recently doubled budget and many women biologists in top posts, took advantage of Hopkins's timely press coverage to expand its existing women's programs into the larger, more ambitious ADVANCE program, with a budget of $9 million, then $17 million, and later $19 million.[26]

Three months after the call for proposals went out, the NSF had seventy-two applications. Evidently university administrators were not embarrassed to ask for help in doing what they might earlier have been expected to do by themselves—getting recalcitrant parts of their universities (engineering, physics, etc.) to hire women faculty. The lure of outside funds (plus overhead) motivated interested faculty, and the NSF's sponsorship legitimized such projects. Those who were already aware of their institutions' problems had an early advantage, as the whole process of gathering the necessary data and putting together a proposal required a high level of coordinated activity, organization, and commitment. But the activity had special appeal to those favorably inclined but possibly frustrated women (and a few men) who were concerned about such issues and had been rising in the administration—heads of units, deans, provosts, and higher. They were eager to promise to the NSF that with its funds they would do more to hire, tenure, and retain women faculty in science and engineering and to put in place new practices and behaviors that would continue after the end of the NSF grant and thus change their institutions permanently. They also agreed to collect the data needed to document this improvement, which often had been secret and inaccessible.

In October 2001 the NSF made its first awards to eight (later nine) institutions, mostly large state universities in the Midwest and the West—the University of California–Irvine (with the biologist and dean Susan Bryant as principal investigator, or PI, and the oceanographer Ellen Druffel and the mathematician Chuu-Lian Terng as ADVANCE professors), the University of Colorado at Boulder, the University of Wisconsin, the University of Washington (with the engineer Denice Denton as PI), New Mexico State University, the University of Michigan (psychologist Abigail Stewart as PI), the University of Texas at El Paso, the University of Puerto Rico at Humacao, and later Hunter College in New York City (psychologist Virginia Valian as co-PI). At some of these first awardees (Irvine, Colorado, Wisconsin), before long there were noticeable increases in the numbers of women hired. But most awardees, whatever their numbers, were euphoric about the innovative processes they were introducing, the workshops, the mentoring, and other activities. They felt that they were improving the quality of personal and personnel interactions, and indeed many of them focused on improving evaluation processes by sensitizing faculty to the many "gender schemas," as Virginia Valian called them in her book *Why So Slow?* (1998), which lurked everywhere in academic decision making.[27]

At the University of Michigan the psychologist Abigail Stewart formed STRIDE (Strategies and Tactics for Recruiting to Improve Diversity and Excellence), a committee of nine faculty members who visited various search and tenure committees and taught them to be aware of and fight off their own predilections and biases.[28] Another innovation was an interactive theater project at the University of Michigan in which student and professional actors portrayed in a skit standard academic scenes, such as an instructor teaching a class or a department or search-committee meeting, with egregious behavior highlighted. Helping faculty members see their own foibles was an entertaining and memorable way to raise their awareness of biases that had hitherto been nearly invisible and unacknowledged.[29] One clear sign of the success of an ADVANCE project was that at the end of the NSF grant the university would agree to continue the project at its own expense. So far the University of Michigan has agreed to continue STRIDE until 2016.[30]

The Lawrence Summers Affair

The abovementioned events, momentous as they were, were quietly under way in Washington, DC, and on several campuses when in January 2005 Lawrence Summers, the president of Harvard University and an economist by training, startled the Western world by expressing, presumably off the record to a small seminar of experts discussing issues relating to women and minorities in science, his skepticism about women's ability and commitment. Having abolished Harvard's post of vice provost in charge of affirmative action and having personally evaluated the many candidates for tenure at Harvard presented to him in his first three and a half years in office, he had begun to doubt that there were many women who had what he called in his attempt to be "provocative" the "intrinsic aptitude" and willingness to put in the eighty-hour weeks that he felt a tenured post at Harvard entailed. (Here he unwittingly followed in the tracks of the Harvard psychologist Edwin Boring, who fifty-four years earlier had expressed the same idea.)[31] Summers therefore believed that Harvard was justified in offering tenure to a dwindling number of women, which had become the pattern under his administration and which twenty-six women faculty there had protested the previous summer and fall. (Only four women got tenure offers from Harvard in 2003–4, of whom just one accepted, whereas twenty-eight men were offered tenure and twenty accepted. This was down from fourteen women offered tenure in 1999–2000 and thirteen in 2000–2001, before Summers's arrival.)[32] Summers's assessment flew in the face of all the data that indicated that all too often working women were already working extraordinary hours, even what Arlie Hochschild had called "the second shift," as they did housework and cared for their children in addition to their full-time jobs.[33]

Among those present at this private meeting called to discuss such subjects off the record was the ubiquitous MIT biologist Nancy Hopkins, who was appalled at Summers's negative views, felt queasy after eating, and left abruptly. Later in the day, in the course of discussing another matter with a reporter from the *Boston Globe*, she leaked Summers's unsettling views, which then appeared on page 1 of the newspaper and promptly set off a tsunami of outrage and consciousness-raising in academia and beyond. If the nation's top educator, the head of perhaps *the* leading coeducational university in the country, was skeptical, even privately but worse publicly, that women could achieve at the highest levels of science, that was grounds for his resignation or removal. Indeed Summers spent most of the next several months clarifying, recanting, and apologizing for his statements, which he claimed were not his real views after all. He said that he had merely meant to stimulate discussion.[34] Curiously, the tremendous outpouring in the press, especially in the *New York Times*, the *Boston Globe*, and the *Chronicle of Higher Education*, barely mentioned the NSF's new ADVANCE program (just twice in letters to the editor), which had by then already made nineteen large grants, mostly to institutions outside the Northeast.[35]

Unfortunately, Summers's apologies did not stop the "perfect storm" that enveloped him. He was already unpopular among Harvard's faculty of arts and sciences for his belligerent management style and his public criticisms of Cornell West, a popular professor of Afro-American Studies, who had promptly left for Princeton University. When the uproar did not abate, the faculty voted in March 2005 that it had no confidence in him, the first time ever for a Harvard president.[36] Summers was stung by the reaction but did not resign. Instead he appointed two task forces (on women faculty at Harvard and on women in science and engineering in general) to come up with recommendations by springtime, and he promised to spend $50 million over the next decade (which he hoped would be the rest of his time in office) on recruiting and retaining women science faculty. One major step recommended by the first task force was the creation of a new office headed by a special vice provost to initiate and oversee new projects and watch over faculty recruitment and tenure activities.[37] In the summer of 2005 Summers appointed Professor Evelynn Hammonds, a graduate of Spelman College and a historian of science and medicine, to this new post. She then had a busy year meeting with numerous groups and individuals on and off campus and putting in place Harvard's own ADVANCE-like programs.[38]

Meanwhile, Harvard's faculty of arts and science scheduled a second vote on

President Larry Summers for February 2006. By then he had been further discredited by revelations of his support for (and refusal to discipline) a fellow economics faculty member who had been heavily fined for spending U.S. funds corruptly while running a program in Russia. Finally, facing the inevitable, Lawrence Summers resigned before the actual vote,[39] and the Harvard Corporation, which by then included two women, the political scientist and former Duke University president Nannerl Keohane and the Georgetown University legal scholar Patricia King, initiated a search.[40] In February 2007 its members finally chose Drew Faust, a woman historian, dean of the Radcliffe Institute for Advanced Study, as president.[41] She got off to a promising start, but when the economy collapsed in late 2008, most hiring was curtailed and construction on a new campus was halted, though in 2009 she did appoint the physicist Cherry Murray, formerly of Bell Labs and most recently of the Lawrence Livermore National Laboratory, as dean of engineering and applied science.[42] Meanwhile, the newly elected U.S. president, Barack Obama, appointed the deposed Summers to the top economics post in the federal government. The saga continues locally and nationally.

One other result of the Summers affair was that the NRC, in Washington, DC, whose prestigious and recently revamped Committee on Science, Engineering, and Public Policy (COSEPUP) now contained several prominent women scientists, established a special committee chaired by the political scientist Donna Shalala, the former secretary of health and human services and by then the president of the University of Miami, to prepare a report on women on science faculties. Seventeen of its eighteen members were women, the only man being Robert Birgenau, formerly provost at MIT and by then chancellor at the University of California–Berkeley. This fact was noted in the press and used to discredit the report, *Beyond Bias and Barriers,* which appeared in 2007. It recommended more data, more vigilance, more enforcement, and more funds for research on the issue of women in science.[43]

Meanwhile momentous events in Stockholm might be interpreted as a kind of trans-Atlantic response to Lawrence Summers's remarks. Starting in the fall of 2004 and continuing to 2009, the Nobelstiftung awarded five science medals to women, and in 2009 the Sveriges Riksbank (Sweden's central bank) awarded a sixth, the prize in economic sciences, to its first woman since it began awarding the prize in 1969. Four of these six laureates were Americans—Linda Buck, Barbara Blackburn, Carol Greider, and Elinor Ostrom, of whom the first three were in the area of "physiology or medicine."

Not much has been written about the 2004 American woman Nobelist, Linda Buck, a psychologist turned microbiologist and molecular biologist who in the early 1990s, as a 44-year-old postdoctoral fellow at Columbia University, discovered the cluster of about one thousand genes that in mice encode for the olfactory factors in the lining of the nose that detect odors. Later a Howard Hughes Medical Institute investigator, she was on the faculty of the Harvard Medical School for ten years before returning to her alma mater, the University of Washington, in her native Seattle, where she was when she shared the prize with her mentor.[44]

More history was made in October 2009 when two women were among the three Nobel Prize winners in physiology or medicine. Elizabeth Blackburn, born in Australia in 1948, earned undergraduate and master's degrees in biology at the University of Melbourne and then moved on to Cambridge, England, for her PhD in Fred Sanger's lab. After that she followed her husband to Yale University, where she got a postdoc with Joseph Gall, a famed cell biologist. While there, she discovered and worked on the end points of chromosomes, distinct entities that she named *telomeres,* which have important functions. (When DNA copies itself, one strand becomes a little shorter than the other. Over time this can lead to an inability to replicate any further, and so death ensues.) Blackburn then became an assistant professor in the troubled department of molecular biology at the University of California–Berkeley, where its first female faculty member, Ellen Daniel, was refused tenure. But Blackburn fared better. In 1984–85, along with the graduate student Carol Greider, Blackburn found an enzyme that she later named *telomerase* (which adds units to the shorter strand of DNA and thus prevents premature aging). Shortly after that she left for a job across the Bay at the University of California–San Francisco, where she has remained ever since. Greider, born in 1961, moved with Blackburn as a postdoc, taught for a few years at the Cold Spring Harbor Laboratory, and then in 1997 got a faculty post at the Johns Hopkins University Medical School, where she is now.[45]

Also in 2009, Elinor Ostrom was at age 76 the first woman to win the Nobel Prize in Economics. She is a political scientist by training, for her bachelor's and doctoral degrees, both from UCLA, were in that field. This gave her a broader viewpoint than most economists who study the causes and remedies for inequality. In 1965, together with her husband, Vincent Ostrom, also a political scientist, she moved Indiana University because it was one of the very few major universities in the country that would give faculty posts to a married couple at that time. She was promoted to full professor in 1974, shortly after she and her husband started an externally funded "workshop" in political theory and policy analysis at the university. Over time she did fieldwork in Nepal, Nigeria, and Kenya and continued to publish theoretical work on "governing the commons," or how people can work together to retain local control over their changing environmental resources and thus deter distant national or international authorities who claim expertise and express confidence from taking over the decision making.[46]

Conclusion

In 1999, after nearly three decades of activity by a variety of individuals and groups of women scientists, the issue of women in science jumped to national prominence with revelations of inequities and admissions of doubts at top levels. The continuing stream of data-rich articles on the unanswerable question Why so few? were now supplemented by riveting human dramas of marginalization and skepticism in high places. Once these doubts and drawbacks were brought into the open, outraged persons criticized the practices scorchingly and relentlessly. Institute and university trustees responded with new female presidents, whose presence acknowledged the gender issue but guaranteed nothing. Yet the chief impact has been the creation at the NSF of a large federal program to transform university culture. In 2006 it made thirteen new awards, bringing its total to thirty-two campuses nationwide, while Caltech, Stanford, USC, and even Harvard have devised others of their own at varying levels of funding. Thus out of this decade of tumult has come increased awareness and admission of the pervasiveness of differential practices within elite places and the need for vigilant women in powerful posts with significant resources to establish and maintain continued progress.

NOTES

Introduction · The New Era

1. The Senate, led by the southerners Sam Ervin (D-NC) and James Allen (D-AL), filibustered an earlier, stronger version of the Equal Employment Opportunities Act and finally settled for this weaker version. The House included at the time such important women representatives as Martha Griffiths (D-MI), Edith Green (D-OR), Bella Abzug (D-NY), Patsy Mink (D-HI), and Shirley Chisholm (D-NY), all of whom deserve biographies. The immense Patsy Mink Papers (900,000 items), in the Manuscript Division of the Library of Congress, have recently been arranged for use. Bella S. Abzug, *Bella! Ms. Abzug Goes to Washington,* ed. Mel Zieger (New York: Saturday Review, 1972), is a flavorful chronicle of her first year in Congress in 1971.

2. See, e.g., John Herbers, "Nixon Busing Curb May Face Filibuster," *NYT,* Mar. 25, 1972, 13; Robert B. Semple Jr., "President Signs School Aid Bill; Scores Congress," ibid., June 24, 1972, 1, 15; "The Busing Distortion . . . ," editorial, ibid., June 27, 1972, 40; Jonathan Spivak, "New Busing Curbs Are Expected to Slow Integration, Tie Up Rights Leaders' Energy," *Wall Street Journal,* June 26, 1972, 5; and idem, "Money Isn't Everything: New Higher-Education Bill Provides Funds, But Sex-Bias Section Could Spark Controversy," ibid., July 13, 1972, 36. See also Dean J. Kotlowski, *Nixon's Civil Rights* (Cambridge, MA: Harvard Univ. Press, 2001), ch. 8. The single best description of Title IX's implications for higher education is Bernice Sandler, "Admissions and the Law," in *Graduate and Professional Education of Women: Proceedings of AAUW Conference Held at the AAUW Educational Center, May 9–10, 1974* (Washington, DC: AAUW, 1974), 23–31.

3. I later identified and named this phenomenon the "Matilda Effect." See Margaret W. Rossiter, "The ~~Matthew~~ Matilda Effect," *Social Studies of Science* 23 (1993): 325–41.

4. See Margaret W. Rossiter, "Women Scientists in America before 1920," *American Scientist* 62 (1974): 312–23; and idem, *Women Scientists in America: Struggles and Strategies to 1940* (Baltimore: Johns Hopkins Univ. Press, 1982).

5. Margaret W. Rossiter, *Women Scientists in America: Before Affirmative Action, 1940–1972* (Baltimore: Johns Hopkins Univ. Press, 1995).

Chapter 1 · From "Sisterhood" to Interest Group

1. List of charter members, attached to news release, Apr. 16, 1971, of the Federation of American Societies for Experimental Biology, 55th annual meeting, Chicago, IL, Apr. 12–17, 1971 copy in Jeanne R. Lowe Papers, Special Collections, Vassar College

Library, Poughkeepsie, NY; there is also a copy in box 6, folder 7, of Fann Harding Papers, Women in Science and Engineering Archives, Special Collections, Iowa State University Library, Ames. See also "Transcript of speech by Estelle Ramey at AWIS 20th anniversary Reception, 17 February 1991," in ibid., box 7, folder 3; D[orothy] M[.] S[kinner], "AWIS Ten Years Later: G. Virginia Upton," *AWIS Newsletter* 5 (May–June 1976): 8–9; and Barbara Mandula, Anne M. Briscoe, Judith A. Ramaley, Adele E. U. Edisen, Phyllis Harber, and Sheila Pfafflin, "AWIS—The First Ten Years," ibid. 10 (June–July 1981): 5–10.

2. On Pool, see K. M. Brinkhous, "Judith Graham Pool, Ph.D. (1919–1975): An Appreciation," *Thrombosis and Haemostasis* 35 (1976): 269–71; *NYT,* July 15, 1975, 36; "NHF Mourns Passing of Dr. Judith Pool," *Hemofax,* Oct. 1975, 1, in Fann Harding Papers, box 14; and K. M. Brinkhous, "Pool, Judith Graham Pool. June 1, 1919–July 13, 1975," in *Notable American Women: The Modern Period,* ed. Barbara Sicherman and Carol Hurd Green (Cambridge, MA: Harvard Univ. Press, 1980), 553–54. On Schwartz, see *AMWS,* 19th ed., 1995–96, 6:697; Neena B. Schwartz, *A Lab of My Own* (Amsterdam: Rodopi, 2010); and Neena Schwartz Papers, Northwestern University Archives, Evanston, IL. She had many ties to women in physiology, especially in Chicago. In fact, several members of the founding group of the AWIS, including Julia Apter and Naomi Weisstein, could be called "Neena's network" or the "Chicago Circle." See also Neena Schwartz, "Why Women Form Their Own Professional Organizations," *Journal of the American Women's Medical Association* 28 (Jan. 1973): 12–15; and idem, "Women in Science," *Science Teacher* 40 (Dec. 1973): 16–17. Schwartz was a 1940s graduate of Goucher College, as was neuroanatomist and feminist Ruth Bleier (see ch. 2 n2). See Toby A. Appel, "Physiology in American Women's Colleges: The Rise and Fall of a Female Subculture," *Isis* 85 (1994): 26–56.

3. Letters in Fann Harding Papers, box 6, folder 6, through box 7, folder 6.

4. On Briscoe, see *AMWS,* 19th ed., 1994, 1:890; "Oral History of Anne Briscoe, June 23, 2002, Ruskin, FL," by Laura Sweeney, Women in Science and Engineering Archives, Special Collections, Iowa State University Library; Anne Briscoe, "Diary of a Mad Feminist Chemist," *International Journal of Women's Studies* 4 (Sept.–Oct. 1981): 420–30; and B[arbara] F[ilner] and Anne Briscoe, "Meet a Member," *AWIS Newsletter* 11 (Apr.–May 1982): 8–10.

5. Judith G. Pool to Vera Kistiakowski [*sic*], Sept. 7, 1971, and reply, Sept. 10, 1971, in box 3, AWIS, Vera Kistiakowsky Papers, Institute Archives, MIT, Cambridge, MA; Natalia Meshkov, "Survey and Evaluation of Registries of [Women?] in the Professions: A Report of the Registry Committee, Federation for Organizations for Professional Women," Oct. 1973, copy in ibid., box 12, FOPW. A digest of this report is "Recruiting Aids #3: Rosters, Registries and Directories of Women in the Professions," Oct. 1974, published by the Project on the Status and Education of Women, Association of American Colleges, Washington, DC, copy in ibid., box 21, NAS-NRC CEEWISE. See also Fann Harding Papers, esp. box 9, folder 4, on FOPW's proposed registry. The American Society of Biological Chemists registry, which ran from 1974 until 1979, was funded initially by the National Institute of Arthritis and Metabolic Diseases. Loretta Leive, "A Registry of Women and Minority Group Biochemists Is Being Established," *Federation Proceedings* 33, no. 5 (1974): 1122; box 58, folders 5–13, Papers of the American Society

for Biochemistry and Molecular Biology, Special Collections, Albin O. Kuhn Library and Gallery, University of Maryland–Baltimore County (UMBC), Catonsville, MD. One female biochemist at the University of Chicago refused to be listed and considered the new registry unnecessary, even "an atrocity." Lillian Eichelberger to Robert A. Harte, Oct. 11, 1973, ibid., folder 6. The National Science Foundation's National Register of Scientific and Technical Personnel, which originated in the 1950s, was terminated in the spring of 1971 because of federal budget cuts, and the placement register run by the American Association of University Women (AAUW) in the 1960s had also ended. See also Avery Russell, "The Women's Movement and Foundations," *Foundation News* 13 (Nov.–Dec. 1972): 16–22.

6. On Apter, see "Julia Apter, Ophthalmologist and Researcher at U.D., Dies," *Chicago Tribune,* Apr. 18, 1979, 4; "Julia Tutelman Apter: In Memoriam," *AWIS Newsletter* 8 (Nov.–Dec. 1979): 10–11; "In Recognition: Julia Apter," ibid. 10 (Apr.–May 1981): 10; Julia T. Apter to Vera Kistiakowsky, Sept. 5, 1971, Vera Kistiakowsky Papers, box 4, APS-CWP Responses, which is autobiographical; and Apter to Kistiakowsky, Feb. 3, 1972, ibid., box 9, Correspondence, 1972–78, summarizing the AWIS lawsuit to that date. Other materials about the campaign are in Fann Harding Papers, box 6, folder 6, esp. "HEW Bias Charged in Women's Suit," *Washington Post,* Mar. 20, 1972; Apter to AWIS, May 14, 1973, ibid., box 7, folder 2; Helene Guttman to colleagues, and Helene Guttman to Dr. Harry M. Doukas, NIH, May 18, 1973, ibid.; and "Julia Apter Reports on Law Suits," *AWIS Newsletter* 3 (Sept.–Oct. 1974): 1. For more on Sylvia Roberts, see ch. 2. There is a copy of the actual legal complaint (United States District Court for the District of Columbia, Civil Action No. 594-72) in carton 13, folder 311, of the Sharon L. Johnson Papers, Schlesinger Library, Radcliffe Institute for Advanced Study (RIAS), Harvard University, Cambridge, MA. The seven coplaintiffs were the Association for Women in Psychology, the Caucus of Women Biophysicists, Sociologists for Women in Society, the Association for Women in Mathematics, the National Organization for Women, Women's Equity Action League, and the psychologist Naomi Weisstein, then of Chicago. See also the optimistic D[eborah] S[hapley], "HEW Advisory Jobs to Go to Women," *Science* 174 (1971): 271, written after an initial meeting with NIH officials in September 1971. In May 1973 Apter compiled and Helene Guttman sent to the NIH a list of current committee members whose appointments violated the NIH's own conflict-of-interest rules by being from the same institution or an affiliated hospital.

7. On Jones, see *AMWS,* 19th ed., 1994, 4:158. On Caserio, see Harold Goldwhite, "Marjorie Constance Beckett Caserio (1929–)," in *Women in Chemistry and Physics: A Biobibliographic Sourcebook,* ed. Louise Grinstein, Rose K. Rose, and Miriam H. Rafailovich (Westport, CT: Greenwood, 1993), 89. Data for 1975 reported in *Manpower Comments* 13 (Mar. 1976): 16; Patricia Roberts Harris, memorandum, "HEW Advisory Committees," Sept. 7, 1979, Fann Harding Papers, box 12, folder 4. See also Donald S. Frederickson, *The Recombinant DNA Controversy: A Memoir; Science, Politics, and the Public Interest, 1974–1981* (Washington, DC: ASM Press, 2001), 227–35. There is no evidence that Apter ever served on one.

8. Schwartz's letters to the Steering Committee are in Fann Harding Papers, esp. box 6, folders 7 and 9, and box 7, folder 1.

9. On Ramey, see *AMWS,* 19th ed., 1994, 6:48–49; Margalit Fox, "Estelle R. Ramey,

89, Used Medical Training to Rebut Sexism," *NYT,* Sept. 12, 2006, B10; Estelle Ramey, "Well, Fellows, What Did Happen at the Bay of Pigs? And Who Was in Control?" *McCall's,* Jan. 1971, 26, 81–83; Betty Beale, "Outraged Women and a Sexy Medical Tome," *Washington Star and Daily News,* Sept. 24, 1972; Nancy L. Ross, "Dr. Ramey: The Mort Sahl of Women's Lib," *Washington Post,* Nov. 14, 1972, C2; Solveig Eggerz, "Equality for the Female Schlemiel," *Washington Daily News,* Apr. 2, 1973, 38; Estelle Ramey, "A Feminist Talks to Men," *Johns Hopkins Magazine,* Sept. 1973, 7–9; Michael Kernan, "Dr. Estelle Ramey: Why Can't a Man Be More Like a Woman?" *Washington Post,* Sept. 28, 1975, E1 and E3; Michael Kolbenschlag, "Dr. Estelle Ramey: Reclaiming the Feminine Legacy," *Human Behavior* 5 (July 1976): 25–27; and "Dr. Estelle Ramey Is a Woman Who Knows Her Places—Wife, Mother, Scientist, Activist," *People Magazine,* Jan. 19, 1996, 69–70. The Columbia University Oral History Research Office also prepared two 1980 interviews with Ramey, "The Reminiscences of Estelle Ramey," which is available in the Rare Book and Special Collections Room, Butler Library, Columbia University, New York. Records relating to the Berman-Ramey debates, which also involved Congresswoman Patsy Mink (D-HI), are in the Smith College Archives, Northampton, MA. There are more clippings in the Fann Harding Papers, box 1, folder 7.

10. The AAAS Committee on Council Affairs requested the name change in December 1974, but such a step would require a vote of the executive board of the AWIS and ratification by its membership. Since all this could not be accomplished before the full AAAS Council met at the end of January 1975, the committee made the AWIS's affiliation contingent on a subsequent change. Agenda, "Meeting of the Committee on Council Affairs," Dec. 8, 1974, 2 and tab C; Catherine Borras to Anne M. Briscoe, Dec. 11, 1974, and reply, Dec. 18, 1974; Catherine Borras to Committee on Council Affairs, Dec. 20, 1974, and Jan. 8, 1975, all in box 35, AAAS, Margaret Mead Papers, Organizations Series, Manuscript Division, Library of Congress, Washington, DC.

11. The AWP became a unique institution, a feminist nonhierarchical collective whose members do its work through committees and without titles, offices, or staff. Part of its role has been to be a gadfly to the profession, as in 1969, when it demonstrated at the large annual meeting of the American Psychological Association and pressured the organization to start a task force on women, whose report a year later recommended the formation of a standing committee on women. Later both the AWP and the committee got the APA to start both Division 35, on the psychology of women, and an Office of Women, with a paid staff at its headquarters in Washington, DC. The AWP has many lesbian members, and its annual meeting features a dance. It has a website (www.awpsych.org), and its papers are at the Schlesinger Library, RIAS. Its unusual structure and functioning deserve an organizational analysis. See Leonore Tiefer, "A Brief History of the Association for Women in Psychology, 1969–1991," *Psychology of Women Quarterly* 15 (1991): 635–49.

12. On the early days of the AWM, see Alice T. Schafer, "Women and Mathematics," in *Mathematics Tomorrow,* ed. Lynn A. Steen (New York: Springer-Verlag, 1981), 176–80; and Lenore Blum, "A Brief History of the Association for Women in Mathematics: The Presidents' Perspectives," *Notices of the American Mathematical Society* 38 (Sept. 1991): 738–54. On Mary Gray, see Florence Fasanelli, "Mary Gray (1939–)," in *Notable Women in Mathematics: A Biographical Dictionary,* ed. Charlene Morrow and Teri Perl

(Westport, CT: Press, 1998), 71–76, which fails to mention that she was also a mainstay of Committee W of the American Association of University Professors (AAUP) and the Women's Equity Action League (WEAL). From 1973 until 1993, when the AWM's central office moved to the University of Maryland, space for it was provided by Wellesley College. Meanwhile, the association's membership rose to more than four thousand. Other women in mathematical fields preferred to work within existing organizations such as the American Mathematical Society and the Mathematical Association of America. By 1980 the AWM was part of the Joint Committee on Women in the Mathematical Sciences, along with six other major mathematical organizations—the American Mathematical Society (AMS), the Mathematical Association of America (MAA), the National Council of Teachers of Mathematics (NCTM), the Society for Industrial and Applied Mathematics (SIAM), the American Statistical Association (ASA), and the Institute for Mathematical Statistics (IMS). "Joint Committee on Women," *AWM Newsletter* 28 (May–June 1998): 15. There is a full set of *AWM Newsletters* in the Mathematics Library at Cornell University, Ithaca, NY.

13. The SWS papers are at the Schlesinger Library, RIAS, but as of 2008 they were not open for use. There is material about the early days of the SWS in Pamela Roby, "Women and the ASA: Degendering Organizational Structures and Processes, 1964–1974," *American Sociologist* 23 (Spring 1992): 18–48; and in the Jessie Bernard Papers, Historical Collections and Labor Archives, Pattee Library, Pennsylvania State University, University Park, PA. Julia Apter was a vigorous member of the Biophysical Society's Committee on Professional Opportunities for Women. Muriel S. Prouty, comp., "How Did We Get Here? A Brief Anecdotal History of Women's Activities in the Biophysical Society," *Spectrum*, Jan. 1991, 2–3, copy in box 16, folder 21, of the Papers of the Biophysical Society, Special Collections, Albin O. Kuhn Library and Gallery, UMBC; Laura Williams, "The History of the WICB: The Founding and Early Years," *WICB Newsletter* 19 (Aug. 1996); idem, "The History of WICB: The Later Years," ibid. 19 (Oct. 1996), available among its archived newsletters online at www.ascb.org. Some material relating to its early years and newsletters and its subsequent reincarnation in 1993 as an official committee of the ASCB are in box 75 of the Papers of the American Society for Cell Biology, Special Collections, Albin O. Kuhn Library and Gallery, UMBC. On Walbot, see Barbara J. Love, comp., *Feminists Who Changed America, 1963–1975* (Urbana: Univ. of Illinois Press, 2006), 475.

14. Martha T. Mednick and Laura L. Urbanski, "The Origins and Activities of APA's Division of the Psychology of Women," *Psychology of Women Quarterly* 15 (1991): 651–63; Anne M. Briscoe, "Phenomenon of the Seventies: The Women's Caucuses," *Signs* 4, no. 1 (1978): 152–58; Alicia E. Leach and Michele Aldrich, comps., *Associations and Committees of or for Women in Science, Engineering, Mathematics and Medicine*, AAAS Publication 84-6 (Washington, DC: AAAS, 1984); Mariam K. Chamberlain, ed., *Women in Academia: Progress and Prospects* (New York: Russell Sage Foundation, 1988), ch. 12.

15. Jesse Bernard to Arlie Hochschild, May 21, 1972, Jessie Bernard Papers, box 8; Ruth B. Kundsin, ed., *Women and Success: The Anatomy of Achievement* (New York: William Morrow, 1974); Ruth B. Kundsin, ed., "Successful Women in the Sciences: An Analysis of Determinants," special issue, *Annals of the New York Academy of Sciences* 208

(1973). NYAS member Anne Briscoe reviewed both works by Kundsin in "Women Pursuing Careers," *Science* 185 (1974): 1040.

16. Joan Huber, "Forget this career stuff . . . ," *CHE*, May 13, 1974, 12; Cecily Cannan Selby, ed., "Women in Science and Engineering: Choices for Success," special issue, *Annals of the New York Academy of Sciences* 869 (1999).

17. Meanwhile, another ad hoc committee on women, set up by the AAAS executive officer, William Bevan, to advise the board of directors, met in May 1972, and as late as March 1973 it continued to insist on a separate office for women. Melanie J. Hunter, "Women in Science: Turmoil within the American Association for the Advancement of Science," July 1999, 7–16, based on materials at the AAAS Archives, Washington, DC (I thank archivist Amy Crumpton for a copy). There is also considerable material on the early days of the OOS in the Janet Welsh Brown Files at the AAAS Archives; the Fann Harding Papers, boxes 3, 6, 7, 11; the Vera Kistiakowsky Papers, box 1, AAAS, 1971–75; and carton 27 of the Elizabeth L. Scott Papers, Bancroft Library, University of California, Berkeley. See also Mary E. Clutter and Virginia Walbot, "AAAS Meeting," *Science* 175 (1972): 944–45; and Sally Gregory Kohlstedt, Michael M. Sokal, and Bruce V. Lewenstein, *The Establishment of Science in America: 150 Years of the American Association for the Advancement of Science* (New Brunswick, NJ: Rutgers Univ. Press, 1999), 142–43. The first scientific society to establish an office devoted solely to women's issues was the APA, which voted in August 1977 to set up its Women's Program Office, long headed by Nancy Felipe Russo.

18. On Brown, see *AMWS*, Social and Behavioral Sciences, 13th ed., 1978, 154; Rayna Green, "Janet Brown Leaves AAAS," *Science* 204 (1979): 490, 532; program, "Women in Science Reception Honoring the Leadership of Janet Welsh Brown, PhD," Sept. 29, 1980, copies in Fann Harding Papers, box 10, folder 4, and in box 6, Women in Science, Jewel Plummer Cobb Papers, SCM 89-63, Schomburg Center for Research in Black Culture, New York Public Library, with the text of her remarks; and Love, *Feminists Who Changed America*, 60–61. Some of the spirit of this group is captured in Nancy Felipe Russo and Marie M. Cassidy, "Women in Science and Technology," in "Women in Washington: Advocates for Public Policy," ed. Irene Tinker, special issue, *Sage Yearbook in Women's Policy Studies* 7 (1983). See also the annual reports of the Committee on Opportunities in Science to the AAAS Council, in the council agenda and minutes for the late 1970s, also at the AAAS Archives.

19. On Cobb, see F. Elaine De Lancey, "Jewel Plummer Cobb (1924–), Cell Biologist," in *Notable Women in the Life Sciences: A Biographical Dictionary*, ed. Benjamin F. Shearer and Barbara S. Shearer (Westport, CT: Greenwood, 1996), 82–87; *Nominations: Hearings Before the Committee of Labor and Public Welfare, United States Senate, 93rd Congress, Second Session, on Jewel P. Cobb, Norman Hackerman, William Neill Hubbard, Jr., Saunders MacLane, Grover E. Murray, Donald B. Rice, Jr., L. Donald Shield, and James H. Zumberge to be Members of the National Science Board, National Science Foundation, September 1 and October 3, 1974* (Washington, DC: GPO, 1974), 52–61, 104; Jewel Plummer Cobb, "A Life in Science: Research and Service," *Sage* 6 (Fall 1989): 39–43; and Diane Ross, "Meet a Member: Jewel Plummer Cobb," *AWIS Newsletter* 19 (Mar.–Apr. 1990): 10–13. There is also much material in the Archives, Sarah Lawrence College, Bronxville, NY, and in the Jewel Plummer Cobb Papers.

20. On Callanan, see *AMWS*, 14th ed., 1979, 691–92; M. Joan Callanan, "NSF Programs for Women in Science, Presented at Conference on Reentry Women in Science, Stouffer's National Center Hotel, Crystal City, Virginia, December 3, 1979," in box 3, Papers Delivered by M. Joan Callanan, Administrative Records, Records of the Women in Science Project, NSF, RG 307, National Archives and Records Administration, College Park, MD; F. James Rutherford, "The Role of the National Science Foundation," in "Expanding the Role of Women in the Sciences," ed. Anne M. Briscoe and Sheila M. Pfafflin, special issue, *Annals of the New York Academy of Sciences* 323 (1979): 276–82; NSF, "Program Estimates FY 1978–1982, Women in Science Program, 11–26–75," in Jewel Plummer Cobb Papers, SCM 89-63, box 10, NSB Committee on Minorities and Women, 1975–77; "Background and Current Status of the NSF Women in Science Program," Aug. 24, 1976, in Vera Kistiakowsky Papers, box 21, NAS-NRC CEEWISE; "Women in Science Career Facilitation Projects, Summary of Awards (1976 through 1981)," box 3, WIS Career Facilitation Awards, 1974–81, Records of the Women in Science Project, NSF, NARA. A good retrospective is Diane Weisz, "Memorandum of Discussion, Thirtieth Meeting, Committee on Minorities and Women in Science, NSB, October 16, 1980," copy in agenda (binder) for meeting of the Women Chemists Committee of the American Chemical Society, Mar. 21, 1983, tab 7, in box 9, folder 7, Kathleen M. Desmond Trehanovsky Papers, Women in Science and Engineering Archives, Special Collections, Iowa State University Library. See also "Brief History and Current Status of Science Career Facilitation Projects, Women in Science Program, National Science Foundation" (June 1980), Memos Pertaining to WIS Program, Administrative Records, box 3, Records of the Women in Science Project, NSF, NARA, also issued as "Brief History of Science Career Facilitation Projects, National Science Foundation, 1976–1981," appendix 3 to Diane Weisz and Jane Stutsman, "Programs for Women Scientists, Discussion Paper, September 1983," in *Reentry Programs for Female Scientists*, by Alma E. Lantz with Marni C. Whittington et al. (New York: Praeger, 1980), which evaluated the NSF program. Most of the projects have entries in Michele L. Aldrich and Paula Quick Hall, comps., *Programs in Science, Mathematics and Engineering for Women in the United States, 1966–1978*, AAAS Publication 80-11 (Washington, DC: AAAS, 1980).

Journalists and program directors published many enthusiastic articles about the science facilitation workshops, e.g., Sister Mary Kieran McElroy, "A Continuing Education Project for Updating Women in Biochemistry," *Journal of Chemical Education* 55 (Oct. 1978): 649; Edwin S. Weaver, "The Short Course Approach to Reentry Problems for Women in Chemistry," ibid. 56 (Aug. 1979): 509; Sharon K. Hahs, "Interviews with Three Participants in the Visiting Women Scientists Program," *WCC Newsletter*, Apr. 1979, 1–2; Anne C. Roark, "Re-Entry Programs Enable Women to Resume Career in Science," *CHE*, Feb. 11, 1980, 304; Alberta Arthurs, letter to the editor, ibid., Mar. 31, 1980, 15; Beverly Jacobson, "The Sciences Want You," *Savvy*, Oct. 1980, 14–15; Marcia Harrison, "Boosting Women Scientists: Northeastern and BU Offer Free, Partially Funded Graduate Programs," *Boston Sunday Globe*, Sept. 20, 1981, B91–B93, on programs at Northeastern University and Boston University; Nancy Kreinberg, *Ideas for Developing and Conducting a Women in Science Career Workshop*, SE 81-16 (Washington, DC: NSF, 1981); Bonita J. Campbell and Tobi A. Roffman, "Women in Engineering: Programs for

Reentry and Career Change," *IEEE Transactions on Education* E-28 (Nov. 1985): 215–21; and Nina Matheny Roscher, "Returning to Science: It Can Be Done," *Scientist,* July 27, 1987, 29. There is more material in the Nina Matheny Roscher Papers, Women in Science and Engineering Archives, Special Collections, Iowa State University Library.

21. H. Guyford Stever to Senator Edward M. Kennedy, June 27, 1975, reprinted in *Women in Cell Biology Newsletter,* Oct. 1975, inside back cover, in Papers of the American Society for Cell Biology, box 75, folder 23. Kennedy's involvement was a bit unusual, since more often it was the science committee of the House of Representatives that looked most closely at NSF legislation. "AAAS Conference Looks at Women in Science," *C&EN* 55 (Oct. 24, 1977): 5.

22. Lilli S. Hornig to M. Joan Callanan, Feb. 25, 1977, Vera Kistiakowsky Papers, box 26, Wellesley; Lilli S. Hornig, "HERStory," *Grants Magazine* 1 (Mar. 1978): 36–42; "Committee on the Education and Employment of Women in Science and Engineering, National Research Council," booklet (1980–81), carton 113, Biobibliographical Backup, 1980–81, Elizabeth L. Scott Papers; Carlos E. Kruytbosch to Vera Kistiakowski [*sic*], Apr. 15, 1977, and Janet Welsh Brown to Kistiakowsky et al., Apr. 20 and May 12, 1977, enclosing copy of Janet Welsh Brown to Dr. Frank Press, May 12, 1977, all in Vera Kistiakowsky Papers, box 12, Lobbying.

23. The full titles were *Climbing the Academic Ladder: Doctoral Women Scientists in Academe, A Report to the Office of Science and Technology Policy from the Committee on the Education and Employment of Women in Science and Engineering, Commission on Human Resources, National Research Council* (Washington, DC: NAS, 1979); and *Women Scientists in Industry and Government: How Much Progress in the 1970s? An Interim Report to the Office of Science and Technology Policy from the Committee on Human Resources, National Research Council* (Washington, DC: NAS, 1980). The Papers of the Committee on the Education and Employment of Women in Science and Engineering, Archives, National Academy of Sciences, Washington, DC, were closed for the years after 1980 when I used them in 2005. The reports were criticized as methodologically limited in Yu Xie and Kimberlee A. Shauman, *Women in Science: Career Processes and Outcomes* (Cambridge, MA: Harvard Univ. Press, 2003), 9.

The committee was established in December 1974, after a difficult eighteen months between women scientists and NAS officials. In 1973 William Kelly, who had provided data on physics PhDs to Vera Kistiakowsky when she was preparing the APS report on the status of women in 1971–72, had invited her to chair a meeting on rosters at the NAS in June 1973. (The NAS staffer and eminent physiologist Louise Marshall had suggested such a meeting earlier, but nothing had come of it.) One recommendation of the assembled knowledgeable women, including Estelle Ramey and the forceful Julia Apter, was that the National Academy should establish a committee on women. Surprisingly, the executive officer Phillip Handler agreed, and he offered the chairmanship to Kistiakowsky. However, she found the charge to the proposed committee unclear, was disturbed by its lack of funding and staff, and negotiated for an NAS commitment to a "talent bank," since those at her meeting, if not Handler, felt that the academy needed a roster to find talented women. When Handler refused this, insisting that the NRC preferred to handle its annual appointment to hundreds of committees via traditional word of mouth, Kistiakowsky declined the post, and Handler then canceled the committee. By then the NRC's Commission on Human Resources, headed by Robert Alberty, also of MIT, had decided to set up its own committee on women, which would study the overall problems of women in science and not meddle with the academy's governance, elections, or NRC appointments. It was this later committee that Hornig was invited to chair in early 1975. But for various reasons the committee was ineffective until it was threatened with cancellation in 1977: Hornig became ill, committee absenteeism was high, those who came brainstormed but then resigned when their ideas were not taken up, and a proposal to the NSF went unfunded. (Many relevant documents are in Vera Kistiakowsky Papers; the Records of the Office of Scientific Personnel—Conference on Women in Science and Engineering, General, 1973, and Records of the Committee on Participation of Women in the National Research Council, 1974, both in Archives, NAS; and Elizabeth L. Scott Papers, cartons 64 and 94).

24. Elizabeth Baranger to colleagues, Nov. 21, 1972, box 2, APS-CSWP, 1972–73 and 1983–84, and Anne Briscoe to Vera Kistiakowsky, Sept. 10 (two letters) and Sept. 27, 1982, box 6, AWIS 1982, all in Vera Kistiakowsky Papers. Fann Harding was a founder of the group, and much about the FOPW can be found in her papers, esp. boxes 7–10. There are also accounts of its early days in the *AWIS Newsletter,* for example, "Historic Professional Women's Meeting Held in Virginia," in vol. 2 (Winter 1973): 1. In 1983 Ruth Oltman was planning to write a history of the beginnings of the FOPW. Several essays in Irene Tinker's *Women in Washington* describe the FOPW's activities in its heyday.

25. Janet Welsh Brown, Julia Lear, and Donna Shavlik, *Effecting Social Change for Women* (Washington, DC: Federation of Organizations of Professional Women, 1976), 34, 37. The two workshops were held in November 1975 and February 1976.

26. Margaret W. Rossiter, "American Scientific Societies and the Equal Rights Amendment, 1977–1982," in "Women Scholars and Institutions: Proceedings of the International Conference (Prague, June 8–11, 2003)," special issue, *Studies in the History of Science and Humanities, Prague* 13 (2004): 101–14; Hunter, "Women in Science," 20–24, reporting that Carey received 45 letters supporting his decision (4 from men and 41 from women) and 106 in opposition (97 from men and 9 women) (I thank Amy Crumpton for a copy of Hunter's paper). Frances K. Graham (chairwoman of Section J [psychology]) to Margaret Mead, Dec. 9, 1977, box 42, AAAS, Mead Papers, Organization Series; John Walsh, "'79 AAAS Meeting Moved from Chicago to Houston," *Science* 199 (1978): 954; John A. McIntyre and Alberta S. Gilinsky, in "Letters: ERA and AAAS," ibid., 1156; Melvin Calvin et al., letter to the editor of *C&EN,* Apr. 21, 1978, enclosed with Michael Heylin to Anna J. Harrison, June 5, 1978, in box 14, folder 4, Anna Jane Harrison Papers, Mount Holyoke College Archives, South Hadley, MA. There is a also great deal of material about the American Astronomical Society and the ERA in boxes 5, 10, and 18 of the Vera C. Rubin Papers, Manuscript Division, Library of Congress, and about the American Physical Society and the ERA in the Vera Kistiakowsky Papers, especially William D. Carey's letter of Sept. 21, 1979, to APS president Lewis Branscomb explaining his rationale, box 5, APS-POPA—1980. See also Florence L. Denmark, "Women in Psychology in the United States," in Briscoe and Pfafflin, "Expanding

the Role of Women in the Sciences," 70–71; and Bruce V. Lewenstein, "Shifting Science from People to Programs: AAAS in the Postwar Years," in Kohlstedt, Sokal, and Lewenstein, *Establishment of Science in America,* 213n127.

27. See, e.g., Paula Quick Hall and Anne Swartz, comps., *Associations of/for Women Scientists, December 1978* (Washington, DC: OOS, AAAS, 1978); idem, *Committees of/for Women of Associations of Scientists, December 1978* (Washington, DC: OOS, AAAS, 1978); Leach and Aldrich, *Associations and Committees;* and Aldrich and Hall, *Programs in Science, Mathematics and Engineering.*

28. Janet Welsh Brown, Heather Coleman, and Susan Posner, *Rosters of Minority and Women Professionals,* AAAS Miscellaneous Publication 75-1 (Washington, DC: AAAS, 1975), 25–31. The report had several appendixes that described the rosters under way at the time. The year-old FOPW was preparing itself to help individual groups. Meshkov, "Survey and Evaluation of Registries of [Women?] in the Professions."

29. The original proposal to the NSF had included Asians. "Minority Women Scientists, A Preliminary Proposal submitted to the National Science Foundation by the Office of Opportunities in Science, American Association for the Advancement of Science," n.d., 3, in Jewel Plummer Cobb Papers, SCM 89-63, box 10, NSB Committee on Women and Minorities, 1974–75; correspondence and lists of nominees in Fann Harding Papers, box 11, folder 6; related items in Vera Kistiakowsky Papers, box 1, Negro Women in Science and Engineering, 1964–65; Shirley Mahaley Malcom, Paula Quick Hall, and Janet Welsh Brown, *The Double Bind: The Price of Being a Minority Woman in Science,* AAAS Report No. 76-R-3 (Washington, DC: AAAS, Apr. 1976). Malcom succeeded Janet Welsh Brown as OOS program head in 1979. Shirley Malcom, "Increasing the Participation of Black Women in Science and Technology," *Sage* 6 (Fall 1989): 15–17; Wini Warren, "Shirley M. Malcolm [*sic*], Directing Changes in Science," in *Black Women Scientists in the United States* (Bloomington: Indiana Univ. Press, 1999), 185–92. See also Paula Quick Hall, *Problems and Solutions in the Education, Employment and Personal Choices of Minority Women in Science* (Washington, DC: OOOS, AAAS, 1981), published with the support of the Polaroid Corporation; and Yolanda Scott George, "Affirmative Action Programs That Work," in *Women and Minorities in Science: Strategies for Increasing Participation,* ed. Sheila M. Humphreys (Boulder, CO: Westview, 1982), 96–97.

30. "Project on the Handicapped Holds Barrier-Free Workshop," *Science* 197 (1977): 551; Martha Ross Redden, Wayne Fortunato-Schwandt, and Janet Welsh Brown, *Barrier Free Meetings: A Guide for Professional Associations,* AAAS Publication 76-07S (Washington, DC: AAAS, 1976).

31. Cheryl M. Fields, "Women in Science: Breaking the Barriers," *CHE,* Oct. 31, 1977, 7–8; Nancy Cahill Joyce, "Women Researchers Analyze Education, Job Barriers," *Science* 198 (1977): 917–18; [Janet Welsh Brown, Michele L. Aldrich, and Paula Quick Hall], "Recommendations of the Conference on Women in Scientific Research Organized by the American Association for the Advancement of Science, 17–20 Oct. 1977, Washington, DC" (mimeograph, n.d.); NSF, *Increasing the Participation of Women in Scientific Research: Summary of Conference Proceedings, October 1977, and Research Study Project Report, March 1978* (Washington, DC, 1978); Janet Welsh Brown, Michele L.

Aldrich, and Paula Quick Hall, "Report on the Participation of Women in Scientific Research" (typescript, OOOS, AAAS, Washington, DC, Mar. 1978), copy in box 14, Records of the Women in Science Project, NSF, RG 307, NARA. Several of the then "young" women attendees have since become well known or important themselves, such as the statistician Barbara Bailar, the chemist Marye Anne Fox, the physicist Elizabeth S. Ivey, the sociologist Barbara F. Reskin, the mathematician Judith Roitman, the historian of science Margaret W. Rossiter, and the economist Laura D'Andrea Tyson.

32. Deborah Shapley, "Women in Science Legislation Advanced," *Science* 199 (1978): 1320–21; Edward M. Kennedy to Vera Kistiakowsky, Apr. 25, 1975, Vera Kistiakowsky Papers, box 26, Women in Science, Background Material.

33. *National Science Foundation Authorization Act for Fiscal Years 1979 and 1980 and the Women in Science and Technology Equal Opportunity Act: Hearing Before the Subcommittee on Health and Scientific Research of the Committee on Human Resources, United States Senate, 95th Congress, Second Session on S. 2549 . . . [and] S. 2550 . . . , April 10, 1978* (Washington, DC: GPO, 1978); Anna J. Harrison to Hon. Harrison A. Williams, Apr. 28, 1978, Anna Jane Harrison Papers, box 4, folder 4. See also S. Mahoob to Anna J. Harrison, Mar. 29, 1978, ibid.; Lester G. Paldy, "The Advancement of Women in Scientific Careers," *Journal of College Science Teaching* 8 (Nov. 1978): 48–50, 53; and Judith Garelick, "Academia's Reluctant Heroines," *Change* 12 (Jan. 1980): 17–20.

34. The cancer researcher Charlotte Friend was the first woman president of the NYAS, in 1977. See the following in Briscoe and Pfafflin, "Expanding the Role of Women in the Sciences": Anne M. Briscoe, introduction, 1–5; Mary Howell, "The New Feminism and the Medical School Milieu," 210–14; Rutherford, "Role of the National Science Foundation," 276–82; and Sheila Pfafflin, "Equal Opportunity for Women in Science," 341–44. For more on Mary Howell, MD, see ch. 4 n16 below. Anna Harrison later wrote to an official at Mount Holyoke College, "It was a strange conference." Anna J. Harrison to Gwen Glass, Apr. 11, 1979, Anna Jane Harrison Papers, box 27, folder 15.

35. Jonathan Cole, *Fair Science: Women in the Scientific Community* (New York: Free Press, 1979). Among the many reviews of Cole's book were Nan Robertson, "Do Female Scientists Face Bias? Columbia Sociologist Says 'No,'" *NYT*, Dec. 17, 1979, D12; Betty M. Vetter, "Sex Discrimination in the Halls of Science," *C&EN* 58 (Mar. 17, 1980): 37–38; Betty M. Vetter to David M. Kiefer, Feb. 7, 1980, in box 7, folder 6, of the Betty Vetter Papers, Women in Science and Engineering Archives, Special Collections, Iowa State University Library; Jacqueline Macaulay, letter to the editor, *Contemporary Psychology* 25 (1980): 1022; Naomi Weisstein, "Fair Science: Survivors in the Savage Wars," unpublished autobiographical manuscript, box 2, Naomi Weisstein Papers, Schlesinger Library, RIAS; Karen Oppenheim Mason, "Sex and Status in Science," *Science* 208 (1980): 277–78; Betty M. Vetter to Karen Oppenheim Mason, Apr. 10, 1980, Betty Vetter Papers, box 7, folder 6; Gaye Tuchman, "Discriminating Science," *Social Policy* 11 (May–June 1980): 59–64; Patricia Yancey Martin, "'Fair Science': Test or Assertion? A Response to Cole's 'Women in Science,'" *Sociological Review* 30 (1980): 478–508; Helen S. Astin, "Book Review," *Psychology of Women Quarterly* 6 (1981): 253–56; Margaret W. Rossiter, "Fair Enough?" *Isis* 72 (1981): 99–103; Ellen Messer-Davidow to Arlie Hochschild, July 8, 1980, Hochschild to Elizabeth Scott, July 30, 1980, and Scott to Hochschild,

Aug. 16, 1980, all in Elizabeth L. Scott Papers, carton 84, box of materials on NAS career differentials; and Jonathan R. Cole, "Women Scientists and Discrimination," *CHE,* Jan. 7, 1980, 23.

36. Anne M. Briscoe, "The Kennedy Bill: A Pipe Dream?" *AWIS Newsletter* 8 (May–June 1979): 7–8; Russo and Cassidy, "Women in Science and Technology"; *Women in Science and Technology Equal Opportunity Act, 1979: Hearing Before the Subcommittee on Health and Scientific Research of the Committee on Labor and Human Resources, United States Senate, 96th Congress, First Session on S. 568 . . . August 1, 1979* (Washington, DC: GPO, 1979); *Women in Science and Technology Equal Opportunity Act, 1980: Hearing Before the Subcommittee on Health and Scientific Research of the Committee on Labor and Human Resources, United States Senate, 96th Congress, Second Session on S. 568 . . . March 3, 1980* (Washington, DC: GPO, 1980); *National Science Foundation and Women in Science Authorization Act for Fiscal Years 1981 and 1982 . . . Report Together with Additional Views,* S. Rep. No. 96-713 (1980); *National Science Foundation Authorization and Equal Opportunities in Science and Technology,* H.R. Rep. No. 96-1474 (1980) (Conf. Rep.); mailgram, Jewel Plummer Cobb to Senator Harrison Williams, Apr. 22, 1980, copy in Jewel Plummer Cobb Papers, SCM 89-63, box 10, NSF; Lewis M. Branscomb, "Women in Science," *Science* 205 (1979): 751; Harold L. Davis, "Helping Women in Physics Succeed," *Physics Today* 33 (Sept. 1980): 144, which provoked some critical letters and a response by Carol Jo Crannell, "Women in Physics," ibid. 34 (Feb. 1981): 11, 13. In addition, the Association for Women in Computing submitted testimony between hearings that was not published. Nancy Bonney Bryan, "Testimony Regarding S. 568 on Behalf of the Association for Women in Computing," Aug. 6, 1979, in box 1, folder 45, Papers of the Association for Women in Computing, Charles Babbage Institute Archives, University of Minnesota, Minneapolis. Later the leaders of the AWM learned that a Kennedy staff member had lost the folder containing the testimony by mathematicians. Memorandum, Alice Schafer to Judy Roitman, Jan. 4, 1980, box 1, Testimonies before Senate Committee re Bill #568, Papers of the Association for Women in Mathematics, Wellesley College Archives, Wellesley, MA.

37. Marian Koshland, quoted in John Walsh, "Women in Science Cut," *Science* 211 (1981): 1027; Robert H. Hartford, "Women in Physics," *Physics Today* 34 (Feb. 1981): 11, 13; "Statement to the Senate Labor and Human Resources Subcommittee of Health and Scientific Research by Professor Mary Gray, Department of Mathematics, American University, Former Chairperson, Committee on the Status of Women in the Academic Profession of the American Association of University Professors, September 6, 1979," in Papers of the Association for Women in Computing, box 1, folder 45, Charles Babbage Institute Archives, University of Minnesota. See also Judith Roitman, "Testimony Regarding S. 568 on Behalf of the Association for Women in Mathematics," July 1979, which makes many of the same points as Gray above, in Papers of the Association for Women in Mathematics, box 1, Testimonies before Senate Committee re Bill #568, Wellesley College Archives.

38. On Brown, see above, n. 18; R. Jeffrey Smith, "Too Much Congressional Direction?" *Science* 211 (1981): 260–61; and Anne C. Roark, "NSF Backs Women's Programs—with Reservations," *CHE,* Jan. 26, 1981, 13.

Chapter 2 · Taking On Academia

1. Earl F. Cheit, *The New Depression in Higher Education: A Study of Financial Conditions at 41 Colleges and Universities* (New York: McGraw-Hill, 1971). On the PhD glut, see Deborah Shapley, "Job Prospects: Science Graduates Face Worst Year in Two Decades," *Science* 172 (1971): 823–24; and H. William Koch, "On Physics and Employment of Physicists in 1970," *Physics Today* 24 (June 1971): 23–27.

2. See, e.g., "Discrimination against Women at the University of Pittsburgh: A Report Compiled by the University Committee for Women's Rights, November 1970," in Papers of the University of Hawaii Commission on the Status of Women, Special Collections, University of Hawaii at Manoa Library, Honolulu. On events at the University of Wisconsin, see the extensive Ruth Bleier Papers, University Archives, University of Wisconsin, Madison. Bleier had gotten tenure in the tiny Department of Neuroanatomy in the 1960s and by 1970 was the firebrand behind the women's movement at the university. On Bleier, see Barbara J. Love, comp., *Feminists Who Changed America, 1963–1975* (Urbana: Univ. of Illinois Press, 2006), 46. The journalist and litigant Joan Abramson, formerly of the University of Hawaii, ridiculed the HEW and the DOL's Office of Federal Contract Compliance as "reluctant dragons" in *Old Boys, New Women: The Politics of Sex Discrimination* (New York: Praeger, 1979). See also Abramson's *The Invisible Woman: Sex Discrimination in the Academic Profession* (San Francisco: Jossey-Bass, 1974); idem, "When Internal Grievance Procedures Fail: The Search for Justice," in "Resolving Sex Discrimination Grievances on Campus: Four Perspectives," ed. Jennie Farley (mimeograph, distributed by Institute of Labor Relations, Cornell University, Ithaca, NY [1982?]), 42–62; and Deborah Shapley, "University Women's Rights: Whose Feet Are Dragging?" *Science* 175 (1972): 151–54; and Judith Glazer-Raymo, *Shattering the Myths: Women in Academe* (Baltimore: Johns Hopkins Univ. Press, 1999).

3. On Sandler, see esp. "An Uppity Woman," *Time,* July 10, 1972, 92; Liz McMillen, "Women's Groups: Going the Old Boys' Network One Better," *CHE,* Dec. 3, 1986, 1517 (which is also about Donna Shavlik, of the Office of Women in Higher Education of the American Council of Education); Debra E. Blum, "Head of College Association's Project on Women Dismissed after 20 Years in Advocacy Role," ibid., Dec. 5, 1990, A15, A20; and Love, *Feminists Who Changed America,* 403. The importance of such Washington insiders deserves analysis. Sandler's Project on the Status and Education of Women also benefited financially from the sale of its publications, including in later years its newsletter, *On Campus with Women,* which was sent to about 17,000 interested persons, keeping them up to date on the numerous judicial actions, administrative rulings, and other news pouring out of Washington. Sandler may merit a biography, as does Green. The extensive papers of both Sandler herself and her project are at the Schlesinger Library, RIAS, Harvard University, Cambridge, MA.

4. See esp. "Discrimination against Women at the University of Pittsburgh." One member of the University of Hawaii Commission on the Status of Women zealously collected materials from similar groups on the mainland, thus making that university's collection a very useful one for this early period. See also Jo Freeman, *The Politics of Women's Liberation: A Case Study of an Emerging Social Movement and Its Relation to*

the Policy Process (New York: David McKay, 1975), 196–200. On Sandler's visit to the University of Michigan in 1970, see Kathryn Kish Sklar, "The Women's Studies Moment, 1972," in *The Politics of Women's Studies: Testimony from Thirty Founding Mothers*, ed. Florence Howe (New York: Feminist Press, 2000), 136.

5. The situation at the University of Washington is covered well in the Davida Teller Papers, University of Washington Archives, Seattle. At the University of California at Berkeley a League of Academic Women filed complaints and a lawsuit against the UC system. See "Chronicle of League of Academic Women, University of California, Berkeley [May 1970–January 1972]" (typescript), and several newsletters in carton 45, HEW-UCB, Elizabeth L. Scott Papers, Bancroft Library, University of California, Berkeley; and "League of Academic Women and National Organization for Women's Class Action Complaint Against the University of California at Berkeley and the Office of the Vice President of the University of California for Violation of Executive Orders 11246 and 11375, Forbidding Discrimination on the Basis of Sex," in box 4, California, University of, at Berkeley, Papers of the University of Hawaii Commission on the Status of Women.

6. On Oct. 5, 1978, President Jimmy Carter signed an Executive Order 12086, which consolidated many formerly decentralized federal contract-compliance activities under the Office of Federal Contract Compliance Programs in the Department of Labor. On Graham, see "Woman Wins KSU Job Dispute," *Cleveland Plain Dealer*, Dec. 18, 1982, clipping in box 12, Cornell Eleven Papers, Rare and Manuscript Collections, Kroch Library, Cornell University, Ithaca, NY; and Kenneth Mines (HEW) to Shirley A. Graham, Nov. 26, 1976, and clippings in Margaret Bryan Davis Papers, University Archives, University of Minnesota, Minneapolis.

7. After the Supreme Court upheld in the mid-1980s the Grove City College and Hillsdale College cases, the lower courts' approvals of the Department of Education's decisions to limit Title IX to student programs, Congress eventually passed the Civil Rights Restoration Act of 1988 (overriding a veto by President Ronald Reagan). It strengthened enforcement of Title VII of the Civil Rights Act of 1964, which applied to employment. The black mathematician Vivienne Malone Mayes, one of the few of her race employed at Baylor University in Waco, Texas, noticed that attention to her treatment lessened when the federal investigators stopped coming to campus in the 1980s. Etta Z. Falconer and Lee Lorch, "Vivienne Malone-Mayes: In Memoriam," *AWM Newsletter* 25 (Nov.–Dec. 1995): 9. Her papers and a 1987 oral history memoir are in the Texas Collection, Carroll Library, Baylor University, Waco, TX.

8. Alice S. Rossi, "Report of Committee W, 1970–71," *AAUP Bulletin* 57 (1971): 215–20; "Faculty Appointments and Family Relationship," *Liberal Education* 57 (1971): 305; Margaret W. Rossiter, *Women Scientists in America: Before Affirmative Action, 1940–1972* (Baltimore: Johns Hopkins Univ. Press, 1995), 379, 527n43.

9. See "The Status of Women," *Academe* 75 (May–June 1989): 17–20, a seventy-fifth-anniversary collage of snippets from some annual reports. A full history of the committee needs to be written. Many of Committee W's annual reports are printed in the *AAUP Bulletin*, which was renamed *Academe* in 1979. There is material on the TIAA-CREF pension battles in the Thelma Kennedy Papers, University of Washington Archives. Other parts of the AAUP took on other issues that affected women; for ex-

ample, Committee Z published its annual salary survey of American colleges, and Committee A investigated cases of free speech, tenure denials, and dismissals. In the 1970s the AAUP also got involved in unionization and collective bargaining, the needs of part-time faculty, and other bellwether issues of the time. On Bergmann, see http://en.wikipedia.org/wiki/Barbara_Bergmann; and Steven Pressman, *Fifty Major Economists* (London: Routledge, 1999), 181–85; and Love, *Feminists Who Changed America*, 39. On Scott, see *AMWS*, 15th ed., 1982, 6:563; and David Blackwell et al., "Elizabeth Leonard Scott, 1917–1988, Professor of Statistics, Emerita," ed. David Krogh, in *University of California: In Memoriam, 1991* (Oakland: Academic Senate, University of California, 1991), 186–90. On Schafer, see *AMWS*, 19th ed., 1995–96, 6:577. On Gray, see Ellen Morgenstern, "Features Profile: Mary Gray," *Academe* 65 (Sept. 1979): 368–69; and Florence Fasanelli, "Mary Gray (1939–)," in *Notable Women in Mathematics: A Biographical Dictionary*, ed. Charlene Morrow and Teri Perl (Westport, CT: Greenwood, 1998), 71–76, which does not mention Gray's work with the AAUP.

10. Margaret Harlow had been an assistant professor at Wisconsin when she married Professor Harry Harlow in 1948, whereupon she was forced off the faculty because of nepotism rules. In 1965 the university relented a bit and let her be a lecturer in another department. Finally, in 1970, with the waning of the nepotism rules, she became a professor in the educational psychology department but died in 1971. She was bitter that she had been cheated out of twenty years of her career. Susan B. Miller, "Female Academicians Claim Careers Curbed by Male Chauvinists," *Wall Street Journal*, June 30, 1971, 1, 18; Deborah Blum, *Love at Goon Park: Harry Harlow and the Science of Affection* (Cambridge, MA: Perseus, 2002), 126–32, 209–11, 225–28. Mary Ellen Rudin, the wife of Walter Rudin, in the Wisconsin mathematics department, was a lecturer from 1958 until her sudden promotion to full professor in 1971. Claudia Henrion, *Women in Mathematics: The Addition of Difference* (Bloomington: Indiana Univ. Press, 1997), 85.

11. On Tidball, see *AMWS*, 19th ed., 1995–96, 7:155; and Janet Battaile, "Debate on Nepotism Rules Grows with a Rise in Working Couples," *NYT*, May 9, 1978, 81. On Kistiakowsky, see *AMWS*, 19th ed., 1995–96, 4:437–38. Kistiakowsky's title had been "scientist" in MIT's Laboratory for Nuclear Science from 1965 until 1969 and then "senior research scientist" in its Department of Physics until 1972, when she became professor of physics.

12. On Shklar, see Judith N. Shklar, *A Life of Learning*, ACLS Occasional Paper No. 9 (Washington, DC: American Council of Learned Societies, 1989); and *Memorial Tributes to Judith Nisse Shklar, 1928–1992* (Cambridge, MA: Harvard University, 1992). Ruth Hubbard was a longtime lecturer and research associate at Harvard who was a visiting professor at nearby MIT in 1972 and then promoted to full professor at Harvard in 1973. *AMWS*, 18th ed., 1992, 3:921; N[ancy] M. T[ooney], "Meet a Member," *AWIS Newsletter* 5 (Nov.–Dec. 1976): 8–10; Ruth Hubbard, "Reflections on My Life as a Scientist," *Radical Teacher*, no. 30 (Jan. 1986): 3–7, copy in box 25, Ruth Hubbard Papers, University Archives, Harvard, Cambridge, MA. There is also an oral history in the Murray Research Archive Database at the Institute for Quantitative Social Sciences, Harvard. Turner had been a research associate at Harvard for thirty years when she became professor of biology and curator of malacology at its Museum of Comparative

Zoology in 1975. *AMWS,* 19th ed., 1995–96, 7:282; Douglas Martin, "Ruth D. Turner, 85, Expert on the Wood-Eating Mollusk," *NYT,* May 9, 2000.

13. On Pool, see K. M. Brinkhous, "Pool, Judith Graham. June 1, 1919–July 13, 1975," in *Notable American Women: The Modern Period,* ed. Barbara Sicherman and Carol Hurd Green (Cambridge, MA: Harvard Univ. Press, 1980), 553–54. Abbott had been a research associate and lecturer in phycology at Stanford since 1960. *AMWS,* 15th ed., 1982, 1:3.

14. On Ravel, see "Oral History of Joanne M. Ravel, April 27, 2002," by Laura Sweeney, with clippings, in Women in Science and Engineering Archives, Iowa State University Library, Ames. On Freier, see Phyllis Freier to Vera Kistiakowsky, Sept. 14, 1971, in box 4, APS-CWP Responses, Vera Kistiakowsky Papers, Institute Archives, MIT, Cambridge, MA; C. Jake Waddington, "Phyllis S. Freier," *Physics Today* 46 (Dec. 1993): 65; and biographical materials in the Phyllis St. Cyr Freier Papers, University Archives, University of Minnesota. On Hoffman, see "Lois Wladis Hoffman," in *A History of Developmental Psychology in Autobiography,* ed. Dennis Thompson and John D. Hogan (Boulder, CO: Westview, 1996), 105–20. On Duncan, see David Gold et al., "Beverly Duncan, 1929–1988, Professor of Sociology, Santa Barbara," in *University of California: In Memoriam, 1988* (Berkeley: University of California, 1988), 45–47.

15. On Robinson, see Constance Reid, "The Autobiography of Julia Robinson," *College Mathematics Journal* 17 (1986): 20.

16. On Roe, see Cutberto Garza et al., "Daphne A. Roe, January 4, 1923–September 22, 1993," in *Memorial Statements, Cornell University Faculty, 1993–94* (Ithaca, NY: Office of the Dean of Faculty, Cornell University, n.d.), 88–91. On Salpeter, see obituary, *NYT,* Oct. 28, 2000, B9. On Earle, see *AMWS,* 19th ed., 1995–96, 2:995.

17. On Kivelson, see *AMWS,* 19th ed., 1995–96, 4:442; Margaret Kivelson, "Graduate Society Medal: Across Space and Through Time," *Radcliffe Quarterly* 69 (Sept. 1983): 7–8; and oral history memoir at Special Collections UCLA Library. On Estrin, see *AMWS,* 19th ed., 1995–96, 2:1152. Estrin had been a research engineer for the UCLA Health Sciences Center from 1960 to 1970 and director of the computer laboratory for the Brain Research Center, which was supported by grants from the Air Force and the NIH, from 1970 to 1980. See also John D. French, Donald B. Lindsley, and H. W. Magoun, *An American Contribution to Neuroscience: The Brain Research Institute, 1959–1984* (Los Angeles: UCLA-BRI, 1984), 103–4.

18. On Ervin-Tripp, see Love, *Feminists Who Changed America,* 136; and http://Socrates.berkeley.edu/~ervintrp/.

19. On Wu, see Richard L. Garwin and Tsung-Dao Lee, "Chien-Shiung Wu," *Physics Today* 50 (Oct. 1997): 120, 122; and Ursula Allen, "Chien-Shiung Wu (1912–1997)," in *Notable Women in the Physical Sciences: A Biographical Dictionary,* ed. Benjamin F. Shearer and Barbara S. Shearer (Westport, CT: Greenwood, 1997), 423–29. The Wu Papers are in the Columbiana Room, Low Library, Columbia University, New York. Thomas Gieryn, personal communication, 1997. Wu merits a full biography.

20. On Weber, see *AMWS,* 15th ed., 1982, 7:465. On Ajzenberg-Selove, see her *A Matter of Choices: Memoirs of a Female Physicist* (New Brunswick, NJ: Rutgers Univ. Press, 1994).

21. On Weisstein, see Love, *Feminists Who Changed America,* 486; and http://

en.wikipedia.org/wiki/Naomi_Weisstein. On Sager, see Eric Pace, "Dr. Ruth Sager, 79, Researcher on Location of Genetic Material," *NYT*, Apr. 4, 1997, A28; and *Current Biography Yearbook, 1967* (New York: H. W. Wilson, 1967), 367–70. Sager's papers are at the Marine Biological Laboratory Library, Woods Hole, MA. On Blake, see Linda B. Bourque and Valerie Oppenheimer, "Judith Blake, 1926–1993," *University of California: In Memoriam, 1994* (Berkeley: University of California, 1994), 32–34; and Linda B. Bourque, "A Biographical Essay on Judith Blake's Professional Career and Scholarship," *Annual Review of Sociology* 21 (1995): 449–77.

22. On Cohen, see *AMWS*, 19th ed., 1995–96, 2:365. On Garvey, see ibid., 3:63. On Garmire, see ibid., 3:53; and "Elsa Garmire—Transcript of a Tape Recorded Interview by Joan Bromberg, February 4, 1985," Niels Bohr Library, Center for the History of Physics, American Institute of Physics, College Park, MD. On Satter, see *AMWS*, 18th ed., 1992–93, 6:472. On Stadler, see finding aid, Joan Stadler Papers, Women in Science and Engineering Archives, Iowa State University Library. On Cowley, see *AMWS*, 19th ed., 1995–96, 2:514; Ruth Bordin, *Women at Michigan: "The Dangerous Experiment," 1870s to the Present* (Ann Arbor: Univ. of Michigan Press, 1999), 82; Sue [Wyckoff] to Vera Rubin, Feb. 17, 1983, box 9, Vera C. Rubin Papers, Manuscript Division, Library of Congress, Washington, DC; and Anne Cowley to Margaret [Bryan Davis], May 24, 1973, enclosing Anne Cowley, "Comments on the Complaint-Appeal Hearing in the Case of Anne Cowley and the University of Michigan," May 16, 1973, in Margaret Bryan Davis Papers. One former research associate who fared well in the federal government was Mary Clutter at the NSF, mentioned in chapter 9.

23. On Steitz, see *AMWS*, 19th ed., 1995–96, 6:1239; and Neal Atebara, "An Insider's Look at Science: YSM Interview with Joan Steitz," *Yale Scientific* 58 (Spring 1984): 20–23. On Hopkins, see Nancy Hopkins, "The High Price of Success in Science," *Radcliffe Quarterly* 66 (June 1976): 16–18; and "Oral History of Nancy Hopkins, April 13, 2002," by Laura Sweeney, Women in Science and Engineering Archives, Special Collections, Iowa State University Archives. On Graham, see *AMWS*, 19th ed., 1995–96, 3:328; and items in Elizabeth L. Scott Papers, cartons 12 and 45, Nepotism.

24. Ann Roscoe, "New Women Faculty Hired, 7/1/80–10/1/80," Nov. 1, 1980, in Cornell Eleven Papers, box 12, folder 12. On Haynes, see *AMWS*, 19th ed., 1995–96, 3:703. On Jordan, see William R. Brice, *Cornell Geology through the Years* (Ithaca, NY: College of Engineering, Cornell University, 1989), 99, 184. On Cooper, see *AMWS*, 19th ed., 1995–96, 2:457.

25. On Wallace, see Julianne Malveaux, "Tilting against the Wind: Reflections on the Life and Work of Phyllis Ann Wallace," *American Economic Review* 84 (May 1984): 93–97. On Widnall, see *AMWS*, 19th ed., 1995–86, 7:739; and Jack Kendall, "WOMEN: Top-level Science: Getting There a Battle," *Boston Herald-American*, Dec. 31, 1979, clipping in Vera Kistiakowsky Papers, box 10, Correspondence, 1979–80. In 1983 the solid-state physicist Mildred Dresselhaus, who had long been a member of MIT's Department of Electrical Engineering and Computer Science, took on a second appointment in the MIT physics department. *AMWS*, 19th ed., 1995–96, 2:916.

26. The University of Massachusetts topped the list with a total of seven at its several campuses. See, e.g., "Few Women in Academia," *C&EN* 49 (May 10, 1971): 21. The full list, "Women on the Faculties of Chemistry Departments Offering Studies in

Chemistry," goes to eight pages. There is a copy of a 1973 update that includes the 1971 data, "Women on the Chemistry Faculties of Institutions Granting the Ph.D. in Chemistry," in the Vera Kistiakowsky Papers, box 28, Women in Science—Background Material. Sister Agnes Ann Green was one of the few PhDs in her order, which was so progressive in the late 1960s that the archconservative James Cardinal McIntyre ousted its sisters from parochial school teaching in the Archdiocese of Los Angeles. Anita M. Caspary, IHM, *Witness to Integrity: The Crisis of the Immaculate Heart Community of California* (Collegeville, MN: Liturgical Press, 2003), 32–33 (I thank Mary Oates for this reference).

27. "Chemistry Faculties Still Have Few Women," *C&EN* 52 (July 22, 1974): 19; Rebecca Rawls and Jeffrey Fox, "Women in Academic Chemistry Find Rise to Full Status Difficult," ibid. 56 (Sept. 11, 1978): 29; "Chemistry Faculties Gain Women Slowly," ibid. 62 (Feb. 13, 1984): 26; Kenneth R. Everett, Will S. DeLoach, and Stephanie E. Bressan, "Women in the Ranks: Faculty Trends in the ACS-Approved Departments," *Journal of Chemical Education* 73 (Feb. 1996): 139–41; Alison Byrum, "Women's Place in Ranks of Academia," *C&EN* 79 (Oct. 1, 2001): 98–99. The *C&EN* was quite slow to take up gender issues, but this changed after 1978, when Anna Harrison, the first woman president of the ACS, responded to complaints and the editors broadened their coverage.

28. Vera Kistiakowsky, "Women in Physics: Unnecessary, Injurious and Out of Place?" *Physics Today* 33 (Feb. 1980): 35, 37; "Letters: Women in Physics," ibid. 33 (June 1980): 15, 74, and reply by Kistiakowsky, ibid., 74, 76; Arthur Schawlow to Dr. Edmund D. Pellegrino, President, Catholic University of America, Oct. 16, 1981 (I thank Carol Crannell for a copy).

29. "Survey of Universities Finds Few Women on Senior Staff," *Physics Today* 35 (Feb. 1982): 99. Eisenstein and Baranger's 1981 full report was titled "Survey Compiled for the Committee on the Status of Women in Physics" (1981, copy in author's possession). See also Hans Frauenfelder and Peter G. Debrunner, "Laura Eisenstein," *Physics Today* 39 (June 1986): 109–10. The physics department at the University of Illinois and the CSWP of the APS created an award in Eisenstein's memory.

30. On Gaillard, see *AMWS*, 19th ed., 1995–96, 3:11; and interview by Elga Wasserman, Mar. 14, 1995, in box 1, Gaillard, Elga Wasserman Collection, Schlesinger Library, RIAS.

31. The women geoscientists collected data from the American Geological Institute's *Directory of Geoscience Departments* but were content with overall averages (i.e., less than 2% in 1972 and 3.9% in 1982). They were less fixated on the top departments, perhaps because earth sciences were not as dependent on academia for jobs as were other fields and perhaps because the geosciences were made up of widely different subfields (meteorology, oceanography, paleontology, geophysics, geochemistry, petroleum geology, etc.). See Virginia Murphy Sand and Bonnie Butler Bunning, "Ten Years of Progress for AGI's Women Geoscientists Committee," *Journal of Geological Education* 33 (1985): 212–15. Political scientists published quantitative data on their departments alphabetically in *PS*, as in "Committee on the Status of Women in the Profession," *PS* 12, no. 2 (1979): 233–38, and department totals (not broken down by rank) for 1973–78 in Doris-Jean Burton, "Ten Years of Affirmative Action and the Changing Status of Women in Political Science," ibid., no. 1 (1979): 18–22.

32. Janet L. Henderson and Barbara E. Cooper, "The Representation of Women Scientists in Land Grant Colleges of Agriculture," *National Association of College Teachers of Agriculture* 31 (Jan. 1987): 14–17. See also Maxime H. Thompson, "Women in Horticulture," *HortScience* 8 (Apr. 1973): 77–78; Christine T. Stephens, "American Women in Plant Pathology," *Plant Diseases* 66 (Feb. 1982): 95; A. K. Vidaver, "Women in Plant Pathology: An Assessment," *Phytopathology* 78, no. 1 (1988): 27–31; and Katherine Albro Houpt and M. Lois Calhoun, "Women in Veterinary Medicine," *Cornell Veterinarian* 67 (1977): 1–23. Janice Bahr, for example, was the first woman faculty member in animal sciences at the University of Illinois when she was hired in 1974. Janice Bahr, "Reproductive Endocrinology: An Understanding of the Basic Sciences Provides the Groundwork for Improved Animal Production" (autobiographical), *Paul A. Funk Recognition Program,* Special Publication 79 (Urbana-Champaign: University of Illinois College of Agriculture, Sept. 1991), iv, 1–8.

33. Roger Sanjek, "The American Anthropological Association Resolution on the Employment of Women: Genesis, Implementation, Disavowal and Resurrection," *Signs* 7, no. 4 (1982): 845–68; Naomi Quinn and Carol A. Smith, "A New Resolution of Fair Employment Practices for Women Anthropologists: Fresh Troops Arrive," ibid., 869–77. There are additional items in the Elizabeth L. Scott Papers, carton 109. See also Cynthia M. Webster and Michael L. Burton, "Summary Report of the Academic Employment of Women in Anthropology," *Anthropology Newsletter* 33 (Feb. 1992): 1, 21, 23; and Michael Burton et al., "Academic Employment of Women in Anthropology," ibid. 35 (Oct. 1994): 11–12.

34. Phrase taken from Judith P. Vladeck, "Litigation: Strategy of the Last Resort," in *Sex Discrimination in Higher Education: Strategies for Equality,* ed. Jennie Farley (Ithaca: New York State School of Industrial and Labor Relations, Cornell University, 1981), 1–22.

35. The upsurge had started in the 1960s with judicial decisions requiring "due process" in student personnel matters, but it had greatly expanded with the coming of federal regulations on research grants after that. Cheryl M. Fields, "Academe's Increased Reliance on Legal Advice Documented by College Attorney's Association," *CHE,* July 17, 1985, 15–16. The "Buckley Amendment" was officially the Family Educational Rights and Privacy Act of 1974, or FERPA, which prohibited schools from releasing information about student grades and behavior to anyone but the student.

36. On Roberts, see *Who's Who of American Women,* 2 vols., 4th ed. (Wilmette, IL: Marquis Who's Who, 1966–67), 2:975; Judith Berger: "Who's Afraid of Sylvia Roberts?" *Synthesis* (University of Pittsburgh Women's Center) 4 (Jan. 1977), copy in carton 13, folder 304, Sharon L. Johnson Papers, Schlesinger Library, RIAS; Jane Howard, *A Different Woman* (New York: E. P. Dutton, 1973), 293–97; and Love, *Feminists Who Changed America,* 386–87.

37. See, e.g., Terry L. Leap, *Tenure, Discrimination, and the Courts* (Ithaca, NY: ILR Press, 1993); George R. LaNoue and Barbara A. Lee, *Academics in Court: The Consequences of Faculty Discrimination Litigation* (Ann Arbor: Univ. of Michigan Press, 1987), which has a chapter on the Rajender case; Farley, *Sex Discrimination in Higher Education;* and idem, "Resolving Sex Discrimination Grievances on Campus." On Tomlin, see Love, *Feminists Who Changed America,* 464.

38. "Controversy Heats Up over Tenure Policies," *C&EN* 51 (Apr. 2, 1973): 9; "Injunction Granted in Sexist Tenure Case," ibid. 51 (June 11, 1973): 2–3; Cheryl M. Fields, "Woman Wins Injunction against Pitt," *CHE,* June 18, 1973, 2. Additional materials are in the Sharon L. Johnson Papers, which I used with Johnson's permission.

39. See, e.g., "Suit on Sex Bias at Pitt Continues after Five Years," *NYT,* Dec. 12, 1976, 31; "Woman Loses Pittsburgh U Suit," ibid., Aug. 3, 1977, 14; Rebecca L. Rawls, "Female Biochemist Loses Tenure Case," *C&EN* 55 (Aug. 15, 1977): 22; and Constance Holden, "Court Rules against Woman Biochemist," *Science* 197 (1977): 743. The Pittsburgh newspapers also covered the case widely. See also the voluminous materials relating to the case in the Sharon L. Johnson Papers. Led by the NOW treasurer, Sheila Tobias, the NOW Legal Defense and Education Fund raised money, and Anne Briscoe's mailing to AWIS members raised $7,000. Sheila Tobias et al. to "Friends and Colleagues," Sept. 18, 1974, in box 6, Jessie Bernard Papers, Historical Collections and Labor Archives, Pattee Library, Pennsylvania State University, University Park; Anne Briscoe to colleagues, Apr. 25, 1975, copy in box 4, folder 7, Charlotte Friend Papers, Archives, Mount Sinai Medical Center, New York; AMB [Anne Briscoe], "President's Remarks: AWIS in Action; Fund Drive for the Legal Defense Fund of Dr. Sharon Johnson," *AWIS Newsletter* 4 (June 1975): 3; "Oral History of Anne Briscoe, June 23, 2002, Ruskin, FL," by Laura Sweeney, Women in Science and Engineering Archives, Special Collections, Iowa State University Archives. The ACS gave Johnson a loan, which was vehemently protested by the University of Pittsburgh's chemistry department. The ACS's Women Chemists Committee, to which Johnson spoke in August 1973, evidently did little officially.

40. Janet Myers and Peter Kovacs, "Lamphere Pact Jolts Brown," *Index,* Sept. 1977, 3, copy in Cornell Eleven Papers, box 3 folder 21; "The Lamphere Case: A Settlement Out of Court," *Brown Alumni Magazine,* [1977?], 3–4, copy in Cornell Eleven Papers, box 3, folder 5; "What Price, Justice?" *On Campus with Women,* no. 26 (1980): 6. See also Anne Fausto-Sterling, "Working with a Court-Ordered Consent Decree," in Farley, "Resolving Sex Discrimination Grievances on Campus," 66–84, with the Lamphere consent decree reprinted in app. 5. The decree was lifted in 1992, when the number of tenured women on the Brown University faculty reached the stipulated sixty-seven. Carolyn J. Mooney, "Court Lifts Decree on Brown U. Hiring and Promotions," *CHE,* June 17, 1992, A15. See also Lyde Cullen Sizer, "A Place for a Good Woman? The Development of Women Faculty at Brown," in *The Search for Equity: Women at Brown University, 1891–1991,* ed. Polly Welts Kaufman (Hanover, NH: Brown Univ. Press, 1991), 208–15. On Cserr, see death notice, *Harvard Magazine,* Nov.–Dec. 1994, 111.

41. On Atwater, see *AMWS,* 19th ed., 1995–96, 1:257. On Surko, see ibid., 6:1392.

42. Barbara B. Reagan, "Stocks and Flows of Academic Economists," *American Economic Review* 69 (May 1979): 143–47. Jessie Bernard wrote numerous letters for young sociologists coming up for tenure in the mid-1970s. Only a few were successful. Jessie Bernard Papers.

43. David Remnick, "Sexism in the Ivy League; Princeton Faculty Women Say Tenure Is Denied Them," *Washington Post,* Dec. 2, 1979, A10. Catherine Clinton, "Women's Graves of Academe," *NYT,* Nov. 5, 1980, 23. "Diane Ruble Support and Liaison Group, STATEMENT, 21 March 1980"; memorandum, Diane Ruble to Marius Janson, "Appeal of Negative Promotion Decision," May 23, 1979; and "Ad Hoc Faculty Committee on

Women and Discrimination in Employment [at Princeton University], Preliminary Report, 27 May 1980," all in Cornell Eleven Papers, box 14. "Ruble Drops Sex-Bias Suit," *Princeton Alumni Weekly,* Apr. 5, 1982, 18. See also "Coeducation at Princeton, Report to the President, Princeton University, April 1980," which stressed that Joan Girgus, a professor of psychology, was serving as dean of the college, and that it was very difficult to find women worthy of a post on the Princeton faculty (20–21, copy in Cornell Eleven Papers, box 4, folder 5). In 1980 five women got tenure at Princeton, raising its total from nine to fourteen.

44. On Judith Moody, see *AMWS,* 18th ed., 1992–93, 5:503; Mark McDonald, "Tenure Process Said 'Flawed,' Woman Professor Denied Promotion," *Chapel Hill (NC) Newspaper,* May 10, 1979, and other clippings and correspondence in Cornell Eleven Papers, box 14, folder 11.

45. Both found opportunities in the fledgling biotechnology industry. On Daniell, see *AMWS,* 19th ed., 1995–96, 2:638; Paul Rabinow, *Making PCR: A Story of Biotechnology* (Chicago: Univ. of Chicago Press, 1996), 105; and Ellen Daniell, *Every Other Thursday: Stories and Strategies from Successful Women Scientists* (New Haven, CT: Yale Univ. Press, 2006), her remarkable account of the long-running group of Bay Area women scientists who sustained one another during difficult career situations after 1977. On Panem, see Sandra Panem to Anna J. Harrison, Oct. 2, 1980, in box 40, folder 11, Anna Jane Harrison Papers, Mount Holyoke College Archives, South Hadley, MA; "An Interview with Sandra Panem," in *The Outer Circle: Women in the Scientific Community,* ed. Harriet Zuckerman, Jonathan R. Cole, and John T. Beckwith (New York: Norton, 1991), 127–54; and letters of recommendation in the Charlotte Friend Papers, box 25, folder 5, and box 27, folder 5.

46. LaNoue and Lee, *Academics in Court,* ch. 7. See also William J. Broad, "Ending Sex Discrimination in Academia," *Science* 208 (1980): 1120–22; Ward Worthy, "Woman Chemist Wins Discrimination Case," *C&EN* 58 (June 30, 1980): 23; Suzanne Perry, "Sex Bias in Academe: A Sweeping Decree Helps Minnesota Women Press Claims," *CHE,* Aug. 31, 1983, 15–16; *Minnesota Alumni Association Magazine,* July–Aug. 1988; and Sally Gregory Kohlstedt and Suzanne M. Fischer, "Unstable Networks among Women in Academe: The Legal Case of Shyamala Rajender," *Centaurus* 51 (2009): 37–62. "Few Women in Academia," *C&EN* 49 (May 10, 1971): 21–22, describes Rajender's work on the ACS's 1971 job survey. The only Minnesota faculty person to support Rajender initially was Charlotte Striebel, an assistant professor of statistics, significantly a student of Elizabeth Scott's at UC–Berkeley, who helped with salary data. On Striebel, see Love, *Feminists Who Changed America,* 448–49. See also Clara Bingham and Laura Leedy Gansler, *Class Action: The Story of Lois Jensen and the Landmark Case That Changed Sexual Harassment Law* (New York: Doubleday, 2002), for an account of another victory by Paul Sprenger. There are also relevant items in the Cornell Eleven Papers, box 3, folder 5. The Sarah Rhoads Papers at the American Heritage Center, University of Wyoming, Laramie, have a few items on Rajender.

47. The best summaries of the Cornell case are Joe Kolman, "Women Professors Sue for Their Jobs, Saying the Tenure System is Sexist," *Cornell Alumni News* 85 (May 1981): 2–9; and Peg Downey and Daryl Endy, "A Struggle for Academic Equity," *Graduate Woman* 75 (Nov.–Dec. 1981): 10–15. PEER news release, May 11, 1979, in Cornell

Eleven Papers, box 14, folder 48. See also Alice H. Cook, *A Lifetime of Labor: The Autobiography of Alice H. Cook* (New York: Feminist Press, 1998), 252–53; and Judy Long, "Paradigm Lost: The Journey from Normal Science to Permanent Marginality," in *Individual Voices, Collective Visions: Fifty Years of Women in Sociology,* ed. Ann Goetting and Sarah Fenstermaker (Philadelphia: Temple Univ. Press, 1995), 130n3. On Long, see *Feminists Who Changed America,* 283–84. The anthropologist Rada Dyson-Hudson was not one of the Cornell Eleven, as in April 1979 she had signed a separate agreement with Cornell that accorded her an appointment as a continuing untenured associate professor.

The AAUW had been considering a legal defense fund since 1977, but it was the Cornell case that caused its board of directors to start one provisionally in 1981. In 1983 it was made permanent. "AAUW's New Legal Advocacy Fund," *Graduate Journal* (AAUW), Fall 1981, 13; Susan Levine, *Degrees of Equality: The American Association of University Women and the Challenge of Twentieth-Century Feminism* (Philadelphia: Temple Univ. Press, 1995), 166. As of 1999, the fund had contributed more than $500,000 to fifty-four cases. It also has a network of volunteer lawyers and social scientists who advise litigants. The AAUW posts a list of litigants aided by the fund on its web page, www.aauw.org/7000/laf/caseplbd.html. See also AAUW Educational Foundation, *Tenure Denied: Cases of Sex Discrimination in Academia* (Washington, DC: AAUW, 2004).

48. See Farley, *Sex Discrimination in Higher Education;* and idem, "Resolving Sex Discrimination Grievances on Campus." The latter has several interesting appendixes, including such hard-to-find documents of the time as the Brown-Lamphere consent decree of 1977, Harvard's 1978 grievance procedures, and Alison Dundes, "RUS vs. Harvard," of April 1981.

49. AAUW Educational Foundation, *Tenure Denied;* Robin Wilson, "Report Shows Difficulty of Sex-Discrimination Lawsuits," *CHE,* Oct. 29, 2004, A12.

50. See, e.g., Michael Knight, "Harvard Panel Says Sex Bias Led to Denial of Tenure for a Woman," *NYT,* Apr. 9, 1981, A16; "A Misunderstanding in Harvard Dispute," ibid., Apr. 26, 1981, 33; Fran R. Schumer, "A Question of Sex Bias at Harvard," *New York Times Magazine,* Oct. 18, 1981; David E. Sanger, "Harvard Offers Tenure to Woman," *NYT,* Jan. 8, 1985, A11; Liz McMillen, "Woman Who Won Tenure Fight Gets Cool Reception at Harvard," *CHE,* Oct. 1, 1986, 1, 16; and the blistering David S. Landes, letter to the editor, ibid., Nov. 5, 1986, 50. There was much other protest at Harvard/Radcliffe at the time, as two other women were denied tenure and a student group filed protests with the Office of Civil Rights of the Department of Education and the Department of Labor. Alison Dundes, "RUS vs. Harvard," *Second Century Radcliffe News,* Apr. 1981, 16–17, copy in Cornell Eleven Papers, box 1, folder 20, and reprinted in Farley, "Resolving Sex Discrimination Grievances on Campus," 90–94, which also reprinted "Harvard University Faculty of Arts and Sciences Grievance Procedures Approved by the Faculty Council, January 11, 1978," on pages 86–89; Ilana DeBare, "Ten Thousand Men of Harvard," *Equal Times,* Aug. 17, 1980, 11–14, copy in Cornell Eleven Papers, box 14.

51. Constance Holden, "Women in Michigan: Academic Sexism under Siege," *Science* 178 (1973): 842. Correspondence, timeline, and clippings in the Margaret Bryan Davis Papers, esp. Kathleen Hampton, "Sex Discrimination: The Case of Margaret

Bryan Davis," [*Ann Arbor News*?], n.d.; Mary Kramer, "'U' to Award Back Pay to Prof. Following Charges of Sex Bias," n.d.; and Margaret Bryan Davis, "Application to the Rajender Settlement Committee." See also D[eborah] S[hapley], "Michigan Approves Equal Pay," *Science* 177 (1972): 336, on another case at Michigan.

52. See Mary W. Gray and Elizabeth L. Scott, "A 'Statistical' Remedy for Statistically Identified Discrimination," *Academe* 66 (May 1980): 174–81; and materials in the Elizabeth L. Scott Papers, cartons 106, 109.

53. The CUNY case attracted interest because it relied solely on statistics. Paul Desruisseaux, "Federal Judge Upholds Women's Claim of Salary Discrimination at CUNY," *CHE*, Mar. 30, 19783, 30. The case against SUNY took thirteen years, 1976–89, and was finally won on appeal. "4 Women Win Bias Case against SUNY Campus," ibid., Nov. 22, 1989; "Legal Notice," ibid., May 30, 1984, 34; "7.5 Million Settlement Reached in CUNY Bias Suit," ibid., June 6, 1984; Karen W. Arenson, "Flood of Data in Women's Suit Fills a Sea of Complexity," *NYT*, Nov. 16, 1980, Education section, 36; Beverly T. Watkins, "U. of Rhode Island Agrees to Pay $1.24 Million to End 8-Year-Old Sex-Discrimination Dispute," *CHE*, Jan. 8, 1986, 25–26.

Chapter 3 · Taking Advantage of Undergraduate Openings

1. See Margaret W. Rossiter, *Women Scientists in America: Before Affirmative Action, 1940–1972* (Baltimore: Johns Hopkins Univ. Press, 1995), ch. 3.

2. See J. McGrath Cohoon and William Aspray, "A Critical Review of the Research on Women's Participation in Postsecondary Computing Education," in *Women and Information Technology: Research on Underrepresentation*, ed. Cohoon and Aspray (Cambridge, MA: MIT Press, 2006), 137–80; and Thomas J. Misa, ed., *Gender Codes: Why Women Are Leaving Computing* (New York: Wiley–IEEE Computer Society Press, 2010) (I thank Misa for a copy).

3. These categories are not mutually exclusive, as Purdue and MIT were more than engineering schools; Caltech and Princeton, here treated as former men's colleges, had engineering programs; and Spelman, one of the HBCUs, was also a liberal arts (and a women's) college.

4. Starting places for this immense literature are Christy Roysdon, *Women in Engineering: A Bibliography on Their Progress and Prospects* (Monticello, IL: Council of Planning Librarians, 1975); and Amy Sue Bix, "From 'Engineeresses' to 'Girl Engineers' to 'Good Engineers': A History of Women's U.S. Engineering Education," *NWSA Journal* 16 (Spring 2004): 27–49, reprinted in *Removing Barriers: Women in Academic Science, Technology, Engineering, and Mathematics*, ed. Jill M. Bystydzienski and Sharon R. Bird (Bloomington: Indiana Univ. Press, 2006), 46–65. See also George Bugliarello et al., eds., *Women in Engineering: Bridging the Gap between Society and Technology; Proceedings of an Engineering Foundation Conference, July 12–16, 1971, New England College, Henniker, New Hampshire* (n.p., 1972); Nancy D. Fitzroy and Sandford S. Cole, *Career Guidance for Women Entering Engineering: Proceedings of an Engineering Foundation Conference, New England College, Henniker, New Hampshire, August 19–24, 1973* (n.p., n.d.); and Mary Diederich Ott and Nancy A. Reese, eds., *Women in Engineering—Beyond Recruitment: Proceedings of the Conference held June 22 to 25, 1975, Cornell*

University, Ithaca, New York (n.p., n.d.), including Robert E. Gardner, "Women—The New Engineers," 158–70, which was influenced by Harold G. Kaufman's recent, "Young Women in Engineering: A Little Bit Better Than the Men," *New Engineer,* Feb. 1975, 31–36, which was based on 1971 data from the American Council on Education.

5. Jane Z. Daniels and William K. LeBold, "Women in Engineering: A Dynamic Approach," in *Women and Minorities in Science: Strategies for Increasing Participation,* ed. Sheila M. Humphreys (Boulder, CO: Westview, 1982), 139–63; Jane Zimmer Daniels, "Women in Engineering: A Program Administrator's Perspective," *Engineering Education* 78 (May 1988): 766–68; idem, "Purdue's Commitment to Women in Engineering: Strategies That Work," *Initiatives* 55, no. 3 (1992): 61–65. Daniels later became head of the Clare Boothe Luce Program.

The SWE had been formed in 1950, but its central office was overwhelmed in the 1970s, when numerous new student chapters were established nationwide. Since its members, whether students or alumnae, come from all branches of engineering, they have formed a convenient sample for many studies of women engineers over the years. See also Marta Navia Kindya, *Four Decades of the Society of Women Engineers* (n.p.: Society of Women Engineers, 1990), 44. The voluminous Society of Women Engineers Papers are at the Walter Reuther Library, Wayne State University, Detroit, Michigan. On the Princeton chapter of the SWE, see Yvonne Ng and Jennifer Rexford, eds., *She's an Engineer? Princeton Alumnae Reflect* (Princeton, NJ: Princeton University, 1993).

6. On MIT, see Amy Sue Bix, "Feminism Where Men Predominate: The History of Women's Science and Engineering Education at MIT," *Women's Studies Quarterly* 28 (Spring–Summer 2000): 24–45; and Mary P. Rowe, "What Actually Works? The One-to-One Approach," in *Educating the Majority: Women Challenge Tradition in Higher Education,* ed. Carol S. Pearson, Donna L. Shavlik, and Judith G. Touchton (New York: Macmillan, 1989), 375–83, and bioblurb on 464. On MacVicar, see "Margaret MacVicar, MIT Dean and School Innovator, Dies at 47," *NYT,* Oct. 2, 1991, D23. Her path at MIT was not easy, as she was not a professor in the Department of Physics but the Cecil and Ida Green Professor of Education, with tenure in the provost's office. See the booklet *Cecil and Ida Green: Philanthropists Extraordinary,* by Robert R. Shrock (Cambridge, MA: MIT Press, 1989). MacVicar's papers are at the Institute Archives, MIT, Cambridge, MA. On McBay, see "Shirley M. McBay," in Clarence G. Williams, *Technology and the Dream: Reflections on the Black Experience at MIT, 1941–1999* (Cambridge, MA: MIT Press, 2001), 764–81; and Nadine Brozan, "MIT Attracts More Women, As Other Schools Try to Do So," *NYT,* Mar. 4, 1987, C14. In 1972 the psychiatrist Carola Eisenberg, MD, the new head of MIT's health services, became the first woman to sit on the institute's Academic Council. In 2007 Marilee Jones, MIT's longtime dean of admissions who had done much to diversify the student body and recruit women in science after 1979, resigned after admitting that she had falsified her credentials. Eric Hoover, "Truth and Admissions: Former MIT Dean Seeks to Reclaim Her Name," *CHE,* Jan. 8, 2010, A17–A19.

7. See, e.g., Humphreys, *Women and Minorities in Science.* Humphreys's vita is on the web at www.eecs.berkeley.edu/~humphreys/Resume/. See also n. 40 below; and ch. 5 for programs at women's colleges.

8. *National Science Foundation Authorization Act for Fiscal Years 1979 and 1980 and the Women in Science and Technology Equal Opportunity Act: Hearing Before the Subcommittee on Health and Scientific Research of the Committee on Human Resources, United States Senate, 95th Congress, Second Session on S. 2549 . . . [and] S. 2550 . . . , April 10, 1978* (Washington, DC: GPO, 1978), 101; Naomi McAfee, "Women in Engineering Revisited," in "Expanding the Role of Women in the Sciences," ed. Anne M. Briscoe and Sheila M. Pfafflin, special issue, *Annals of the New York Academy of Sciences* 323 (1979): 94–99.

9. Emily M. Wadsworth, "Women's Activities and Women Engineers: Expansions over Time," *Initiatives* 55, no. 2 (1992): 59–65; Jane Zimmer Daniels, ed., *Women in Engineering Conference: A National Initiative, Conference Proceedings May 30–June 1, 1990* (n.p., n.d.); idem, ed., *Women in Engineering Conference: A National Initiative, Conference Proceedings, Washington, D.C., June 2–4, 1991* (n.p., n.d.); Robin Wilson, "Colleges Start Programs to Encourage Women Who Are Interested in Engineering Careers," *CHE,* June 12, 1991, A27, A29; Karla Haworth, "Mentor Programs Provide Support via E-mail to Women Studying Science," ibid., Apr. 17, 1998, A29–A30. See also www.mentornet.net.

10. Engineering Workforce Commission of the American Association of Engineering Societies, *Engineering and Technology Degrees, 2000* (New York, 2001), 22.

11. Most notable was the anonymous gift of $20 million to Stanford University in 2000. James Robinson, news release, May 5, 2000, http://news.stanford.edu/pr/00/anonymousgift55.html. On Baum, see *AMWS,* 19th ed., 1995–86, 1:450. On Denton, see Angela Y. Davis, "In Memoriam: Denice Denton, Chancellor, Professor of Electrical Engineering, UC Santa Cruz, 1959–2006," www.universityofcalifornia.edu/senate/inmemoriam/denicedenton.htm. On Klawe, see http://en.wikipedia.org/wiki/Maria_Klawe. On Eibeck, see Emma L. Carew, "Mom's the President, and I'm Outta Here," *CHE,* Nov. 27, 2009, A6.

12. Wilson Smith, "'Cow College' Mythology and Social History: A View of Some Centennial Literature," *Agricultural History* 44 (July 1970): 299–310; Jim Hightower, *Hard Tomatoes, Hard Times: A Report of the Agribusiness Accountability Project on the Failure of America's Land Grant College Complex* (Cambridge, MA: Schenkman, 1973); Stéphane Castonguay, *Protection des Culture, Construction de la Nature: Agriculture, Foresterie et Entomologie au Canada, 1884–1954* (Sillery, QC: Septentrion, 2004).

13. U.S. Department of Agriculture, Research, Education, and Economics Information System (REEIS), "Total FAEIS Enrollments by Gender by Degree: Baccalaureate, 1993–2000" (database), www.reeis.usda.gov/discoverer.app/grid?event=displayData&stateStr=eNrtUktvozAQ. On CALS name change in 1971, see www.cals.cornell.edu/cals/about/overview/index.cfm. The feminization of the agricultural college might be a possible topic for further work. On Susan Henry, see "Susan Armstrong Henry," in *Journeys of Women in Science and Engineering: No Universal Constraints,* ed. Susan A. Ambrose et al. (Philadelphia: Temple Univ. Press, 1997), 208–12; and http://mbg.cornell.edu/faculty-staff/faculty/henry.cfm. On Wintersteen, see http://www.ent.iastate.edu/dept/faculty/wintersteen.html. On Jahn, see http://news.cals.wisc.edu/newsDisplay.asp?id=1416.

14. Leslie Miller-Bernal and Susan L. Poulson, eds., *Going Coed: Women's Experiences in Formerly Men's Colleges and Universities, 1950–2000* (Nashville: Vanderbilt Univ. Press, 2004). See also "25 Years of Women at Holy Cross: The History, the Memories, the People," special issue, *Holy Cross Magazine* 32 (Nov.–Dec. 1997); and Jolane Baumgartenn Solomon, "A Woman Scientist at a Catholic College," *Initiatives* 54, no. 4 (1952): 57–62.

15. See Elizabeth A. Duffy and Idana Goldberg, *Creating a Class: College Admissions and Financial Aid, 1955–1994* (Princeton, NJ: Princeton Univ. Press, 1998); "Looking Back on 30 Years of Student Financial Aid," supplement, *CHE,* [2003]; and Constance E. Cook, *Lobbying for Higher Education: How Colleges and Universities Influence Federal Policy* (Nashville: Vanderbilt Univ. Press, 1998).

16. Elaine Yaffe, *Mary Bunting: Her Two Lives* (Savannah, GA: Frederic C. Beil, 2005), 273–80. Unfortunately, Bunting claimed that as she neared retirement she was more eager to learn about issues other than women. She was followed by the historians Patricia Graham and Adele Simmons. See Adele Simmons, "Princeton's Women," *Change* 9 (Dec. 1977): 42–44. On Yale, see Elga Wasserman, *The Door in the Dream: Conversations with Eminent Women in Science* (Washington, DC: Joseph Henry, 2000), 6–7. On Wasserman, see Barbara J. Love, comp., *Feminists Who Changed America, 1963–1975* (Urbana: Univ. of Illinois Press, 2006), 480.

For more on the Yale transition, see Janet Lever and Pepper Schwartz, *Women at Yale: Liberating a College Campus* (Indianapolis: Bobbs-Merrill, 1971); Joseph A. Soares, *The Power of Privilege: Yale and America's Elite Colleges* (Stanford, CA: Stanford Univ. Press, 2007), 101–12; and Deborah L. Rhode, "The Woman Question," in *Gender Matters: Women and Yale in Its Third Century* (New Haven, CT: Women's Faculty Forum, 2004), 156–63. One could also add the physician Mary Howell's unhappy experience as the first female associate dean at the Harvard Medical School (1972–75) and her forthright letter of resignation, printed in a women's health newsletter: Mary Howell, "An Open Letter to the Women's Health Movement," *Healthright* 1 (Spring 1975): 2, copy in box 10, folder 2, Fann Harding Papers, Women in Science and Engineering Archives, Special Collections, Iowa State University Library, Ames, and reprinted in *AWIS Newsletter* 4 (Jan.–Feb. 1976): 4–5. For her obituary, see Wolfgang Saxon, "Mary Howell a Leader in Medicine, Dies at 65," *NYT,* Feb. 6, 1998, D19. Her papers are at the Schlesinger Library, RIAS, Harvard University, Cambridge, MA. She may merit a biography, especially as she was also an abused wife and mother of seven. Leigh Marlowe to Ms. Waite, Mar. 22, 1982, carton 1, Newsletter, AWP Papers, Schlesinger Library, RIAS.

17. Lee A. Daniels, "Women at Princeton, 20 Years Later," *NYT,* Apr. 5, 1989, B13. See also special supplement to the *Princeton Weekly Bulletin,* Sept. 25, 1989, S1–S4 (I thank Anne Simon Moffat for a copy).

18. Alessandra Stanley, "Court Tells Princeton Clubs They Must Admit Women," *NYT,* July 4, 1990, 33, 36; Michele N.-K. Collison, "20 Years Later, Women on Formerly All-Male Campuses Fight to Change Their Institutions' 'Old Boy' Images," *CHE,* Dec. 12, 1990, A23–A24; Kirsten Bibbins, Anne Chiang, and Heather Stephenson, *Women Reflect about Princeton* (Princeton, NJ: Princeton University Office of Communications, 1989); Gina Barreca, *Babes in Boyland: A Personal History of Co-Education in the*

Ivy League (Hanover, NH: Univ. Press of New England, 2005). Barreca was an English major in the class of 1979 who took a few science classes.

19. Larry Gordon, "Caltech Chemistry Improves," *Los Angeles Times*, Aug. 6, 2007, B1, B6; Alycia J. Weinberger, "WISE (Women in Science and Engineering) at Caltech," in *Women at Work: A Meeting on the Status of Women in Astronomy, Held at the Space Telescope Science Institute, September 8–9, 1992*, ed. C. Megan Urry, Laura Danly, Lisa E. Sherbert, and Shireen Gonzaga (Baltimore: STSI, [1993]), 165–67; *Manpower Comments* 17 (June 1980): 17. See also Jeffrey Selingo, "Science-Oriented Campuses Strive to Attract More Women," *CHE*, Feb. 20, 1998, A53–A54; "Services and Programs," www.womenscenter.caltech.edu/Services_and_Programs.html; and R[ebecca] R[osenberg], "Twenty-Five Years of Women at Caltech—But Who's Counting?" *Engineering & Science* 60, no. 1 (1997): 5 (I thank Karen Rader for a copy).

20. Liesl Schillinger, "Yale's Women Science Students Face Insecurity in a Male-Dominated Field," *Yale Daily News*, Apr. 24, 1986, 1, 4; Edward B. Fiske, "In Search of Accountability in Campus Courses," *NYT*, Feb. 3, 1988, B5; Collison, "20 Years Later." Women at Yale did better in the biological sciences. Naomi Pierce, a 1976 graduate of Yale and later a biology professor at Harvard and world expert on butterflies, had initially been interested in art. D. Boswell Lane, "Naomi E. Pierce (1954–), Biologist," in *Notable Women in the Life Sciences: A Biographical Dictionary*, ed. Benjamin F. Shearer and Barbara S. Shearer (Westport, CT: Greenwood, 1996), 320–24.

21. "At Princeton, a Bid to Draw More Women," *NYT*, Nov. 5, 1986, B1, B9; Paul J. Korshin, "Women in the Sciences," ibid., Nov. 27, 1986, A24; Barbara Wilson, "Attracting Women to Princeton," *CSWP Gazette* 7 (Apr. 1987): 4. Neta Bahcall went to Princeton in 1990. *AMWS*, 19th ed., 1995–96, 1:304.

22. Carol B. Muller and Mary L. Pavone, "The Women in Science Project at Dartmouth: One Campus Model for Support and Systemic Change," in *Women in Science: Meeting Career Challenges*, ed. Angela M. Pattatucci (Thousand Oaks, CA: SAGE, 1998), 247–63; Michele Montejo, "From Eden to the Seashore," ibid., 265–74; Natalie M. Bachir, "Oh, the Places You'll Go . . . ," ibid., 271–79.

23. Dartmouth had employed some women scientists, such as the botanist Hannah Croasdale and the biochemist Lucile Smith, in various roles long before it went coed. Shelby Grantham, "Women at the Top (almost)," *Dartmouth Alumni Magazine*, May 1977, 34–38. On Robinson, see her obituary in *NYT*, June 1, 1988, B8. On Smith, see oral history at Women in Science and Engineering Archives, Special Collections, Iowa State University Library. The March 1997 issue of the *Dartmouth Alumnae [sic] Magazine* is devoted to female topics. Similarly, the sociologist Suzanne Keller had been at Princeton since 1968. Suzanne Keller, "Bridging Worlds: A Sociologist's Memoir," in *Individual Voices, Collective Visions: Fifty Years of Women in Sociology*, ed. Ann Goetting and Susan Fenstermaker (Philadelphia: Temple Univ. Press, 1995), 164–66. On Tilghman, see www.princeton.edu/president/biography/; "Tilghman on Women in Science and Being a Role Model," in special supplement, *Princeton Weekly Bulletin*, Sept. 25, 1989, S4 (I thank Anne Simon Moffat for a copy); and Claudia Dreifus, "Career That Grew from an Embryo: A Conversation with Shirley Tilghman," *NYT*, July 8, 2003, F2. On Navrotsky, see *AMWS*, 19th ed., 1995–96, 5:767–68. On Warren, see www.brown

.edu/Departments/Anthropology/people/facultypage.php?id=210182. On Girgus, see http://psych.princeton.edu/psychology/research/girgus/cv.php. On the Clare Boothe Luce Program, see ch. 11.

24. Mary Catherine Bateson, *Composing a Life* (New York: Atlantic Monthly, 1989), 53–55, 197–207, 222–23.

25. See Duffy and Goldberg, *Creating a Class.*

26. Betty L. Pollack [*sic*] and Lee K. Little, "Experimental Project in Physics Education or New Avenues for Women," *Physics Teacher* 11 (Oct. 1973): 391–99.

27. Ann L. Fuller, "The Status of Women at Oberlin," *Oberlin Alumni Magazine,* Sept.–Oct. 1972, 8–11; idem, "The Status of Women at Oberlin," ibid., Mar.–Apr. 1973, 6–13, which summarizes the "Final Disposition of the Twenty-two Recommendations made by the Ad Hoc Committee on the Status of Women at Oberlin in its Reports and Recommendations dated December 8, 1972," copy in box 6, Committee on the Status of Women, 1967–83, Anna Ruth Brummett Papers, Oberlin College Archives, Oberlin, OH. See also Brummett's "The Status of Women Legislation at Oberlin: A Personal Reflection" (typescript, May 4, 1981), also in box 6, Committee on the Status of Women, 1967–83.

28. Conrad Stanitski, Frank Frankfort, and Mary Muir, "Science-Active Liberal Arts Colleges and the Future of Basic Science," *Change* 18 (Nov.–Dec. 1986): 52–53; Thomas R. Cech, "Science at Liberal Arts Colleges: A Better Education?" *Daedalus* 128 (Winter 1999): 195–216; "A Hothouse for Female Scientists," *CHE,* May 5, 2006, A12. Every few years the NAS or the NSF publishes retrospective data on the "baccalaureate origins" of recent PhDs, as in "Baccalaureate Sources of 1984 Doctorate Recipients," in Susan L. Coyle and Peter D. Syverson, *Doctorate Recipients from United States Universities: Summary Report 1984* (Washington, DC: National Academy Press, 1986), 10–24, and NSF, *Undergraduate Origins of Recent (1991–1995) Science and Engineering Doctorate Recipients: Detailed Statistical Tables,* NSF 96-334 (Arlington, VA, 1996). It is hard to know from the data presented which students were women and at which institutions (other than women's colleges) they fared best, yet these rankings matter, since some foundations—the Howard Hughes Medical Institute, the Alfred P. Sloan Foundation, the Pew Trust, and others—give large grants to these "hothouses" to produce more scientists.

29. Research Corporation, *Academic Excellence: The Sourcebook; A Study of the Role of Research in the Natural Sciences at Undergraduate Institutions* (Tucson, AZ, 2001), executive Summary and tables 5.2–5.8. See also Norean Radke Sharpe and Carol H. Fuller, "Baccalaureate Origins of Women Physical Science Doctorates: Relationship to Institutional Gender and Science Discipline," *Journal of Women and Minorities in Science and Engineering* 2 (1995): 1–15, which is on women in the physical sciences from 1976 to 1986.

30. Robin Wilson, "At Carleton, Working Closely with Professors Leads Women to Careers in Science," *CHE,* May 5, 2006, A13–A14. Receiving less attention but also perennially on any list of universities "producing" the most undergraduate women who later earned PhDs in the sciences were large, long-coed institutions such as UC–Berkeley, Harvard-Radcliffe, MIT, the University of Chicago, and Cornell.

31. Roberta M. Hall and Bernice R. Sandler, *The Classroom Climate: A Chilly One*

for Women? (Washington, DC: Association of American Colleges, 1982). On Sandler, see www.bernicesandler.com; Debra E. Blum, "Head of College Association: Project on Women Dismissed after 20 Years in Advocacy Role," *CHE,* Dec. 5, 1990, A15, A20; and David Potter and Helen Rippier Wheeler, letters to the editor, ibid., Jan. 9, 1991, B5. Her papers and those of her Project on the Status and Education of Women are at the Schlesinger Library, RIAS.

32. See ch. 4, n. 17.

33. Undated (1988?) leaflet in Mathematical Association of America Records, box F/6 2000–01/1, Archives of American Mathematics, Center for American History, University of Texas, Austin; Barbara F. Sloat and Catherine M. De Loughry, *Summer Internships in the Sciences for High School Women: A Model Program at the University of Michigan* (Ann Arbor: University of Michigan, 1985); Pattatucci, *Women in Science,* 246; Cinda-Sue Davis et al., eds., *The Equity Equation: Fostering the Advancement of Women in the Sciences, Mathematics, and Engineering* (San Francisco: Jossey-Bass, 1996), xi; Russel S. Hathaway, Sally Sharp, and Cinda-Sue Davis, "Programmatic Efforts Affect Retention of Women in Science and Engineering," *Journal of Women and Minorities in Science and Engineering* 7 (2001): 107–24; Jane Zimmer Daniels and Jane Butler Kahle, eds., *Girls and Science and Technology: Proceedings and Contributions of the GASAT Conference (4th, Ann Arbor, Michigan, July 24–29, 1987),* 4 vols. (1987), www.eric.ed.gov/ERICWebPortal/search/detailmini.jsp?_nfpb=true&_&ERICExtSearch_SearchValue_0=ED384486&ERICExtSearch_SearchType_0=no&accno=ED384486.

34. For a brief history of WIS programs from 1986 to 1996, see NSF, Committee for the Review of Undergraduate Education, *Shaping the Future: New Expectations for Undergraduate Education in Science, Mathematics, Engineering, and Technology; A Report on its Review of Undergraduate Education by the Advisory Committee to the NSF Directorate for Education and Human Resources,* NSF 96-139 (Arlington, VA, 1996), ch. 2. Some of the courses on women in science included the then new, provocative writings of the historian and philosopher of science Carolyn Merchant, the biologist Ruth Bleier, the physicist turned biologist turned philosopher Evelyn Fox Keller, and others. The biologist and women's-studies proponent Sue Rosser started several programs and wrote about them.

35. Charlene M. Hoffman, Thomas D. Snyder, and Bill Sonnenberg, National Center for Education Statistics, *Historically Black Colleges and Universities, 1976–1994,* NCES 96-902 (Washington, DC: U.S. Department of Education, National Center for Education Statistics, 1996), 8, 17, 40. See also Henry N. Drewry and Humphrey Doerrman, *Stand and Prosper: Private Black Colleges and Their Students* (Princeton, NJ: Princeton Univ. Press, 2001); and Peter Schmidt, "At the University of Puerto Rico, Ambitions Meet Ambiguity," *CHE,* Apr. 15, 2005, A23, A26.

36. Rachel Ivie and Kim Nies Ray, *Women in Physics and Astronomy, 2005,* AIP Report No. R-430.02 (College Park, MD: American Institute of Physics, Feb. 2005), 8; Cheryl Leggon and Willie Pearson Jr., "The Baccalaureate Origins of African American Women Female Ph.D. Scientists," *Journal of Women and Minorities in Science and Engineering* 3 (1997): 213–22. See also Wini Warren, *Black Women Scientists in the United States* (Bloomington: Indiana Univ. Press, 1999); and Diann Jordan, *Sisters in Science: Conversations with Black Women Scientists on Race, Gender, and Their Passion for Science*

(West Lafayette, IN: Purdue Univ. Press, 2006). Although not one of the HBCUs, the University of Puerto Rico also graduated large numbers of women with BS degrees in science and engineering.

37. Kathleen Crane, *Sea Legs: Tales of a Woman Oceanographer* (Boulder, CO: Westview, 2003), 13–15, 26–27; "Geology: The Women Are Coming!" *U-M, Ann Arbor, MI, News,* Sept. 11, 1980, clipping in box 21, Geosciences and Women, Records of the Women in Science Project, 1974–82, NSF, RG 307, NARA, College Park, MD. The American Geological Institute also kept track of student enrollments and degrees by sex, with results published occasionally in the "Women Geoscientists Committee Newsletter."

38. "Institute for Advanced Study and Princeton University Program for Women in Mathematics," *Institute Letter,* Summer 2003, 3. See also Karen Uhlenbeck, "The Mentoring Program for Women in Mathematics," in NAS, Committee on Women in Science and Engineering (CWSE), Office of Scientific and Engineering Personnel, NRC, *Who Will Do the Science of the Future? A Symposium on Careers of Women in Science* (Washington, DC: National Academy Press, 2000), 49–54. A visiting journalist was somewhat skeptical: Anne Matthews, *Bright College Years: Inside the American Campus Today* (New York: Simon & Schuster, 1997), 163–70. See also Tami Worner, "Programs for Women in Math at Purdue," *AWM Newsletter* 28 (July–Aug. 1998): 19–20. Freshmen women at Purdue can reside in the same dorm as other female math and science majors and have access to special tutoring. Another program for potential math graduate students, EDGE (Enhancing Diversity in Graduate Education), was held at Spelman and Bryn Mawr Colleges. See also ch. 6, n. 18.

39. Betty Vetter, cited in two articles by Ronald Rosenberg, "Engineering Losing Its Luster among Students" and "More Women Shun Engineering as Other Careers Beckon," *Boston Globe,* Mar. 13, 1989, both on p. 1.

40. Leaflets, *Computer Science Reentry Program* and *Why Study Electrical Engineering and Computer Science at Berkeley?,* both published by UC–Berkeley and undated; Sheila Humphreys, comp., "Milestones: Women in Engineering at the University of California at Berkeley" (typescript, 1995); newsletters, "Excellence and Diversity Student Programs" (Department of Electrical Engineering and Computer Science, UC–Berkeley, Fall 1994 and Fall 1995) (I thank Sheila Humphreys for copies).

41. Nancy Leveson, "Women in Computer Science: A Report for the NSF CISE Cross Disciplinary Activities Advisory Committee" (typescript, Dec. 1989), 3–4, 39–41, copy in Cornell Engineering Library, Cornell University, Ithaca, NY; Amy Pearl et al., "Becoming a Computer Scientist: A Report by the ACM Committee on the Status of Women in Computer Science," *Communications of the A[ssociation for] C[omputer] M[achinery]* 33 (Nov. 1990): 47–58.

42. Jane Margolis and Alan Fisher, *Unlocking the Clubhouse: Women in Computing* (Cambridge, MA: MIT Press, 2002). Later Lenore Blum, long of Mills College and the Bay Area Math/Science Network, moved to Carnegie Mellon, where her son was on the faculty, and began to take up the subject of women and computers once again.

43. Steve Lohr, "Microsoft, amid Dwindling Interest, Talks Up Computing as a Career," *NYT,* Mar. 1, 2004, C1, C2; Andrea L. Foster, "Student Interest in Computer Science Plummets," *CHE,* May 27, 2005, A31–A32; "Letters—The Decline and Fall of Computer Science," ibid., June 24, 2005, A39; Scott Carlson, "Wanted: Female Computer-

Science Students," ibid., Jan. 13, 2006, A35–A38; John Fischman, "Robots to the Rescue," ibid., June 1, 2007, A29–A30; NSF website for Broadening Participation in Computing, Division of Computer and Network Systems, www.nsf.gov/funding/pgm_summ.jsp?pims_id=13510.

Chapter 4 · Innovative Outreach

1. See, e.g., the following by Lucy Watson Sells: "High School Mathematics as the Critical Filter in the Job Market," in *Developing Opportunities for Minorities in Graduate Education,* ed. R. T. Thomas, "Proceedings of the Conference on Minority Graduate Education, University of California at Berkeley, May 11–12, 1973" (Berkeley: Graduate Minority Program, University of California, 1973), 37–39; "Sex, Ethnic, and Field Differences in Doctoral Outcomes" (PhD diss., University of California at Berkeley, 1975); "Mathematics—A Critical Filter," *Science Teacher* 45 (Feb. 1978): 28–29; and "The Mathematical Filter and the Education of Women and Minorities," in *Women and the Mathematical Mystique,* ed. Lynn H. Fox, Linda Brody, and Dianne Tobin (Baltimore: Johns Hopkins Univ. Press, 1980), 66–75.

2. Nancy Kreinberg, "EQUALS: Working with Educators," in *Women and Minorities in Science: Strategies for Increasing Participation,* ed. Sheila Humphreys (Boulder, CO: Westview, 1982), 39–54; Teri Perl, *Women and Numbers: Lives of Women Mathematicians Plus Discovery Activities* (San Carlos, CA: Wide World /Tetra, 1993), 169–86. On Sells, see Barbara J. Love, comp., *Feminists Who Changed America, 1963–1975* (Urbana: Univ. of Illinois Press, 2006), 416–17.

3. Nancy Kreinberg, *"I'm Madly in Love with Electricity" and Other Comments about Their Work by Women in Science and Engineering* (Berkeley, CA: Lawrence Hall of Science, 1977), a resource directory that grew out of the 1976 conference; Jean Fetter, "Caught in the Network," *AWIS Newsletter* 6 (Sept–Oct. 1977): 4–5; Nancy Kreinberg, "The Development and Growth of the Math/Science Network" (paper presented at the Fourth International Congress on Mathematical Education, University of California at Berkeley, Aug. 11, 1980), copy in box 25, Kreinberg, Nancy, NSF, Records of the Women in Science Program, NARA, College Park, MD; Lenore Blum, Nancy Kreinberg, and Joan Koltnow, "The Math/Science Network and Resource Center," position paper, in ibid., box 24, Bay Area Math, with a detailed chronology in an appendix; Ruth C. Cronkite and Teri Hoch Perl, "A Short-Term Intervention Program: Math-Science Conferences," in Humphreys, *Women and Minorities in Science,* 65–85; Nancy Angle, "From Nancy Angle, President-elect," *Women and Mathematics Education Newsletter* 5 (Oct. 1982): 3, copy in Eugenie V. Mielczarek Papers, Special Collections and Archives, George Mason University, Fairfax, VA; Michelle Levander, "Expanding the Science Set," *San Jose Mercury News,* Mar. 22, 1992, 1E–2E; Joanne Pugh, "Expanding Your Horizons," in Perl, *Women and Numbers,* 193–201; www.expandingyourhorizons.org. The Berkeley statistician Elizabeth Scott evaluated several EYH programs for foundations in the 1970s. Cartons 64, 108, and 113, Elizabeth L. Scott Papers, Bancroft Library, University of California, Berkeley.

4. There is a considerable amount of material on Women and Mathematics in the Mathematical Association of America Records, Archives of American Mathematics,

Center for American History, University of Texas, Austin. See esp. Mary W. Gray to Alice J. Kelly, Sept. 24, 1986, and "Women and Mathematics Report, July 1, 1986 through December 31, 1986," 2–3, both in ibid., box F/6, 2000-01 2/1. See also Carole B. Lacampagne, "An Evaluation of the Women and Mathematics (WAM) Program and Associated Sex-Related Differences in the Teaching, Learning and Counseling of Mathematics" (EdD diss., Teachers College, Columbia University, 1979); and Eileen Poiani, "The Real Energy Crisis," in *Mathematics Tomorrow,* ed. Lynn A. Steen (New York: Springer-Verlag, 1981), 155–63. Mary Gray's early involvement in WAM later surprised others, as she and the MAA executive staff had differences. Sheila Tobias's article "Math Anxiety: Why Is a Smart Girl Like You Still Counting on Your Fingers?" *Ms.,* Sept. 1976, 56–59, 92, generated a lot of queries to WAM.

5. Sheila Tobias, *Overcoming Math Anxiety* (New York: Norton, 1978); idem, "Math Anxiety"; idem, "Who's Afraid of Math and Why?" *Atlantic Monthly,* Sept. 1978, 63–65; Sheila Tobias and Carol Weissbrod, "Anxiety and Mathematics: An Update," *Harvard Educational Review* 50 (Feb. 1980): 63–70; Fred M. Hechinger, "Curing Math Anxiety," *NYT,* Dec. 22, 1987, C12. On Tobias, see Love, *Feminists Who Changed America,* 463. Tobias's papers are at the Schlesinger Library, RIAS, Harvard University, Cambridge, MA, and the Newcomb College Center for Research on Women, Tulane University, New Orleans, LA.

6. See Lynn H. Fox, Elizabeth Fennema, and Julia Sherman, *Women and Mathematics: Research Perspectives for Change,* NIE Papers in Education and Work No. 8 (Washington, DC: NIE, HEW, 1977); Fox, Brody, and Tobin, *Women and the Mathematical Mystique,* for articles by Edith Luchins, Elizabeth Fenema, John Ernest, Lynn Fox, and others; and Lynn H. Fox, *The Problem of Women and Mathematics: A Report to the Ford Foundation* (New York: Ford Foundation, 1981).

7. Jean E. Taylor and Sylvia M. Wiegand, "AWM in the 1990s: A Recent History of the Association for Women in Mathematics," *Notices of the American Mathematical Society* 46 (Jan. 1999): 31–32. The AWM Papers from 1973 through the early 1980s are at the Wellesley College Archives, Wellesley, MA.

8. Gina Bari Kolata, "Math and Sex: Are Girls Born with Less Ability?" *Science* 210 (1980): 1234–35; Camilla Persson Benbow and Julian Stanley, "Sex Differences in Mathematical Ability: Fact or Artifact?" ibid., 1262–64; Alice T. Schafer and Mary W. Gray, "Sex and Mathematics," editorial, ibid. 211 (1981): 231. See also Nancy Tooney, "The 'Math Gene' and Other Symptoms of the Biology Backlash," *Ms.,* Sept. 1981, 56, 59; Jacquelynne S. Eccles and Janis E. Jacobs, "Social Forces Shape Math Attitudes and Performance," *Signs* 11, no. 2 (1986): 367–80; Anne Leggett, "Barbie," *AWM Newsletter* 22 (Nov.–Dec. 1992): 12; and "AAUW in the Media," *AAUW Outlook* 87 (Spring 1993): 24.

9. Association for the Promotion of the Mathematics Education of Girls and Women, undated announcement, Eugenie V. Mielczarek Papers, box 21, folder 2, Women in Women and Mathematics Education, Science and Technology Conference; correspondence between Judith Jacobs, Alice Schafer, Judith Roitman, and Dora Skypek, 1978–79, in box 2, WME, 1978–, Dora Skypek Papers, Manuscripts, Archives, and Rare Book Library, Emory University, Atlanta; WME website, www.wme-usa.org/home.html. One of the early members of the WME, Skypek had also been involved in bringing coeducation to Emory. See her "Girls Need Mathematics Too," *Arithmetic Teacher,*

Feb. 1980, 5–7; and E. Marie Robertson, "Strength in Numbers: Dora Helen Skypek is a Role Model for Women and Mathematicians," *Emory Magazine,* Feb. 1983, 26–29.

10. Sue E. Berryman, *Who Will Do Science? Minority and Female Attainment of Science and Mathematics Degrees: Trends and Causes; A Special Report [to] The Rockefeller Foundation* (n.p., [1981]), 84.

11. Marcia Linn and Janet S Hyde, "Gender, Mathematics, and Science," *Educational Researcher* 18 (Nov. 1989): 17–27; C. M. Reese et al., *N[ational] A[ssessment of] E[ducational] P[rogress], 1996 Mathematics Report Card for the Nation and the States,* NCES 97-488 (Washington, DC: U.S. Department of Education, 1997), as cited in NSF, *Women, Minorities, and Persons with Disabilities in Science and Engineering: 1998* (Arlington, VA, 1999), 14–15, 147; Yupin Bae et al., *Trends in Educational Equity of Girls and Women,* NCES 2000-030 (Washington, DC: National Center for Education Statistics, 2000), 5, 20–21; Janet S. Hyde et al., "Gender Similarities Characterize Math Performance," *Science* 321 (2008): 494–95.

12. Steven Olson, *Count Down: Six Kids Vie for Glory at the World's Toughest Math Competition* (Boston: Houghton Mifflin, 2004); Titu Andreescu et al., "Cross-Cultural Analysis of Students with Exceptional Talent in Mathematical Problem Solving," *Notices of the American Mathematical Society* 55 (Nov. 2008): 1248–60. On Wood, see "An Interview with Melanie Wood, Olympiad Team Member," by Sylvie Wiegand, *AWM Newsletter* 28 (July–Aug. 1998); "A Conversation with Melanie Wood," interview by Joseph A. Gallan, *Math Horizons,* Sept. 14, 2004, 13–14, 31 (also at www.maa.org/math horizons); and Polly Shulman, "The Girl Who Loved Math," *Discover* 21 (June 2000): 66–71, with peculiar pictures.

13. For an extensive compendium of projects from elementary to graduate school, see Michele L. Aldrich and Paula Quick Hall, comps., *Programs in Science, Mathematics and Engineering for Women in the United States, 1966–1978,* AAAS Publication 80-11 (Washington, DC: AAAS, 1980), with indexes of fields, funders, and sponsors. IBM and Ford funded the most projects. The Carnegie Corporation was also an early supporter, and the Alfred P. Sloan Foundation supported several projects, especially in the physical sciences. By the 1990s several other foundations, such as the Howard Hughes Medical Institute, Hewlett Packard, and the biggest of them all, the Melinda and Bill Gates Foundation, were also funding undergraduate science-education projects and programs at colleges around the nation.

14. See, e.g., Timothy M. Phelps, "Top Winner in Westinghouse Science Search is Hunter High Pianist, 17," *NYT,* Mar. 3, 1981, A1, B1; Sam Howe Verhovek, "Two Girls Win Westinghouse Competition," ibid., Mar. 3, 1987, C1, C3; "Connecticut Girl Wins $100,000 in Intel Contest," ibid., Mar. 13, 2001, B4; Joseph Berger, *The Young Scientists: America's Future and the Winning of the Westinghouse* (Reading, MA: Addison-Wesley, 1994), 116 and ch. 8; Rena F. Subotnik, Richard A. Duschl, and Eric H. Selmon, "Retention and Attrition of Science Talent: A Longitudinal Study of Westinghouse Science Talent Search Winners," *International Journal of Science Education* 15, no. 1 (1993): 61–72; and Allan Richter, "Six Decades of Science Contest Prowess," *NYT,* Mar. 9, 2003, LI 1, LI 8. There is a box of Rena Subotnik's paper at the Women in Science and Engineering Archives, Special Collections, Iowa State University Library, Ames. For other high-school prizewinners, see Stephanie Zaharoudis, "Girls Capture Top Honors at

Science Fair," *Washington Post,* Mar. 24, 1982, A1, on when four of the five top winners of the Montgomery Area Science Fair were girls; and Karen W. Arenson, "A: One, From Romania; Q. How Many Women Have Won the Top Math Contest?" *NYT,* May 1, 1997, B1, B4, on the first woman to win the William Lowell Putnam Mathematical Competition, first awarded in 1941.

15. Roger Ricklefs, "Bronx School of Science Offers Bright City Kids Top-Notch Education," *Wall Street Journal,* May 17, 1977, 1, 21; "Bronx Science, Despite Changes, Still a Top School," *NYT,* Oct. 22, 1979, B3; Gene I. Maeroff, "An Emphasis on Excellence," ibid., Feb. 23, 1982, C1, C3; Jane Perlez, "50 Years of Nurturing Excellence in Science," ibid., Aug. 30, 1988, 33, 35; http://alumni.bxscience.edu/?page=Notable Alumni. On Stuyvesant High, see Alyse Reckson, "Girl in a 'Boys' School: The Way It Was," ibid., Apr. 24, 1983, ESS 76; Robert D. McFadden, "Finally, a Façade to Fit Stuyvesant," ibid., Sept. 8, 1992, B1, B6; and www.shsaa.org/index.php?option=com_directory&page=viewcat&catid=38&Itemid=46. On special high schools, see Fred M. Hechinger, "Is It Fair to Have Selective Schools?" ibid., Mar. 19, 1985, C10; Joseph Berger, "States Planting Seeds to Grow Crop of Scientists," ibid., Mar. 15, 1989, B10; and idem, *Young Scientists,* chs. 5–7.

16. See, e.g., Margaret Anne Rogers, "A Different Look at Word Problems—Even Mathematics Texts Are Sexist," *Mathematics Teacher* 68 (1975): 285–88; Lois Arnold, "Madame Curie Was Great, But . . . ," *School Science and Mathematics* 75 (1975): 577–84; idem, "Florence Bascom and the Exclusion of Women from Earth Science Curriculum Materials," *Journal of Geological Education* 23 (1975): 110–13; Alleen Pace Nilsen, "Three Decades of Sexism in School Science Materials," *School Library Journal,* Sept. 1987, 117–22; Willa Ramsay to AAPT Committee on Women in Physics, "Report of Textbook Publishers' Guidelines to Authors Regarding Feminine Roles in Physics," Jan. 20, 1983, in Eugenie V. Mielczarek Papers, box 24, folder 16, APS (misfiled).

17. AAUW Educational Foundation, *The AAUW Report: How Schools Shortchange Girls* (Washington, DC: AAUW Educational Foundation/NEA, 1992). Its findings were widely publicized, for example, in Susan Chira, "Bias against Girls Is Found Rife in Schools, With Lasting Damage," *NYT,* Feb. 12, 1992, A1, and Richard N. Ostling and S. Urquhart, "Is School Unfair to Girls?" *Time,* Feb. 24, 1992, 62. See also Myra Sadker and David Sadker, *Failing at Fairness: How America's Schools Cheat Girls* (New York: Charles Scribner's Sons, 1994). The widespread publicity revitalized many girls-only high schools, some of which added more math and science to their curricula. Ilana DeBare, *Where Girls Come First: The Rise, Fall, and Surprising Revival of Girls' Schools* (New York: Jeremy P. Tarcher/Penguin, 2004), 191–93, 202–13.

18. Margaret E. Law et al., "A Study of Women High School Physics Teachers," *Bulletin of the American Physical Society,* 2nd ser., 21 (June 1976): 888–92; Eugenie V. Mielczarek, "Survey of Women in Physics," *Physics Teacher* 14 (Nov. 1976): 511–12. Correspondence about this article is in box 13, folder 3, APS-CSWP, of the Eugenie V. Mielczarek Papers.

19. Judith Roitman, "Careers for Women in Mathematics," copy in AWM Papers, box 2, AWM History; APS, CSWP, "Annual Report, February 1977–February 1978," 2, in box 5, folder 22, C. S. Wu Papers, Columbiana Room, Low Library, Columbia University, New York; [Laura Roth and Nancy O'Fallon], *Women in Physics* (New York:

APS, [1977]); "Pamphlets Encourage Young Women to Study Science," *Physics Today* 34 (June 1981): 77; a copy of the proposal for preparing new pamphlets is in box 3, Physics, Records of the Women in Science Project, Alphabetical Reference File for Scientific Organizations and Fields, NSF, NARA; APS, CSWP, *Wanted: More Women in Science and Technology: A Packet of Information and Suggestions for Junior and High School Counselors* (New York: APS, 1981).

20. Iris R. Weiss to colleague, Sept. 14, 1977, enclosing leaflet, in Papers of the Association for Women in Mathematics, Correspondence, Sept.–Oct. 1977, Wellesley College Archives.

21. Judith E. Parker, "Visiting Women Scientists Program," *Journal of College Science Teaching* 11 (May 1982): 363–64; "Visiting Women Scientists Program," *Women Chemists* (newsletter of the WCC of the ACS), Mar. 1983, 2; Lois Fritts, "What Employers Can Do in the Community: 3M's Visiting Technical Women Program," ibid., Aug. 1989, 2.

22. Melitta Rorty, "The Founding of the Association for Women Geoscientists," box 45, folder 13; program, "Careers in the Geosciences: A Conference for Students and Professionals," Nov. 18, 1978, box 5, folder 2; and program, "Careers in the Geosciences: A Conference for Students and Professionals," Oct. 20, 1979, box 5, folder 5, all in Papers of the Association for Women Geoscientists, Women in Science and Engineering Archives, Special Collections, Iowa State University Library.

23. Wendy Katkin, "Project WISE: A Community of Wise Women," in *Women in Science: Meeting Career Challenges*, ed. Angela M. Pattatucci (Thousand Oaks, CA: SAGE, 1998), 219–41.

24. Betty M. Windham, "Training Women in Science & Technology (TWIST) Project Progress Report: October 1983," in Eugenie V. Mielczarek Papers, box 24, folder 16, APS (misfiled); Jack M. Wilson, "Women in Physics," *AAPT Announcer* 14 (May 1984): 27–28.

25. Betsy Hammond, "Gail Whitney, Renowned for Saturday Academy, Dies at 65," *Oregonian*, Sept. 26, 2008, online at www.saturdayacademy.org/Default.aspx?tabid=163; John Travis, "Making Room for Women in the Culture of Science," *Science* 260 (1993): 412.

26. Women's College Coalition, "Math and Science Graduates: Women's Colleges Are Answering the Call; A Working Package of Information Developed by the Women's College Coalition, September 1989," in Jewel Plummer Cobb Papers, SCM 90-42, box 1, Schomburg Center for Research in Black Culture, New York Public Library; Ellen F. Mappen, "The Douglass Science Institute for High School Women," in *Women in Engineering Conference: A National Initiative, Conference Proceedings, Washington, D.C., June 2–4, 1991*, ed. Jane Zimmer Daniels (n.p., n.d.), 93–96 (three of New Jersey's chemical companies provided partial support for the program at Douglass: Merck & Co, Hoechst Celanese, and Schering Plough); leaflet, *Summermath* (South Hadley, MA: Mount Holyoke College, 1984), copy in author's possession (the Mount Holyoke program charged $1,750); Nancy Costello, "Budding Women Scientists Test Careers in Radcliffe Program," *Harvard Gazette*, Aug. 9, 1985, 8; Cindy Maxwell and Linda Maxwell, "The Radcliffe Summer Program in Science," *Radcliffe Quarterly* 80 (Mar. 1994): 15–17; Susan Staffin Metz, "ECOES: A Summer Engineering and Science Program for

High School Women," *Gifted Child Today* 14 (May–June 1991): 40–43; Bill Steele, "CURIE Academy Encourages High School Girls Pursue Engineering," *Cornell Chronicle,* July 27, 2000, 7; idem, "CU Helps CURIE-ous High School Girls Explore Engineering Careers," ibid., July 25, 2002, 7; Cinda-Sue Davis, "Summerscience: An Enrichment Program for Eighth Grade Girls," in *Women in Engineering Conference: A National Initiative, May 30–June 1, 1990* (n.p., n.d.), 125–28.

27. Dorrit Hoffleit, *Misfortunes as Blessings in Disguise: The Story of My Life* (Cambridge, MA: American Association of Variable Star Observers, 2002), 98–99. Her successor was Emilia Pisani Belserene, who served from 1978 until 1990. *AMWS,* 19th ed., 1995–96, 1:524. For more on the Maria Mitchell Observatory, see ch. 10.

28. Erinn C. Howard, "Not Your Ordinary Summer Camp," *AWIS Newsletter* 33 (Summer 2004): 6–9; leaflet, *Women in Technology, Vermont Technical College* ([Randolph Center, VT], n.d.); Peter Schmidt, "State Notes—Vermont Panel Approves College's Girls-Only Technology Camp," *CHE,* Mar. 15, 1996, A30.

29. Travis, "Making Room for Women in the Culture of Science," 412–15; Diane Sentementes Serpina, "Girl Scouts: From Arts and Crafts to Issues," *NYT,* Dec. 5, 1993, CN1, CN31; Marsha Lakes Matyas, "Weaving a Network of Early Support for Girls in Science: Empowering Girl Scout Leaders," in *Women in Engineering Conference . . . May 30–June 1, 1990,* 51–55; "Footnotes," *CHE,* June 7, 1989, A4; "Girl Scouts and Science Museums," *A[ssociation for] W[omen in] M[athematics] Newsletter* 23 (Mar.–Apr. 1993): 23. On the Girls Clubs of America, see Ellen Wahl Sullivan to Carole B. Lacampagne, Feb. 26, 1986, and attachment, in Mathematical Association of America Records, box F/6, 2000-01/1; leaflet, *Operation SMART, Encouraging girls in science, math and technology, girls, inc., Formerly Girls Clubs of America,* [ca. 1988], in box 30, Women in Science (published), Vera Kistiakowsky Papers, Institute Archives, MIT, Cambridge, MA.

30. *My Daughter, the Scientist: A Guide to the Exhibit* (Chicago: Museum of Science and Industry, 1985); "Urban Camping, Amid the Dinosaurs," *NYT,* Mar. 23, 1989, C11; Deborah L. Jacobs, "A Slumber Party Where the Wild Things Are," ibid., Mar. 12, 2008, 24. See also José Luis Sanz, *Starring T. Rex! Dinosaur Mythology and Popular Culture,* trans. Philip Mason (Bloomington: Indiana Univ. Press, 2002); and Mary Sladek, *A Report on the Evaluation of the National Science Foundation's Informal Science Education Program,* NSF 98-65 (Arlington, VA: NSF, 1998). On the founding of children's television, 1966–70, see Richard M. Polsky, *Getting to Sesame Street: Origins of the Children's Television Workshop* (New York: Praeger, 1974).

31. On the Commission on Excellence in Education, see Terrel H. Bell, *The Thirteenth Man: A Reagan Cabinet Memoir* (New York: Free Press, 1988); and David Pierpoint Gardner, *Earning My Degree: Memoirs of an American University President* (Berkeley and Los Angeles: Univ. of California Press, 2005), ch. 4, which has an extensive bibliography.

32. The immunologist Marian Koshland, of UC–Berkeley, was also at the Stanford hearing, representing the National Science Board, of which she was then a member. National Commission on Excellence in Education, *A Nation at Risk: The Imperative for Educational Reform; A Report to the Nation and the Secretary of Education, United States Department of Education* (Washington, DC: GPO, 1983), 49.

33. National Science Board, Task Committee on Undergraduate Science and Engi-

neering Education, *Undergraduate Science, Mathematics and Engineering Education: Role for the National Science Foundation and Recommendations for Action by Other Sectors to Strengthen Collegiate Education and Pursue Excellence in the Next Generation of U.S. Leadership in Science and Technology,* NSB 86-100, 2 vols. (Washington, DC: NSF, Mar. 1986–Nov. 1987), 1:58–59; Betty M. Vetter, "Issues in Undergraduate Education in Science and Engineering," ibid., 2:67–73; Carolyn K. Rozier, "Women's Underrepresentation in Science," ibid., 2:163–66.

34. NSF, *Decade of Achievement: Educational Leadership in Mathematics, Science and Engineering,* NSF 92-94 (Washington, DC, 1992), 8.

35. Sladek, *Report on the Evaluation of the National Science Foundation's Informal Science Education Program;* Urban Institute, Education Policy Center, *Summary Report on the Impact Study of the National Science Foundation's Program for Women and Girls* (Arlington, VA: NSF, Dec. 2000); Katherine Darke, Beatriz Clewell, and Ruta Seva, "Meeting the Challenge: The Impact of the National Science Foundation's Program for Women and Girls," *Journal of Women and Minorities in Science and Engineering* 8 (2002): 285–303; *NSF's Program for Gender Equity in Science, Technology, Engineering, and Mathematics: A Brief Retrospective, 1993–2001,* NSF 02-107 (Arlington, VA: NSF, 2002). The AAUW evaluated 416 research and intervention projects supported by the NSF and the AAUW between 1993 and 2001 in American Association of University Women Educational Foundation, *Under the Microscope: A Decade of Gender Equity Projects in the Sciences* (Washington, DC, 2004).

Chapter 5 · Using Science to Fight Back

1. Margaret W. Rossiter, *Women Scientists in America: Before Affirmative Action, 1940–1972* (Baltimore: Johns Hopkins Univ. Press, 1995), ch. 10 and pp. 543–44. Of course, these accusations were not entirely accurate.

2. Elizabeth A. Duffy and Idana Goldberg, *Crafting a Class: College Admissions and Financial Aid, 1955–1994* (Princeton, NJ: Princeton Univ. Press, 1998), 52–53. This informative study includes Wellesley, Smith, Wheaton, Amherst, and Williams Colleges among its sixteen institutions. On the women's colleges in these years, see Irene Harwarth, *Women's Colleges in the United States: History, Issues, and Challenges* (Washington, DC: GPO, 1997); and Mariam K. Chamberlain, ed., *Women in Academia: Progress and Prospects* (New York: Russell Sage Foundation, 1988), ch. 6. For more on these newly coed institutions see ch. 3.

3. Zoe Ingalls, "Colleges Run by Nuns Forced to Change as Fewer Women Enter Religious Orders," *CHE,* Mar. 25, 1987, 13–15.

4. Cedar Crest College, *Report of Conference on the Undergraduate Education of Women, July 8–10, 1969* (Allentown, PA, 1969); Pauline Tompkins, "What Future for the Women's Colleges?" *Liberal Education* 58 (1972): 298–303; Auden D. Thomas, "Preserving and Strengthening Together: Collective Strategies of U.S. Women's College Presidents," *History of Education Quarterly* 48 (Nov. 2008): 565–89; Barbara W. Newell, "Statement to Department of Health, Education and Welfare Regarding the Proposed Title IX Regulations," Oct. 8, 1974, in Title IX: Statement and Comments, 1974, Papers of the President's Office, Wellesley College Archives, Wellesley, MA. Besides people at

Wellesley, she consulted a small group of "contacts" committee, made up of Sheila Tobias, then at Wesleyan; Adele Simmons, then at Princeton; Lilli Hornig; and Elsa Wasserman, then at Yale. In the early 1970s the presidents of women's colleges (often nuns or other females) were among the few women on any ACE boards or commissions. Later the women came from a broader range of institutions.

5. In 1979–81 the Women's College Coalition had a two-year grant from the Ford Foundation to assemble and publish data. This resulted in several publications about the colleges and their recent graduates, summarized in Susan Nall Bales and Marcia Sharp, "Women's Colleges—Weathering a Difficult Era with Success and Stamina," *Change* 13 (Oct. 1981): 53–56. See also Jadwiga S. Sebrechts, "Cultivating Scientists at Women's Colleges," *Initiatives* 55, no. 2 (1993): 45–52. The AAC, which sponsored Bernice Sandler's Project on the Status and Education of Women, was more receptive in the early 1970s to the women's movement than were some other educational associations in Washington, DC (e.g., the ACE), that fought the new governmental regulations). Some of this is mentioned briefly in Mark H. Curtis, *Enhancing, Promoting, Extending Liberal Education: The Association of American Colleges at Seventy-Five* (Washington, DC: AAC, 1988). The AAC may merit a full history.

6. Bernice Segal, quoted in "Why a Women's College?" *Barnard Reporter,* Jan.–Feb. 1982, clipping in box 20, folder 17, Eugenie V. Mielczarek Papers, Special Collections and Archives, George Mason University, Fairfax, VA. For more on Barnard, see Rosalind Rosenberg, *Changing the Subject: How the Women of Columbia Shaped the Way We Think about Sex and Politics* (New York: Columbia Univ. Press, 2004). Segal is described in "Sally Chapman, Transcript of an Interview Conducted by Hillary Domosh at Barnard College, New York, New York, on 5 and 6 January 2009," 43, 44, 58, Chemical Heritage Foundation, Philadelphia.

7. Katherine Bishop, "Women's College Rescinds Its Decision to Admit Men," *NYT,* May 19, 1990, 7; Mary S. Hartman, "Mills Students Provided Eloquent Testimony to the Value of Women's Colleges," *CHE,* July 5, 1990, A40.

8. Betty Friedan's book *The Feminine Mystique* started as a data-collection project for her Smith College class's fifteenth reunion. See Daniel Horowitz, *Betty Friedan and the Making of The Feminine Mystique* (Amherst: Univ. of Massachusetts Press, 1998). The history of the women's colleges underwent a great change at about this time, as the whole field of women's history exploded. Some lively books that were better than their titles were Ellen Kendall, *"Peculiar Institutions": An Informal History of the Seven Sister Colleges* (New York: Putnam, 1975); and Liva Baker, *I'm Radcliffe, Fly Me! The Seven Sisters and the Failure of Women's Education* (New York: Macmillan, 1976). Charlotte Conable, a Cornell graduate and trustee, was so shaken by student criticisms that she wrote *Women at Cornell: The Myth of Coeducation* (Ithaca, NY: Cornell Univ. Press, 1977).

9. The best bibliography of Tidball's writings is in her book with autobiographical details, M. Elizabeth Tidball et al., *Taking Women Seriously: Lessons and Legacies for Educating the Majority* (Phoenix: Oryx, 1999), 15–18. See also M. Elizabeth Tidball, "Baccalaureate Origins of Entrants into American Medical Schools," *Journal of Higher Education* 56 (1985): 385–402; and idem, "Baccalaureate Origins of Recent Natural Science Doctorates," ibid. 57 (1986): 606–20.

10. M. Elizabeth Tidball and Vera Kistiakowsky, "Baccalaureate Origins of American Scientists and Scholars," *Science* 193 (1976): 646–52. See also oral history of Vera Kistiakowsky, as well as many items in box 8, Baccalaureate Origins Study (9 folders), and data printouts in box 35, in Vera Kistiakowsky Papers, all in Institute Archives, MIT, Cambridge, MA.

11. Mary J. Oates and Susan Williamson, "Women's Colleges and Women Achievers," *Signs* 3, no. 4 (1978): 795–806. For a more recent analysis of this topic, see Daryl G. Smith, Lisa E. Wolf, and Diane E. Morrison, "Paths to Success: Factors Related to the Impact of Women's Colleges," *Journal of Higher Education* 66 (1995): 245–66; Norean Radke Sharpe and Carol H. Fuller, "Baccalaureate Origins of Women Physical Science Doctorates: Relationship to Institutional Gender and Science Discipline," *Journal of Women and Minorities in Science and Engineering* 2 (1995): 1–15; and Norean Radke Sharpe, "Sisters Sell Success in Science," *Mount Holyoke Alumnae Quarterly,* Winter 1995, 26–28. See also Mary J. Oates, CSJ, "Professor Susan Williamson, 1936–2008," *Regis Today,* Fall 2008, 28–29 (I thank Mary Oates for a copy).

12. "Chatham College Legacies," advertisement, *USAir Magazine,* Nov. 1991, 46; Linda Lear, *Rachel Carson, Witness for Nature* (New York: Henry Holt, 1997); *Women in Science: Mount Holyoke College,* booklet (n.p.: Davis Press, n.d.); Elizabeth Becker, "Family History Forges Labor Secretary's Convictions," *NYT,* Feb. 26, 2001, A10; Women's College Coalition, "Women Can't Do Science?" full-page advertisement, ibid., Mar. 20, 2005 (I thank Joy Harvey for bringing this ad to my attention).

13. "Interview with Carolyn Shaw Bell, Katharine Coman Professor of Economics Emerita, Professor of Economics, 1950–89, November 24, 1997, by Millie Rahn," 68, 82, Wellesley College Archives; Judith H. Dobrzynski, "How to Succeed? Go to Wellesley: Its Graduates Scoff at Glass Ceilings," *NYT,* Oct. 29, 1995, F1, F9.

14. "The College of St. Catherine," advertisement, *CHE,* Mar. 19, 1999, B71.

15. Nancy J. Vickers, "Reflecting on Women, Coeducation, and Cultural Change," in *Gender Matters: Women and Yale in its Third Century* (New Haven, CT: Women's Faculty Forum, 2004), 62.

16. Women's College Coalition, *A Profile of Women's College Presidents: Summary Report* (Washington, DC, 1982), 6. In New York State the financial support was called "Bundy Aid," in honor of McGeorge Bundy, who chaired a commission that recommended in 1968 that such Catholic colleges be made eligible if they reorganized their governance so as not to be sectarian institutions. See also Beth McMurtrie, "Report Examines Role of Lay Leaders at Catholic Colleges," *CHE,* July 4, 2003, A22. Boards of trustees increasingly sought presidents with business skills and management experience.

17. Don Wyclif, "Women as Presidents: Shalala Takes Charge at U. of Wisconsin . . . ," *NYT,* Aug. 15, 1990, B6; Julianne Basinger, "How Nan Keohane Is Changing Duke," *CHE,* Nov. 3, 2000, A35–A37; Glenn Collins, "Spelman College's First 'Sister President,'" *NYT,* July 20, 1987, C20; Kit Lively, "Discrimination or Compensation? Questions Raised over Right of Texas Woman's University to Bar Men from Some Majors," *CHE,* Oct. 12, 1994, A23; Jacques Steinberg, "Brown U Breaks Ground in Picking Black as Chief," *NYT,* Nov. 9, 2000, A18; idem, "Proud Daughter of a Janitor Reaches an Academic Peak," ibid., Nov. 11, 2000, 30.

18. Beverly Guy-Sheftall and Jo Moore Stewart, *Spelman: A Centennial Celebration, 1881–1981* (Atlanta: Spelman College, 1981), 88, 92, 119. See also Albert E. Manley, *A Legacy Continues: The Manley Years at Spelman College, 1953–1976* (Lanham, MD: Univ. Press of America, 1995); Henry N. Drewry and Humphrey Doermann, *Stand and Prosper: Private Black Colleges and Their Students* (Princeton, NJ: Princeton Univ. Press, 2001), 167–73; Gaynelle Evans, "Spelman President to Head the College Board," *CHE,* Sept. 3, 1986, 3; Michele N-K. Colison, "At Spelman, Jubilation Greets Its First Black Woman President," ibid., Oct. 7, 1987, A3; Ronald Smothers, "President's Door Open for Spelman Students," *NYT,* Jan. 20, 1988, B9; Johnnetta B. Cole, "The Long Road through Gendered Questions," in *The Politics of Women's Studies: Testimony from Thirty Founding Mothers,* ed. Florence Howe (New York: Feminist Press, 2000), 327–33; idem, *Conversations: Straight Talk with America's Sister President* (New York: Doubleday, 1993); and Mary Catherine Bateson, *Composing a Life* (New York: Atlantic Monthly, 1989).

19. On Cobb, see F. Elaine De Lancey, "Jewel Plummer Cobb (1924–), Cell Biologist," in *Notable Women in the Life Sciences: A Biographical Dictionary,* ed. Benjamin F. Shearer and Barbara S. Shearer (Westport, CT: Greenwood, 1996), 82–87; Samuel Weiss, "Scientists' Group Backs Dr. Cobb for Hunter Post," *NYT,* Sept. 16, 1979, 47 (she was supported by Anne Briscoe and Janet Welsh Brown, among others); and idem, "Hunter Search for President Is Ugly but May End Soon," ibid., Nov. 11, 1979, E6. See also ch. 1, n. 19. The Jewel Plummer Cobb Papers, Schomburg Center for Research in Black Culture, New York Public Library, include many clippings about her academic career.

20. On Smith, see "Briefly Noted," *CHE,* Feb. 21, 1990, A35. On Barnard, see "15 Colleges Announce Update on Campaigns," ibid., Jan. 12, 2001, A30. On Chatham, see Audrey Williams June, "Remaining the Province of Women," ibid., Aug. 1, 2003, A28. On Wellesley, see "Give and Take," ibid., Apr. 28, 2000, A40; and "Bond Upgrades," ibid., Feb. 15, 2008, A19.

21. On Spelman, see Lee A. Daniels, "A Black College Gets 20 Million from Bill Cosby," *NYT,* Nov. 5, 1988, A1, A20; "The Cosbys' $20 Million Challenge," editorial, ibid., Nov. 11, 1988, A30; Liz McMillen, "Bill Cosby, Urging Support for Black Colleges, Gives $20 Million to Spelman," *CHE,* Nov. 16, 1988, A29, A31; Julie L. Nicklin, "Fund Gives Spelman College Stock Valued at $37 Million," ibid., May 13, 1992, A32; and "Spelman Sets Record for Black Colleges by Raising $114 Million in Campaign," ibid., July 12, 1996, A34. On the College of St. Catherine, see "Gifts and Bequests," ibid., Sept. 15, 2000, A35. On Agnes Scott, see Julie L. Nicklin, "The Terms of a Widow's Gift Bottle Up an Endowment That's Awash in Coke," ibid., Nov. 19, 1996, A35–A36.

22. On Simmons, see the booklet *The Simmons College Science Center* ([Boston?], Oct. 1972) and "A Case for Simmons and Its Science Center" (typescript, n.d.), both in box 85 of the Dixy Lee Ray Papers, Library and Archives, Hoover Institution, Stanford University, Stanford, CA; Eleanor Webster, "Chemistry at Wellesley College," *Nucleus* (Northeastern Section of the ACS) 5 (Nov. 1977): 8–11, copy in box 4, folder 3, Anna Jane Harrison Papers, Mount Holyoke College Archives, South Hadley, MA; Nancy Harrison Kolodny, "The New Science Center," *Wellesley Alumnae Magazine,* Winter 1977, 16–17; "Wellesley's New Science Center," photo collage, ibid., 18–29; Harold E.

Andrews, "Report on the Sciences at Wellesley College," Fall 1977, in "Self-Study Report," Feb. 1, 1979, submitted to Commission on Institutions of Higher Education, New England Association of Schools and Colleges, copy in Wellesley College Archives. The coming of the new science center at Wellesley is also mentioned in the oral histories of Janet Brown Guernsey (pp. 31–32), Elizabeth J. Rock (24–31), Dorothea J. Widmayer (29–30), and trustee Anne Cohen Heller (23–24), as well as in the college's self-study for reaccreditation in 1989, all in Wellesley College Archives. See also *Wellesley Alumnae Magazine,* Autumn 1973, Winter 1975, and Winter 1987; Duffy and Goldberg, *Crafting a Class,* 56 (on flyer); and Paula Rayman and Belle Brett, *Pathways for Women in the Sciences: The Wellesley Report, Part I* (Wellesley, MA: Wellesley College Center for Research on Women, 1993), app. A.

23. The Mount Holyoke chemistry department had a lot of scientific equipment paid for by a variety of sources. See, e.g., "Report to the President of Mount Holyoke College from the Chemistry Department, 1992–1993," app. 15, "Major Instruments and Equipment, Carr Laboratory, Department of Chemistry, Mount Holyoke College," Chemistry Department Records, Series B, Mount Holyoke College Archives. On the Office of Sponsored Research, see "Interview with Elizabeth J. Rock, June 25, 1998," by Millie Rahn, 32–33, Wellesley College Archives; awards database, HHMI website, www.hhmi.org/grants/reports/awards.

24. The data for Rutgers had always included Douglass. "Women on the Chemistry Faculties of Institutions Granting the Ph.D. in Chemistry," Vera Kistiakowsky Papers, box 28, Women in Science—Background Material; "Chemistry Faculties Still Have Few Women," *C&EN* 52 (July 22, 1974): 19; Rebecca Rawls and Jeffrey Fox, "Women in Academic Chemistry Find Rise to Full Status Difficult," ibid. 56 (Sept. 11, 1978): 29; "Chemistry Faculties Gain Women Slowly," ibid. 62 (Feb. 13, 1984): 26; Kenneth R. Everett et al., "Women in the Ranks: Faculty Trends in the ACS-Approved Departments," *Journal of Chemical Education* 73 (Feb. 1996): 139–41; Allison Byrum, "Women's Place in Ranks of Academia," *C&EN* 79 (Oct. 1, 2001): 98–99.

25. Ellen F. Mappen, "Expanding the Field," *Douglass Alumnae Bulletin,* Spring 1993, 1–4; "Crucial Experiments," *Women's Review of Books* 12 (Feb. 1995): 20.

26. Kathleen Teltsch, "Douglass Reserves a Dorm for Women in Science," *NYT,* Oct. 11, 1989, B6 ("Douglass" was changed to "Rutgers" in the late edition); Patricia B. Campbell, "The Douglass Project for Rutgers Women in Math, Science and Engineering Longitudinal Research Study," and "Final Report to the Exxon Education Foundation," both in Jewel Plummer Cobb Papers, SCM 98-32, box 2, Personal Papers, 1996.

27. Jennie Farley, *Academic Women and Employment Discrimination: A Critical Annotated Bibliography,* Cornell Industrial and Labor Relations Bibliography Series, No. 16 (Ithaca: New York State School of Industrial and Labor Relations, Cornell University, 1982), 8; Margaret W. Rossiter, *Women Scientists in America: Before Affirmative Action, 1940–1972* (Baltimore: Johns Hopkins Univ. Press, 1995), ch. 10; "Interview with Phyllis Fleming, Professor of Physics, September 12, 1997," 42, and "Interview with Dorothea J. Widmayer, '52, William R. Kenan Professor of Biological Sciences Emerita, Science Faculty, 1961–1996, Nov. 3, 1998," 20, both by Millie Rahn, in Wellesley College Archives.

28. Wade Roush, "No Women Chemists at Women's College," *Science* 272 (1996):

24. Ironically, the tenured chemistry faculty had voted unanimously for tenure for Sharon Palmer, but she was turned down by a collegewide committee. She sued, claiming that her qualifications were almost the same as the most recently tenured man's, and as one of the few women in the department she was expected to do a lot more advising and counseling. She won a lot of publicity but not her case.

29. "Faculty Profiles: Sheila Ewing Browne," www.mtholyoke.edu/offices/comm/profile/sbrowne.html.

30. On Keller, see "Sister Mary Kenneth Keller," *Communications of the A[ssociation] of C[omputer] M[achinery]* 38 (Jan. 1995): 47 (I thank Mary Oates for a copy). On Blum at Mills, see below, n. 33, and ch. 4, n. 3. On Wellesley, see Rayman and Brett, *Pathways for Women in the Sciences . . . Part I,* 131–32.

31. Agnes Scott ad, "Science, Technology and the Liberal Arts," *NYT,* Nov. 16, 2000, A29 (Op-Ed page); "Gifts and Bequests," *CHE,* Sept. 15, 2000, A35. On Smith, see Jill Ker Conway, *A Woman's Education* (New York: Knopf, 2001), 82, 103, 126; Samuel Florman, "Engineering and the Female Mind," *Harper's,* Feb. 1978, 57–58, 60, 62–63; Elizabeth S. Ivey, "Engineering at Smith College," *Engineering Education* 73 (Dec. 1982): 235–37; Ethan Bronner, "Women's College to Diversify via Engineering," *NYT,* Feb. 20, 1999, A1, A11; and Elizabeth F. Farrell, "Smith College's First Engineers Feel Like 'Rock Stars,'" *CHE,* May 28, 2004, A34.

32. On Simmons, see Sally Gregory Kohlstedt, personal communication, 2009. On the lack of interest in women's studies at the women's colleges, see Baker, *I'm Radcliffe, Fly Me!,* 115. On the Barnard meeting, see box 11 of Ruth Hubbard Papers, University Archives, Harvard, Cambridge, MA. On Douglass, see Assembly Bill No. 3702, introduced Nov. 23, 1981, State of New Jersey, copy in Jewel Plummer Cobb Papers, SCM 91-4, box 1.

33. Lenore Blum and Steven Givant, "Increasing the Participation of Women in Fields That Use Mathematics," *American Mathematics Monthly* 87 (Dec. 1980): 785–93; idem, "Increasing the Participation of College Women in Mathematics-Related Fields," in *Women and Minorities in Science: Strategies for Increasing Participation,* ed. Sheila M. Humphreys (Boulder, CO: Westview, 1982), 119–37. There is a chapter on Lenore Blum in *Women in Mathematics: The Addition of Difference,* by Claudia Henrion (Bloomington: Indiana Univ. Press, 1997), 145–64. See also *AMWS,* 19th ed., 1995–96, 1:716; and Teri Perl, "Lenore Blum (1942–)," in *Notable Women in Mathematics: A Biographical Dictionary,* ed. Charlene Morrow and Teri Perl (Westport, CT: Greenwood, 1998), 11–16.

34. Charlene Morrow, "Harriet Pollatsek (1942–)," in Morrow and Perl, *Notable Women in Mathematics,* 168; "Math Education Programs That Work," *Focus: The Newsletter of the Mathematical Association of America* 15 (Dec. 1995): 4.

35. On Falconer, see *AMWS,* 19th ed., 1995–96, 2:1190; Ulrica Wilson Parker, "Etta Zuber Falconer," in Morrow and Perl, *Notable Women in Mathematics,* 43–47; and "Etta Falconer, a Story of Success: The Sciences at Spelman College," *Sage* 6 (Fall 1989): 36–38. On McBay, see Clarence G. Williams, *Technology and the Dream: Reflections on the Black Experience at MIT, 1941–1999* (Cambridge, MA: MIT Press, 2001), 764–81, which has some comments on her Spelman years. On Bozeman, see Ulrica Wilson Parker, "Sylvia Bozeman (1947–)," in Morrow and Perl, *Notable Women in Mathematics,* 17–21. See also Mona T. Philips, "'I Bring the History of My Experience': Black Women

Professors at Spelman College Teaching Out of Their Lives," in *Black Women in the Academy: Promises and Perils,* ed. Lois Benjamin (Gainesville: Univ. Press of Florida, 1997), 302–14, which includes the psychologist Myra Burnett.

36. Sharpe and Fuller, "Baccalaureate Origins of Women Physical Science Doctorates"; Mary K. Campbell, Edwin Weaver, and Sheila Ewing Browne, conversations with author, Jan. 2001; Peter Beckmann, "Physics at Bryn Mawr," *CSWP Gazette* 5 (Dec. 1985): 1–2.

37. Marcia Sharp, "The Women's Colleges Go to Washington," *Radcliffe Quarterly* 66 (Mar. 1980): 14–17.

38. Oral histories of Elizabeth J. Rock, Carolyn Shaw Bell, and Janet Brown Guernsey, Wellesley College Archives. On Witkin, see *AMWS,* 19th ed., 1995–96, 7:855.

39. On Harrison, see *AMWS,* 19th ed., 1995–96, 3:639; Anna Jane Harrison Papers; and Shirley M. Malcom et al., *Science, Technology and Women: A World Perspective; Proceedings of the Ad Hoc Panel of Specialists of the United Nations Advisory Committee on Science and Technology for Development Held at South Hadley, Massachusetts, USA, 12–16 September 1983,* AAAS Publication 85-14 (Washington, DC: AAAS, 1985). On Cobb, see above, n. 19, and ch. 1, n. 19.

40. Duffy and Goldberg, *Crafting a Class,* ch. 5.

41. See www.mtholyoke.edu/offices/comm/profile/sbrowne.html.

42. See Afro-American Students/Alumnae, Mount Holyoke College Archives, with clippings and directories; and Kate Zernike, "They're Back and Proud: Black Alumnae Revisit Smith," *Boston Globe,* May 16, 1999, A1, D17.

43. Cheryl B. Leggon and Willie Pearson Jr., "The Baccalaureate Origins of African American Female Ph.D. Scientists," *Journal of Women and Minorities in Science and Engineering* 3 (1997): 213–24. Data on Hispanic or Asian American science students at women's colleges are hard to find. (Perhaps there is an undergraduate project in finding more.) See, however, "Changes in Demography at Wellesley College, 1978–1988," fig. 1 in "Wellesley College: Self-Study for Reaccreditation," submitted to the Commission on Institutions of Higher Education, New England Association of Schools and Colleges, Inc., Feb. 1989, Wellesley College Archives. The percentage of Asian American students jumped from 4 percent in 1978 to 15 percent in 1988. In the same period the percentage of Hispanics doubled, from 2 percent to 4 percent; that of blacks dropped from 8 percent to 6 percent; and that of foreign students dropped from 5 percent to 4 percent.

44. Jewel Plummer Cobb, "SAT Scores and Douglass College," in Jewel Plummer Cobb Papers, SCM 91-4, box 1; Ben Gose, "Second Thoughts at Women's Colleges," *CHE,* Feb. 10, 1995, A22–A24; Duffy and Goldberg, *Crafting a Class,* 100–101; Casey Clark et al., "A Peer Mentoring Program for Underrepresented Students in the Sciences," in *Women Succeeding in the Sciences: Theories and Practices across Disciplines,* ed. Jody Bart (West Lafayette, IN: Purdue Univ. Press, 2000), 149–67.

45. Honorary Degree Lists, Mount Holyoke College Archives; Honorary Degree Files, boxes 80–87, Smith College Archives, Northampton, MA; Dixy Lee Ray Papers, box 7. On Harrison, see www.mtholyoke.edu/offices/comm/press/releases/annaharrison.shtml. On Shalala, see www.miami.edu/index.php/about_us/leadership/office_of_the_president/president_donna_e_shalalas_biography/.

46. Suzanne Varimbi and Sr. Mary Leo Bryan, SHCJ, "A Reentry Program for Women Chemists," *Journal of Chemical Education* 68 (July 1991): 544; Sister Mary Kieran McElroy, "A Continuing Education Project for Updating Women in Biochemistry," ibid. 55 (Oct. 1978): 649; Edwin S. Weaver, "The Short Course Approach to Reentry Problems for Women in Chemistry," ibid. 56 (Aug. 1979): 509; "The Women in Science Program: A Diary," *Mount Holyoke Alumnae Quarterly,* Fall 1978, 36–37; Jamilah Evelyn, "Making the Leap: With Support, Community College Graduates Succeed to Competitive Private Colleges," *CHE,* Feb. 28, 2003, A36–A37; Eric Hoover, "A College's Near-Death Experience," ibid., June 18, 2004, A37–A38.

47. On CROW, see "Interview with Carolyn Elliott, January 8, 1980" (typescript), Wellesley College Archives; Rayman and Brett, *Pathways for Women in the Sciences . . . Part I;* and Paula Rayman, Belle Brett, and Janet T. Civian, *Pathways for Women in the Sciences: The Wellesley Report, Part II* (Wellesley, MA: Wellesley College Center for Research on Women, 1997).

48. "Interview with Alice Turner Schafer," Nov. 6, 1998, 15, 31, and 32, in Wellesley College Archives. The Wellesley Archives also house a collection of Schafer's papers and those of the AWM. See ch. 4 on the AWM's activities while its headquarters were at Wellesley. See also Women's College Coalition, "Math and Science Graduates: Women's Colleges Are Answering the Call; A Working Package of Information Developed by the Women's College Coalition, September 1989," in Jewel Plummer Cobb Papers, SCM 90-42, box 1, for data on science enrollments and special programs. On Sylvia Bozeman and Rhonda Hughes, see ch. 6, n. 18; and Lenore Blum, "A Brief History of the Association for Women in Mathematics: The Presidents' Perspectives," *Notices of the American Mathematical Society* 38 (Sept. 1991): 747, 749.

49. Andrew Brownstein, "Enrollment Falls, and a Small College Debates Its Future," *CHE,* May 4, 2001, A39–A40; idem, "Bare Naked Protesters," ibid., May 15, 2001, A10. David W. Breneman considered Hollins's future uncertain in *Liberal Arts Colleges: Thriving, Surviving, or Endangered?* (Washington, DC: Brookings Institution, 1994), 125–26.

50. "Needed: Men," *CHE,* Sept. 29, 2000, A47.

51. Correspondence about Emmanuel College, 1980–82, in Anna Jane Harrison Papers, box 21; Molly Honan, "Emmanuel Welcomes Its Largest Class in History," *Pilot* (Boston), Oct. 10, 2003, special section, 6–7; Peter Schworm, "College Finds Its Angel," *Boston Globe,* Oct. 1, 2007, A1, A8.

52. Bill Pennington, "Small Colleges, Short of Men, Embrace Football," *NYT,* July 10, 2006, A1, A14. On Newcomb, see advertisement for executive director in *CHE,* Nov. 21, 2008, A94. On Washington University, see leaflet, *The Mr. and Mrs. Spencer T. Olin Fellowships for Women in Graduate Study: A Joint Undertaking of the Monticello College Foundation and Washington University in St. Louis* (n.p., 1974). On Radcliffe, see ch. 7.

Chapter 6 · Surviving the "Minefields" in Graduate School

1. Over the years parasitology, pathology, and pharmacology moved from the applied "medical" category to the "biological sciences." "Specialties List," in *Summary Report, 1975: Doctorate Recipients from United States Universities* (Washington, DC:

NAS, 1976), inside back cover, and later editions. The report for 1972 was the first to provide data on women doctorates in all its tables.

2. Institutions vary on how they count noncompleters. See NRC, Office of Scientific and Engineering Personnel, Ad Hoc Panel on Graduate Attrition Advisory Committee, *The Path to the Ph.D.: Measuring Graduate Attrition in the Sciences and Humanities* (Washington, DC: National Academy Press, 1996). In 1995 it was reported that more women than men in economics were dropping out of graduate school, but the reason for this was not known. Shulamit Kahn, "Women in the Economics Profession," *Journal of Economic Perspectives* 9 (Autumn 1995): 200.

3. John F. Ohles and Shirley M. Ohles, comps., *Private Colleges and Universities,* 2 vols. (Westport, CT: Greenwood, 1982), 1:170. Wellesley and Smith, which had had small master's-degree programs for women only, quietly discontinued them in the early 1970s rather than provoke the attention of federal investigators.

4. On Geller, see "Video Portraits: Margaret Geller," videotape interview by Matthew Schreps, Feb. 5, 1989, tape 4, Smithsonian Institutions Archives, Washington, DC; and *AMWS,* 19th ed., 1995–96, 3:93. On Farrar, see http://as.nyu.edu/object/Glennys Farrar.html. On Max, see *AMWS,* 19th ed., 1995–86, 5:344.

5. P. J. Bickel, E. A. Hammel, and J. W. O'Connell, "Sex Bias in Graduate Admissions: Data from Berkeley," *Science* 187 (1975): 398–404. See also Saul D. Feldman, *Escape from the Doll's House: Women in Graduate and Professional School Education* (New York: McGraw-Hill, 1974), based on a questionnaire sent to almost thirty-three thousand graduate students at 158 institutions in the spring of 1969 that included many questions about the perceived masculinity and femininity of various fields; and Bernice Sandler, "Admissions and the Law," in *Graduate and Professional Education of Women: Proceedings of AAUW Conference Held at the AAUW Educational Center, May 9–10, 1974* (Washington, DC: AAUW, 1974), 23–31.

6. Lucy Watson Sells, "Sex, Ethnic, and Field Differences in Doctoral Outcomes" (PhD diss., University of California at Berkeley, 1975); idem, "Preliminary Report on the Status of Graduate Women: University of California, Berkeley" (report prepared for the Graduate Assembly's Committee on the Status of Women, Mar. 30, 1973); idem, "Sex Differences in Graduate School Survival" (paper presented at the annual meeting of the American Sociological Association, New York, Aug. 28, 1973).

7. [Donald E. Stokes], *The Higher, the Fewer: Report and Recommendations of the Committee to Study the Status of Women in Graduate Education and Later Careers to the Executive Board of the Graduate School, The University of Michigan, March 1974* ([Ann Arbor: University of Michigan, 1974]). Few of the other status-of-women reports of the time had much to add. See Lewis C. Solomon, *Male and Female Graduate Students: The Question of Equal Opportunity* (New York: Praeger, 1976), app. B. On Larimore, see *AMWS,* 13th ed., 1976, 4:2506. On Davis, see ibid., 19th ed., 1995–96, 2:686.

8. The dean of women was a respected figure in the university administration until the end of World War II. By the 1950s the post had been recast as "dean of students" and was held by a man, possibly with a female assistant dean on his staff, whose role had been demoted in the hierarchy. See Margaret W. Rossiter, *Women Scientists in America: Struggles and Strategies to 1940* (Baltimore: Johns Hopkins Univ. Press, 1982), 64–65, 71–72; idem, *Women Scientists in America: Before Affirmative Action, 1940–1972*

(Baltimore: Johns Hopkins Univ. Press, 1995), 33, 35; and Kathryn Nemeth Tuttle, "What Became of the Dean of Women? Changing Roles for Women Administrators in American Higher Education, 1940–1980" (PhD diss., University of Kansas, 1996).

9. Audrey N. Slate, *AGS: A History* (Austin, TX: Association of Graduate Schools in the Association of American Universities, 1994), 180–81. On Fromkin, see *AMWS*, 19th ed., 1995–96, 2:1492; "Victoria Fromkin, Linguistics: Los Angeles," *University of California: In Memoriam, 2001* (Berkeley: University of California, 2001), 53–54; Council of Graduate Schools in the United States, Committee on Women, "A Study of the Graduate Deanship: Does Gender Make a Difference? Report of the Committee on Women, 1983–84" (mimeograph, 1984), 2, 3, 14; and Greg Winter, "Rockefeller Foundation Names a New President," *NYT*, Aug. 25, 2004, B2. Starting in the 1970s some graduate schools also hired an African American assistant dean. For a forthright discussion of this and related matters, see Kenneth Manning's interview with Clarence Williams in *Technology and the Dream: Reflections on the Black Experience at MIT, 1941–1999*, by Clarence G. Williams (Cambridge, MA: MIT Press, 2001), 523–24, 528.

10. Few records remain of these graduate student clubs. On Stanford, see [Judith Lemon, ed.], *Stanford Women in Science and Engineering, September 1981*, booklet (Stanford, CA, 1981), 44. On MIT, see Amy Sue Bix, "Feminism Where Men Predominate: The History of Women in Science and Engineering Education at MIT," *Women's Studies Quarterly* 28 (Spring–Summer 2000): 40. For Caltech, see ch. 3, n. 19. The existence of a club of women graduate students at Princeton is mentioned in Madeleine Jacobs, "Reasons Sought for Lack of Diversity," *C&EN* 79 (Oct. 1, 2001): 103. See also Maresi Nerad and Joseph Cerny, "From Facts to Action: Expanding the Educational Role of the Graduate Division," in *Council of Graduate Schools Communicator*, special ed., May 1991; and Maresi Nerad and Debra Sands Miller, "Increasing Student Retention in Graduate and Professional Programs," in *New Directions for Institutional Research*, no. 92 (Winter 1996): 61–76.

11. Robert R. Shrock, *Cecil and Ida Green: Philanthropists Extraordinary* (Cambridge, MA: MIT Press, 1989), 198, 230–33; Vincent A. Fulmer, *Dedication of Ida Flansburgh Green Hall* (Cambridge, MA: MIT, 1983).

12. Rachel Ivie and Kim Nies Ray, *Women in Physics and Astronomy, 2005*, AIP Report No. R-430.02 (College Park, MD: American Institute of Physics, Feb. 2005), 6. For data on doctorates by field, gender, and citizenship for 2005, see http://webcaspar.nsf.gov. Unfortunately, the extensive annual NAS *Summary Reports* do not break their data down by sex, citizenship status, and field.

13. Renata Frank de Verthelyi, "International Female Graduate Students in Engineering at a U.S. University: Survival of the Fittest?" *Journal of Women and Minorities in Science and Engineering* 3 (1997): 245–64. Verthelyi interviewed twenty women graduate students at Virginia Polytechnic Institute and State University. The NSF published occasional reports on foreign scientists, but very little of its data was broken down by sex. See, e.g., NSF, *Foreign Citizens in U.S. Science and Engineering: History, Status and Outlook*, NSF 86-305, rev. ed. (Washington, DC, 1987). On Ruiz, see W. Patrick McCray, *Giant Telescopes: Astronomical Ambition and the Promise of Technology* (Cambridge, MA: Harvard Univ. Press, 2004), 239–40.

14. There still is not much available on minority women's experiences in graduate

school, but see Anne J. MacLachlan, "Careers of Minority Women Scientists from the University of California, Berkeley," *Making Strides* (newsletter of the Directorate for Education and Human Resources Programs, AAAS) 3 (July 2001): 1–5; Jayni Flores, "Chicana Doctoral Students: Another Look at Educational Equity," in *Ethnolinguistic Issues in Education,* ed. Herman S. Garcia and Rudolfo Chavez Chavez (Lubbock: Texas Tech Press, 1988), 72–79; and "Lydia Villa-Komaroff," in *Journeys of Women in Science and Engineering: No Universal Constraints,* ed. Susan A. Ambrose et al. (Philadelphia: Temple Univ. Press, 1997), 405–8. The NAS's annual Survey of Earned Doctorates began to break its published data down by race and ethnicity in 1974 but not further by sex and field, so that one cannot document minority women's numbers by field.

15. Peter D. Syverson, *Summary Report: Doctorate Recipients from United States Universities, 1981* (Washington, DC: National Academy Press, 1982), table 5 (p. 40). See also Shirley Mahaley Malcom, "Women in Science and Engineering: An Overview," *IEEE Transactions on Education,* E-28 (Nov. 1985): 192–93.

16. Interviews with Shirley Jackson, Jenny Patrick, and Evelynn Hammonds are included in Williams, *Technology and the Dream,* 220–30, 626–39, 937–44. On Malcom, see Wini Warren, *Black Women Scientists in the United States* (Bloomington: Indiana Univ. Press, 1999), 185–92. See also box 16, MIT, Physics Department of, Committee for Minority Graduate Students in Physics, Vera Kistiakowsky Papers, Institute Archives, MIT, Cambridge, MA. For more on Hammonds, see Aimee Sands, "Never Meant to Survive: A Black Woman's Journey; An Interview with Evelynn Hammonds," *Radical Teacher,* no. 30 (Jan. 1986): 8–15; and Evelynn Hammonds and Banu Subramaniam, "A Conversation on Feminist Science Studies," *Signs* 28, no. 3 (2003): 923–44. See also Alex Kellogg, "A University Beats the Odds to Produce Black Ph.D.'s in Math," *CHE,* Feb. 16, 2001, A14–A15.

17. On the Minority Women Scientists' Network, see ch. 1, n. 29.

18. Sylvia T. Bozeman and Rhonda J. Hughes, "Improving the Graduate School Experience for Women in Mathematics: The EDGE Program," *Journal of Women and Minorities in Science and Engineering* 10 (2004): 243–53; Robin Wilson, "Strength in Numbers: A Summer Program Gives a Boost to Women Going for Ph.D's in Mathematics," *CHE,* July 18, 2003, A10–A12.

19. Lilli Hornig, comment in *Women in Engineering: Beyond Recruitment; Proceedings of the Conference Held June 22 to 25, 1975, Cornell University, Ithaca, New York,* ed. Mary Diedrich Ott and Nancy A. Reese (n.p., n.d.), 153.

20. Malcom, "Women in Science and Engineering," 192, referring to Syverson, *Summary Report . . . 1981,* table D (p. 17).

21. Beverly Jacobson, "Career Development—The Sciences Want You," *Savvy,* Oct. 1980, 14–15; Marcia Harrison, "Boosting Women Scientists: Northeastern and BU Offer Free, Partially Funded Graduate Programs," *Boston Sunday Globe,* Sept. 20, 1981, B91–B93.

22. Joan Sherry and Linda Skidmore Dix, "Promoting Graduate and Postdoctoral Studies in Science and Engineering," in *Science and Engineering Programs: On Target for Women?* ed. Marsha Lakes Matyas and Linda Skidmore Dix (Washington, DC: National Academy Press, 1992), table 5-6 (p. 88). See also Richard B. Freeman, Tanwin Chang, and Hanley Chiang, "Supporting 'The Best and Brightest' in Science and Engineering:

NSF Graduate Research Fellowships," Working Paper 11623 (National Bureau of Economic Research, Cambridge, MA, Sept. 2005); and J. Scott Long, ed., *From Scarcity to Visibility: Gender Differences in the Careers of Doctoral Scientists and Engineers* (Washington, DC: National Academy Press, 2001), 51–57.

23. Freeman, Chang, and Chiang, "Supporting 'The Best and Brightest,'" 6.

24. Sherry and Dix, "Promoting Graduate and Postdoctoral Studies," table 5-7 (p. 90). See also Freeman, Chang, and Chiang, "Supporting 'The Best and Brightest'"; Jeffrey Mervis, "NIH Settles, NSF Sued in Student Cases," *Science* 279 (1998): 22; and Colleen Cordes, "NSF Agrees to Settle Reverse-Bias Suit," *CHE*, July 3, 1998, A24. Meanwhile, in 1987 the Department of Energy, the Lawrence Livermore National Laboratory, the NSF, and eventually about one hundred graduate schools formed the National Physical Science Consortium to provide six-year graduate fellowships in the physical sciences and engineering to women and minorities. By 2004 it had awarded more than three hundred fellowships, three-fourths of these to women, but the consortium may have been terminated, as the website provides no more recent information. See www.npsc.org.

25. Sherry and Dix, "Promoting Graduate and Postdoctoral Studies," table 5-3 (p. 82).

26. Ibid., 84. Barbara Filner, longtime head of the AWIS Foundation, headed the HHMI fellowship program in the 1990s and may have had some impact in raising consciousness at the HHMI.

27. For more on Bell Labs see ch. 10. In 1996, when Lucent Technologies took over part of Bell Labs, the fellowships' name changed to Lucent Graduate Research Fellowship Program for Women. Women in Engineering Programs and Advocates Network website (www.wepan.org); Vera Kistiakowsky Papers, box 8, Bell Labs. On Cooper, see *AMWS*, 19th ed., 1995–96, 2:457; and Neil Ashcroft, Douglas Fitchen, and Wilson Ho, "Barbara Hope Cooper," *Physics Today* 53 (Apr. 2000): 84–85. The Barbara Hope Cooper Papers are in the Rare and Manuscript Collections, Kroch Library, Cornell University, Ithaca, NY.

28. Shrock, *Cecil and Ida Green*, 229–30, 418–25; leaflet, *The Mr. and Mrs. Spencer T. Olin Fellowships for Women in Graduate Study, A Joint Undertaking of the Monticello College Foundation and Washington University in St. Louis* (n.p., 1974).

29. See, e.g., the long list in Sandra Johnson Baylor, "Graduate Fellowship Programs," *Communications of the A[ssociation for] C[omputer] M[achinery]* 38 (Jan. 1995): 37–42. The astrophysicist Margaret Geller held a Zonta International Fellowship in her fourth year at Princeton in the 1970s.

30. Susan Levine, *Degrees of Equality: The American Association of University Women and the Challenge of Twentieth-Century Feminism* (Philadelphia: Temple Univ. Press, 1995), 146–50; Doris C. Davies, *Idealism at Work: AAUW Educational Foundation Programs, 1967–1981* (Billings, MT: AAUW Educational Foundation, 1981); AAUW website, www.aauw.org; "Annual Report, 2001–2002, AAUW Educational Foundation," *AAUW Outlook* 97 (Spring–Summer 2003): 34–35. The AAUW's Educational Foundation also held occasional meetings and issued reports in the 1970s on the continuing disparate treatment of women in academia. It also changed the name of its magazine to *Graduate Woman* in 1978 and then to *Outlook* in 1989. AAUW, *Graduate and Profes-*

sional Education of Women; Suzanne Howard, *But We Will Persist: A Comparative Research Report on the Status of Women in Academe* (Washington, DC: AAUW, 1978). In the 1990s it turned its attention to a series of reports on the inequities facing schoolgirls. See ch. 4, n. 17.

31. Robin Wilson and Karen Birchard, "Looking for Gender Equity in the Lab," *CHE,* Jan. 20, 2006, A19; Piper Fogg, "New Maternity Policy for Stanford Ph.D.'s," ibid., Feb. 10, 2006, A11; Robin Wilson, "Dartmouth Students to Get Parental Leave," ibid., June 6, 2006, A12; Sierra Millman, "Princeton Expands Family-Friendly Benefits for Graduate Students with Children," ibid., Apr. 13, 2007, A13; J.C., "Policy Benefits Grad Students," *Yale Medicine,* Autumn 2007, 7.

32. "Interview with Susan Solomon, 5 September 1997," by Dale Kellogg, American Meteorological Society/University Corporation for Atmospheric Research Tape Recorded Interview Project, transcript, at National Center for Atmospheric Research Archives, Boulder, CO (I thank Diane Rabson for a copy). On Denton, see www.universityofcalifornia.edu/senate/inmemoriam/denicedenton.htm. On Greider, see Catherine Brady, *Elizabeth Blackburn and the Story of Telomeres: Deciphering the Ends of DNA* (Cambridge, MA: MIT Press, 2007), ch. 4. Although there were beginning to be a few female research professors, especially in the biological and social sciences—e.g., Ruth Sager, Charlotte Friend, Mildred Dresselhaus, Margaret Bryan Davis, Susan Ervin-Tripp, Judith Blake, Shirley Tilghman, and Joan Steitz—it is hard to generalize about their treatment of students, except to say that Davis was beloved and highly successful and that Carol Greider, one of Elizabeth Blackburn's graduate students at UC–Berkeley, later shared her Nobel Prize. On Davis, see "Statement of Nomination, Regents' [*sic*] Professor Margaret B. Davis, Award for Outstanding Contribution to Post-Baccalaureate, Graduate, and Professional Education, January 29, 1999," copy in box 5, Desert Research Institute, University of Nevada 1, 1993 [*sic*], Margaret Bryan Davis Papers, University Archives, University of Minnesota, Minneapolis.

33. Edith H. Luchins and Abraham S. Luchins, "Female Mathematicians: A Contemporary Appraisal," in *Women and the Mathematical Mystique,* ed. Lynn H. Fox, Linda Brody, and Dianne Tobin (Baltimore: Johns Hopkins Univ. Press, 1980), 11–14. The average age of the respondents was 37, so their experience in graduate school could have been at least a decade earlier.

34. Kathleen Crane, *Sea Legs: Tales of a Woman Oceanographer* (Boulder, CO: Westview, 2003), esp. 31–39, 118–22.

35. Scott Smallwood, "Bitter Aftertaste: After a Professor Took Credit for a Graduate Student's Research, Cornell Found Little Amiss," *CHE,* Apr. 12, 2002, A10–A12.

36. Jean L. Marx, "Lasker Award Stirs Controversy," *Science* 203 (1979): 341; Joan Arehart-Treichel, "Winning and Losing: The Medical Awards Game," *Science News,* Feb. 24, 1979, 120, 126; William Pollin, letter to the editor, *Science* 204 (1979): 8; [Eugene Garfield], "Controversies over Opiate Receptor Research Typify Problems Facing Awards Committees," *Current Contents,* no. 20 (May 14, 1979): 141–55, reprinted in his *Essays of an Information Scientist,* vol. 4 (Philadelphia: ISI Press, 1981), 141–55; Candace B. Pert, "This Week's Citation Classic—The Naloxone Methodology and the Discovery of the Opiate Receptors," *Current Contents,* no. 40 (Oct. 2, 1989): 16–17; idem, *Molecules of Emotion: Why You Feel the Way You Feel* (New York: Scribner, 1997);

Solomon H. Snyder, *Brainstorming: The Science and Politics of Opiate Research* (Cambridge, MA: Harvard Univ. Press, 1989); Robert Kanigel, *Apprentice to Genius: The Making of a Scientific Dynasty* (New York: Macmillan, 1986); Jeff Goldberg, *Anatomy of a Scientific Discovery* (Toronto: Bantam Books, 1988); Susan E. Cozzens, *Social Control and Multiple Discovery in Science: The Opiate Receptor Case* (Albany: State Univ. of New York Press, 1989).

37. Deborah C. Fort, ed., *A Hand Up: Women Mentoring Women in Science* (Washington, DC: AWIS, 1993).

38. Linda Grant, Kathryn B. Ward, and Carrie Forshner, "Mentoring Experiences of Women and Men in Academic Physics and Astronomy," in *Women at Work: A Meeting on the Status of Women in Astronomy, Held at the Space Telescope Science Institute, September 8–9, 1992,* ed. C. Megan Urry et al. (Baltimore: STSI, [1993]), 81–86. On the Argonne Lab meeting in 1987, see Ellen A. Keiter and Elizabeth A. Piocos, "Marion Thurnauer, Senior Scientist: A View from a Government Laboratory," in *Successful Women in Chemistry: Corporate America's Contribution to Science,* ed. Amber S. Hinkle and Jody A. Kocsis, ACS Symposium Series 907 (Washington, DC: ACS, 2005), 111–16. N. K. Meshkov, comp., *Graduate School and Beyond: A Panel Discussion from the Conference "Science Careers in Search of Women," Held at Argonne National Laboratory, Waterfall Glen Conference, April 6–7, 1989* ([Argonne, IL: Argonne National Laboratory], 1990).

39. Joy L. Frestedt, "Mentoring Women Graduate Students: Experience of the Coalition of Women Graduate Students at the University of Minnesota, 1993–1995," *Journal of Women and Minorities in Science and Engineering* 2 (1995): 151–70. See also *Mentoring for the 1990s and Beyond: New Perspectives on an Old Way to Move Ahead* (Minneapolis: University of Minnesota, Commission on Women, 1994), written and compiled by the anthropologist Janet Spector (I thank Sally Gregory Kohlstedt for a copy).

40. Laraine T. Zappert and Kendyll Stansbury, "In the Pipeline: A Comparative Analysis of Men and Women in Graduate Programs in Science, Engineering and Medicine at Stanford University," Working Paper 20 (Institute for Research on Women and Gender, Stanford University, 1984). See also "Making It in the Sciences: Susan Groh, '74, Graduate Student," *Douglass Alumnae Bulletin,* Spring 1973, 8, 15, on Groh's skeptical reception in the Stanford chemistry department; and for an evaluation of the topic to 1987, see Lilli S. Hornig, "Women Graduate Students: A Literature Review and Synthesis," in *Women: Their Underrepresentation and Career Differentials in Science and Engineering; Proceedings of a Workshop, Office of Scientific and Engineering Personnel, National Research Council,* ed. Linda S. Dix (Washington, DC: National Academy Press, 1987), 103–22.

41. This was in part because fewer women there had initially intended to complete a PhD; many had planned to get only a master's degree. Mildred Dresselhaus, "Reflections on Women Graduate Students in Engineering," *IEEE Transactions on Education,* E-28 (Nov. 1985): 202. A report a year later on a similar survey of the MIT physics department (297 students, 14% women) was less informative but similarly upbeat. Idem, "Women Graduate Students," *Physics Today* 39 (June 1986): 74–75. On Dresselhaus, see Barbara J. Love, comp., *Feminists Who Changed America, 1963–1975* (Urbana: Univ. of Illinois Press, 2006), 124.

42. Sheila E. Widnall, "AAAS Presidential Lecture: Voices from the Pipeline," *Science* 241 (1988): 1740–45.

43. Carol S. Hollenshead, Patricia Soellner Younce, and Stacy A. Wenzel, "Women Graduate Students in Mathematics and Physics: Reflections on Success," *Journal of Women and Minorities in Science and Engineering* 1 (1994): 63, 71. For use of the term *minefield,* see Paul Selvin, "Mathematics: Heroism Is Still the Norm," *Science* 255 (1992): 1383–84.

44. Maria M. Ferreira, "The Research Lab: A Chilly Place for Graduate Women," *Journal of Women and Minorities in Science and Engineering* 8 (2002): 85–98; Carol S. Hollenshead, Stacy A. Wenzel, Barbara B. Lazarus, and Indira Nair, "The Graduate Experience in the Sciences and Engineering: Rethinking a Gendered Institution," in *The Equity Equation: Fostering the Advancement of Women in the Sciences, Mathematics, and Engineering,* ed. Cinda-Sue Davis et al. (San Francisco: Jossey-Bass, 1996), 122–62; *The Women, Gender, and Science Question: [Symposium] May 12–145, 1995, Earle Brown Continuing Education Center, St. Paul Campus, St. Paul Minnesota; Abstracts* (St. Paul: University of Minnesota, 1995).

45. Candace Sidner, "Surviving as a Female Graduate Student," *Sojourner* 4 (Aug. 1979): 19, 26.

46. "Barriers to Equality in Academia: Women in Computer Science at MIT, Prepared by Female Graduate Students and Research Staff in the Laboratory for Computer Science and the Artificial Intelligence Laboratory at MIT, February 1983."

47. Roberta Brawer, "Open Letter to the Department of Physics," Apr. 10, 1986, copy in Vera Kistiakowsky Papers, box 20, Women at MIT, 1970–78 (misfiled), coinciding with Mildred Dresselhaus's more upbeat articles in n. 41 above; Judith Fleischman to Vera Kistiakowsky, Apr. 20, 1992, enclosing Judith Fleischman, "Student Sound-Off: The Visible and the Isolated: The Story of Women Students in Physics," in *APS [American Physical Society] News* 1 (Apr. 1992): 17–19, in Vera Kistiakowsky Papers, box 3, APS-CSWP, 1991–92.

48. The best summary of this effort is Jean M. Curtin, Geneva Blake, and Christine Cassagnau, "The Climate for Women Graduate Students in Physics," *Journal of Women and Minorities in Science and Engineering* 3 (1997): 95–117. See also Mildred S. Dresselhaus, Judy R. Franz, and Bunny C. Clark, "Interventions to Increase Participation of Women in Physics," *Science* 263 (1994): 1392–93; Meg Urry, "Affecting the Climate for Women in Physics: The CSWP Site Visit Program," *Status: A Report on Women in Astronomy,* June 2007, 13–19; and Zhenya Gallon, "The Atmosphere Inside: Surveying the Climate for Women at UCAR," *Staffnotes Monthly,* Apr. 2000, 1, available online at www.ucar.edu/communications/staffnotes/0004/women.html.

49. See esp. Mary Frank Fox, "Organizational Environments and Doctoral Degrees Awarded to Women in Science and Engineering Departments," *Women's Studies Quarterly* 28 (Spring–Summer 2000): 47–61. In the 1990s some graduate schools, such as MIT in 1993, also began to provide written guidelines in a graduate-student handbook that spelled out much of what was expected.

50. Debbie Franzblau and Susan Landau to "Graduate Student," Jan. 5, 1982, box 4, Production, Papers of the Association for Women in Mathematics, Wellesley College Archives, Wellesley, MA.

51. Debbie Franzblau to Linda Rothschild, Dec. 19, 1982, AWM Fund-Raising Committee (1981), Alice T. Schafer Accession, A88-060, Wellesley College Archives.

52. Marjorie Farnsworth, *The Young Woman's Guide to an Academic Career* (New York: Richards Rosen, 1974); Susan Goldhor, "How to Get a Job," *Women in Cell Biology Newsletter,* [1976], in box 75, folder 23, Papers of the American Society for Cell Biology, Special Collections, Albin O. Kuhn Library and Gallery, UMBC, Catonsville, MD, revised by the Steering Committee of Women in Cell Biology and reprinted in 1989 (copy in ibid., folder 18), later published as Women in Cell Biology, "Landing Your First Job," in Fort, *A Hand Up: Women Mentoring Women in Science,* 313–31; Virginia Walbot, "On Choosing a Postdoc," *Women in Cell Biology Newsletter,* Oct. 1979, copy in box 75, folder 23, Papers of the American Society for Cell Biology; [Cornell University, Faculty Advisory Committee, Conference on Women in Science], "Career Guide for Women in Science" (mimeograph, 1988); Clarice M. Yentsch and Carl J. Sindermann, *The Woman Scientist: Meeting the Challenge for a Successful Career* (New York: Plenum, 1992). The special issues on women in *Science* were 255 (Mar. 13, 1992), 260 (Apr. 16, 1993), 263 (Mar. 11, 1994), and 269 (Apr. 11, 1995).

53. Jean E. Taylor, "AWM in the 1990s: A Recent History of the Association for Women in Mathematics: Part 3," *AWM Newsletter* 29 (May–June 1999): 28; Ruth Strathearn Dickie, *Time of Transition: Women in Science* (n.p., 1986), 4, www.gwis.org/about/history.html. Starting in 1970 Vera Kistiakowsky, Vera Pless, Elizabeth Urey Baranger, Margaret Laws, and others ran a Boston-based Women in Science and Engineering group that held potluck dinners at members' homes until 1980, when it was replaced by the Boston chapter of the AWIS. "Women in Science and Engineering, Boston," in Vera Kistiakowsky Papers, box 29, WISE—Meetings and Speakers.

54. Suzanne Rice, "The 'Discovery' and Evolution of Sexual Harassment as an Educational Issue," *Initiatives* 57, no. 1 (1995): 1–14; Karen W. Bauer and Kelly E. Green, "Graduate Student Sexual Harassment: Do Personal Perceptions Make a Difference?" ibid. 57, no. 4 (1996): 43–50.

55. K. S. Pope, H. Levenson, and L. Schover, "Sexual Intimacy in Psychology Training: Results and Implications of a National Survey," *American Psychologist* 34 (Aug. 1979): 682–89; Robert D. Glaser and Joseph S. Thorpe, "Unethical Intimacy: A Survey of Sexual Contact and Advances between Psychology Educators and Female Graduate Students," ibid. 41 (Jan. 1986): 43–51.

56. Kim A. McDonald, "Many Female Astronomers Say They Face Sex Harassment and Bias," *CHE,* Feb. 13, 1991, A11, A15; Faye Flam, "Still a 'Chilly Climate' for Women?" *Science* 252 (1991): 1604–6. Interestingly, the Committee on the Status of Women in Astronomy of the American Astronomical Society did not wish to be associated with Bonner's survey, which the society's staff circulated anyway. Annual reports of the committee's activities are published most years in the *Bulletin of the American Astronomical Society:* see, related to the sexual-harassment survey, 21, no. 3 (1989): 883–84; 23, no. 3 (1991): 1091; 24, no. 3 (1992): 890–91 (there was no report in 1990). Only 10 percent of 350 departments polled responded. Their findings were presented in Geoff Marcy and Debra Meloy Elmegren, "University Approaches to Affirmative Action," in Urry et al., *Women a Work,* 107–15.

57. Rice, "'Discovery' and Evolution of Sexual Harassment as an Educational Issue," 7.

58. Karla Hayworth, "Professor Loses a Round in Battle against Cornell," *CHE*, Apr. 10, 1998, 20.

Chapter 7 · Postdoctoral Pathways

1. Daniel S. Greenberg, *Science, Money, and Politics: Political Triumph and Ethical Erosion* (Chicago: Univ. of Chicago Press, 2001), table 4 (p. 484); Eugene Russo, "Victims of Success," *Nature* 422 (Mar. 20, 2003): 354–55.

2. This mobility is most visible in the annual yearbooks of the Carnegie Institution of Washington, in which numerous asterisks after names indicate address changes. For a history of postdoctoral fellowships, see NRC, *Postdoctoral Appointments and Disappointments* (Washington, DC: National Academy Press, 1981).

3. Virginia Walbot, "On Choosing a Postdoc," *Women in Cell Biology Newsletter*, Oct. 1979, copy in box 75, folder 23, Papers of the American Society for Cell Biology, Special Collections, Albin O. Kuhn Library and Gallery, UMBC, Catonsville.

4. Margaret W. Rossiter, *Women Scientist in America: Before Affirmative Action, 1940–1972* (Baltimore: Johns Hopkins Univ. Press, 1995), 317–19; "Energy-Related Postdoctoral Fellowships," *NSF 25th Annual Report for Fiscal Year 1975* (Washington, DC: GPO, [1976]), 84; Carter Kimsey, "Outcomes and Impacts of the National Science Foundation's Minority Postdoctoral Research Fellowships Program," in Committee for the Evaluation of the Lucille P. Markey Charitable Trust Programs in Biomedical Sciences, Board on Higher Education and Workforce, NRC, *Enhancing Philanthropy's Support of Biomedical Scientists: Proceedings of a Workshop on Evaluation* (Washington, DC: National Academies Press, 2006), 106–10. The names of the awardees are listed in the NSF's annual *Grants and Awards* for most fiscal years into the mid-1980s, but in some years the names of the winners of all types of fellowships were grouped together in a single, list making it impossible to know who received which award.

The Ford Foundation and the NIH also had postdoctoral programs for underrepresented minorities. See Connie L. McNeely and Christine O'Brien, "Exploring Program Effects on Life Sciences Faculty Diversity: Assessing the Ford Foundation Postdoctoral Fellowships for Minorities," ibid., 89–98, stating that an evaluation was under way; and Committee for the Assessment of NIH Minority Research Training Programs, Oversight Committee for the Assessment of NIH Minority Research Training Programs, Board on Higher Education and Workforce, NRC, *Assessment of NIH Minority Research and Training Programs, Phase 3* (Washington, DC: National Academies Press, 2005), ch. 5. In 1997 there were 1,242 underrepresented minority postdoctoral fellows in the United States. Committee on Science, Engineering, and Public Policy of the National Academy of Sciences, National Academy of Engineering, and Institute of Medicine, *Enhancing the Postdoctoral Experience for Scientists and Engineers: A Guide for Postdoctoral Scholars, Advisers, Institutions, Funding Organizations, and Disciplinary Societies* (Washington, DC: National Academy Press, 2000), 39.

Gerhard Sonnert, assisted by Gerald Holton, published the results of their highly quantitative study of male and female NSF postdocs, 1952–85, and NRC associateships, 1959–86, in two volumes: *Gender Differences in Science Careers: The Project Access Study* (New Brunswick, NJ: Rutgers Univ. Press, 1995), where a lot depends on small differences in coefficients; and *Who Succeeds in Science? The Gender Dimension* (New Brunswick, NJ: Rutgers Univ. Press, 1995), aimed at policymakers and women. Unfortunately, neither were useful here, as those they interviewed and occasionally quoted are unnamed. But the records from Project Access are at the Murray Research Archive Database, Institute for Quantitative Social Sciences, Harvard University, Cambridge, MA, http://dvn.iq.harvard.edu/dvn/dv/mra/faces/study/StudyPage.xhtml?globalid=hdl:1902.1/00994.

5. Cynthia L. Attwood, *Women in Fellowship and Training Programs* (Washington, DC: Association of American Colleges, Nov. 1972); Judith Nies, "Fellowships and Women," in *Graduate and Professional Education of Women: Proceedings of AAUW Conference Held at the AAUW Educational Center, May 9–10, 1974* (Washington, DC: AAUW, 1974), 34–37; idem, *Women and Fellowships, 1974* (Washington, DC: WEAL, 1974); idem, *Women and Fellowships, 1976* (Washington, DC: WEAL, 1976); idem, *Women and Fellowships, 1981* (Washington, DC: WEAL, 1981), which reported the complicated and protracted reform of the Rhodes Trust, which required an act of Parliament. See also Nies's autobiography, *The Girl I Left Behind: A Narrative History of the Sixties* (New York: HarperCollins, 2008), 248–58, 284–85.

6. See, e.g., *AWM Newsletter* 1 (Sept. 1971): 3; *AWIS Newsletter* 15 (Sept.–Oct. 1986): 8; *WCC Newsletter*, May 1988, 1–2; "We Hear That . . . ," *CSWP Gazette* 6 (July 1986): 2; and "Three Women Physicists Receive Alfred Sloan Research Fellowship," ibid., 9 (May 1989): 7–8.

7. Nan Trent, "Computer Tallies Success Story," *Christian Science Monitor*, Mar. 2, 1964, 6.

8. The annual reports of the Alfred P. Sloan Foundation list the awardees, but it is hard to determine the gender and ethnicity of foreign recipients from their names. On Goldman-Rakic, see Eric J. Nestler, "Obituary: Patricia S. Goldman-Rakic (1937–2003)," *Nature* 425 (Oct. 2, 2003): 471. At some point the Sloan fellowships modified their age restrictions. Unfortunately, some other fellowship programs have not. In 2005 Committee W (on women) of the American Association of University Professors (AAUP) denounced such age limits and prepared a list of (still) offending programs. Robin Wilson, "Giving Junior Professors a Voice," *CHE*, Sept. 23, 2005, A24.

9. Amy L. Francis, "Analysis of the Damon Runyon Cancer Research Foundation Fellowship Program (1947–2003)," in Committee for the Evaluation of the Lucille P. Markey Charitable Trust Programs in Biomedical Sciences, Board on Higher Education and Workforce, NRC, *Enhancing Philanthropy's Support of Biomedical Scientists: Proceedings of a Workshop on Evaluation* (Washington, DC: National Academies Press, 2006), 6–72.

10. Committee for the Evaluation of the Lucille P. Markey Charitable Trust Programs in Biomedical Sciences, Board on Higher Education and Workforce, NRC, *Evaluation of the Markey Scholars Program* (Washington, DC: National Academies Press, 2006), 12.

11. G. Thomas Tanselle, Peter F. Kardon, and Eunice R. Schwager, eds., *The John*

Simon Guggenheim Memorial Foundation, 1925–2000: A Seventy-Fifth Anniversary Record (New York: John Simon Guggenheim Memorial Foundation, 2001); and Elizabeth M. Gurl, comp., *Reports of the President and the Treasurer, 2000, John Simon Guggenheim Memorial Foundation* (New York: John Simon Guggenheim Memorial Foundation, 2001). The numbers for 1971–2000 include Latin American and Caribbean as well as Canadian women. For earlier years, see Margaret W. Rossiter, *Women Scientists in America: Struggles and Strategies to 1940* (Baltimore: Johns Hopkins Univ. Press, 1982), 272–75; and idem, *Women Scientists in America: Before Affirmative Action, 1940–1972,* 320–22.

12. Rada Dyson-Hudson to Gordon Ray, May 1, 1980, box 4, folder 26, Cornell Eleven Papers, Rare and Manuscript Collections, Kroch Library, Cornell University, Ithaca, NY.

13. *MacArthur Fellows: The First 25 Years, 1981–2005* (Chicago: John D. and Catherine T. MacArthur Foundation, 2005); Kathleen Teltsch, "20 Get Cash Prizes in 'Genius Search,'" *NYT,* Jan. 19, 1983, A9, mentioning criticism; idem, "MacArthur Trust, a Philanthropic Maverick, Faces Divestiture and Discord," ibid., Aug. 17, 1983, A17. One of the difficulties was that even women nominated mostly men. Kenneth Hope to Vera Rubin, May 26, 1987, box 7, MacArthur, Vera C. Rubin Papers, Manuscript Division, Library of Congress, Washington, DC.

14. On Taylor, see *AMWS,* 19th ed., 1995–96, 7:52. On McDuff, see ibid., 5:66. On Roitman, see ibid., 6:317. See also [Juliet A. Mitchell, ed.,] *A Community of Scholars: The Institute for Advanced Study; Faculty and Members, 1930–1980* (Princeton, NJ: Institute for Advanced Study, 1980); and the institute's website, www.math.ias.edu/include/history_chron.

15. Susan Ervin-Tripp to Betty Scott, May 7, 1975, carton 32, Woman's Faculty Group, Elizabeth L. Scott Papers, Bancroft Library, University of California, Berkeley; CASBS website, www.casbs.org. See also Jessie Bernard's notes on the center's "Notice of Nondiscriminatory Policy" in *NYT,* Dec. 28, 1975, and Jessie Bernard to Dr. Lindzey, Nov. 28, 1977 (denying that she had pressured anyone for an invitation to the center), in boxes 5 and 6, Jessie Bernard Papers, Historical Collections and Labor Archives, Pattee Library, Pennsylvania State University, University Park; and Harriet B. Presser, "The Personal Is Political and Professional," in *Gender and the Academic Experience: Berkeley Women Sociologists,* ed. Kathryn P. Meadow Orlans and Ruth A. Wallace (Lincoln: Univ. of Nebraska Press, 1994), 156.

16. On Hillis, see *AMWS,* 19th ed., 1995–96, 3:859; and Llewellya Hillis-Colinvaux, "Ocean in My Backyard," *Radcliffe Quarterly* 72 (June 1986): 23–24, special issue on the Bunting Institute's first twenty-five years. On Ruskai, see *AMWS,* 19th ed., 1995–96, 6:440. On Grosz, see Jillian Lokere, "Science at Radcliffe: Drawing People Together," *Radcliffe Quarterly* 92 (Winter 2007): 10–16; and Pat Harrison, "Rebel, Builder, Boundary Crosser: Dean Barbra J. Grosz," ibid. 93 (Summer 2008): 10–11.

17. Doris C. Davies, *Idealism at Work: AAUW Educational Foundation Programs, 1967–1981* (Billings, MT: AAUW Educational Foundation, 1981), 126, 129, 193, 204. There were other private postdocs specifically for women at particular places.

18. Christine Wennerds and Agnes Wold, "Nepotism and Sexism in Peer-Review," *Nature* 387 (May 22, 1997): 341–43. These authors had unusual access to confidential

data on the applicants, because a Swedish court had ruled that although the Medical Research Council dossiers involved were official documents, they were not state secrets. To my knowledge, there have been no similar studies of American peer-review practice.

19. NRC, Committee on National Needs for Biomedical and Behavioral Scientists, *Addressing the Nation's Changing Needs for Biomedical and Behavioral Scientists* (Washington, DC: National Academy Press, 2000), ch. 1.

20. IOM, Committee on National Needs for Biomedical and Behavioral Research Personnel, *Personnel Needs and Training for Biomedical and Behavioral Research, 1983 Report of the Committee* (Washington, DC: National Academy Press, 1983).

21. Susanne D. Ellis, "Initial Employment of Physics Doctorate Recipients: Class of 1992," *Physics Today* 46 (Dec. 1993): 30, 32, 33; Kate Kirby and Roman Czujko, "The Physics Job Market: Bleak for Young Physicists," ibid., 22–27.

22. NIH, Division of Research Grants, Statistics, Analysis and Evaluation, *Women in NIH Extramural Grant Programs: Fiscal Years 1984–1993,* NIH Publication No. 95-3876 (Bethesda, MD, 1994), 116, 122, 126; Karen Kreeger, "Special Report," Naturejobs, *Nature* 417 (May 9, 2002); Howard K. Schachman and Marvin Cassman, "Ruth L. Kirschstein (1926–2009)," *Science* 326 (2009): 947.

23. "Notes on the Women Faculty Meeting of October 25, 1988," box 20, Women at MIT, 1970–78 (misfiled); Marie K. Huffman to Vera Kistiakowsky, Nov. 20, 1990, box 13, MIT Misc. 1990; and *Handbook for Entering Postdoctoral Associates and Fellows at MIT,* 2nd ed. (Cambridge, MA: MIT, Oct. 1990), box 30—all in Vera Kistiakowsky Papers, Institute Archives, MIT, Cambridge, MA.

24. Dan Ferber, "Postdoc Activism: Getting to the Front of the Bus," *Science* 285 (1999): 1514–17. This was part of a special section of an issue devoted to the concerns of postdoctoral fellows, itself a sign of the expanding coverage.

25. On the Salk Institute, see the institute's website, www.salk.edu/srf/; and Bruno Latour and Steve Woolgar, *Laboratory Life: The Social Construction of Scientific Facts* (Thousand Oaks, CA: Sage, 1979), based on Latour's fieldwork at Roger Guillemin's laboratory at Salk in 1975–77 (the people are unnamed, but some were women). On the NYAS, see "The Academy's Science Alliance Brings Together the Area's Top Young Investigators," Naturejobs, *Nature* 429 (June 17, 2004).

26. On Hanson, see David G. Cassel, "Gail Hanson (1947–)," in *Out of the Shadows: Contributions of Twentieth-Century Women to Physics,* ed. Nina Byers and Gary Williams (Cambridge: Cambridge Univ. Press, 2006), 427–39. On Blackburn, see Catherine Brady, *Elizabeth Blackburn and the Story of Telomeres: Deciphering the Ends of DNA* (Cambridge, MA: MIT Press, 2007), ch. 3. On Tilghman, see Claudia Dreifus, "Career That Grew from an Embryo: A Conversation with Shirley Tilghman," *NYT,* July 8, 2003, F2. On Buck, see Lawrence K. Altman, "Unraveling Enigma of Smell Wins Nobel for Two Americans," ibid., Oct. 5, 2004, A18; and Alison Abbott, "Science of Smell Wins Medicine Nobel," *Nature* 430 (Oct. 7, 2004): 616. On Bargmann, see Melissa Marino, "Inaugural Article: Biography of Cornelia I. Bargmann," *Proceedings of the National Academy of Sciences (USA)* 102 (Mar. 1, 2005): 3181–83.

27. On Chang, see www.awm-math.org/noetherbrochure/Chang01.html. On Tinsley, see Sandra Faber, "Beatrice Tinsley," *Physics Today* 34 (Sept. 1981): 110, 111; and

Edward Hill, *My Daughter Beatrice: A Personal Memoir of Dr. Beatrice Tinsley, Astronomer* (New York: APS, 1986). See also the quite insightful unpublished essay by Jill C. Bonner, "Life and Death in the Holding Pattern (A Short Outline Study of the Problems of People Holding Temporary Positions in Physics)" (typescript, June 1976), copy in box 13, folder 3, APS-CSWP, Eugenie V. Mielczarek Papers, Special Collections and Archives, George Mason University, Fairfax, VA.

28. Barbara Mandula, "Perpetual Postdocs," *AWIS Newsletter* 6 (Jan.–Feb. 1977): 4–7.

29. Peter Monaghan, "U. of Idaho Looks Into Researcher's Complaint about Schedule," *CHE,* Nov. 16, 1994, A26.

30. One early effort to effect some changes came from Yale University in 1976. Katherine Finseth, MD, to Federation of Organizations for Professional Women, Jan. 19, 1976, Vera Kistiakowsky Papers, box 12, FOPW. For another effort, see Sidney Altman and Maxine Singer (Yale trustee) to Bernadine Healy, Oct. 21, 1991, and Healy's supportive reply, Dec. 12, 1991, both in box 54, folder 15, NIH Correspondence, 1988–91, Maxine Singer Papers, Manuscript Division, Library of Congress.

31. "Donna J. Nelson, Transcript of Interview Conducted by Hilary Domush and Leah Webb-Halpern at University of Oklahoma, Norman, Oklahoma, on 21 and 22 July 2008," 15–20, Chemical Heritage Foundation, Philadelphia; Robin Wilson, "The Laws of Physics: A Postdoc's Pregnancy Derails Her Career," *CHE,* Nov. 11, 2005, A10–A12. In 2005 the National Institute of Allergy and Infectious Diseases quietly started a small program ($500,000) to provide help to young parents. Robin Wilson, "Family Science," ibid., July 22, 2005, A8.

32. Jeffrey Brainard, "NIH's Minority-Training Programs Lack Data on Success, Report Says," ibid., July 1, 2005, A18; Committee for the Assessment of NIH Minority Research Training Programs, Oversight Committee for the Assessment of NIH Minority Research Training Programs, Board on Higher Education and Workforce, NRC, *Assessment of NIH Minority Research and Training Programs,* 129.

33. Eliot Marshall, "Dispute Slows Paper on 'Remarkable' Vaccine," *Science* 268 (1995): 1712–15; Philip J. Hilts, "U.S. Scientist Says Superiors Stole Credit for Her Vaccine," *NYT,* Dec. 18, 1993, 10; letters to the editor, *Science* 269 (1995): 1029–35, 1119 (two of the letters mention an earlier, similar case involving the NIH fellow Jane Rosen).

34. Reva Kay Williams, "A Word from a Black Female Relativistic Astrophysicist: Setting the Record Straight on Black Holes," Apr. 6, 2004, http://arxiv.org/abs/physics/0404029.

35. See, e.g., Alisa Solomon, "Snake Pit: When Dr. Margaret Jensvold Dared to Challenge the All-Male Pecking Order at the National Institute of Mental Health, the Men in White Coats Came after Her—with a Vengeance," *Mirabella,* Apr. 1993, 140–44; Scott S. Greenberger, "Science Friction: The Struggle of Female Researchers at NIH," *Washington Post,* July 11, 1993, C3; "Women Doctors Lose NIH Sex Discrimination Lawsuit," *Feminist Majority Newsletter* 8, no. 1, n.d., online at http://feminist.org/research/report/eight.html; and Clippings in box 11, folder 11, Fann Harding Papers, Women in Science and Engineering Archives, Special Collections, Iowa State University Library, Ames.

36. Philip Hilts, "Hero in Exposing Science Hoax Paid Dearly," *NYT,* Mar. 22, 1991,

A1, B6; Judy Sarasohn, *Science on Trial: The Whistle Blower, the Accused and the Nobel Laureate* (New York: St. Martin's, 1993); Daniel Kevles, *The Baltimore Case: A Trial of Politics, Science, and Character* (New York: Norton, 1998), 139; Margot O'Toole, "Scientists Must Be Able to Disclose Colleagues' Mistakes without Risking Their Own Jobs or Financial Support," *CHE,* Jan. 25, 1989, A44. Allegra Goodman's *Intuition* (New York: Dial, 2006) is a novel about postdocs and a possible case of scientific fraud.

37. Committee on Dimensions, Causes, and Implications of Recent Trends in the Careers of Life Scientists, Board on Biology, Commission on the Life Sciences, Office of Scientific and Engineering Personnel, NRC, *Trends in the Early Careers of Life Scientists* (Washington, DC: National Academy Press, 1998); NRC, *Bridges to Independence: Fostering the Independence of New Investigators in Biomedical Research* (Washington, DC: National Academy Press, 2005); Jeffrey Brainard, "New Grant Program Unveiled by NIH," *CHE,* Feb. 10, 2006, A24.

38. Committee on Science, Engineering, and Public Policy of the National Academy of Sciences, National Academy of Engineering, and Institute of Medicine, *Enhancing the Postdoctoral Experience for Scientists and Engineers.* Mildred Dresselhaus chaired COSEPUP's Postdoctoral Guidance Group. Ron Southwick, "NIH Offers Postdoctoral Scientists a Raise and Prospects for Better Treatment," *CHE,* Apr. 27, 2001, A29–A30.

39. See also Richard Freeman, Eric Weinstein, Elizabeth Marincola, James Rosenbaum, and Frank Solomon, "Competition and Careers in Biosciences," *Science* 294 (2001): 2293–94; and Sally Goodman and Karen Kreeger, "Getting Organized," *Nature* 423 (May 1, 2003): 98–99.

40. On Jean Fort, see Yudhijit Bhattacharjee, ed., "Random Samples: People," *Science* 307 (2005): 843.

41. The report is available at http://postdoc.sigmaxi.org. See also Jeffrey Brainard, "Postdoctoral Researchers Value Structured Training over Pay, Survey Says," *CHE,* Apr. 15, 2005, A21.

42. Maresi Nerad and Joseph Cerny, "Postdoctoral Patterns, Career Advancement, and Problems," *Science* 285 (1999): 1533–35.

43. Frederick H. Buttel and Jessica R. Goldberger, "Gender and Agricultural Science: Evidence From Two Surveys of Land-Grant Scientists," *Rural Sociology* 67, no. 1 (2002): 33–34. The men and women predominated in different fields. The classic study on the value of postdoctoral fellowships is Barbara Reskin, "Sex Differences in the Professional Life Chances of Chemists" (PhD diss., University of Washington, 1973), which was based on a study of 450 male and female chemists who earned their doctorates between 1955 and 1961. She found that holding a prestigious postdoctoral fellowship had more impact on the women's subsequent careers than it did on the men's. See also Barbara F. Reskin, "Scientific Productivity, Sex, and Location in the Institution of Science," *American Journal of Sociology* 83, no. 5 (1978): 1235–43.

44. Dean O. Smith, "Women in Neuroscience," *Science* 257 (1992): 1611.

Chapter 8 · Industrial and Self-Employment

1. The historic AT&T case started in November 1970, a year and a half before Congress passed, and President Nixon signed, the two pieces of equal-opportunity legisla-

tion that affected academia, the Equal Opportunity Act of 1972 and the Educational Amendments Act of 1972, which did not apply to AT&T. The case grew out of AT&T's request as a federally regulated utility to the Federal Communications Commission (FCC) for its first rate increase in many years. The lawyers on the staff of the FCC asked the Equal Employment Opportunity Commission to join the case and prove that as a racially and sexually discriminatory employer, AT&T was inefficient and did not deserve the price hike. For an absorbing account of the case, see Lois Kathryn Herr, *Women, Power and the AT&T: Winning Rights in the Workplace* (Boston: Northeastern Univ. Press, 2003). Herr was one of several NOW members involved in the case. Another was the sociologist Sally Hacker, whose papers are at the Schlesinger Library, RIAS, Harvard University, Cambridge, MA. Phyllis Wallace, an African American economist and at the time a researcher at the EEOC, later published statistical data on the case: *Equal Employment Opportunity and the AT&T Case* (Cambridge, MA: MIT Press, 1976). Several women social scientists involved in the case in one way or another were the economists Barbara Bergmann, Valerie Oppenheimer, and Wallace; the sociologists Judy Long Laws and Hacker; and the psychologists Sandra Bem, Leona Tyler, and JoAnn Evansgardner.

2. A useful survey of the topic with a good bibliography is Paula Rayman and Jennifer S. Jackson, "Women Scientists in Industry," in *The Equity Equation: Fostering the Advancement of Women in the Sciences, Mathematics, and Engineering,* ed. Cinda-Sue Davis et al. (San Francisco: Jossey-Bass, 1996), 290–320. Another good starting place on the issues is CWSE, Office of Scientific and Engineering Personnel, NRC, *Women Scientists and Engineers Employed in Industry: Why So Few? A Report Based on a Conference* (Washington, DC: National Academy Press, 1994).

3. NSF, *Research and Development in Industry, 1979: Funds, 1979; Scientists and Engineers, January 1980,* NSF 82-304 (Washington, DC, 1982), table B-52; NSF, Division of Science Resources Studies, *Research and Development in Industry, 1998: Funds, 1998; Scientists and Engineers, January 1999,* NSF 01-305 (Arlington, VA, 2000), table A-42, which gives data for January 1999; NSF, *Guide to NSF Science and Engineering Resources Data,* NSF 95-318 (Arlington, VA, 1995), 22–24, 38. Although seemingly the best data available, these numbers should be regarded skeptically, since the NSF's surveys were based on samples that changed frequently.

4. Carol Hymowitz and Timothy D. Schellhardt, "The Glass Ceiling: Why Women Can't Seem to Break the Invisible Barrier That Blocks Them from the Top Jobs," *Wall Street Journal,* Mar. 24, 1986, special report on corporate women, 1, 4. Judith Glover, *Women and Scientific Employment* (Basingstoke, UK: Macmillan, 2000), calls for new and better data.

5. Susan Fitzpatrick, "Successfully Creating Your Own Business," *AWIS Magazine* 31 (Autumn 2002): 37; Esmeralda Santiago, "Women-Owned Business: The Fastest Growing Segment of the Economy," *Radcliffe Quarterly* 73 (Dec. 1987): 13, claiming that 2 million women started small businesses in 1986 alone. Exec. Order 12138, May 18, 1979, creating a national women's business enterprise policy, was amended by Exec. Order 12608, Sept. 9, 1987, www.archives.gov/federalregister/execuiveorders/1987.html .#12608.

6. CWSE, Office of Scientific and Engineering Personnel, NRC, *Women Scientists*

and Engineers Employed in Industry; Catalyst, Inc., *Women in Engineering: An Untapped Resource* (New York, 1992); idem, *Cracking the Glass Ceiling: Strategies for Success* (New York, 1994), with an extensive bibliography; idem, *Women Scientists in Industry: A Winning Formula for Companies* (New York, 1999); idem, *Women of Color in Corporate Management: Opportunities and Barriers* (New York, 1999).

7. Eric Pooley, "Leaders of the Pack," *Fortune,* Oct. 16, 2006, 25.

8. Margaret W. Rossiter, *Women Scientists in America: Struggles and Strategies to 1940* (Baltimore: Johns Hopkins Univ. Press, 1982), 117–18. See also Bureau of Vocational Information, *Women in Chemistry: A Study of Professional Opportunities* (New York, 1922).

9. ACS, Office of Manpower Studies, *1975 Report of Chemists' Salaries and Employment Status Supplement, Economic Status of Women in the ACS* (Washington, DC: ACS, Apr. 1976). See also Committee on the Education and Employment of Women in Science and Engineering, NRC, *Women Scientists in Industry and Government: How Much Progress in the 1970s? An Interim Report to the Office of Science and Technology Policy from the Committee on Human Resources, National Research Council* (Washington, DC: NAS, 1980), which used the National Academy's Survey of Earned Doctorates for data on both doctorates of the 1970s and a sample of all past doctorates.

10. *Women Chemists 1990: A Supplementary Report to the 1990 Survey of Members' Salary and Employment* (Washington, DC: ACS, Dec. 1990), 52. See also Roscher's obituary in the *Washington Post,* Sept. 27, 2001, B6; and "Oral History of Nina Matheny Roscher, April 24, 1998," by Tanya Zanish-Belcher, Women in Science and Engineering Archives, Special Collections, Iowa State University Library, Ames.

11. *Women Chemists 1990,* ix, 66.

12. Michael Heylin, "ChemCensus 2000," *C&EN* 78 (Aug. 14, 2000): 47; *Women Chemists 2000* (Washington, DC: ACS, 2001), 31.

13. Lillian H. Sello, "Corporate Networks," *CHEMTECH* 12 (Dec. 1982): 721–23; Susan Struthers, "Improvements, but Not Quite Parity for Women Chemists," *Industrial Chemical News* 4 (Sept. 1983): 1, 29–31; Nancy M. Tooney, "Meet a Member: Floie Vane [of Hoffman-La Roche]," *AWIS Newsletter* 13 (Feb.–Mar. 1984): 12–14. See also Hoffman-La Roche's web page, www.rocheusa.com. There was a sartorial side to the coming of women in business in the 1970s. See John T. Molloy, *Dress for Success* (New York: Wyden, 1975); idem, *Women: Dress for Success* (London: Foulsham, 1980); and M. R. Montgomery, *In Search of L. L. Bean* (Boston: Little, Brown, 1984). In 1975 women made up one-quarter of L.L. Bean's customers, but by 1984 they constituted 70 percent, buying mostly women's clothing. This disconcerted the Maine owners, who preferred a masculine, outdoors profile.

14. *Working Mother* started its list in 1986 with thirty companies, increasing it to forty in 1987 and then fifty in 1988. Sherrie Clinton, "Best Firms for Working Moms," *Newsday,* Nov. 2, 1988. Bayer was on the list by 2004. Clipping in box 17, Dora Skypek Papers, Manuscript, Archives, and Rare Book Library, Emory University, Atlanta.

15. "Metro Women Chemists Committee," in agenda binder for meeting of the Women Chemists Committee of the American Chemical Society, June 5, 1988, tab 4, in box 12, folder 4, Kathleen M. Desmond Trehanovsky Papers, Women in Science and

Engineering Archives, Special Collections, Iowa State University Library, which is historical; Valerie Schlemann, "FemChem—One Year Old," *Women Chemists,* Mar. 1984, 2; Raquel Diaz-Sprague, "The Columbus Section, Women Chemists Committee," ibid., Mar. 1987, 3.

16. Roseanne Savol, "The Women's Committee of Miles Laboratories," *Women Chemists,* July 1985, 7.

17. "Oral History of Jeannette E. Brown, December 19, 2001," by Laura Sweeney, Women in Science and Engineering Archives, Special Collections, Iowa State University Library.

18. On DuPont, see "Women in Research," *Better Living* 14 (Jan.–Feb. 1960): 12–15. On Kaiser, see *AMWS,* 19th ed., 1995–96, 4:210. On Larsen, see ibid., 4:744. On Issler, see "The Drive to Be More Competitive," *DuPont World* 6, no. 5 (1991) (I thank Issler for a copy). On Mazur, see "Barbara Mazur," *Nature Biotechnology* 29 (Mar. 2011): 195. In 1965 DuPont had hired a female African American chemistry graduate of Fisk University as a patent searcher. "Peggy Joplin, Patent Searcher," *Better Living* 26 (Jan. 1972): 10–21.

19. U.S. Patent and Trademark Office, *Buttons to Biotech: U.S. Patenting by Women, 1977 to 1988* (Alexandria, VA: NTIS, Jan. 1990), 16, 25. On Walworth, see "Officers," *Photographic Science and Engineering* 25 (July–Aug. 1981), front matter; and "Officers," *Journal of Imaging Science* 29 (May–June 1985), front matter.

20. On Grasselli, see *AMWS,* 19th ed., 1995–96, 3:338; "Oral History of Jeanette Grasselli Brown, December 27, 2001," by Tanya Zanish-Belcher, transcript in Women in Science and Engineering Archives, Special Collections, Iowa State University Library; and Francie Bauer, "Jeanette Grasselli (Brown) (1929–), Analytical Chemist," in *Notable Women in the Physical Sciences: A Biographical Dictionary,* ed. Benjamin F. Shearer and Barbara S. Shearer (Westport, CT: Greenwood, 1997), 154–58.

21. "Oral History of Helen Free, October 21, 2001," by Tanya Zanish-Belcher, transcript in Women in Science and Engineering Archives, Special Collections, Iowa State University Library; James J. Bohning, "Diagnosing Disease: A Story from the CHF Oral History Program," *Chemical Heritage,* Fall 2003, 12–13, 42–44; Raymond B. Seymour, "Helen Murray Free (1923–)," in *Women in Chemistry and Physics: A Biobibliographic Sourcebook,* ed. Louise S. Grinstein, Rose K. Rose, and Miriam H. Rafailovich (Westport, CT: Greenwood, 1993), 201–6; Heather Martin, "Helen M. Free (1923–), Chemist," in Shearer and Shearer, *Notable Women in the Physical Sciences,* 110–12. Free deserves a biography.

22. On Taylor, see *AMWS,* 19th ed., 1995–96, 6:53; and Mairin B. Brennan, "Eminent Women Chemists Share Success Stories," *C&EN* 71 (Apr. 26, 1993): 30. Also at GM were the physicist Betsy Ancker-Johnson and the economist Marina von Neumann Whitman.

23. On Newman, see "United States Court of Appeals for the Federal Circuit, Judicial Biographies," www.fedcir.gov/judgbios.html.

24. On Hopkins, see "Esther A. H. Hopkins," *Journeys of Women in Science and Engineering: No Universal Constraints,* ed. Susan A. Ambrose et al. (Philadelphia: Temple Univ. Press, 1997), 213–16. For more on Polaroid, see James E. Post and Marilyn

Mellis, "Corporate Responsiveness and Organizational Learning," *California Management Review* 20 (Spring 1978): 57–63; and Lisa Birnbach, *Going to Work* (New York: Villard Books, 1988), 78–89.

25. Amber S. Hinkle and Jody A. Kocsis, eds., *Successful Women in Chemistry: Corporate America's Contribution to Science,* ACS Symposium Series 907 (Washington, DC: ACS, 2005), chs. 3, 11, 19. See also Frankie Wood-Black, "From Bench Researcher to Business Director," *Chemical Engineering Progress* 99 (Aug. 1, 2003): 45S.

26. L. Shannon Davis, "An Oddity No Longer: Women Scientists in Industry," in NRC, Commission on Physical Sciences, Mathematics and Applications, Board on Chemical Sciences and Technology, Chemical Sciences Roundtable, *Women in the Chemical Workforce: A Workshop Report to the Chemical Sciences Roundtable* (Washington, DC: National Academy Press, 2000), 61.

27. On Simon, see *AMWS,* 19th ed., 1995–96, 6:932; and Betty Smith, "Woman Scientist Earns Respect by Performance," *Bridgeport Sunday Post,* Apr. 1977, clipping in Records of the Women in Science Project, NSF, NARA, College Park, MD. See also Margaret W. Rossiter, *Women Scientists in America: Before Affirmative Action, 1940–1972* (Baltimore: Johns Hopkins Univ. Press, 1995), 264–65, 340, 366. Journalists were respectful but disbelieving of her tales of her charmed life and career.

28. On Berkowitz, see *AMWS,* 19th ed., 1995–96, 1:576; and Susan Klarreich, "Joan Berkowitz (1931–)," in Grinstein, Rose, and Rafailovich, *Women in Chemistry and Physics,* 50–56.

29. On Good, see *AMWS,* 19th ed., 1995–96, 3:266; Margaret A. Cavanaugh, "Mary Lowe Good (1931–)," in Grinstein, Rose, and Rafailovich, *Women in Chemistry and Physics,* 218–29; Eileen Horn Stanley, "Mary Lowe Good (1931–), Chemist," in Shearer and Shearer, *Notable Women in the Physical Sciences,* 148–53; and "Mary L. Good, Transcript of an Interview Conducted by James G. Traynham at Little Rock, Arkansas, on 2 June 1998 (With Subsequent Corrections and Additions)," Chemical Heritage Foundation, Philadelphia.

30. "Polaroid Elects 4 Officers," *Boston Globe,* Apr. 5, 1987, clipping in Polaroid Corporation Archives, Cambridge, MA.

31. Doug Henze, "Things Have Been 'Different' for Female Dow Executive," *Midland Daily News,* May 1992, in Clippings File, Post Street Archives, Midland, MI.

32. "DuPont CEO to Step Down, Company Names Successor," *Boston Globe,* Sept. 24, 2008, C2; *1999 Catalyst Census of Women Corporate Officers and Top Earners* (New York: Catalyst, 1999), 24, 34; "Meet the Executives," www2.dupont.com/Our_compan/en_US/executives/kullman.html; Alexander H. Tullo, "Women in Industry," *C&EN* 86 (Aug. 11, 2008): 330–31. On Burns, see Amy Zipkin, "The Boss: Delight in Science, Then and Now," *NYT,* Jan. 28, 2007, B18. Burns's big break came in 1994, when she was named director of women's health to deal with the barrage of lawsuits from angry women who claimed that they had been harmed by their silicone breast implants from Dow Corning. The company declared bankruptcy, but Burns handled the situation and a subsequent foreign assignment so well that she was made president at age 49.

33. For more on the Garvan-Olin Medal, see ch. 12. On Helen Free, see above, n. 21.

34. On Sherman, see "Hall of Fame/Inventor Profile," www.invent.org/hall_of_fame/160.html. On Kwolek, see "Interview with Stephanie Kwolek," by Laura Sweeney,

Women in Science and Engineering Archives, Special Collections, Iowa State University Library, Ames.

35. On Elion, see *AMWS*, 16th ed., 1989, 2:867; Richard Kent and Brian Huber, "Obituary: Gertrude Belle Elion (1918–99), Pioneer of Drug Discovery," *Nature* 398 (Apr. 1, 1999): 380; Lawrence K. Altman, "Gertrude Elion, Drug Developer, Dies at 81," *NYT*, Feb. 23, 1999, A21; Marilyn McKinley Parrish, "Gertrude Belle Elion (1918–), Biochemist," in Shearer and Shearer, *Notable Women in the Physical Sciences*, 84–88; and her autobiographical article "The Quest for a Cure," *Annual Review of Pharmacology and Toxicology* 33 (1993): 1–23. It may have been Elion's salary (for a woman chemist with an MS and more than forty years' experience) that sent the salary data in *Women Chemists 1990*, fig. 3.20 (p. 57), skyrocketing.

36. Leticia Kent, "Leigh Anderson, Chemical Engineer," *Chemistry* 52 (Feb. 1979): 27–29.

37. J. Gregg Robinson and Judith S. McIlwee, "Women in Engineering: A Promise Unfulfilled?" *Social Problems* 36 (Dec. 1989): 455–72, based on a mid-1980s study of 328 male and 79 female mechanical and electrical engineers in southern California in the aerospace and high-tech industries.

38. [Laura I. Langbein], *Profile of IEEE Women Members: Their Salaries, Demographics, Attitudes toward the Workplace and Professional Status* (Washington, DC: IEEE, 1984), 3–7.

39. SWE, *A National Survey of Women and Men Engineers: A Study of the Members of 22 Engineering Societies* (New York, 1993). On McAfee, see Nancy Tooney, "Meet a Member," *AWIS Newsletter* 6 (Sept.–Oct. 1977): 10.

40. Patrick has mellowed over the years. See Jennie Patrick, "Trials, Tribulations, Triumphs," *Sage* 6 (Fall 1989): 51–53; Wini Warren, *Black Women Scientists in the United States* (Bloomington: Indiana Univ. Press, 1999), 219–22; Clarence G. Williams, *Technology and the Dream: Reflections on the Black Experience at MIT, 1941–1999* (Cambridge, MA: MIT Press, 2001), 626–39; and Diann Jordan, *Sisters in Science: Conversations with Black Women Scientists on Race, Gender, and Their Passion for Science* (West Lafayette, IN: Purdue Univ. Press, 2006), 163–74 (I thank Jordan for a copy).

41. Esther M. Conwell, "Promoting Science and Engineering Careers in Industry," in *Science and Engineering Programs: On Target for Women?* ed. Marsha Lakes Matyas and Linda Skidmore Dix (Washington, DC: National Academy Press, 1992), 119–39. On the CWSE, see also France A. Córdova, "Projects of the National Academies on Women in Science and Engineering," in *Gendered Innovations in Science and Engineering*, ed. Londa Schiebinger (Stanford, CA: Stanford Univ. Press, 2008), 198–211.

42. CWSE, Office of Scientific and Engineering Personnel, NRC, *Women Scientists and Engineers Employed in Industry: Why So Few? A Report Based on a Conference* (Washington, DC: National Academy Press, 1994), ch. 3. The book also included the Aerospace Corporation, of El Segundo, California, but as that is a federally financed nonprofit corporation, its remarkable turnaround is considered in ch. 10.

43. A later version was published as Anne E. Preston, "Why Have All the Women Gone? A Study of Exit of Women from the Science and Engineering Profession," *American Economic Review* 84 (Dec. 1994): 1446–62; and then idem, *Leaving Science: Occupational Exit from Scientific Careers* (New York: Russell Sage Foundation, 2004).

44. Ann M. Thayer, "Family Leave Law to Have Little Impact on Major Chemical Firms," *C&EN* 71 (May 10, 1993): 9–10.

45. Cecily Cannan Selby, ed., "Women in Science and Engineering: Choices for Success," special issue, *Annals of the New York Academy of Sciences* 869 (1999); Mairin B. Brennan, "Reflections of Women in Science," *C&EN* 76 (Apr. 6, 1998): 37–41. It probably is not a coincidence that Madeleine Jacobs returned to *C&EN* as managing editor in 1998 (see ch. 10).

46. Scanning the cores for foraminifera or signs of oil was a kind of "women's work" in the 1920s. See Rossiter, *Women Scientists in America: Struggles and Strategies to 1940,* 259.

47. Dorothy Jung Echols, "Memorial to Doris M. Curtis, 1914–1991," *Geological Society of America, Memorials* 23 (1993): 175–83.

48. There is not much available on women scientists in the petroleum industry, although some worked on the Alyeska Pipeline, in Alaska, in the early 1970s. Duane R. Packer, William R. Dickson, and Kathryn M. Nichols, "Memorial to Marjorie K. Korringa, 1943–1974," ibid. 6 [1977]: 1–3. There is material on other women workers on the pipeline in the Helen L. Atkinson Collection, Alaska and Polar Regions Collections, Rasmuson Library, University of Alaska, Fairbanks.

49. On Schwartzer, see *AMWS,* 19th ed., 1995–96, 6:701; Martha J. Bailey, *American Women in Science 1950 to the Present: A Biographical Dictionary* (Santa Barbara, CA: ABC-CLIO, 1998), 351–52; Theresa Flynn Schwartzer, "Women Geoscientists in Petroleum Industry," *American Association of Petroleum Geologists Bulletin* 60 (Apr. 1976): 720; and idem, "Career Opportunities for Women Geoscientists in the Petroleum Industry," *Geology* 5 (1977): 500–501. See also Virginia Murphy Sand and Bonnie Butler Bunning, "Ten Years of Progress for AGI's Women Geoscientists Committee," *Journal of Geological Education* 33 (1985): 212–15; untitled article, *Geotimes* 22 (Apr. 1977): 20–23, for a condensed version of the 1974 report; and Theresa F. Schwartzer, "The Changing Status of Women in the Geosciences," in "Expanding the Role of Women in the Sciences," ed. Anne M. Briscoe and Sheila M. Pfafflin, special issue, *Annals of the New York Academy of Sciences* 323 (1979): 48–64. There was some friction in the oil industry between women who had started as secretaries and inched their way up to more responsible positions and the new degreed professional women of the 1970s. Mary Kay Ritz, "Oil Industry Women Say Advancement a Real Fight," *Rocky Mountain Business Journal,* Nov. 8, 1982, 4–5.

50. On Landon, see *AMWS,* 19th ed., 1995–96, 4:712; "WOMEN: Success Abounds in 1983's Oil Fields," *Rocky Mountain News,* Feb. 15, 1983, 1B, 12B; Lori Roll, "'Gutsy' Women Geoscientists Log Career Successes," *AAPG Explorer,* Oct. 1981, 26–27 (I thank Michele Aldrich for a copy); and "AGI to Induct Susan M. Landon as 1998 President," *AGI Spotlight,* Oct. 20, 1997, www.agiweb.org/news/landon97.html.

51. "Barbara Sue McBride (1956–2000)," *AAPG Bulletin* 85 (Feb. 2001): 360–61.

52. On Lindstedt-Siva, see *AMWS,* 19th ed., 1995–96, 4:978. On Moody, see ibid., 18th ed., 1992–93, 5:503. On Wood-Black, see above, n. 25. On Price, see "Breaking Through," *Working Woman,* Dec. 1992, 21.

53. Suzanne Neuschatz, "Association of Women Geoscientists National Survey of Geoscientists, 1983–1984," copy in box 6, folder 6, Papers of the Association for Women

Geoscientists, Women in Science and Engineering Archives, Special Collections, Iowa State University Library. See also Sigrid Asher and Suzanne Webel, "Women Geoscientists Report Survey Results," *Geotimes* 30 (Nov. 1985): 10–12; and "Survey of Women Geoscientists: Results," *Eos*, Dec. 24, 1985, 1358–59.

54. Mary J. Repar to Vera Kistiakowski [*sic*], July 6, 1989, box 29, Women in Federal Government, Vera Kistiakowsky Papers, Institute Archives, MIT, Cambridge, MA; Mary J. Repar, personal communication to author, Jan. 12, 1991.

55. Katharine M. Donato, "Programming for Change? The Growing Demand for Women Systems Analysts," in *Job Queues, Gender Queues*, ed. Barbara F. Reskin and Patricia A. Roos (Philadelphia: Temple Univ. Press, 1990), 167–82; Myra Strober and Carolyn L. Arnold, "Integrated Circuits/Segregated Labor: Women in Computer-Related Occupations and High-Tech Industries," in *Computer Chips and Paper Clips: Technology and Women's Employment*, ed. Heidi Hartmann (Washington, DC: National Academy Press, 1987), 136–82. Some data are available in NSF, *Profiles—Computer Sciences: Human Resources and Funding*, NSF 88-324 (Washington, DC, 1988); and Larry Drake, "The Outlook for Computer Professions: 1985 Rewrites the Program," *Occupational Outlook Quarterly* 30 (Winter 1986): 2–11. See also Rosemary Wright, *Women Computer Professionals: Progress and Resistance* (Lewiston, ME: Edwin Mellen, 1997).

56. Preston Galla, "High-Tech Gypsies," *Boston Magazine*, Sept. 1987, 79, 81, 83, 85–87.

57. Box 1, Officer Resignations, Papers of the Association of Women in Computing, Charles Babbage Institute Archives, University of Minnesota, Minneapolis.

58. On Sammet, see "Sammet, Jean Elaine (1928–)," in Bailey, *American Women in Science 1950 to the Present*, 345–46. On Allen, see *AMWS*, 19th ed., 1995–96, 1:102. See also www-03.ibm.com/ibm/history/witexhibit/wit_hall_allen.html; and "Pfeiffer, Jane (Cahill) (1932–)," in Bailey, *American Women in Science 1950 to the Present*, 306–8.

59. On Wu, see "Lilian Shiao-Yen Wu," in Ambrose et al., *Journeys of Women in Science and Engineering*, 437–40. On Ferrante, see www.jacobsschool.ucsd.edu/news/news_releases/release.sfe?id=654. On Klawe, see www.hmc.edu/headline/Klawe.html; and Maria M. Klawe, "Computer Science," *AWM Newsletter* 15 (July–Aug. 1985): 7–9, reprinted in *Complexities: Women in Mathematics*, ed. Bettye Anne Case and Anne M. Leggett (Princeton, NJ: Princeton Univ. Press, 2005), 228–30.

60. Ron Wolf, "IBM Eases Policies for Its Employees," *Miami Herald*, Oct. 19, 1988, clipping in Dora Skypek Papers, box 17.

61. Katie Hafner, "Anita Borg, 54, Trailblazer for Women in Computer Field," *NYT*, Apr. 10, 2003, A25. See also Anita Borg, "Why Systers?" at http://anitaborg.org/projects/systers/whysysters/html, which reproduces a version of the original essay that appeared in *Computing Research News* in 1993; and idem, "What Draws Women to and Keeps Women in Computing?" in Selby, "Women in Science and Engineering," 102–5, 247. After Borg's death, an Anita Borg Institute for Women and Technology was established in California with support from industry, foundations, and individuals. Borg may merit a biography.

62. Anita Borg and Telle Whitney, "The Grace Hopper Celebration," *Communications of the Association for Computer Machinery* 38 (Jan. 1995): 50–51. See also http://grace hopper.org/2010/about/history-of-the-conference/.

63. In 2006 Klawe became the first woman president of Harvey Mudd College, which has an engineering program.

64. "Chairman's Daughter Gets Post," *NYT,* Nov. 8, 1984, D2; "Kim Polese, Web Entrepreneur," *Time,* Apr. 21, 1997, 42. On Carly Fiorina, see her autobiography, *Tough Choices: A Memoir* (New York: Penguin Group, 2006); and the Hewlett Packard website, www.hp.com/hpinfo/execteam/bios/fiorina.html. In 2010 she ran unsuccessfully as the Republican nominee for a U.S. Senate seat for California.

65. The NSF publication *Biotechnology Research and Development Activities in Industry: 1984 and 1985,* NSF 87-311 (Washington, DC, 1987), tried to set definitions, but a proposed follow-up report was never published. See also Potter Wickware, "The Birthplace of Biotech—California," Naturejobs, *Nature* 411 (May 17, 2001); and Marcia Barinaga, "The Attractions of Biotech Careers over Academia," *Science* 257 (1992): 1718–21, 1765.

66. John Timpane, "Unfinished Business: Careers for Women in Biotechnology and Pharmaceuticals," supplement, *Science* 272 (1996): 573–74, 578, 582; Susan C. Eaton, "Surprising Opportunities: Gender and the Structure of Work in Biotechnology Firms," in Selby, "Women in Science and Engineering," 175–88. See also Laurel Smith-Doerr, *Women's Work: Gender Equality vs. Hierarchy in the Life Sciences* (Boulder, CO: Lynne Rienner, 2004), which has an extensive bibliography; and Lorna D. McLeod, "Women and Biotech in the 21st Century," *Biopharm* 14 (June 2001): 32–41 (I thank Sarah K. Benz for this reference).

67. Belinda Martineau's *First Fruit: The Creation of the Flavr Savr™ Tomato and the Birth of Genetically Engineered Food* (New York: McGraw-Hill, 2001), is semiautobiographical.

68. On Chilton, see *AMWS,* 19th ed., 1995–96, 2:243; and "Career Stories: Those That Have Been There and Back," *Science* 256 (1992). On Panem, see http://people.forbes.com/profile/sandra-panem/90062; and "Interview with Sandra Panem," in *The Outer Circle: Women in the Scientific Community,* ed. Harriet Zuckerman, Jonathan R. Cole, and John T. Breuer (New York: Norton, 1991), ch. 5.

69. On Daniell, see *AMWS,* 20th ed., 1998–99, 2:630; and "Interview: Ellen Daniell," in *Making PCR: A Story of Biotechnology,* by Paul Rabinow (Chicago: Univ. of Chicago Press, 1996), 105–7. See also Ellen Daniell, *Every Other Thursday: Stories and Strategies from Successful Women Scientists* (New Haven, CT: Yale Univ. Press, 2006).

70. On Naughton, see http://awis.npaci.edu/wib_website/speakers/gail_bio.html.

71. On Sato, see Ronald Rosenberg, "Sato Promoted to President of Vertex Pharmaceuticals, Cambridge Firm's Stock Soars," *Boston Globe,* Dec. 6, 2000, D5; and Vicki L. Sato, "Women in Biotechnology: Finding a Place," in Selby, "Women in Science and Engineering," 189–90, 258–59. See also Barry Werth, *The Billion-Dollar Drug: One Company's Quest for the Perfect Molecule* (New York: Simon & Schuster, 1994), which describes Sato as "a quick street-smart no-nonsense former biology professor at Harvard" (411); Ann R. Earls, "Women Find Room at the Top in Biotech Sector; Boasts Many Females at Senior Executive Levels in Mass. Companies," *Boston Globe,* Oct. 20, 2002, G1; and Gareth Cook, "The Gender Gap," *Boston Sunday Globe,* May 6, 2007, K9 (I thank Mary Oates for a copy).

72. See the Immusol website, www.immusol.com.

73. On WEST, Inc., see the website, www.westorg.org/about.whoweare.php.

74. For a list of industrially employed women physicists about 1971, see Committee on Women, APS, "Report on the Status of Women," app. F (1972). Actual questionnaires are in the Vera Kistiakowsky Papers, boxes 32 and 33, but are restricted. See also "Survey of Industrial Labs," *CSWP Gazette*, no. 9 (Mar. 1, 1983): 3–5 (copy in Vera Kistiakowsky Papers, box 2, APS-CSWP, 1972–73 and 1983–84); and Rosalie G. Genovese and Sylvia F. Fava, "Women Physicists: Where Are They?" *Current Research on Occupations and Professions* 10 (1998): 259–82.

75. On Ancker-Johnson, see "Contributions of 20th Century Women to Physics," at http://cwp.library.ucla.edu. On Stearns, see *AMWS*, 19th ed., 1995–96, 6:1214. On Conwell's career in electronics, see ibid., 2:441; and "Esther Marly Conwell," in Ambrose et al., *Journeys of Women in Science and Engineering*, 88–90. Jean Dickey Apker worked for many years at the General Electric Research Laboratory in Schenectady, NY. "David Heckermann Wins Apker Memorial Award," *Physics Today* 32 (Jan. 1979): 101.

76. Ann Branigar Hopkins, *So Ordered: Making Partner the Hard Way* (Amherst: Univ. of Massachusetts Press, 1996).

77. On Earle, see Wallace White, "Profiles—Her Deepness," *New Yorker*, July 3, 1989, 41–65.

78. On Hamilton, see www.klabs.org/home_page/hamilton.html (I thank Joy Harvey, of the Cambridge Women's Historical Commission, for telling me about Hamilton).

79. On Osmer, see "Osmer, Judith Ann (1940–)," in *Feminists Who Changed America, 1963–1975*, comp. Barbara J. Love (Urbana: Univ. of Illinois Press, 2006), 348.

80. See Arlene Blum, *Breaking Trail: A Climbing Life* (New York: Scribner, 2005).

81. On Whitman, see Marshall Loeb, "Rise of the Role Model," *Time*, Nov. 20, 1978, 92; Marina von Neumann Whitman, "Economics from Three Perspectives," *Business Economics* 18 (Jan. 1983): 20–24; "Making It in the Corporate World: What's Good for General Motors; A Vice President of General Motors Discusses Her Role as an Economist in the Corporate World, An Interview with Marina Whitman by Aida K. Press," *Radcliffe Quarterly* 69 (June 1983): 11–13; and Love, *Feminists Who Changed America*, 492–93. Some of Whitman's papers are at the Richard Nixon Presidential Library, Yorba Linda, CA.

82. Kimberly Mayer, "Women Economists Make Denver Mark," *Rocky Mountain News*, Dec. 2, 1979, 93, 96; Anne B. Fisher, *Wall Street Women* (New York: Knopf, 1990), 128–29; Judith H. Dobrzynski, "How to Succeed? Go to Wellesley: Its Graduates Scoff at Glass Ceilings," *NYT*, Oct. 29, 1995, F1, F9.

83. "Sandra Wood Scarr," in *Models of Achievement: Reflections of Eminent Women in Psychology*, vol. 3, ed. Agnes N. O'Connell (Mahwah, NJ: Lawrence Erlbaum Associates, 2001), 108–9; Bailey, *American Women in Science 1950 to the Present*, 349–50.

Chapter 9 · Federal Employment

1. U.S. Civil Service Commission, Bureau of Manpower Information Systems, *Occupations of Federal White-Collar Workers, October 31, 1974 and 1975* (Washington, DC: GPO, 1976), table C-2; Office of Personnel Management, *Occupations of Federal*

White-Collar and Blue-Collar Workers, As of September 1997 (Washington, DC, 1998), the latest volume available when I started counting in July 2000. For later data sets, see www.opm.gov/feddata/.

2. See Margaret W. Rossiter, *Women Scientists in America: Before Affirmative Action, 1940–1972* (Baltimore: Johns Hopkins Univ. Press, 1995), ch. 13.

3. Elaine Einarson, *Woods-Working Women: Sexual Integration in the U.S. Forest Service* (Tuscaloosa: Univ. of Alabama Press, 1984), is based on interviews with women firefighters and forest technicians and their male supervisors in Oregon in 1978–79.

4. Diane K. Winokur, "History and Summary of Consent Decree: U.S. Forest Service," *Women in Forestry* 8 (Spring 1986): 66–68.

5. Mary Albertson, "Progress of Women in the Forest Service," *Women in Natural Resources* 15 (Spring 1994): 5.

6. James J. Kennedy and Joseph A. Mincolla, "Women and Men Natural Resources Managers in Early Stages of Their Professional Forest Service Careers," *Women in Forestry* 8 (Spring 1986): 27.

7. "U.S. Forest Service Ordered to Promote California Women," *NYT*, May 10, 1988, A22; Andrea Warner, "Forest Service Women Professionals *Begin* the Journey," *Women in Natural Resources* 15 (Spring 1994): 8–10; Gary Gerhardt, "Woman Climbs to Key Forestry Post," *Rocky Mountain News*, May 7, 1992, 10. See also Vincent Y. Dong, "Profile of Jacqueline Robertson," *Women in Natural Resources* 9 (Summer 1988): 31–32, on an entomologist who was the second woman at the USFS to attain the high rank of GS-15.

8. Impressions gained from the Fann Harding Papers, Women in Science and Engineering Archives, Special Collections, Iowa State University Library, Ames. Harding was at the National Institute of Heart and Lung for a long time, was involved in many of the protests at the NIH, and fortunately kept "everything," making her collection an invaluable source on the history of women at the NIH.

9. Beverlee A. Myers, "Homemaking in the Federal Government: What It Is Like to Be a Female Fed," supplement, *A[merican] A[ssociation of] H[ospital] A[dministrators] Newsletter* 24, no. 2 (June 1975), copy in Fann Harding Papers, box 10, folder 2.

10. On SHER, see Billy Goodman, "Controversial Group Marks Quarter-Century of Fighting for NIH Women Scientists' Rights," *Scientist*, Jan. 6, 1997, 1, 10. There is a great deal on SHER in the Fann Harding Papers and a little bit in box 10 of the Midge Costanza Papers, Jimmy Carter Presidential Library, Atlanta. SHER also took an active role in the NIH postdoctoral fellows' complaints and lawsuits, discussed in ch. 7, and is mentioned in Scott Huler, "Woman Scientist Victorious in Discrimination Case," *Scientist*, Apr. 27, 1992, 1, 6, which is about Sharon Johnson's later case against the NIH. On Ramey, see ch. 1, n. 9.

11. On Harding, see *AMWS*, 19th ed., 1995–96, 3:604–5. Her case is well documented in the Fann Harding Papers, boxes 1–3.

12. On Marimont, see *AMWS*, 19th ed., 1995–96, 5:249. The Rosalind Marimont Papers are at the Women in Science and Engineering Archives, Special Collections, Iowa State University Library; and she appears often in the Fann Harding Papers there. See also Gail Robinson, "Women Sue NIH over Alleged Bias," *Washington Post*, July 18, 1974, A27; A. O. Sulzberger Jr., "Sex Bias Inquiry Seeks to Alleviate Tense Situation at

Health Agency," *NYT,* Nov. 20, 1978, D11; Barbara B. Davis, letter to the editor, *Washington Post,* Apr. 25, 1974, A1; and Goodman, "Controversial Group Marks Quarter-Century," 1, 10. There are also legal documents and correspondence in box 55, folder 7, Marimont, Rosalind, Discrimination Complaint, 1973, in Maxine Singer Papers, Manuscript Division, Library of Congress, Washington, DC.

13. Goodman, "Controversial Group Marks Quarter-Century," 10. Kirschstein's own account differs. Carla Garnett, "Intramural Women Scientists Speak Out on Status at NIH," *NIH Record* 44 (Mar. 17, 1992): 6. See also the obituary by Howard K. Schachman and Marvin Cassman, "Ruth L. Kirschstein (1926–2009)," *Science* 326 (2009): 947.

14. See, e.g., Christopher Anderson, "NIH Aims at 'Glass Ceiling,'" *Nature* 356 (Mar. 5, 1992): 6; Garnett, "Intramural Women Scientists Speak Out," 1, 6–8; Traci Watson, "Women at NIH: Task Force; Level the Playing Field," *Science* 260 (1993): 888–89; Hynda K. Kleinman, "NIH Women and Men Scientists: Still Not Equal in Pay, Tenure, Promotion and Visibility," *AWIS Magazine* 22 (July–Aug. 1993): 8–9; and Lisa Seachrist, "Women in Science: Disparities Detailed in NCI Division," *Science* 264 (1994): 340.

15. "Dolores J. Copeland, Individually and on Behalf of the Class of All Others Similarly Situated v. F. Ray Marshall, Secretary of Labor et al., Appellants—641 F.2d 880," http://cases.justia.com/us-court-of-appeals/F2/641/880/25648/.

16. On Ray, see below, n. 64. On Chewning, see "June Chewning, Plaintiff, v. Robert C. Seamans, Jr., Defendant, U.S. District Court for the District of Maryland, Civil Action No. 76-34F," in box 5, folder 27, Cornell Eleven Papers, Rare and Manuscript Collections, Kroch Library, Cornell University, Ithaca, NY; Timothy S. Robinson, "U.S. Admits Sex Bias by Agency," *Washington Post,* July 15, 1978, A1, A5; Judy Mann, "Sex Bias Complaint Based on Experience," ibid., July 19, 1978, B1, B5; Janet Raloff, "The Unequal Scientists," *Science News,* Aug. 5, 1978, 92–95; June S. Chewning, "Surviving a Class Action Suit: A Matter of Attitude," *Federal Times,* Jan. 15, 1979, 12; Laura A. Kiernan, "130 Women to Get $2.2 Million in Pay in DOE Bias Suit," *Washington Post,* Aug. 2, 1982, A11; Joan Abramson, *Old Boys, New Women: The Politics of Sex Discrimination* (New York: Praeger, 1979), 137–42; and Barbara J. Love, comp., *Feminists Who Changed America, 1963–1975* (Urbana: Univ. of Illinois Press, 2006), 81–82. The statistics were even worse at ERDA's brother agency, the Nuclear Regulatory Commission. *Manpower Comments* 15 (May 1978): 18.

17. Robert E. Davis, "Career Status and Opportunities for Women Earth Scientists in the Geological Division of the U.S. Geological Survey," *Geology* 5 (1977): 495–97. The complaint filed in 1978 may have been related to a more grisly saga that had unfolded in Alaska in August 1977. Because most women geologists had long had difficulties getting opportunities to do fieldwork, some did it alone or with other women. This had been the practice for years without incident. For example, the USGS geologist Cornelia Cameron had for decades taken her mother, who held a PhD in botany, into the field with her. Maxine J. Levin, "Opening Opportunities: Women in Soil Science and the Soil Survey," in *Profiles in the History of the U.S. Soil Survey,* ed. Douglas Helms, Anne B. W. Effland, and Patricia J. Durana (Ames: Iowa State Univ. Press, 2002), 155–56 (I thank Douglas Helms for a copy). In the summer of 1977 the longtime fieldworker Florence Weber, of the Alaska Division, who prided herself on not carrying a gun, was

in the field with a younger woman geologist, Cynthia Dusel-Bacon. They split up at one point, whereupon a black bear appeared from nowhere and attacked and mangled the younger woman, who, following the traditional practice, also had no gun, but she did have a radio transmitter. Fortunately, a helicopter arrived in time to scare the bear off, but Dusel-Bacon suffered the loss of both arms. She continues to work for the USGS. On Weber, see *AMWS*, 19th ed., 1995–96, 7:58; and tape recording, "Interview with Florence Weber, February 2000," by Karen Brewster, in Alaska and Polar Regions Collections, Rasmuson Library, University of Alaska, Fairbanks; "Cynthia Dusel-Bacon," in *Journeys of Women in Science and Engineering: No Universal Constraints*, ed. Susan A. Ambrose et al. (Philadelphia: Temple Univ. Press, 1997), 120–25. The two accounts differ.

18. "Career Patterns of Women Scientists in the Geological Division of the U. S. Geological Survey," Open File 81-0246 (1980), a remarkably candid unpublished report, is available through interlibrary loan. The committee was chaired by the longtime USGS geologist Elaine Greening Ames Weed, Mount Holyoke, class of 1953. *AMWS*, 18th ed., 1992–93, 7:515. A USGS newsletter for women, called *WInGS*, is available only in USGS libraries. The Denver office of the USGS also employed the accomplished African American chemist Vicki Smith to recruit other minorities. Celeste G. Engel, *Rocks in My Head* (New York: Vantage, 1987), 87, 196. Smith does not appear in *AMWS* or in Wini Warren, *Black Women Scientists in the United States* (Bloomington: Indiana Univ. Press, 1999).

19. Penny Hanshaw to Dr. Lochman-Balk, May 11, 1981, Christina Lochman-Balk Papers, American Heritage Center, University of Wyoming, Laramie. Hanshaw was a founder of the AWG's Potomac branch, which named a fellowship for her after her retirement in 1990. She was also the first woman president of the Geological Society of Washington.

20. On Lucchitta, see http://astrogeology.usgs.gov/About/People/Baerbell Lucchitta/. On Kieffer, see "Kieffer, Susan Werner (1942–)," in *American Women in Science 1950 to the Present: A Biographical Dictionary*, by Martha J. Bailey (Santa Barbara, CA: ABC-CLIO, 1998), 211–12. On Zoback, see *AMWS*, 19th ed., 1995–96, 7:1085.

21. "Older Women Scientists Fight USGS over Layoffs," *Nature* 404 (Mar. 16, 2000): 219; "Mary A. Carey, William C. Butler v. Department of Interior," 2006 MSPB 295, Sept. 27, 2006.

22. Molly Gleiser, "The Glass Wall," in *Women in Science: Meeting Career Challenges*, ed. Angela M. Pattatucci (Thousand Oaks, CA: SAGE, 1998), 204–18; correspondence and legal documents about Gleiser's loan from the ACS in box 3 of the Anna Jane Harrison Papers, Mount Holyoke College Archives, South Hadley, MA; Sally Lehrman and Jonathan King, "Lab Beats Charge of Discrimination," *Daily Californian* (UC–Berkeley), Mar. 3, 1980, 1, 11; and Gleiser's response, "Battling with Hiring Policies," ibid., Mar. 20, 1980, 4. In the 1980s Gleiser became interested in the topic of women chemists who took their own lives and became head of an organization called Scientists Against Suicide.

23. "Sex Discrimination Suit at Lawrence Livermore," *Nature* 397 (Jan. 7, 1999): 12; Mary F. Singleton, "Gender Equity: It's a Class Action at Lawrence Livermore National Laboratory," *AWIS Magazine* 30 (Autumn 2001): 11–16; "Oral History of Mary Frances

Singleton," June 14, 2002, by Tanya Zanish-Belcher, Archives of Women in Science and Engineering, Special Collections, Iowa State University Library. Singleton's papers, also at Iowa State, include materials about the case. See also Warren E. Leary, "Report Finds Disparities in Hiring at U.S. Labs," *NYT*, May 22, 2002, A20, on a General Accounting Office report documenting underemployment of women and minorities at the LBL, the LLNL, and the Sandia National Laboratory, in Albuquerque, New Mexico.

24. Betsy Mason, "Women Join Supervisory Ranks at Lab," *Valley Times* (Pleasanton, CA), Sept. 25, 2004, 1, 23, clipping in box 1, folder 14, Lawsuit Clippings, LLNL and LANL, Mary Singleton Papers, Women in Science and Engineering Archives, Special Collections, Iowa State University Library.

25. The one exception was a short account in a 1976 SHER newsletter of Dolores Copeland's victory at the DOL: "Class Action Victory for Women at the Labor Department," *SHER Newsletter*, Sept.–Oct. 1976, 1–2, copy in box 4, folder 10, SHER Newsletter, Rosalind Marimont Papers.

26. See Constance Holden, "NASA: Sacking of Top Black Woman Stirs Concern for Equal Employment," *Science* 183 (1973): 804–7; Kim McQuaid, "Race, Gender, and Space Exploration: A Chapter in the Social History of the Space Age," *Journal of American Studies* 41, no. 2 (2007): 405–34; correspondence, Nov. 1973–Jan. 1974, in box 12, FOPW, Vera Kistiakowsky Papers, Institute Archives, MIT, Cambridge, MA; *Hearings Before the Subcommittee on Civil Rights and Constitutional Rights of the Committee on the Judiciary, House of Representatives, 93rd Congress, Second Session on NASA's Equal Opportunity Program, March 13 and 14, 1974*, No. 56 (Washington, DC: GPO, 1975); obituary of Ruth Bates Harris McKenzie, *Washington Post*, Feb. 3, 2004; and Harris's autobiography, Ruth Bates Harris, *Harlem Princess: The Story of Harry Delaney's Daughter* (New York: Vantage, 1991).

27. Margaret Weitekamp, *Right Stuff, Wrong Sex: America's First Women in Space Program* (Baltimore: Johns Hopkins Univ. Press, 2004), 177; Rossiter, *Women Scientists in America: Before Affirmative Action*, 293–95, 370; Isaac Asimov, "No Space for Women," *Ladies' Home Journal*, Mar. 1971, 115, 201–2, 204; Carl Konkel and William G. Holder, "Sputnik to Mutnik to—Picnic!" *Space World*, Dec. 1972, 46–48; Joan McCullough, "13 Who Were Left Behind," *Ms.*, Sept. 1973, 41–45; "Ladies on the Pad?" *Time*, Oct. 22, 1973, 73; "Space for Women," ibid., Nov. 5, 1973, 71; "Study of Women as Space Flight Candidates Completed," *Space World*, Feb. 1974, 4–10, 27–29.

28. Weitekamp, *Right Stuff, Wrong Sex*, 187–88; "New Crop of Astronauts," *Science* 199 (1978): 36–37. On Ride, see http://en.wikipedia.org/wiki/Sally_Ride. On Sullivan, see http://en.wikipedia.org/wiki/Kathryn_Dwyer_Sullivan. On Resnik, see http://en.wikipedia.org/wiki/Judith_Resnik. On Lucid, see http://en.wikipedia.org/wiki/Shannon_Lucid. On Jemison, see "Jemison, Mae Carol (1956–)," in Bailey, *American Women in Science 1950 to the Present*, 194–95. On Ochoa, see "Ochoa, Ellen (1958–)," ibid., 292–93. See also Carolyn Huntoon, "Opening Doors for Women in Space: A Perspective from the National Aeronautics and Space Administration," in "Women in Science and Engineering: Choices for Success," ed. Cecily Cannan Selby, special issue, *Annals of the New York Academy of Science* 869 (1999): 204–6; and Amy Elizabeth Foster, "Sex in Space: The Politics and Logistics of Sexually Integrating NASA's Astronaut Corps," (PhD diss., Auburn University, 2005).

29. One early exception was the anthropologist Jean McWhirt Pinckley, at Mesa Verde National Park from 1943 to 1966, the widowed daughter-in-law of a park official. Rossiter, *Women Scientists in America: Before Affirmative Action,* 283–84; Polly Welts Kaufman, *National Parks and the Woman's Voice: A History* (Albuquerque: Univ. of New Mexico Press, 1996), 140, 141, 169–72. Devising an acceptable uniform for women rangers was difficult. Early versions included a blouse and skirt that made them look like airline stewardesses, as they were then called, or Girl Scout leaders. In the 1970s pantsuits became more popular, but the women were not allowed to wear the Stetson hat, and their badges were smaller than the men's. With the coming of the modern "visitor center" after 1956, more rangers worked indoors, behind a counter. Sarah Allaback, *Mission 66 Visitor Centers: The History of a Building Type* (Washington, DC: U.S. Department of the Interior, National Park Service, 2000). The passage of the Archaeological Resources Protection Act of 1979 increased the Park Service's responsibilities in the area of antiquity theft. See Charles D. Rafkind, "Prosecution under the Archaeological Resource [*sic*] Protection Act," *Women in Forestry* 8 (Fall 1986): 8–9.

30. Toby A. Appel, *Shaping Biology: The National Science Foundation and American Biological Research, 1945–1975* (Baltimore: Johns Hopkins Univ. Press, 2000); *Interviews of National Science Foundation Program Officers, Final Report, Prepared by the Subcommittee on Science, Research and Technology of the Committee on Science and Technology, U.S. House of Representatives, 96th Congress, 1st Session, Serial N* (Washington, DC: GPO, 1979); NSF, *Celebrating 50 Years: Resource Guide 2000,* NSF 00-87 (Arlington, VA, 2000). On James, see *AMWS,* 15th ed., 1982, 4:30. On Doherty, see Forest Stearns, "Distinguished Service Citation," *Bulletin of the Ecological Society of America* 67 (Mar. 1986): 42–43. On Litman, see "Rose Litman Is Dead at 49; Research Aide at Colorado U.," *NYT,* Sept. 24, 1981, D27. On Clark, see *AMWS,* 19th ed., 1995–96, 2:308; and Name File, White House Records, Gerald Ford Presidential Library, Bentley Library, University of Michigan, Ann Arbor. On Callanan, see *AMWS,* 19th ed., 1995–96, 2:19–20. On Bautz, see ibid., 1:458. On Sunley, see ibid., 6:1388–89. On Dornan, see "Elvira Dornan," *NSF Annual Report for 1986* (Washington, DC: GPO, 1987), 54; and Diann Jordan, "Elvira Doman [*sic*]: A Class Act," in *Sisters in Science: Conversations with Black Women Scientists on Race, Gender, and Their Passion for Science* (West Lafayette, IN: Purdue Univ. Press, 2006), 81–90. On Bryant, see *AMWS,* 19th ed., 1995–96, 1:971. On Willis, see ibid., 7:801.

31. Vera Kistiakowsky to Sheila Pfafflin, Feb. 3, 1983, enclosing Vera Kistiakowsky to Dr. Edward A. Knapp, Jan. 10, 1983, in Vera Kistiakowsky Papers, box 6, AWIS, 1982–83.

32. Betty Vetter, "Is There a Federal Role in Developing Women Chemists?" (paper delivered at ACS meeting, St. Louis, May 1984), in box 9, folder 19, Status of Women in Science in 1983, Betty Vetter Papers, Women in Science and Engineering Archives, Special Collections, Iowa State University Library.

33. Mary Gray, "The NSF and Women in Mathematics," *AWM Newsletter* 10 (Sept.–Oct. 1980): 12–13, typescript in box 1, AWM Correspondence (1979–80); the article criticized the fact that Sunley had been hired as an associate program director rather than a full one like the men with her credentials. On Sunley, see *AMWS,* 19th ed., 1995–96, 6:1388–89; and "Judith Sunley Appointed NSF Director," *AWM Newsletter* 17

(July–Aug. 1987): 2. See also Margaret Cozzens, "Some Thoughts on Recruiting Women for NSF Division Director Positions," ibid., 23 (Nov.–Dec. 1993): 25–27. On Clutter, see *AMWS*, 19th ed., 1995–96, 2:346; "ASCB Profile: Mary Clutter," www.ascb.org/ascb/profiles/Clutter.html; and B[arbara] F[ilner], "Meet a Member," *AWIS Newsletter* 10 (Winter 1980–81): 6–10. On Clutter's refusal to fund certain meetings, see Marcia Clemmitt, "Toughest Federal Science Jobs Elude Women," *Scientist,* Oct. 15, 1990, 13. See also ch. 12.

34. On Osteryoung, see *AMWS,* 19th ed., 1995–96, 5:995.

35. On Leverton, see Jeffrey S. Hampl and Marilynn I. Schnepf, "Ruth M. Leverton (1908–1982)," *Journal of Nutrition* 129 (1999): 1769–72; Lena Williams, "Agriculture Department: Latest Charge of Racism Prompts a Debate," *NYT,* June 30, 1986, 16; and idem, "Agriculture Dept. Criticized over Civil Rights," ibid., Sept. 30, 1986, 23.

36. On Carter, see *AMWS,* 19th ed., 1995–86, 2:102; and "Action Lines," *Women in Action* 4 (Spring 1974): 4, copy in Vera Kistiakowsky Papers, box 29, Women in Federal Government.

37. Levin, "Opening Opportunities," ch. 5; Douglas Helms, "Women in the Soil Conservation Service: A History and Current Events," *Women in Natural Resources* 14 (Spring 1993): 9–12.

38. Andrea Warner, "In the Beginning, There was the Forest Service's Federal Women's Program," *Women in Natural Resources* 15 (Sept. 1994): 10; Elaine J. Zieroth, "The Benefits of an Active Federal Women's Program," ibid. 9 (Spring 1988): 31–32, 36.

39. Bonnie G. Mani, "Gender and the Federal Senior Executive Service: Where Is the Glass Ceiling?" *Public Personnel Management* 26 (Winter 1997): 545–58; U.S. General Accounting Office, *Senior Executive Service: Diversity Increased in the Past Decade; Report to Congressional Requesters* (Washington, DC, 2001); U.S. Merit Systems Protections Board, *A Ten-Year Retrospective of the MSPB, 1978–1988* (Washington, DC: GPO, 1989). On Sengers, see *AMWS,* 19th ed., 1995–96, 6:761; "Part-Time Workers: Five Success Stories," *Women in Action* 4 (Spring 1974): 8, copy in Vera Kistiakowsky Papers, box 29, Women in Federal Government; and Johanna M. H. Levelt Sengers to Elga Wasserman, July 25, 1996, in box 2, Sengers, Elga Wasserman Collection, Schlesinger Library, RIAS, Harvard University, Cambridge, MA. On Zoback, see *AMWS,* 19th ed., 1995–96, 7:1085; and Mary Lou Zoback to Elga R. Wasserman, June 8, 1995, in Elga Wasserman Collection, box 2.

40. Sonia Pressman Fuentes, "Covering Women's Awards," *Washington Post,* Mar. 18, 1971, A21; "Woman's Award," ibid., Oct. 4, 1972, B2; Jeannette Smyth, "A Paradox for Women," ibid., Mar. 7, 1973, B2. The chemist B. Jean Apgar, of the USDA's Plant, Soil and Nutrition Laboratory in Ithaca, NY, was one of ten winners of the Arthur S. Flemming Award in 1973, possibly the first female. "Chemist Receives Second Honor," *Women in Action* 3 (Spring 1973): 5.

41. Barbara Mandula and Nancy M. Tooney, "Meet Joan Humphries," *AWIS Newsletter* 15 (May–June 1986): 8–10; "Goals and Objectives for FWP/IAG, Subcommittee on Women in Science and Engineering," with list of members, n.d., in box 2, FWP/IAG Subcommittee on WISE, Alphabetical Reference File for Scientific Organizations and Fields, Records of the Women in Science Project, NSF, NARA, College Park, MD (another copy, along with Office of Personnel Management, Federal Women's Program,

Intra-agency Subcommittee, *WISE Directory of Federally Employed Women in Science and Engineering* [1980], is contained in the Fann Harding Papers, box 5, folder 1).

42. NIH, *Women in Science* ([Bethesda, MD, 1988]). On Kirschstein, see *AMWS,* 19th ed., 1995–96, 4:433; Jeffrey Brainard, "Acting Chief of the NIH Stresses Stability, but Not the Status Quo," *CHE,* May 26, 2000, A40–A41; and her obituary, Schachman and Cassman, "Ruth L. Kirschstein (1926–2009)."

43. On Goldman-Rakic, see Eric J. Nestler, "Obituary: Patricia S. Goldman-Rakic (1937–2003)," *Nature* 425 (Oct. 2, 2003): 471. On Neufeld, see *AMWS,* 18th ed., 1992, 5:703; and Elizabeth Neufeld, notes on a telephone interview by Elga Wasserman, July 25, 1995, box 2, Elga Wasserman Collection. On Singer, see *AMWS,* 19th ed., 1995–96, 6:950; and ch. 10. On Pert, see *AMWS,* 19th ed., 1995–96, 5:1183; and ch. 6. On Wong-Staal, see *AMWS,* 19th ed., 1995–96, 7:893; and ch. 8. On Silbergeld, see *AMWS,* 19th ed., 1995–96, 6:912; and Robin Marantz Henig, "Ellen Silbergeld: The Making of a Biochemist," *SciQuest* 54 (Mar. 1981): 22–24.

44. On Goldhaber, see Wolfgang Saxon, "Gertrude Scharff Goldhaber, 86, Crucial Scientist in Nuclear Fission," *NYT,* Feb. 6, 1998, D18; and Peter D. Bond and Chellis Chasman, "Gertrude Scharff Goldhaber," *Physics Today* 512 (July 1998): 82–83. Her papers exist but have not yet found a repository. She may merit a biography. The microbiologist Marian "Bunny" Koshland, later at UC–Berkeley, also worked at Brookhaven in the 1950s and 1960s.

45. On Edwards, see *AMWS,* 19th ed., 1995–96, 2:1030; and John Peoples Jr., "Helen Thom Edwards (1936–)," in *Out of the Shadows: Contributions of Twentieth-Century Women to Physics,* ed. Nina Byers and Gary Williams (Cambridge: Cambridge Univ. Press, 2006), 385–98.

46. On Skinner, see *AMWS,* 19th ed., 1995–96, 6:977; vita in "Misc. Files, 1969–82," box 10, Women in Science, 1976–78, Anna Ruth Brummett Papers, Oberlin College Archives, Oberlin, OH; Linda H. Mantel, "Contributions of Dorothy M. Skinner to the Development of Crustacean Biology," *American Zoologist* 39 (1999): 465–70; and Susan A. Gerbi, "In Memory, Dorothy Skinner," *ASCB Newsletter,* Mar. 2005, 19. On Russell, see *AMWS,* 19th ed., 1995–96, 6:445.

47. On Bissell, see *AMWS,* 19th ed., 1995–96, 1:662; Alison Abbott, "Biology's New Dimension," *Nature* 424 (Aug. 21, 2003): 370–71; Mina J. Bissell, "Viewpoint—A Female Problem?" *LBL Newsmagazine,* Summer 1979, 3; and correspondence about nomination for the LBL's Ernest O. Lawrence Award, box 14, folder 16, Ruth Sager Papers, Marine Biological Laboratory Library, Woods Hole, MA. See also Office of Biological and Environmental Research, DOE, *A Vital Legacy: Biological and Environmental Research in the Atomic Age* (Washington, DC, 1997), a celebration of 50 years of biological research by energy agencies.

48. On Hoffman, see *AMWS,* 19th ed., 1995–96, 4:911; and oral history in Women in Science and Engineering Archives, Special Collections, Iowa State University Library.

49. Louise A. Raphael, "Science Careers in Search of Women: Argonne National Laboratory," *AWM Newsletter* 17 (July–Aug. 1987): 10–11; N. K. Meshkov, comp., *Graduate School and Beyond: A Panel Discussion from the Conference "Science Careers in Search of Women," Held at Argonne National Laboratory, Waterfall Glen Conference, April 6–7, 1989* (Argonne, IL: Argonne National Laboratory, 1990), 5–10; "The Maria

Goeppart-Mayer Distinguished Scholar Program Announcement," Bulletin Board, *CHE*, June 21, 1989, p. B-11; Linda Skidmore Dix, "Promoting Careers in the Federal Government," in *Science and Engineering Programs: On Target for Women?* ed. Marsha Lakes Matayas and Linda Skidmore Dix (Washington, DC: National Academy Press, 1992), 150–52, citing DOE, Office of Energy Research, *Department of Energy Review of Laboratory Programs for Women (DOE/E-0510P), November 16, 1990* (Washington, DC, 1990). On Meyer-Schützmeister, see John P. Schiffer, "Luise Meyer-Schützmeister," *Physics Today* 35 (June 1981): 74–75; Suzanne Gronemayer, "In Memoriam: Luise Meyer-Schütmeister," *AWIS Newsletter* 10 (1981): 10; and "Contributions of 20th Century Women to Physics," at http://cwp.library.ucla.edu. On Hartline, see "Promoting Success in the Department of Energy," *Annals of the New York Academy of Sciences* 869 (1999): 200–203, 252–53; and Beverly Karplus Hartline and Dongqi Li, *Women in Physics: The IUPAP International Conference on Women in Physics* [Paris, Mar. 2002], AIP Conference Proceedings 628 (College Park, MD: American Institute of Physics, 2002). See also Jack M. Holl, *Argonne National Laboratory, 1946–96* (Urbana: Univ. of Illinois Press, 1997), 378, 491–93.

50. On Holberton, see "Betty Holberton, 84, Computing Pioneer," *Boston Globe*, Dec. 13, 2001 (I thank Mary Oates for a copy). On Jacox, see Lester Andrews, Bruce Ault, and Zakya Kafafi, "A Brief Scientific Biography," in "Marilyn Jacox Festschrift," special issue, *Physical Chemistry A* 104, no. 16 (2000): 3431–34. On Gallagher, see Katharine B. Gebbie and John R. Rumble, "Jean Weil Gallagher," *Physics Today* 50 (Aug. 1997): 78. On Gebbie, see *AMWS*, 19th ed., 1995–96, 3:82. See also Raymond G. Kammer, ed., *NIST at 100: Foundations for Progress*, NIST Special Publication 956 (Darby, PA: Diane, 2000).

51. Margaret A. LeMone, Joan V. Frisch, and Lesley T. Julian, "Tracking Women and the Weather: Their Growing Role in Meteorology," *Weatherwise* 37 (Aug. 1984): 176–81; Edward F. Taylor, "Joanne Simpson: Pathfinder for a Generation," ibid., 182–83, 206–7; "Interview of Joanne Simpson, September 6, 1989," by Margaret LeMone, American Meteorological Society / University Corporation for Atmospheric Research Tape Recorded Interview Project, transcript, National Center for Atmospheric Research Archives, Boulder, CO (I thank archivist Diane Rabson for a copy). Patricia Sullivan, "Joanne Malkus Simpson, Famed Female Meteorologist," *Boston Globe*, Mar. 10, 2010, B11 (I thank Mary Oates for a copy). The Joanne Simpson Papers are at the Schlesinger Library, RIAS.

52. On Solomon, see *AMWS*, 20th ed., 1998–99, 6:1094; and "Interview with Susan Solomon, 5 September 1997," by Dale Kellogg, American Meteorological Society/University Corporation for Atmospheric Research, Tape Recorded Interview Project, transcript, National Center for Atmospheric Research Archives, Boulder, CO (I thank archivist Diane Rabson for a copy). On Antarctica, see also Rossiter, *Women Scientists in America: Before Affirmative Action*, 377–78; and, more recently, Irene C. Peden, "If You Fail, There Won't Be Another Woman on the Antarctic Continent for a Generation," and Sister Mary Odile Cahoon, "If Women Are in Science and Science Is in the Antarctic, Then Women Belong There," in *Women in the Antarctic*, ed. Esther D. Rothblum, Jacqueline Weinstock, and Jessica F. Morris (New York: Harrington Park, 1998), 17–29 and 31–39.

53. Bill Workman, "Substitute Science Teacher Is a Meteorology Legend," *San Francisco Chronicle,* Mar. 23, 2000.

54. On Roman, see *AMWS,* 19th ed., 1995–96, 6:323. Nancy Roman's papers are at the Center for the History of Physics, American Institute of Physics, College Park, MD, as is an oral-history interview, "Interview with Nancy G. Roman by David DeVorkin at Her Office, NASA, August 19, 1980." On Underhill, see *AMWS,* 19th ed., 1995–96, 7:308; and Theresa A. Nagy, "Anne Barbara Underhill (1920–)," in *Women in Chemistry and Physics: A Biobibliographic Sourcebook,* ed. Louise S. Grinstein, Rose K. Rose, and Miriam H. Rafailovich (Westport, CT: Greenwood, 1993), 562–71. On Townsend, see Constance Holden, "NASA Satellite Project: The Boss Is a Woman," *Science* 179 (1973): 48–49. On Shirley, see Kenneth Change, "Making Science Fact, Now Chronicling Science Fiction," *NYT,* June 15, 2004, D3. On Wilson, see www.mtholyoke.edu/acad/physics/Wilson.html. In 2005 her alma mater, Mount Holyoke College, awarded her an honorary ScD.

55. See, e.g., *Women at Work in NASA* (Washington, DC: GPO, 1979), copy in Vera Kistiakowsky Papers, box 27, Women in the Federal Government; Wini Warren, "Patricia Suzanne Cowings" and "Christine Voncile Mann Darden," in Warren, *Black Women Scientists in the United States,* 67–71, 77–80 (Warren also includes three others at NASA). See also "Linda C. Shackelford," in Ambrose et al., *Journeys of Women in Science and Engineering,* 340–44.

56. On Fischer, see Irene K. Fischer, *Geodesy? What's That? My Personal Involvement in the Age-Old Quest for the Size and Shape of the Earth* (New York: iUniverse, 2005) (I thank her son Michael Fischer for a copy); and J. M. Lawrence, "Irene K. Fischer; Measured Earth; at 102," *Boston Globe,* Oct. 28, 2009, 13 (I thank Mary Oates for a copy). Fischer's papers are at the Schlesinger Library, RIAS. On Karle, see ch. 12.

57. On Geller, see *AMWS,* 19th ed., 1995–96, 3:93. On Dupree, see ibid., 2:967. On Marvin, see ibid., 5:297. The rather scathing report, "Center for Astrophysics Gender Equity Report," is summarized in *STATUS: A Report on Women in Astronomy* (American Astronomical Society), June 2007, 1, 7–8. The Harvard-Smithsonian Astrophysical Observatory committee had been chaired by Giuseppina "Pepi" Fabbiano.

58. On Bailar, see *AMWS,* 19th ed., 1995–96, 1:307; and Sandra Stinnett et al., "Women in Statistics: Sesquicentennial Activities," *American Statistician* 44 (May 1990): 75. Perhaps as a result of the Native American movement, the Census Bureau hired its first full-time Native American, Edna Paisano, in 1976 to reach and help devise a questionnaire for Native Americans in the 1980 and 1990 censuses. Teri Perl, *Women and Numbers: Lives of Women Mathematicians Plus Discovery Activities* (San Carlos, CA: Wide World / Tetra, 1993), 127.

59. See materials in the Anna Jane Harrison Papers. See also *The National Science Board: A History in Highlights, 1950–2000* (Washington, DC: NSB, n.d.).

60. On Cobb, see ch. 1. On McBay, see Clarence G. Williams, *Technology and the Dream: Reflections on the Black Experience at MIT, 1941–1999* (Cambridge, MA: MIT Press, 2001), 764–81. On MISIP, see *Minority Institutions Science Improvement Program (MISIP): A Brief History, 1972–1980* (Washington, DC: National Science Foundation, Directorate for Science and Engineering Education, 1981), available at ERIC ED210149.

61. Dix, "Promoting Careers in the Federal Government," 156–58. See also National

Research Council, Office of Scientific and Engineering Personnel, Committee on Scientists and Engineers in the Federal Government, *Recruitment, Retention, and Utilization of Federal Scientists and Engineers: A Report to the Carnegie Commission on Science, Technology, and Government* (Washington, DC: National Academy Press, 1990).

62. On Eshoo, see "Science Committee Hears from NSF on Reauthorization," *C[onsortium] O[f] S[ocial] S[cience] A[ssociations] Washington Update,* June 28, 1993, 1; and Renee Twombley, "Congresswomen Take NIH and NSB to Task over Gender Bias," *Scientist,* Nov. 15, 1993, 1. Eshoo also criticized the low number of women at the top of the NSF, which may have led to the presidential appointments of Anne Petersen and Rita Colwell later on. The five other women on the House committee at the time were Connie Morella (R-MD), Jennifer Gunn (R-WA), Marilyn Lloyd (D-TN), Jane Harman (D-CA), and Eddie-Bernice Johnson (D-TX).

63. On Ancker-Johnson, see *AMWS,* 19th ed., 1995–96, 1:146–47; and ch. 8. On Whitman, see ch. 8, n. 81.

64. On Ray, see Eric Pace, "Dixy Lee Ray, 79, Ex-Governor; Led Atomic Energy Commission," *NYT,* Jan. 3, 1994, A24; Janet Newland Bower, "Dixy Lee Ray (1914–1994)," in *Women in the Biological Sciences: A Biobibliographical Sourcebook,* ed. Louise S. Grinstein, Carol A. Biermann, and Rose K. Rose (Westport, CT: Press, 1997), 424–32; Robert Gillette, "Ray's Shift to State Department Will Test Kissinger's Interest in Science," *Science* 186 (1974): 612–13; idem, "A Conversation with Dixy Lee Ray," ibid. 189 (1975): 124–27; and a campaign biography by her speechwriter, Louis R. Guzzo, *Is It True What They Say about Dixy? A Biography of Dixy Lee Ray* (Mercer Island, WA: Writing Works, 1980). Most of Ray's papers are at the Hoover Institution, Stanford University, Stanford, CA, but Ray's gubernatorial papers are in the Washington State Archives in Olympia.

65. Juanita Kreps File, Biographical Files, Duke University Archives, Durham, NC. See also Leonard Silk, "A Candid Academic at Commerce," *NYT,* May 8, 1977, sec. 3, pp. 1, 9; and obituary, "Juanita Kreps, 89, Dies, Headed Commerce Dept.," by Robert D. McFadden, ibid., July 8, 2010, 22. On Bingham, see http://en.wikipedia.org/wiki/Eula_Bingham. On Newell, see chs. 5 and 11. On Shalala, see www.miami.edu/index.php/about_us/leadership/office_of_the_president/president_donna_e_shalalas_biography.

66. On Payton, see "A Very Special Volunteer," *Ebony,* September 1978, 64–66, 70; and "Carolyn Robertson Payton," in *Models of Achievement: Reflections of Eminent Women in Psychology,* ed. Agnes N. O'Connell and Nancy Felipe Russo (Hillsdale, NJ: Erlbaum, 1981), 228–41. On Norwood, see Philip Shabecoff, "She Takes Her Computers Home," *NYT,* July 22, 1979; and Robert D. Hershey Jr., "An Even-Handed Collector of Labor Statistics Retires," ibid., Dec. 31, 1991, D1, D4. Jessica Mathews, a Caltech PhD, was also on the staff of the National Security Council at the Carter White House. Sarah Weddington's office prepared a large chart showing Jimmy Carter's two hundred women appointees, a copy of which is in the Fann Harding Papers at Iowa State. See also Office of Sarah Weddington, The White House, Washington, DC, *Women in Government: Your Guide to More Than 600 Top Women in the Federal Government* (Aug. 1980), available online at www.twu.edu/library/wedd/. It lists women at the GS-16 level and above but omits federal judges. There is also some related material in the Midge Costanza Files at the Jimmy Carter Presidential Library.

67. On Earle, see Susan A. Ambrose et al., eds., "Earle, Sylvia Alice (1935–)," in Bailey, *American Women in Science 1950 to the Present,* 85–87. On Healy, see, for starters, Larry Thompson, "NIH Gets Its First Woman Director," *Washington Post,* Mar. 26, 1991, 8; Joseph Palca, "Bernadine Healy: A New Leadership Style at NIH," *Science* 253 (1991): 1087–89; Erik Eckholm, "A Tough Case for Dr. Healy," *Washington Post,* Sunday suppl., *Outlook,* Dec. 8, 1991, 67–68, 122–23; Maggie Garb, "Opening Doors for Women in Science?" *American Medical News,* Feb. 17, 1992; and Malcolm Gladwell, "The Healy Experiment," *Washington Post Magazine,* June 21, 1992, 8–13, 23–25. Healy may merit a biography. She was married to Fred Loop, MD, who became the head of the Cleveland Clinic in 1989. There is a folder, "Swearing In of Dr. Healy, NIH Director 6/24/91," on her appointment, in box 107, Backup Files, 1989–93, Speech File, Office of White House Speechmaking, George H. W. Bush Presidential Library, College Station, TX. Two items were withdrawn from the file on the grounds that they would disclose confidential advice.

68. On Kreps and Shalala, see above, n. 65. On Widnall, see *AMWS,* 19th ed., 1995–96, 7:739. On Good, see Christopher Anderson, "Good to Take Top Commerce Post," *Science* 260 (1994): 1421.

69. On Petersen, see "Ways and Means," *CHE,* Mar. 9, 1994, A20. She left for a post at the Kellogg Foundation in 1996. On Colwell, see *AMWS,* 19th ed., 1995–96, 2:414–15; Andrew Lawler, "Clinton Names Adviser, NSF Chief," *Science* 279 (1998): 1122–23; Paulette Walker Campbell, "Clinton Revamps His Science-Policy Team," *CHE,* Feb. 27, 1998, A33–A34; Jeffrey Mervis, "Keeping Up With Rita Colwell," *Science* 279 (1998): 1622–23; Claudia Dreifus, "'Always, Always, Going against the Norm': A Conversation with Rita Colwell," *NYT,* Feb. 16, 1999, F3; Kathleen Gavin, "Letters: Questioning the Questions," ibid., Feb. 23, 1999, F3; Jeffrey Mervis, "NSF Shines Brightest in New Good-Government Scorecard," *Science* 295 (2002): 953; and Anne Marie Borrego and Alyson Klein, "Director of National Science Foundation, A Clinton Appointee, Resigns," *CHE,* Feb. 20, 2004, A24. Colwell has given more than 130 boxes of her papers to the Manuscript Division of the Library of Congress in Washington, DC, and to the American Society for Microbiology Archives at the University of Maryland–Baltimore County, Catonsville.

70. On Tyson, see "Tyson, Laura (D'Andrea) (1947–)," in Bailey, *American Women in Science 1950 to the Present,* 394–95; and Susan F. Rasky, "A Rare Glimpse of Sexism in Economics," *CHE,* Jan. 27, 1993, B1–B2. On Krebs, see *AMWS,* 20th ed., 1998–99, 4:583; "Key Research Official to Leave Energy Department," *CHE,* Oct. 15, 1999, A37; and Lynn Yarris, "Martha Krebs of LBL Nominated to DoE Office of Energy Research," July 16, 1993, www.lbl.gov/Science-Articles/Archive/krebs-DOE-nomination.html.

71. On Beattie, see William Dicke, "Mollie Beattie, 49; Headed Wildlife Service," *NYT,* June 29, 1996, 26; and "Mollie Beattie, An Interview," by Daina Dravnieks Apple, *Women in Natural Resources* 15 (June 1994): 28–32. On Graham, see "Rhea L. Graham," in Ambrose et al., *Journeys of Women in Science and Engineering,* 189–93; and "The Journal Talks with U.S. Bureau of Mines Director Rhea L. Graham," *J[ournal] O[f] M[etals]* 47 (Apr. 1995): 42–43. In 1995 the Republican Congress voted to close the Bureau of Mines, and its functions were either terminated or sent to other agencies. Subsequent mine disasters have led to calls to reestablish the bureau.

72. On Cordova, see *AMWS*, 19th ed., 1995–96, 2:476; and "Cordova, France Anne-Dominic (1947–)," in Bailey, *American Women in Science 1950 to the Present*, 57–58. On Jackson, see ch. 10.

73. On Prabhakar, see *AMWS*, 19th ed., 1995–96, 5:1348; Ambrose et al., *Journeys of Women in Science and Engineering*, 304–7; and "NIST Head to Step Down," *Science* 275 (1997): 1745.

74. Joan Steitz and Shirley Tilghman to "Dear Colleague," and Tilghman and Seitz to Dr. Donna Shalala, both May 7, 1993, both in box 55, folder 2, NIH Correspondence, 1993–96, Maxine Singer Papers.

75. On Olsen, see AMWS, 19th ed., 1995–96, 5:948; and Yudhijit Bhattacharjee, "People," *Science* 308 (205): 1407.

Chapter 10 · Nonprofit Alternatives

1. Defining *nonprofit* is a continuing challenge. Harvy Lipman, "Congress Urged to Define 'Nonprofit,'" *CHE*, May 6, 2005, A26. The NSF's definition has varied over the years, as explained in the "Technical Notes" to its data surveys, *Federal Support to Universities, Colleges, and Selected Nonprofit Institutions* (Washington, DC: GPO, 1969–93), renamed and continued as *Federal Science and Engineering Support to Universities, Colleges and Nonprofit Institutions* (Arlington, VA: NSF, Directorate for Social, Behavioral, and Economic Sciences, Division of Science Resources Studies, 1994–). In the late 1970s the NSF broadened its coverage of the nonprofits to include more hospitals, scientific societies, museums, zoos, botanical gardens, and other entities, but not the Federally Funded Research and Development Centers (FFRDCs). See below, n. 3. Despite being somewhat dated, Harold Orlans, *The Non-Profit Research Institute: Its Origin, Operation, Problems and Prospects* (New York: McGraw-Hill, 1972), is still of value.

2. Donald D. Brown, "The Department in the Second Half of the Twentieth Century," in *The Department of Embryology*, ed. Jane Maienschein, Marie Glitz, and Garland E. Allen, vol. 5 of *Centennial History of the Carnegie Institution of Washington*, 5 vols. (Cambridge: Cambridge Univ. Press, 2004–5), 189; Louis Brown, *The Department of Terrestrial Magnetism*, vol. 2 of *Centennial History of the Carnegie Institution of Washington*, 205.

3. NSF, "Master Government List of Federally Funded Research and Development Centers," Feb. 2005, www.nsf.gov/statistics/ffrdclist/. The status of individual centers changes from time to time.

4. There are historical entries for most of these institutions in Joseph C. Kiger, ed., *Research Institutions and Learned Societies* (Westport, CT: Greenwood, 1982). Otherwise the historical literature on nonprofits is eclectic. See, e.g., David Micklos, *A History of Man and Science at Cold Spring Harbor: The First Hundred Years* (Cold Spring Harbor, NY: Cold Spring Harbor Laboratory, 1988), 29–32; Leslie Roberts, "Cold Spring Harbor Turns 100," *Science* 250 (1990): 496–98; Victor K. McElheny, *Watson and DNA: Making a Scientific Revolution* (Cambridge, MA: Perseus, 2003); James D. Watson, *Avoid Boring People: Lessons from a Life in Science* (New York: Knopf, 2007), 279–80; Kenneth Paigen, *The Jackson Laboratory* (New York: Newcomen Society of the United States, 1999); Jean Holstein, *The First Fifty Years at the Jackson Laboratory* (Bar Harbor,

ME: n.p., 1979); and Vicky Cullen, *Down to the Sea for Science: 75 Years of Ocean Research, Education, and Exploration at the Woods Hole Oceanographic Institution* (Woods Hole, MA: WHOI, 2005). Otherwise, Boyce Thompson, Salk, Dana-Farber, Whitehead, The Institute for Genomic Research (TIGR), and others all merit serious historical treatments.

5. NSF, *R&D Activities of Independent Nonprofit Institutions, 1973* (Washington, DC: GPO, 1975), includes the numbers of persons employed in 444 institutions spending more that $100,000 on research and development (by field, degree level, salary, and state but not by sex), but the NSF's later surveys were of funding only. Thus there are no good comparative data on scientific personnel employed in the nonprofit sector broken down by sex after the American Science Manpower series was terminated in the early 1970s.

6. One exception was Charlotte Friend, long a staff scientist at Sloan-Kettering Memorial Hospital, who left for a faculty position at Mount Sinai Medical School, also in New York City, when it opened in the late 1960s.

7. The much-honored mathematician Karen Uhlenbeck was, however, a frequent visitor. The lack of a woman scientist on the faculty of the IAS was in fact one of the few steps backward at nonprofit institutions in this period, as the archaeologist Hetty Goldman had been there from 1936 to 1947. Joan Wallach Scott, in its School of Social Sciences after 1985, was a historian. See [Juliet A. Mitchell, ed.], *A Community of Scholars: The Institute for Advanced Study; Faculty and Members, 1930–1980* (Princeton, NJ: Institute for Advanced Study, 1980).

8. On Peggy LeMone, see *AMWS*, 19th ed., 1995–96, 4:858. I thank the NCAR archivist, Diane Rabson, for sending materials from the early days of its first women's committee, including Diane Rabson, "It Happened Here: 'Don't Call Me a Girl,'" *[UCAR] Staff Notes Monthly*, Apr. 1998. See also www.ucar.edu/org/history.shtml; Zhenya Gallon, "The Atmosphere Inside: Surveying the Climate for Women at UCAR," *[UCAR] Staff Notes Monthly*, Apr. 2000, www.ucar.edu/communications/staffnotes/0004/women.html.; and Kiger, *Research Institutions and Learned Societies*, 473–76. A new women's committee was formed in 2007. "November 2007, NCAR Launches Women In Science Committee," www.ucar.edu/communications/staffnotes/0711/women.shtml.

9. On Quinn and Hanson, see www.cwp.library.ucla.edu/. On Drell, see "Movers," NatureJobs, *Nature* 417 (Mar. 28, 2002); Charles Seife, "Shakeup at SLAC," *Science* 308 (2005): 1393; and Adrian Cho, "For a Famous Physics Laboratory, A Quick and Painful Rebirth," ibid. 326 (2009): 221–23.

10. Cullen, *Down to the Sea for Science*, 138, 159–60. For example, Mary Silver of UC–Santa Cruz and Ellen Druffel of UC–Irvine both spent leaves at WHOI in the 1980s, and Mary Sears and Betty Bunce had been on the staff since the 1930s and 1940s, respectively. Interestingly, Cullen saw the perception at WHOI of the increasing presence of women in oceanography as a series of waves (no pun intended), first in the number of undergraduate applications for its summer programs, then in the number of applications for its graduate program, then among postdocs, and finally at the staff level.

11. Irwin Goodwin, "Down and Out in Murray Hill," *Nature* 412 (Aug. 9, 2001): 578–79. On Cladis, see *AMWS*, 19th ed., 1995–96, 2:300. On Hu, see www.engineering

.ucsb.edu/faculty/profile/207. On Jackson, see *AMWS,* 19th ed., 1995–96, 4:12. On Reichmanis, see ibid., 6:130; and "Elsa Reichmanis, Transcript of an Interview Conducted by David C. Brock at Murray Hill, New Jersey, on 1 August 2001," Chemical Heritage Foundation, Philadelphia. On Murray, see *AMWS,* 19th ed., 1995–96, 5:707; and Babak Ashrafi, "Interview with Cherry Ann Murray," Jan. 17, 2006, Niels Bohr Library, Center for the History of Physics, American Institute of Physics, College Park, MD, available online at www.aip.org/history/ohilist/29936-1.html. On Wilson, see *AMWS,* 19th ed., 1995–96, 7:807. Also at Bell Labs, from 1958 to 1983, was the applied mathematician Florence Jessie MacWilliams. On MacWilliams, see obituary by Vera Pless in *SIAM News* 23 (Nov. 1990) and www.awm-math.org/noetherbrochure/MacWilliams80.html. Part of the Sally Hacker Papers at the Schlesinger Library, RIAS, Harvard University, Cambridge, MA, relate to Bell Labs in the 1970s. On Hacker, see Barbara J. Love, comp., *Feminists Who Changed America, 1963–1975* (Urbana: Univ. of Illinois Press, 2006), 194.

12. On Hoover, see "Inventor of the Week: Erna Schneider Hoover, Computerized Telephone Switching System," http://web.miit.edu/invent/iow/hoover.html; and "Outstanding Alumni," *Yale Graduate School Newsletter* 10 (Mar. 2008) (I thank Mary Oates for a copy). On sexual harassment, see memoranda, box 14, folder 5, Bell Labs, Eugenie V. Mielczarek Papers, Special Collections and Archives, George Mason University, Fairfax, VA; and *CSWP Gazette,* no. 6 (July 1, 1982).

13. On Surko, se *AMWS,* 19th ed., 1995–96, 6:1392. On Reichmanis, see www.bell-labs.com/org/physicalsciences/profiles/reichmanis.html. She also served as president of the ACS in 2003. On Wright, see *AMWS,* 19th ed., 1995–96, 7:934; and www.awm-math.org/noetherbrochure/Wright00.html. On Murray, see Valerie Jameson, "The Industrial Physicist Who Has It All," *Physics World* 14 (May 2001): 9, available online at http://physicsweb.org/articles/world/1/5/7; and Betsy Mason, "Women Join Supervisory Ranks at Lab," *Valley Times* (Pleasanton, CA), Sept. 25, 2004, 1, 23, clipping in box 1, folder 14, Lawsuit Clippings, LLNL and LANL, Mary Singleton Papers, Women in Science and Engineering Archives, Special Collections, Iowa State University Library, Ames. On Chan, see *AMWS,* 20th ed., 1998–99, 2:157.

14. The Office of Scientific Programs of the institute was directed by the astrophysicist Neta Bahcall until 1990, when she left for a professorship at Princeton University. On Bahcall, see *AMWS,* 19th ed., 1995–96, 1:304.

15. Debra Elmegreen, "Committee on the Status of Women in Astronomy," *Bulletin of the American Astronomical Society* 23, no. 3 (1991): 1091; idem, "Committee on the Status of Women in Astronomy," ibid. 24, no. 3 (1992): 890–91; idem, "Committee on the Status of Women in Astronomy," ibid., 25, no. 3 (1993): 991; Jill S. Price, "Gender Issues in Astronomy: Results of a Survey of Women AAS Members," ibid. 22, no. 4 (1990): 1236; Kim A. McDonald, "Many Female Astronomers Say They Face Sex Harassment and Bias," *CHE,* Feb. 13, 1991, A11, A15; letters to the editor, ibid., Mar. 20, 1991, B4; Faye Flam, "Still a 'Chilly Climate' for Women?" *Science* 252 (1991): 1604–6; Jill S. Price, "Discrimination in the Workplace: Results of Two Recent Surveys and Some Recommendations," in "Women in Astronomy," ed. Sally Stephen and Andrew Fraknoi, special issue, *Mercury: The Journal of the Astronomical Society of the Pacific* 21 (Jan.–Feb. 1992): 29–31. Price died suddenly at age 38 in 1996. Kathleen DeGioia East-

wood, "Jill S. Price, 1957–1996," *Bulletin of the American Astronomical Society* 26, no. 4 (1996): 1462–63. See also ch. 6, n. 56.

16. "The Baltimore Charter for Women in Astronomy," and numerous other articles in *Women at Work: A Meeting on the Status of Women in Astronomy, Held at the Space Telescope Science Institute, September 8–9, 1992*, ed. C. Megan Urry et al. (Baltimore: STSCI, [1993]), esp. preface, also online at www.stsci.edu/meetings/WiA/Balto Charter.html. See also Diana Steel, "Astronomers Fight Sexism," *New Scientist* 135 (Sept. 26, 1992): 8; and "Astronomers Explore Equal Rights Universe," *Science* 258 (1992): 19. A follow-up meeting, "Women in Astronomy II: Ten Years After," held in Pasadena, California, in June 2003, came up with "Equity Now: The Pasadena Recommendations for Gender Equality in Astronomy," available online at www.aas.org/cswa/pasadenarecs .html. The meeting was partially subsidized by an NSF grant.

17. Diane L. Fowlkes, "Workshop on Sexual Harassment," in Urry et al., *Women at Work*, 93–97; Anne Kinney, "Astronomizing at STScI," ibid., 193–95; Andrew Lawler, "Institute Faulted on Attitudes toward Women," *Science* 299 (2003): 993; "Steven Beckwith Named UC VP for Research," www.universityofcalifornia.edu/news/article/16816.

18. On Berman, see *AMWS*, 19th ed., 1995–96, 1:581; and "Helen M. Berman, Transcript of an Interview by David N. Berol at New Brunswick, New Jersey, on 11 February 2000," Chemical Heritage Foundation. She was part of a female protégée chain in twentieth-century Anglo-American crystallography. While an undergraduate at Barnard College she had worked with the crystallographer Barbara Low, a student of the British Nobelist Dorothy Hodgkin, and later with another, Jenny Glusker, at the Institute for Cancer Research, as the Fox Chase Cancer Center was then known.

19. On Hanson, see http://cwp.library.ucla.edu/Phase2/Hanson_Gail_Gulledge @8412345. Hanson later moved back to California with a distinguished professorship at UC–Riverside, possibly replacing the physicist Anne Kernan. On Kernan, see *AMWS*, 19th ed., 1995–96, 4:356.

20. On Klinman, see *AMWS*, 19th ed., 1994, 4:470; notes on a telephone interview by Elga Wasserman, Mar. 13, 1995, in box 1, Elga Wasserman Collection, Schlesinger Library, RIAS; and "Oral History of Judith Klinman, June 13, 2002," by Tanya Zarnish-Belcher, Women in Science and Engineering Archives, Special Collections, Iowa State University Library. See also Ellen Daniell, *Every Other Thursday: Stories and Strategies from Successful Women Scientists* (New Haven, CT: Yale Univ. Press, 2006), 26–28.

21. On Adelman, see Irma Adelman, "My Life Philosophy," *American Economist* 34 (Fall 1990): 3–13; and "Explorations: Irma Adelman: A Pioneer in the Expansion of Economics, An Interview by Shoshana Grossbard-Schechtman, Introduced by Regenia Gagnier," *Feminist Economics* 8, no. 1 (2002): 101–16. Adelman coauthored many works with Cynthia Taft Morris, later at Smith College. Not much has been written about collaborations between women.

22. On Fedoroff, see *AMWS*, 19th ed., 1995–96, 2:1229; D. Brown, "Department in the Second Half of the Twentieth Century," 190–91; and Nina Fedoroff, "Two Women Geneticists," *American Scholar* 65 (Autumn 1996): 587–92, which includes information about her marriages and children.

23. On Hajek, see http://bti.cornell.edu/pdfs/BTIviews_Faculty_Bios.pdf.

24. On Orr-Weaver, see www.wi.mit.edu/research/faculty/orrweaver.html; and Nat-

alie Angier, *Natural Obsessions: Striving to Unlock the Deepest Secrets of the Cancer Cell* (Boston: Houghton Mifflin, 1999), 174. See also the website of the Whitehead Institute, www.wi.mit.edu/about/index.html. The twenty-fifth-anniversary issue of its in-house newsletter, *Discovery* (Winter 2008), online at www.wi.mit.edu/news/discovery/index.html, had celebratory articles. Its origins were more controversial. Colin Norman, "MIT Agonizes over Links with Research Unit," *Science* 214 (1981): 416–17; clippings and other materials in box 10, folder 46, Margaret A. MacVicar Papers, Institute Archives, MIT, Cambridge, MA. See also David. F. Noble, "The Selling of the University," *Nation*, Feb. 6, 1982, 129, 143–48.

25. On Moody, see *AMWS*, 18th ed., 1992–93, 5:503. Also at Battelle in Columbus was the botanist Ann Waterman Rudolph, who wrote many reports on radioactivity and the environment. Ronald L. Stuckey, "Necrology—Ann (Waterman) Rudolph (1934–1991)," *Ohio Journal of Science* 93 (1993): 163; idem, "Ann Waterman Rudolph (1934–1991): Her Life with Plants, Books, and Buttons," *Michigan Botanist* 34 (Jan. 1995): 24–45. At Battelle's Pacific Northwest Laboratories was the statistician Ethel Schaefer Gilbert, who in the 1970s completed award-winning studies of the causes of death of workers exposed to ionizing radiation at Hanford. *Radcliffe Quarterly* 66 (Dec. 1980): 32.

26. Watson, *Avoid Boring People*, 283. There were a few women scientists at Cold Spring Harbor. Susan Hockfield spent four years there (1980–84) before accepting a post at the Yale Medical School. See http://en.wikipedia.org/wiki/Susan_Hockfield; and McElheny, *Watson and DNA*, 213.

27. See the Fox Chase web page, www.fccc.edu/information/mission.html. See also Margaret W. Rossiter, *Women Scientists in America: Before Affirmative Action, 1940–1972* (Baltimore: Johns Hopkins Univ. Press, 1995), 237–38. On Mintz, see *AMWS*, 19th ed., 1995–96, 5:520; and "Awards and Honors," *Journal of Dairy Science* 60 (Apr. 1977): 4.

28. On Glusker, see *AMWS*, 20th ed., 1998–99, 3:209; and Miriam Rossi, "Jenny Pickworth Glusker (1931–): Crystallographer, Cancer Researcher," in *Notable Women in the Physical Sciences: A Biographical Dictionary*, ed. Benjamin F. Shearer and Barbara S. Shearer (Westport, CT: Greenwood, 1997), 132–36.

29. On Tilghman, see http://en.wikipedia.org/wiki/Shirley_M._Tilghman. On Skalka, see *AMWS*, 19th ed., 1995–96, 6:972; and A. M. Skalka to Maxine Singer, Apr. 9, 1987, box 47, folder 6, Maxine Singer Papers, Manuscript Division, Library of Congress, Washington, DC.

30. Judith Klinman, notes on a telephone interview by Elga Wasserman, Mar. 13, 1995, in box 1, Elga Wasserman Collection.

31. On Freedman, see Ella N. Strattis, "Wendy Laurel Freedman (1957–): Astronomer," in Shearer and Shearer, *Notable Women in the Physical Sciences*, 113–17. See also Andrew Watson, "The Universe Shows Its Age," *Science* 279 (1998): 981–83; Kim A. McDonald, "Astronomers Obtain a More Exact Measurement of the Age of the Cosmos," *CHE*, June 4, 1999, A17–A18; and "Transitions," *Nature* 421 (Jan. 30, 2003): 558.

32. On Rubin, see *AMWS*, 19th ed., 1995–96, 6:418; and Joann Eisberg, "Vera Cooper Rubin (1928–): Astronomer," in Shearer and Shearer, *Notable Women in the Physical Sciences*, 350–55. See also Robert Irion, "The Bright Face behind the Dark Sides of Galaxies," *Science* 295 (2002): 960–61; "Vera Rubin," in *Origins: The Lives and Worlds of*

Modern Cosmologists, ed. Alan Lightman and Roberta Brawer (Cambridge, MA: Harvard Univ. Press, 1990), 285–305; Vera Rubin, *Bright Galaxies, Dark Matters* (Woodbury, NY: American Institute of Physics, 1997); and L. Brown, *Department of Terrestrial Magnetism,* ch. 28. The Vera C. Rubin Papers are in the Manuscript Division of the Library of Congress.

33. On Sager, see *AMWS,* 19th ed., 1995–96, 6:480; and Eric Pace, "Dr. Ruth Sager, 79, Researcher on Location of Genetic Material," *NYT,* Apr. 4, 1997, A28. Her papers are in the Marine Biological Laboratory Library at Woods Hole, MA.

34. On Marrack, see *AMWS,* 19th ed., 1995–96, 5:265.

35. Patsy A. McLaughlin and Sandra Gilchrist, "Women's Contributions to Carcinology," in *History of Carcinology,* ed. F. Truesdale (Rotterdam: A. A. Balkema, 1993), 165–207 (I thank Patsy McLaughlin for a copy).

36. William E. Metcalf, "Margaret Thompson (22 February 1911–29 February 1992)," *American Journal of Archaeology* 96 (1992): 547–49.

37. Ann Lindell, "Ruth Patrick (1907–): Limnologist," in *Notable Women in the Life Sciences: A Biographical Dictionary* (Westport, CT: Greenwood, 1996), ed. Benjamin F. Shearer and Barbara S. Shearer, 313–16; Ruth Patrick, "Water Pollution," in *Life Stories: World-Renowned Scientists Reflect on Their Lives and the Future of Life on Earth,* ed. Heather Newbold (Berkeley and Los Angeles: Univ. of California Press, 2000), 85–92.

38. On Bliss, see "Dorothy Bliss, Ex-Curator of Museum, Dies," *NYT,* Jan. 2, 1988, 28; Linda H. Mantel, "Dorothy E. Bliss (1916–1987)," *Journal of Crustacean Biology* 8 (1988): 706–9; and McLaughlin and Gilchrist, "Women's Contributions to Carcinology," 201. On Stiassny, see Robin Finn, "Small Fry with a Scent Only a Mother Could Love," *NYT,* May 6, 2003, B2.

39. "G. & G. Collective History," box 3, folder 17, Genes and Gender Group Papers, Schlesinger Library, RIAS; Ethel Tobach and Betty Rosoff, eds., *Genes and Gender* (New York: Gordian, 1978). See also N[ancy] M[.] T[ooney], "Genes and Gender," *AWIS Newsletter* 6 (Mar.–Apr. 1977): 8–9; idem, "Meet a Member: Ethel Tobach," ibid., 10–12; "Kurt Lewin Memorial Award, 1993: Ethel Tobach," *Journal of Social Issues* 50, no. 1 (1994): 213–44; and "Gold Medal Award for Life Achievement in Psychology in the Public Interest," *American Psychologist* 58 (Aug. 2003): 551–53.

40. On Sanford, see *AMWS,* 16th ed., 1986, 6:429. There are items on Sanford's resignation and the 1989 fire in box 44, Jackson Lab, Maxine Singer Papers. Singer was on the laboratory's board of directors at the time. On Morawetz, see Gina Bari Kolata, "Morawetz to Direct Math Institute," *Science* 223 (1984): 1274. See also idem, "Cathleen Morawetz: The Mathematics of Waves," ibid. 206 (1979): 206–7; and "Cathleen S. Morawetz," in *More Mathematical People: Contemporary Conversations,* by Donald J. Albers, Gerald L. Alexanderson, and Constance Reid (Boston: Harcourt Brace Jovanovich, 1990), 221–38. Morawetz was also a trustee of Princeton University and of the Alfred P. Sloan Foundation.

41. On Healy, see *AMWS,* 19th ed., 1995–96, 3:714. See also ch. 9, n. 67. On Colwell, see *AMWS,* 19th ed., 1995–96, 2:414–15; and Stefanie Buck, "Rita Rossi Colwell (1934–): Microbiologist," in Shearer and Shearer, *Notable Women in the Life Sciences,* 87–93.

42. MacVicar served until 1987. In 1990 she developed breast cancer, and she died in 1991 of lung cancer. "A Resolution on the Death of Margaret L. A. MacVicar, For

Presentation at the Meeting of the Faculty of MIT, November 20, 1991," in box 11, folder 1, Margaret A. MacVicar Papers; CIW materials are in box 9, folders 7–9, and box 11, folders 4 and 23.

43. Singer may merit a biography or an autobiography.

44. On Fraser, see Elizabeth Pennisi, "TIGR's Chief: Results without the Roar," *Science* 296 (2002): 1957–58; and J. Craig Venter, *A Life Decoded* (New York: Viking, 2007), passim.

45. On Reid, see P.W., "Access to Education Provides a Way Out," *Nature* 405 (June 8, 2000): 718; and Wini Warren, "Yvonne A. Reid, American Type Culture Collection Researcher," in *Black Women Scientists in the United States* (Bloomington: Indiana Univ. Press, 1999), 247–49. At Howard, Reid had been a student of Margaret Collins in a kind of black protégée chain. Ibid., 52–66.

46. On Lindquist, see http://en.wikipedia.org/wiki/Susan_Lindquist. Materials relating to her selection as the new director in 2001 are in box 46, folder 7, Whitehead Institute—Director Search Committee, 2000–01, Maxine Singer Papers. Singer was especially displeased at the unprofessional comments made by some about Elizabeth Blackburn and Shirley Tilghman, two other candidates. Maxine Singer to Susan Whitehead, Feb. 20, 2001, ibid. On Villa-Komaroff, see *AMWS*, 19th ed., 1995–96, 7:409; and "Lydia Villa-Komaroff," in *Journeys of Women in Science and Engineering: No Universal Constraints*, ed. Susan A. Ambrose et al. (Philadelphia: Temple Univ. Press, 1997), 405–8. By 2007 Villa-Komaroff was the chief executive officer of Cytonome. See also ch. 6.

47. Ad hoc Panel on Industry, CWSE, Office of Scientific and Engineering Personnel, NRC, *Women Scientists and Engineers Employed in Industry: Why So Few? A Report on a Conference* (Washington, DC: National Academy Press, 1994), ch. 3. On McCarty, see *Marquis Who's Who Directory of Computer Graphics* (Chicago: Marquis Who's Who, 1984), 203; Carol Truxal, "The Woman Engineer," *IEEE Spectrum*, Apr. 1983, 58, 61–62; and Andrea Adelson, "Women Still Finding Bias in Engineering," *NYT*, Mar. 9, 1988, D6. By 1988 McCarty, age 53, was a division director supervising 120 software engineers and the highest-ranking woman in a technical position at the huge corporation. See her video at www.youtube.com/watch?V=LalhPPFe0jo. On Judith Osmer's lawsuit, see Karen Kaplowitz to Vera Kistiakowsky, Oct. 27, 1982, box 6, AWIS—1982, Vera Kistiakowsky Papers, Institute Archives, MIT; "Elsa Garmire: Transcript of a Tape-Recorded Interview by Joan Bromberg, February 4, 1985," Niels Bohr Library, Center for the History of Physics, 32; and Love, *Feminists Who Changed America*, 348. On Austin, see www.aero.org/corporation/corporateofficers/Austin.html.

48. "Museums, in Shift, Aim at Illiteracy in Science," *NYT*, Mar. 8, 1987, 48; Steve Fiffer, *Tyrannosaurus Sue: The Extraordinary Saga of the Largest, Most Fought Over T. Rex Ever Found* (New York: W. H. Freeman, 2000).

49. On Futter, see William H. Honan, "Barnard's President to Head Museum of Natural History," *NYT*, June 29, 1993, C13, C22. W. T. Golden was chair of the AMNH's board of trustees at the time.

50. "Breaking Through," *Working Woman*, Dec. 1992, 21; Fernbank website, www.fernbank.edu/museum/history.html. There are materials on the Fernbank Museum in box 4 of the Augusta S. Cooper Papers, Manuscript, Archives, and Rare Book Library, Emory University, Atlanta.

51. On Sullivan, see "Sullivan, Kathryn D. (1951–)," in *American Women in Science 1950 to the Present: A Biographical Dictionary,* by Martha J. Bailey (Santa Barbara, CA: ABC-CLIO, 1998), 381–82.

52. Leighton Taylor, *Aquariums: Windows to Nature* (New York: Prentice Hall General Reference, 1993); John Adam, "Piloting through Uncharted Seas," *Scientific American* 284 (June 2001): 38–39. On Earle, see Susan A. Ambrose et al., eds., "Earle, Sylvia Alice (1935–)," in Bailey, *American Women in Science 1950 to the Present,* 85–87. On McNutt, see Adam, "Piloting through Uncharted Seas."

53. Noreen Grice, "Women's Roles in America's Planetariums," in Urry et al., *Women at Work,* 143–44. Grice's summary described the other unpublished study, by David Menke, of the Buehler Planetarium, Broward Community College, Davie, FL. For earlier times, see Jordan Marché, *Theaters of Time and Space: American Planetaria, 1930–1970* (New Brunswick, NJ: Rutgers Univ. Press, 2005).

54. Dorrit Hoffleit, "The Maria Mitchell Observatory—For Astronomical Research and Public Enlightenment," *Journal of the American Association of Variable Star Observers* 30 (2001): 62–93, is a comprehensive history. On Belserene, see *AMWS,* 19th ed., 1995–96, 1:524. On Friel, later an executive officer for astronomy at the NSF and briefly director of the Lowell Observatory, in Flagstaff, Arizona, see "Lowell Observatory Director Resigns," June 11, 2010, http://azdailysun.com/news/local/article_afd7c0c0-75a0-11df-94e4-001cc4c002e0.html. There is material about the Nantucket Maria Mitchell Association, previously mentioned in ch. 4, in box 13, Correspondence, June '82–, Vera C. Rubin Papers.

55. Tony Reichhardt, "Cuts Force Telescope Closures at Kitt Peak," *Nature* 401 (Sept. 16, 1999): 199; "Wolff Resigns from Troubled Arizona Observatories," ibid. 402 (Nov. 25, 1999): 339. On Wolff, see *AMWS,* 16th ed., 1989, 7:725; "Oral History Interview with Sidney Carne Wolff, 1999 October 28 and 29," by Patrick McCray, Niels Bohr Library, Center for the History of Physics; Sidney C. Wolff, "Cautions for Astronomy's Golden Age," *Scientist,* Nov. 2, 1987, 18; and Kim A. McDonald, "Financial Squeeze at U.S. Observatories Forces Staff Cuts, Deferred Maintenance," *CHE,* Jan. 11, 1989, A1, A8.

56. On Cole, see David J. Hoff, "ETS President Cole Announces Retirement," www.edweek.org/ew/articles/2000/01/19/19ets.h19.html.

57. On Vetter, see *AMWS,* 19th ed., 1995–96, 7:399; and obituary, *Scientist,* Jan. 9, 1995, 23. See also the commemorative booklet *Celebrating Fifty Years: Commission on Professionals in Science and Technology, 1953–2003* (Washington, DC: CPST, 2003) (I thank Michele Aldrich for a copy). The Betty Vetter Papers are in the Women in Science and Engineering Archives, Special Collections, Iowa State University Library.

58. Wini Warren, "Shirley M. Malcolm [*sic*], Directing Changes in Science," in *Black Women Scientists in the United States,* 185–92; "Linkages," in *Linkages* 1 (Spring 1986): 1, 11; AAAS, *Linking Science and Technology to Woman's Needs* (Washington, DC, 2000). Malcom's acceptance speech for the NAS medal is on the web at www.nasonline.org/site/PageServer?pagenames=AARDS_pwm. She was nominated in 2001 by William T. Golden, box 43, Malcom, Shirley, 2001, Maxine Singer Papers. Also at the AAAS was Priscilla Reining, an anthropologist, who was hired in 1974 by the then AAAS president Margaret Mead to work in its Office of International Science, especially on arid lands.

She later became an expert on the role of circumcision in the spread of AIDS in Africa. Matt Schudel, "Priscilla Reining, 84; Anthropologist Broke Ground on AIDS in Africa," *Boston Globe,* Aug. 3, 2007, D13 (I thank Mary Oates for a copy).

59. On Filner, see Adam Bernstein, "Barbara Filner, 67, Senior Program Officer at Hughes Medical Institute," *Washington Post,* Mar. 1, 2009, B5; and "Barbara Filner, Biographical Note," in *Graduate School and Beyond: A Panel Discussion from the Conference "Science Careers in Search of Women," Held at Argonne National Library, Waterfall Glen Conference, April 6–7, 1989,* comp. N. K. Meshkov (Argonne, IL: Argonne National Laboratory, 1990), 27. On Didion, see www.anitaborg.org/about/who-we-are/catherine-didion/.

60. On Sandler's project, see Debra E. Blum, "Head of College Association's Project on Women Dismissed after 20 Years in Advocacy Role," *CHE,* Dec. 5, 1990, A15, A20. See also ch. 2, n. 3. The project's extensive papers are now at the Schlesinger Library, RIAS.

61. On Rivlin, see http://clinton1.nara.gov/White_House/EOP/OMB/html/amrbio.html. She lived near and knew the biochemist Maxine Singer. On Sawhill, see http://en.wikipedia.org/wiki/Isabel_Sawhill.

62. On Searle, see vita in box 57, folder 17, Searle, Barbara, Maxine Singer Papers. On Mansfield, see "Jane Elliott *Mansfield* Holt," in *Harvard and Radcliffe Class of 1966, Thirtieth Anniversary Report* (Cambridge, MA: Harvard University, 1996), 288–89. On the World Bank, see Judith H. Dobrzynski, "How to Succeed? Go to Wellesley: Its Graduates Scoff at Glass Ceilings," *NYT,* Oct. 29, 1995, F1, F9; and data on women at the World Bank (as of June 30, 1980) attached to Rayna Dyson-Hudson to Jenny Farley, Sept. 20, 1981, box 6, folder 14, Cornell Eleven Papers, Rare and Manuscript Collections, Kroch Library, Cornell University, Ithaca, NY; and Jerri Dell to Martijn Paijmans, Dec. 21, 1979, attached to memorandum from Norman Brown to Dr. Shirley Malcolm [*sic*], Jan. 28, 1980, Shirley Malcom Correspondence, 1979–80, OOS, Janet Welsh Brown Files, 1975–80, AAAS Archives, Washington, DC. On Krueger, see www.imf.org/external/np/omd/bios/ak.html.

63. On Hartmann, see Ellen K. Coughlin, "Policy Researchers Shift the 'Terms of the Debate' on Women's Issues," *CHE,* May 31, 1989, A5–A6; Heidi Hartmann, "Policy Research for Social Change," in *Gender Matters: Women and Yale in Its Third Century* (New Haven, CT: Women's Faculty Forum, 2004), 128–42; and Love, *Feminists Who Changed America,* 203. She was also one of the several Yale graduate students in economics behind the founding of the Committee on the Status of Women in the Economics Profession, or CSWEP, in the early 1970s and later a MacArthur Prize fellow. See Carolyn Shaw Bell, "The Reasons for CSWEP," *Journal of Economic Perspectives* 12 (Fall 1998): 191–95; www.iwpr.org/About/HHBio.html; and Alice H. Cook, *A Lifetime of Labor: The Autobiography of Alice H. Cook* (New York: Feminist Press, 1998), 243.

64. On Croker, see Malene L. Cohen et al., "Women in ASPET: A Centennial Perspective," *Pharmacologist* 49 (Dec. 2007): 127. On Brown, see *AMWS,* 13th ed., Social and Behavioral Sciences, 1978, 154; and Rayna Green, "Janet Brown Leaves AAAS," *Science* 204 (1979): 490, 532.

65. On Jacobs, see "An Interview with ACS Executive Director Madeleine Jacobs, Part I," by E. Thomas Strom, *Southwest Retort* [newsletter of the Dallas–Fort Worth,

Texas, section of the ACS], Nov. 2004, 5–9; and "An Interview with ACS Executive Director Madeleine Jacobs, Part 2," ibid., Dec. 2004, 5–15.

66. On Newman, see Susan Newman, "Opinion: Looking Backward, Facing Forward," *S[eismographic] R[esearch] L[etters]* 77 (Nov.–Dec. 2006): 643–45, available online at www.seismosoc.org/publications/SRL/SRL_77/srl_77_6_op.html.

67. On Sward, see http://en.wikipedia.org/wiki/Marcia_P_Sward.

68. On Bailar, see *AMWS,* 19th ed., 1995–96, 1:307; and Sandra Stinnett et al., "Women in Statistics: Sesquicentennial Activities," *American Statistician* 44 (May 1990): 75.

69. On Franz, see *AMWS,* 19th ed., 1995–96, 2:1432; "Judy R. Franz," in Ambrose et al., *Journeys of Women in Science and Engineering,* 173–76; and "Franz to Become New APS Executive Officer in April," *APS News* 3 (Mar. 1994): 1–2. On Kirby, see "Kirby to Become Executive Officer of APS," *Physics Today* 67 (July 2009): 27–28. Nancy Felipe Russo, director of the women's program at the APA from 1974 to 1985, then held a professorship at Arizona State University. Felice J. Levine served as executive officer of the American Sociological Association from 1991 to 2002. On Russo, see Love, *Feminists Who Changed America,* 400.

70. See J. J. Hermes, "Chemical Society Defends Policy on Open-Access Publishing," *CHE,* Nov. 2, 2007, A9.

71. Catherine Didion to Jaleh Daie et al., "Salary Review," n.d, mentioning *The 1994 Association Salary Review,* published by the Greater Washington Society of Association Executives, and enclosing a clipping, "Directors Manage Society Affairs," *Scientist,* May 2, 1994, 1, 5–6, which included recent salary data from "Non Profit Paychecks: SGR's Sixth Annual Survey," from *Science and Government Report* 23 (Apr. 1, 1994): 1–4, in box 17, AWIS Board Meeting, 1994, Vera C. Rubin Papers. Rubin was on the AWIS board at the time. The highest-paid executive director of a scientific society in 1994 was reportedly John Crum, of the ACS, at $258,455.

72. Wolfgang Saxon, "Janet Akyuz Mattei, Astronomer, Dies at 61," *NYT,* Apr. 2, 2004, B9; Gloria Negri, "Janet A. Mattei, Astronomer with a Passion for Flowers," *Boston Globe,* Apr. 7, 2004, F12; Lee Anne Willson and Elizabeth O. Waagen, "Janet Akyuz Mattei," *Physics Today* 58 (Jan. 2005): 66–67. See also Dorrit Hoffleit, *Misfortunes as Blessings in Disguise: The Story of My Life* (Cambridge, MA: American Association of Variable Star Observers, 2002), and "Variable Star Symposium; History, Science, Associations; Dedication of AAVSO Headquarters; 75th Annual Meeting," special issue, *Journal of the American Association of Variable Star Observers* 15, no. 2 (1986). For more on Mattei and the search for her successor, see the Lee Anne Willson Papers, Women in Science and Engineering Archives, Special Collections, Iowa State University Library, Ames.

73. On Marincola, see www.ascb.org/index.cfm?id=1311&navid=110&tcode=nws3; and Catherine Brady, *Elizabeth Blackburn and the Story of Telomeres: Deciphering the Ends of DNA* (Cambridge, MA: MIT Press, 2007), 173–78, 265, 280–81, 283, 299, 301.

74. See Avery Russell, "The Women's Movement and Foundations," *Foundation News* 13 (Nov.–Dec. 1972): 16–22; and esp. "Statement of Franziska P. Hosken, Na-

tional Organization for Women," *Public Hearings Before the Committee on Ways and Means, House of Representatives, Ninety-Third Congress, First Session on the Subject of General Tax Reform, Part 16 of 18 (April 13, 1973)* (Washington, DC: GPO, 1973), 6424–57. NOW urged Congress to deprive the foundations of their tax-free status, since they discriminated heavily against women and lavishly funded prestigious institutions that also did so. See also May Jean Tully, "Funding the Feminists," *Foundation News* 16 (Mar.–Apr. 1975): 24–33. After about 1973 several foundations specifically for women, including the Wonder Woman Foundation and the Susan G. Komen Breast Cancer Foundation, were established. Andrew Yarrow, "Feminist Philanthropy Comes Into Its Own," *NYT*, May 21, 1983, A7. In 1975, interested staff and trustees formed Women and Foundations, a forum to help women get more grants. Susan Calhoun, "New Ways to Lead," *Foundation News* 28 (Nov.–Dec. 1987): 24–29.

75. On Anderson, see obituary in *NYT*, Dec. 20, 1985, D26; and Isabel Grossner, "Carnegie Corporation Project—Florence Anderson," 1966 and 1967, transcript at Rare Book and Special Collections Room, Butler Library, Columbia University, New York.

76. The extensive records of the Carnegie Corporation of New York are in the Rare Book and Special Collections Room, Butler Library, Columbia University.

77. On the Ford Foundation, see Susan M. Hartmann, *The Other Feminists: Activists in the Liberal Establishment* (New Haven, CT: Yale Univ. Press, 1998), ch. 5. On Chamberlain at Ford, see Mariam K. Chamberlain, "There Were Godmothers, Too," in *The Politics of Women's Studies: Testimony of Thirty Founding Mothers*, ed. Florence Howe (New York: Feminist Press, 2000), 353–64. See also Mariam K. Chamberlain and Alison Bernstein, "Philanthropy and the Emergence of Women's Studies," *Teachers College Record* 93 (Spring 1992): 556–68; and Rosa Proietto, "The Ford Foundation and Women's Studies in American Higher Education," in *Philanthropic Foundations: New Scholarship, New Possibilities*, ed. Ellen Condliffe Lagemann (Bloomington: Indiana Univ. Press, 1999), ch. 12.

78. On Simmons, see *Who's Who in America, 1994*, 48th ed. (New Providence, NJ: Marquis Who's Who, 1993), 2:3175; and Love, *Feminists Who Changed America*, 424. See also the commemorative booklet *30 Years of the John D. & Catherine T. MacArthur Foundation* (Chicago: MacArthur Foundation, 2008).

79. On Bond, see http://forms.bwfund.org/forms/summer_2007/bond.html. On Davis, see "Karen Davis, Ph.D.," www.commonwealthfund.org/Content/Bios/D/Davis_Karen.aspx. On Berresford, see Mitch Naufitts, "Featured Newsmaker Interview with Susan V. Berresford: September 11 and Beyond," *Philanthropy News Digest*, May 28, 2002, available online at www.fordfound.org/news/views_news_detail.cfm?news_index=122; and idem, "Newsmakers—Susan V. Berresford, Former President, Ford Foundation: A Life in Philanthropy," *Philanthropy News Digest*, Jan. 9, 2008, online at http://foundationcenter.org/pnd/newsmakers/nwsmkr.jhtml?id=20000008. On Rodin, see Greg Winter, "Rockefeller Foundation Names a New President," *NYT*, Aug. 25, 2004, B2; and E.S., "Judith Rodin to Head Rockefeller Fund," *CHE*, Sept. 10, 2004, A29. On Aiello, see "Wenner-Gren Board of Trustees Appoints D. Leslie Aiello as Next President," ibid., Jan. 14, 2005, A39. A few women headed small foundations with a

particular cause; for example, the neurologist Nancy Wexler headed the Hereditary Disease Foundation of New York City, which is concerned with Huntington's disease. See http://en.wikipedia.org/wiki/Nancy_Wexler.

Chapter 11 · Academia after Rajender

1. See, e.g., Jennie Farley, ed., *Sex Discrimination in Higher Education: Strategies for Equality* (Ithaca: New York State School of Industrial and Labor Relations, Cornell University, 1981); and idem, ed., "Resolving Sex Discrimination Grievances on Campus: Four Perspectives" (mimeograph, distributed by Institute of Labor Relations, Cornell University, Ithaca, NY, 1981).

2. See, e.g., Anne C. Roark, "Women in Science: Unequal Pay, Unsold Ideas, and, Sometimes Unhappy Marriages," *CHE,* Apr. 21, 1980, 3–4; Natalie Angier, "Women Swell Ranks of Science, But Remain Invisible at the Top," *NYT,* May 21, 1991, C1, C12; Anthony DePalma, "Rare in Ivy League: Women Who Work as Full Professors," ibid., Jan. 24, 1993, 1, 23; and, later, Sara Rimer, "For Women in the Sciences, the Pace of Progress at Top Universities is Slow," ibid., Apr. 15, 2005, A15.

3. "Awards and Honors," *AWM Newsletter* 14 (July–Aug. 1984): 3–4.

4. NSF, *NSF—Increasing the Visibility of Women in Science, Visiting Professorship for Women,* NSF 93-144 (Arlington, VA, 1994), has an incomplete list of the awardees through 1993.

5. The physicist Carol Jo Crannell said that the idea for the visiting professorships grew out of an idea that she had while on the Committee on the Status of Women in Physics of the American Physical Society. She could not get funding for it, but it was incorporated into the second version of the Women in Science Bill. Joan Ruskus and Cynthia Williamson, *The Visiting Professorships for Women Program: Lowering the Hurdles for Women in Science and Engineering, Final Report,* NSF 93-159 (Arlington, VA: NSF, 1993), app. E, 8. Materials relating to the Ruskus evaluation are at the Murray Research Archive Database, Institute for Quantitative Social Sciences, Harvard University, Cambridge, MA, http://dvn.iq.harvard.edu. There was a suggestion for NSF funding for visiting appointments for women in Clare Rose, Sally Ann Menninger, and Glenn F. Nyre, *The Study of the Academic Employment and Graduate Enrollment Patterns and Trends of Women in Science and Engineering, Summary* (Los Angeles: Evaluation and Training Institute, 1978), 34.

6. Mirjam Kempf, "EmPOWREment and ADVANCEment for Women: NSF Programs for Women in Science," Next Wave, *Science* 297 (2002), online only at http://sciencecareetrs.sciencemag.org/career_magazine/previous_issues/articles/2002_09_20/noDol.159438463419.

7. Kathleen Teltsch, "Mrs. Luce Left $70 Million for Women's Science Education," *NYT,* July 2, 1989, 15; Liz McMillen, "Clare Boothe Luce Fund to Spend $3.5 Million a Year to Encourage Women to Study and Teach Science," *CHE,* July 5, 1989, A23–A24; Lois R. Ember, "Luce Foundation Program Helps Women Develop Science Careers," *C&EN* 67 (July 24, 1989): 23–25; C[onstance] H[olden], "New Support for Women Scientists," *Science* 245 (1989): 126. Four awards had been made to a more eclectic group in 1987. Marilyn Hoffman, "Women Scholars Get a Boost," *Christian Science*

Monitor, Oct. 23, 1987, 19. Vera Kistiakowsky advised Luce Foundation officials setting up the program. Melissa S. Topping to Vera Kistiakowsky, Jan. 24, 1989, box 13, MIT, Misc. 1985, Vera Kistiakowsky Papers, Institute Archives, MIT, Cambridge, MA.

8. See also Sue V. Rosser and Jane Z. Daniels, "Widening Paths to Success: Improving the Environment, and Moving toward Lessons Learned from the Experiences of POWRE and CBL Awardees," *Journal of Women and Minorities in Science and Engineering* 10 (2004): 131–48. See also the web page of the Clare Boothe Luce Program, a part of the Henry Luce Foundation, www.hluce.org/cblprogram.aspx.

9. Ethan J. Schrier, "A Snapshot Survey of Women in Astronomy," in *Women at Work: A Meeting on the Status of Women in Astronomy, Held at the Space Telescope Science Institute, September 8–9, 1992,* ed. C. Megan Urry et al. (Baltimore: STSI, [1993]), 59, 62. There is a certain amount of material on the Harvard department in the Vera C. Rubin Papers, Manuscript Division, Library of Congress, Washington, DC, as she was on its visiting committee from 1975 to 1983 and was almost offered a professorship there. See also Faye Flam, "Still a 'Chilly Climate' for Women?" *Science* 252 (1991): 1604–6.

10. Louis Uchitelle, "In Economics, a Subtle Exclusion," *NYT,* Jan. 11, 1993, D1, D3. See also Randy Albelda, *Economics and Feminism: Disturbances in the Field* (New York: Twayne, 1997), 42–45; and Marianne A. Ferber and Julie A. Nelson, eds., *Feminist Economics Today: Beyond Economic Man* (Chicago: Univ. of Chicago Press, 2003).

11. Maria Luisa Crawford, Judith B. Moody, and Jan Tullis, "Women in Academia: Students and Professors Revisited," *Geology* 15 (1987): 771–72; Lois K. Ongley, Matthew W. Bromley, and Katherine Osborne, "Women Geoscientists in Academe: 1996–1997," *GSA Today,* Nov. 1998, 12–14, which was dedicated to the memory of Judith Moody, whose early death stilled a strong voice for gender equity in geoscience. See Judith B. Moody to Vera Kistiakowsky, Oct. 24, 1982, with enclosures, Vera Kistiakowsky Papers, box 6, AWIS—1982.

12. Arthur R. M. Nowell and Charles D. Hollister, "Graduate Students in Oceanography: Recruitment, Success, and Career Prospects," *Eos,* Sept. 6, 1988, 834–35, 840–43; M. Grant Gross, "Women in Ocean Sciences," ibid., Sept. 26, 1989, 857. See also Ben McKelway, "Women in Oceanography," *Oceanus* 25 (Winter 1982–83): 75–79.

13. Margaret A. LeMone, Joan V. Frisch, and Lesley T. Julian, "Tracking Women and the Weather: Their Growing Role in Meteorology," *Weatherwise* 37 (Aug. 1984): 179; Julie A. Winkler, Donna Tucker, and Anne K. Smith, "Salaries and Advancement of Women Faculty in Atmospheric Science: Some Reasons for Concern," *Bulletin of the American Meteorological Society* 77 (Mar. 1996): 475, 477–78. Two of the twenty-eight women were significant grantswomen, bringing in two-thirds of the nearly $3 million awarded to all the women faculty (483).

14. Paul Selvin, "Profile of a Field: Mathematics; Heroism Is Still the Norm," *Science* 255 (1992): 1382–83; "Women in Math Update" (chart) ibid. 257 (1992): 323 (the chart lists five full professorships held by women in the top ten mathematics departments but has some errors, as Joan Birman, of Barnard, who had never been given the usual courtesy appointment at Columbia, was nevertheless counted as one of them; someone else had resigned; and Berkeley counted as full time Sun-Yung Alice Chang, who had a joint appointment with UCLA and lived in Los Angeles); Alice T. Schafer, "Mathematics and Women: Perspectives and Progress," *Notices of the American Mathematical*

Society 38 (Sept. 1991): 735–37; Judith Axler Turner, "More Women Are Earning Doctorates in Mathematics, but Few Are Being Hired by Top Universities," *CHE,* Dec. 6, 1989, A13–A14. See also Alice T. Schafer, "Women and Mathematics," in *Mathematics Tomorrow,* ed. Lynn A. Steen (New York: Springer-Verlag, 1981), 176–85; Joan S. Birman, letter to the editor, *AWM Newsletter* 22 (Nov.–Dec. 1992): 3–4; and Jean E. Taylor and Sylvia Wiegand, "AWM in the 1990s: A Recent History of the Association for Women in Mathematics: Part 1," ibid. 29 (Jan.–Feb. 1999): 9–10.

15. On the Harrison case, see Jenny Harrison, "The Escher Staircase," *Notices of the American Mathematical Society* 38 (Sept. 1991): 730–34; Paul Selvin, "Does the Harrison Case Reveal Sexism in Math?" *Science* 252 (1991): 1781–83; Brady Kahn, "The Gender Factor," *Express* (East Bay, CA) 13 (July 1991): 1, 15–18, 22–28; Paul Selvin, "Jenny Harrison Finally Gets Tenure in Math at Berkeley," *Science* 261 (1993): 286; Denise K. Magner, "Debate over Woman's Tenure Continues at Berkeley," *CHE,* Oct. 20, 1993, A16; Allyn Jackson, "Fighting for Tenure: The Jenny Harrison Case Opens Pandora's Box of Issues about Tenure, Discrimination, and the Law," *Notices of the American Mathematical Society* 41 (Mar. 1994): 87–94; and Steve Batterson, *Stephen Smale: The Mathematician Who Broke the Dimension Barrier* (Providence, RI: AMS, 2000), 230–36. See also Liz McMillen, "U. of California Professors Launch a Drive to Give Tenure Candidates Access to Files," *CHE,* Mar. 18, 1987, 16–18; Linda Greenhouse, "Universities Lose Shield of Secrecy in Tenure Disputes," *NYT,* Jan. 10, 1990, A1, B7; and James Fristrom, Renee Sung, and Leon Wofsky, "Patricia St. Lawrence, Genetics: Berkeley, 1922–1996," in *University of California: In Memoriam, 1996,* ed. David Krogh (Oakland: Academic Senate, University of California, [1997]), 172–74. In the 1970s the *AWM Newsletter* had printed many revelations about the UC–Berkeley math department's heavy-handed methods of recruitment as a kind of bad-faith operation. See ch. 2.

16. See, e.g., Marina Ratner, Letter to the Editor, *AWM Newsletter* 23 (Sept.–Oct. 1993): 17; and Paul Selvin, "Harrison Case: No Calm after Storm," *Science* 262 (1993): 324–27. See also Nina Toren, *Hurdles in the Halls of Science: The Israeli Case* (Lanham, MD: Lexington Books, 2000), 23.

17. Carol Hollenshead, "Women in the Academy: Confronting Barriers to Equality," in *Equal Rites, Unequal Outcomes: Women in American Research Universities,* ed. Lilli S. Hornig (New York: Kluwer Academic/Plenum, 2003), 214. On Daubieches, see *AMWS,* 19th ed., 1995–96, 2:657; and Pamela Profitt, ed., *Notable Women Scientists* (Detroit: Gale Group, 1999), 126–28. On Chang, see *AMWS,* 19th ed., 1995–96, 2:175–76; and Profitt, *Notable Women Scientists,* 85–86.

18. Taylor and Wiegand, "AWM in the 1990s," 10. On Sally, see www.agnesscott.edu/lriddle/women/sally.html. On Roitman, see *AMWS,* 19th ed., 1995–96, 6:317. On Uhlenbeck, see http://en.wikipedia.org/wiki/Karen_Uhlenbeck. Kopell articulated well the differences she felt in the treatment of women in the mathematics departments at Northeastern and MIT in "Interview with Nancy Kopell by Shirlee Sherkow," Jan. 11, 1977, Institute Archives, MIT, Cambridge, MA.

19. Donna J. Nelson, "Diversity in the Physical Sciences," *AWIS Magazine* 31 (Winter 2002): 28–32; "Survey of Universities Finds Few Women on Senior Staff," *Physics Today* 35 (Feb. 1982): 99; Mildred S. Dresselhaus, Judy R. Franz, and Bunny C. Clark, "Intervention to Increase the Participation of Women in Physics," *Science* 263 (1994):

1392–93. But conditions could be quite unpleasant for the lone woman, as a 2004 article on the Duke physics department revealed. Robin Wilson, "Louts in the Lab," *CHE*, Jan. 23, 2004, A7–A9.

20. See, e.g., "Chemistry Faculties Gain Women Slowly," *C&EN* 62 (Feb. 13, 1984): 26; Ivan Amato, "Profile of a Field: Chemistry; Women Have Extra Hoops to Jump Through," *Science* 255 (1992): 1372–73; Kenneth Everett et al., "Women in the Ranks: Faculty Trends in the ACS-Approved Departments," *Journal of Chemical Education* 73 (Feb. 1996): 139–41; Mairin B. Brennan, "Women Make Slow Progress on Chemistry Faculties of Major Universities," *C&EN* 76 (July 20, 1998): 31; and Allison Byrum, "Women's Place in Ranks of Academia," ibid. 79 (Oct. 1, 2001): 98–99. The focus in these articles was on tenure-track jobs and thus they do not mention the growth in non-tenure-track jobs, a disturbing phenomenon of the time. The data were based largely on ACS directories, such as Bonnie R. Blaser and Jeanann M. Dellantonio, eds., *College Chemistry Faculties*, 4th ed. (Washington, DC: ACS, 1977), which also includes Mexico and Canada. On Minnesota, see Maureen Smith, "Catalysts for Change: Chemistry Department Finds New Formulas for Resolving Conflict," *Items* (University of Minnesota), Winter 1989, 12–15, copy in box 12, folder 3, Printed Materials, 1986–91, Kathleen M. Desmond Trehanovsky Papers, Women in Science and Engineering Archives, Special Collections, Iowa State University Library, Ames. On Friend, see Kathleen Koman, "Catalytic," *Harvard Magazine*, Jan.–Feb. 1995, 50–52. See also Nina Matheny Roscher and Margaret A. Cavanaugh, "Academic Women Chemists in the 20th Century: Past, Present, Projections," *Journal of Chemical Education* 64 (Oct. 1987): 823–27; and idem, "Academic Women Chemists in the 20th Century," ibid. 69 (Nov. 1992): 870–73.

21. Robin Wilson, "Beggar, Psychologist, Mediator, Maid: The Thankless Job of a Chairman," *CHE*, Mar. 2, 2001, A11.

22. Joyce Y. Corey to Eileen Reilley, Apr. 5, 1986, in agenda binder for meeting of the Women Chemists Committee of the American Chemical Society, Sept. 9, 1986, Anaheim National Meeting, tab 6, in box 11, folder 5, Kathleen M. Desmond Trehanovsky Papers (Corey had been a VPW in 1984); Alison Schneider, "Female Scientists Turn Their Backs on Jobs at Research Universities," *CHE*, Aug. 18, 2000, A12–A14; Robin Wilson, "Where the Elite Teach, It's Still a Man's World," ibid., Dec. 3, 2004, A8–A9, A12, A14.

23. Amato, "Profile of a Field: Chemistry," 1372–73; and "Letters: Women in Science; The Response," *Science* 256 (1992): 1610–15.

24. Alison Schneider, "Support for a Rare Breed," *CHE*, Nov. 10, 2000, A12–A14; "Funding Boost for Women Chemists in Fight for Equality," *Nature* 410 (Mar. 15, 2001): 240; Robin Wilson, "The Chemistry between Women and Science," *CHE*, May 26, 2006, A12; http://coach.uoregon.edu/coach/index.php?page. See also "Sally Chapman, Transcript of an Interview Conducted by Hilary Domush at Barnard College, New York, New York, on 5 and 6 January 2009," 51–59, Chemical Heritage Foundation, Philadelphia; and Geraldine L. Richmond, "COAChing Women to Succeed in Academic Careers in the Chemical Sciences," *Journal of Chemical Education* 82 (Mar. 2005): 351–53. At some point the Dreyfus Foundation established awards for those who encouraged or mentored women and minorities.

25. Ruth Sager to Marcus M. Rhoades, Sept. 17, 1971, box 21, folder 6, Ruth Sager Papers, Marine Biological Laboratory Library, Woods Hole, MA. Some of these top women (e.g., Sager) deserve biographies or autobiographies. Already there are Catherine Brady, *Elizabeth Blackburn and the Story of Telomeres: Deciphering the Ends of DNA* (Cambridge, MA: MIT Press, 2007), and Neena B. Schwartz, *A Lab of My Own* (Amsterdam, NY: Rodopi, 2010). Obituaries and archival collections exist for most of them, such as the Charlotte Friend Papers, Archives, Mount Sinai Medical Center, New York; the Margaret Bryan Davis Papers, University Archives, University of Minnesota, Minneapolis; and the Neena Schwartz Papers, Northwestern University Archives, Evanston, IL. On Judith Blake, see Linda B. Bourque and Valerie Oppenheimer, "Judith Blake, 1926–1993," *University of California: In Memoriam, 1994* (Berkeley: University of California, 1994), 32–34; and Linda B. Bourque, "A Biographical Essay on Judith Blake's Professional Career and Scholarship," *Annual Review of Sociology* 21 (1995): 449–77.

26. Lists of current and former investigators of the HHMI are available at www.hhmi.org/research/investigators/. See also *The Howard Hughes Medical Institute, 1953–1993* (Bethesda, MD, 1993); and *The Howard Hughes Medical Institute: A Twentieth Century History* (Bethesda, MD: HHMI, 1999), 37, 46.

27. Ellen Daniell, *Every Other Thursday: Stories and Strategies from Successful Women Scientists* (New Haven, CT: Yale Univ. Press, 2006), 26–28. On Klinman, see ch. 10, n. 20.

28. Debra E. Blum, "Faculty Notes," *CHE*, Sept. 19, 1990, A24; idem, "Faculty Notes—U. of Iowa to Pay $1.1 Million to Settle Sexual Harassment Suit," ibid., Dec. 5, 1990, A17; idem, "Medical Professor, U. of Iowa Face Aftermath of Bitter Sexual-Harassment Case," ibid., Mar. 13, 1991, A15–A16.

29. See Frances K. Conley, MD, *Walking Out on the Boys* (New York: Farrar, Straus, & Giroux, 1998).

30. Shirley Tilghman, "Science vs. the Female Scientist," *NYT*, Jan. 25, 1993, A17; idem, "Science vs. Women—A Radical Solution," ibid., Jan. 26, 1993, A23. See also Natalie Angier, "Scientist at Work: Shirley M. Tilghman; Fighting and Studying Battle of the Sexes with Men and Mice," ibid., June 11, 1996, C1, C11; and Nancy Hopkins, letter to the editor, ibid., June 16, 1996, sec. 4, p. 14.

31. Henry Etzkowitz et al., "The Paradox of Critical Mass for Women in Science," *Science* 266 (1944): 51–54. On Shavlik, see Barbara J. Love, comp., *Feminists Who Changed America, 1963–1975* (Urbana: Univ. of Illinois Press, 2006), 419.

32. On the OWHE, see boxes 1280–84 (annual reports) and 1348–1402, American Council on Education Records, Library and Archives, Hoover Institution, Stanford University, Stanford, CA; and Emily Taylor and Donna Shavlik, "To Advance Women: A National Identification Program," *Educational Record* 58 (1977): 91–100. In 1980 the OWHE ran an executive seminar on and for women administrators of color. "Minority Women Administrators Discuss Major Concerns in Higher Education," *Comment* 12 (June 1980): 4–5, 12.

33. On Scott, see David Blackwell et al., "Elizabeth Leonard Scott, 1917–1988, Professor of Statistics, Emerita," in *University of California: In Memoriam, 1991*, ed. David Krogh (Oakland: Academic Senate, University of California, 1991), 186–90. On David, see "A Conversation with F. N. David," by Nan M. Laird, *Statistical Science* 4 (1989): 235–46; Jessica Utts, "Obituary: Florence Nightingale David, 1909–1993," *Biometrics* 49

(1993): 1289–91; and M. J. Garber, "Florence Nightingale David, Statistics: Riverside," in *University of California: In Memoriam, 1995,* ed. David Krogh (Oakland: Academic Senate, University of California, 1995), 38–39. David's papers are at the Bancroft Library, University of California, Berkeley. On Schwartz, see ch. 1, n. 2. On Hay, see *AMWS,* 19th ed., 1995–96, 3:694; Bryan Marquard, "Elizabeth Hay, at 80; was Pioneer for Women in Science," *Boston Globe,* Sept. 16, 2007, B9 (I thank Mary Oates for a copy); and items in the Papers of the American Society for Cell Biology, box 75, Special Collections, Albin O. Kuhn Library and Gallery, UMBC, Catonsville, MD. At one point Hay wanted the society to cut off funding for the irreverent newsletter. Hay's papers, which are at the Countway Library at Harvard Medical School, are unprocessed and partly restricted for eighty years. On Rodin, see Molly O'Neill, "In an Ivy League of Her Own: On Campus with Judith Rodin" *NYT,* Oct. 20, 1994, C1, C4. On Zuber, see "Peer Review," *CHE,* Aug. 8, 2003, A8.

34. On Cobb at Michigan, see Ruth Bordin, *Women at Michigan: The "Dangerous Experiment," 1870s to the Present* (Ann Arbor: Univ. of Michigan Press, 1999), 84–85; and Sara Rimer and Judy Ruskin, "Black Woman to Be New LSA Dean," *Michigan Daily,* Jan. 9, 1975, 1–2, clipping in box 1, 1974 Personal Letters, Jewel Plummer Cobb Papers, SCM 89-63, Schomburg Center for Research in Black Culture, New York Public Library. On Goodman, see "Madeleine J. Goodman, Dean of Vanderbilt's College of Arts and Sciences, Dies," http:www.vanderbilt.edu/News/news/oct96/nr4.html. On Osborn, see *AMWS,* 19th ed., 1995–96, 5:986. On Baum, see ibid., 1:450. On Denton, see Angela Y. Davis, "In Memoriam, Denice Denton, Chancellor, Professor of Electrical Engineering, UC Santa Cruz, 1959–2006," online at www.universityofcalifornia.edu/senate/inmemoriam/denicedenton.htm.

35. Richard M. Weintraub, "Woman Physics Professor Named New Tufts Provost," *Boston Sunday Globe,* May 29, 1973, 8. On Gonzalez, see Nancie L. Gonzalez, "The Anthropologist as Female Head of Household," *Feminist Studies* 10 (Spring 1984): 97–114. On Reynolds, see below, n. 45. On Calloway, see *AMWS,* 19th ed., 1995–96, 2:22; and Kenneth Carpenter, Sally Fairfax, and Janet C. King, "Doris Howes Calloway, Nutrition: Berkeley," in *University of California: In Memoriam, 2001,* ed. Micki Conklin (Oakland: Academic Senate, University of California, 2001), 23–28. On Clark, see *AMWS,* 19th ed., 1995–96, 2:308.

36. On Caserio, see *AMWS,* 19th ed., 1995–96, 2:111. On Rodin, see Courtney Leatherman, "'Saint Judy' Goes One More unto the Breach," *CHE,* July 22, 1992, A5. Details of Rodin's career are in O'Neill, "In an Ivy League of Her Own." See also "Early Career Awards for 1977—Judith Rodin, Citation," *American Psychologist* 33 (Jan. 1978): 77–80; and Kit Lively, "Women in Charge: More Elite Universities Hire Female Provosts, Creating a New Pool for Presidential Openings," *CHE,* June 16, 2000, A33–A35. Rodin was the second such woman in the Ivy League, for the historian Hanna Holborn Gray had been provost at Yale briefly in the 1970s. On Coleman, see below, n. 43. On Richard, see http://en.wikipedia.org/wiki/Alison_Richard. On Holbrook, see http://en.wikipedia.org/wiki/Karen_Holbrook. On Cantor, see www.syracuse.edu/chancellor/about/index.html. See also Eleanor L. Babco, *Professional Women and Minorities: A Total Human Resources Data Compendium,* 13th ed. (Washington, DC: Commission on Professionals in Science and Technology, 2000), table 5-21 (p. 166), for data on women

college presidents by type of institution (public, private, two- or four-year) from 1975 through 1995 from the Office of Women in Higher Education of the American Council on Education. The page also has data on the women presidents' racial groups and religious orders.

37. On Rogers, see Dennis Hevesi, "Lorene Rogers, 94, President of University of Texas in '70s, Is Dead at 94," *NYT,* Jan. 26, 2009, 15; "Top University Post to Woman Biochemist," *C&EN* 52 (Oct. 21, 1974): 27; "The Regents' Choice," *Time,* Oct. 6, 1975, 50–51; *AMWS,* 19th ed., 1995–96, 6:309; and John Archibald Wheeler, *Geons, Blacks Holes and Quantum Foam: A Life in Physics* (New York: Norton, 1998), 319. See also oral history of Joanne Ravel, Women in Science and Engineering Archives, Special Collections, Iowa State University Library. Ravel was a friend of Rogers's and both belonged to the same department that hired the chemist Marye Anne Fox in the 1970s. As Fox too later held several high administrative posts, the department may have been a kind of nursery for potential leaders.

38. On Newell, see *Who's Who in America, 1994,* 48th ed. (New Providence, NJ: Marquis Who's Who, 1993), 2:2528. On Cobb, see *AMWS,* 19th ed., 1995–96, 2:359. See also "Black Women College Presidents: Diverse Group Heads 14 Colleges across Nation," *Ebony,* Feb. 1986, 108–15; Colleen Jones, "Does Leadership Transcend Gender and Race? The Case of African American Women College Presidents," in *Black Women in the Academy: Promises and Perils,* ed. Lois Benjamin (Gainesville: Univ. Press of Florida, 1987), 201–9; and items in Jewel Plummer Cobb Papers.

39. On Handler, see Don Wyclif, "Head of Brandeis Quits after Troubled Tenure," *NYT,* June 16, 1990, 8.

40. On Shalala, see Don Wyclif, "Women as Presidents: Shalala Takes Charge at U. of Wisconsin," ibid., Aug. 15, 1990, B6. The phrase "master politician" is from the accompanying article by Wyclif on Hanna Gray, ". . . Gray Guards the Ivory Tower at U. of Chicago," ibid.

41. See National Association of State Universities and Land-Grant Colleges, *Assessing Change: A Profile of Women and Minorities in Higher Education Administration at State and Land-Grant Universities* (n.p., 1988), 3 (the report has data on 122 campus posts); and Debra E. Blum, "165 Female College Presidents 'Honor Progress, Connect with Each Other,' and Commiserate," *CHE,* Dec. 19, 1990, A13, A15. See also OWHE, ACE, *Women in Presidencies: A Descriptive Study of Women College and University Presidents* (Washington, DC, 1993).

42. On Rodin, see Courtney Leatherman, "An Ivy League First: Woman to Lead U. of Pennsylvania," *CHE,* Dec. 15, 1993, A20; and O'Neill, "In an Ivy League of Her Own." See also Julie Nicklin, "Few Women Are among the Presidents With the Largest Compensation Packages," *CHE,* Nov. 9, 2001, A30.

43. On Keohane, see Denise K. Magner, "A 'Risk' Worth Taking," *CHE,* Nov. 10, 1993, A16–A17; Julianne Basinger, "How Nan Keohane Is Changing Duke," ibid., Nov. 3, 2000, A35–A37; and Piper Fogg, "Duke's President to Leave Office," ibid., Mar. 14, 2003, A28. On Coleman, see "Mary Sue Coleman, President, University of Michigan," www.umich.edu/pres/about.html; "Peer Review: A Michigan First," *CHE,* June 7, 2002, A10; and "Oral History of Mary Sue Coleman, July 31, 1997," by Tanya Zanish-Belcher, Women in Science and Engineering Archives, Special Collections, Iowa State University

Library. In 2003–4 Coleman was the best-paid president of a public university in the United States. Julianne Basinger and Sarah H. Henderson, "Hidden Costs of High Public Pay," *CHE,* Nov. 14, 2003, S3–S4. On Jackson, see Joshua Rolnick, "A New Challenge for a Theoretical Physicist," ibid., Jan. 8, 1999, A10; and Audrey Williams June, "Shirley Ann Jackson Sticks to the Plan," ibid., June 15, 2007, A24–A27. She was later billed as the best-paid college president in the country. Julianne Basinger, "Closing In on $1-Million," *CHE,* Nov. 14, 2003, S1; and Emma L. Carew and Paul Fain, "Paychecks Top More Than $1-Million for 23 Private-College Presidents," ibid., Nov. 6, 2009, A1, A30. Jackson surpassed Judith Rodin and Nannerl Keohane, the earlier best-paid women presidents. Nicklin, "Few Women Are among the Presidents With the Largest Compensation Packages."

44. On Wilkening, see Rich Elbaum, "UC Irvine Chancellor to Step Down Next Summer," Sept. 3, 1997, www.ucsc.edu/oncampus/currents/97-09-15/Irvine.html. On Moses, see http://leftspot.com/blog/?q=farewellmoses; and Yolanda T. Moses, "Linking Ethnic Studies to Women's Studies," in *The Politics of Women's Studies: Testimony from Thirty Founding Mothers,* ed. Florence Howe (New York: Feminist Press, 2000), 316–24. On Ramaley, see *AMWS,* 19th ed., 1995–96, 6:45; Ana Marie Cox, "U. of Vermont President Quits under Pressure," *CHE,* Feb. 23, 2001, A38; and Audrey Williams June, "An Interim President Can Be More Than a Caretaker," ibid., Nov. 15, 2002, A31–A32, about the wonders wrought by her successor.

45. On Reynolds, see *AMWS,* 19th ed., 1995–96, 6:176; William Trombley, "Governor Played Key Role in Reynolds' Resignation," *Los Angeles Times,* Apr. 27, 1990, A38; Mary Crystal Cage, "California State U. Chief Resigns amid Controversy over Raises for Administrators; System Is Still Plagued by Political Problems," *CHE,* May 2, 1990, A20, A22–A23; Samuel Weiss, "Californian Is Approved as Chancellor of CUNY," *NYT,* June 2, 1990, 30; Michael Lev, "Chancellor Used to Fray, Wynetka Ann Reynolds," ibid., 29, 30; Jeffrey Selingo, "A Battle-Scarred President Finds a Home in a Surprising Setting," *CHE,* Jan. 28, 2000, A36–A37; and Peter Schmidt, "Ann Reynolds to Resign in 2002, Years Early," ibid., Sept. 21, 2001, A24. It may be a coincidence that Reynolds, Jewel Plummer Cobb, and the AWIS founder and Northwestern University biologist Neena Schwartz all spent some time in the biological sciences at the University of Illinois Medical School in Chicago in the 1950s and 1960s. For more on the unusual atmosphere in the school at the time, see "Negro Profs at White Medical School," unidentified clipping, Jewel Plummer Cobb Papers.

46. On Albino, see Courtney Leatherman, "Bitter Feelings in Boulder," *CHE,* Dec. 1, 1995, A26, A28–A29; and, for her own view of the situation, Judith Albino, "Judith E. N. Albino," in *Models of Achievement: Reflections of Eminent Women in Psychology,* vol. 3, ed. Agnes N. O'Connell (Mahwah, NJ: Erlbaum, 2001), 219–37. On Denton, see Paul Fain, "In Apparent Suicide, Chancellor Dies in a Fall," *CHE,* July 7, 2006, A1, A28–A29; idem, "Too Much, Too Fast; Denise [*sic*] Denton made a rapid rise to become a university chancellor. Then she lept to her death. Why?," ibid., Jan. 19, 2007, A24–A27; and "Letters: The Suicide of a Chancellor," ibid., Feb. 23, 2007, A47.

47. On Lyall, see Kit Lively, "Lowering the Decibels: Head of Vast U. of Wis. System Wins Points for Hard Work and Unfailing Courtesy," *CHE,* May 18, 1994, A23–A24.

48. NSF, *Women & Science: Celebrating Achievements, Charting Challenges, Confer-*

ence Report, March 1997 (Arlington, VA, 1997); hrrp://www.ehr.nsf.gov/conferences/women95.htm. Interestingly in light of later events, the former deputy director, Anne Peterson, used the term *transformation* in her plenary address. Sue V. Rosser ran POWRE for three years, 1997–99. Sue V. Rosser, "Balancing: Survey of Fiscal Year 1997, 1998, and 1999 POWRE Awardees," *Journal of Women and Minorities in Science and Engineering* 7 (2001): 9–18. See also Sue V. Rosser and Eliesh O'Neil Lane, "A History of Funding for Women's Programs at the National Science Foundation: From Individual POWRE Approaches to the ADVANCE of Institutional Approaches," *Journal of Women and Minorities in Science and Engineering* 8 (2002): 327–46; and Sue V. Rosser, *The Science Glass Ceiling: Academic Women Scientists and the Struggle to Succeed* (New York: Routledge, 2004).

49. Kempf, "EmPOWREment and ADVANCEment for Women."

50. Ann Gibbons, "Women in Science—Congress Focuses on Job Discrimination," *Science* 257 (1992): 23; Shari Rudavsky, "Recalling Tales of Sex Discrimination," *Washington Post,* July 1, 1992, A21; Judy Green, "Hearing on the House Bill on Women in Science," *AWM Newsletter* 22 (Sept.–Oct. 1992): 9–10; Jenny Harrison, letter to the editor, ibid. 22 (Nov.–Dec. 1992): 4–5.

51. "Sex and Science," *Science* 280 (1998): 367; "US Panel to Advise on Women in Science," *Nature* 395 (Oct. 22, 1998): 736; Jeffrey Mervis, "Efforts to Boost Diversity Face Persistent Problems," *Science* 284 (1999): 1757, 1759; idem, "Diversity: Easier Said Than Done," ibid. 289 (2000): 378–79; *Land of Plenty: Diversity as America's Competitive Edge in Science, Engineering, and Technology; Report of the Congressional Commission on the Advancement of Women and Minorities in Science, Engineering and Technology Development* (n.p., Sept. 2000). Morella's papers are in the Historical Manuscripts, Special Collections, University of Maryland, College Park, and copies of her political advertisements are in the Julian P. Kanter Political Commercial Archive, Department of Communication, University of Oklahoma, Norman. Morella may have known Rita Colwell, also from Maryland.

52. Maxine Singer, "Dynamics of Change: Strategies for the 21st Century," keynote address, NIH Conference on Women in Biomedical Careers, Bethesda, MD, June 11, 1992, *Journal of Women's Health* 1, no. 3 (1992): 239–42.

Chapter 12 · Taking the Scientific Societies beyond Recognition

1. "Affiliated Organizations," in *AAAS Handbook: 2000–01* (Washington, DC: AAAS, 2000), 86–90.

2. See Margaret W. Rossiter, *Women Scientists in America: Struggles and Strategies to 1940* (Baltimore: Johns Hopkins Univ. Press, 1982), ch. 4; and idem, *Women Scientists in America: Before Affirmative Action, 1940–1972* (Baltimore: Johns Hopkins Univ. Press, 1995), ch. 14.

3. On the Chemists' Club, see Rossiter, *Women Scientists in America: Before Affirmative Action, 1940–1972,* 309, 380; and William Z. Lidicker Jr., "An Essay on the History of the Biosystematists of the San Francisco Bay Area," in *Cultures and Institutions of Natural History: Essays in the History and Philosophy of Science,* ed. Michael T. Ghiselin and Alan E. Leviton (San Francisco: California Academy of Sciences, 2000), 322–23. On

the Nuttall Ornithological Club, see Rossiter, *Women Scientists in America: Before Affirmative Action, 1940–1972,* 142; and William E. Davis Jr., *History of the Nuttall Ornithological Club, 1873–1986* (Cambridge, MA: Nuttall Ornithological Club, 1987), 89–92. Of course those women who were excluded were angrier and more bitter than the male members of the clubs, who often later claimed to have been unaware of the exclusionary practices.

4. Josh Barbanel, "Explorers to Admit Women," *NYT,* Apr. 13, 1981, B4; "Explorers Club Discovers Women," ibid., Apr. 19, 1981, E7; George Van B. Coleman, "President's Statement," *Explorers Journal* 60 (Mar. 1982): [27].

5. See, e.g., "Citing Female Bar, Brown Won't Dine at Cosmos," *Science* 179 (1973): 667; Daniel S. Greenberg, "Cosmos Club Battling Again on Women Members," *Science and Government Report* 15 (Jan. 15, 1985): 6–7; and Lawrence Feinberg, "18 Women End Cosmos Club's 110-Year Male Era," *Washington Post,* Oct. 12, 1988, B3. The saga can be followed in the *Washington Post* via keywords on the electronic ProQuest Historical Newspapers database. The astronomer Vera Rubin battled the Cosmos Club's membership policies in various ways at least as early as 1972. Vera C. Rubin Papers, Manuscript Division, Library of Congress, Washington, DC. In July 1988 William Golden offered to nominate her for membership. Golden was a wealthy investment banker and an important philanthropist and trustee in Washington and New York City. Starting about 1988, perhaps under the influence of his second wife, the mathematician Jean E. Taylor, he nominated several women for high honors and top posts. He may deserve a biography. See Dennis Overbye, "William T. Golden, Financier and Key Science Adviser, Is Dead at 97," *NYT,* Oct. 9, 2007, A29.

6. "Scientific Manpower Commission Surveys AAAS Affiliates," *Science* 196 (1977): 1976–77; John D. Hogan and Virginia Staudt Sexton, "Women and the American Psychological Association," *Psychology of Women Quarterly* 15 (1991): 630; "Gender Survey of the APS Membership—1990," *CSWP Gazette* 12 (Jan. 1993): 5–6; "Prime Numbers," *CHE,* July 11, 2003, A91 (on astronomy).

7. APA, Committee on Women in Psychology, *Women in the American Psychological Association* (Washington, DC, 1984), 7; APA, Women's Programs Office, Public Interest Directorate, *Women in the American Psychological Association, 1988* (Washington, DC, 1988), 10.

8. Entomological Society of America web page, http://entsoc.org/awards/honors/fellows_list.htm; Ellen R. M. Druffel, "Looking at Gender Distribution among AGU Fellows," *Eos,* Sept. 13, 1994, 42.

9. Society of Experimental Psychologists web page, www.sepsych.org/fellows.html; Frances K. Graham, "Frances K. Graham," in *Models of Achievement: Reflections of Eminent Women in Psychology,* vol. 2, ed. Agnes N. O'Connell and Nancy Felipe Russo (Hillsdale, NJ: Earlbaum, 1988), 182; Georgine M. Pion et al., "The Shifting Gender Composition of Psychology," *American Psychologist* 51 (May 1996): 515 (table 2).

10. Rossiter, *Women Scientists in America: Struggles and Strategies to 1940,* ch. 11.

11. Rossiter, *Women Scientists in America: Before Affirmative Action, 1940–1972,* 352–53; [E.] Margaret Burbridge, "Moving to Washington, the Cannon Prize, and Other Thoughts," in *The American Astronomical Society's First Century,* ed. David H. DeVorkin (Washington, DC: American Astronomical Society, 1999), 146–47; E. Margaret

Burbridge to Dr. Laurence W. Frederick, May 27, 1971, and I. R. King to Women in Berkeley Astronomy, Oct. 26, 1972, enclosing a copy of the brief report of the Special Annie J. Cannon Prize Committee, both in carton 45, Astronomy, Elizabeth L. Scott Papers, Bancroft Library, University of California, Berkeley; "The Stargazer," *Time,* Mar. 20, 1972, 38.

12. "The Annie Jump Cannon Award," in *Idealism at Work: AAUW Educational Foundation Programs, 1967–1981,* by Doris C. Davies (Billings, MT: AAUW Educational Foundation, 1981), 368. For a list of Cannon prizewinners, see www.aas.org/grants/awards.html. On Tinsley, see Edward Hill, *My Daughter Beatrice: A Personal Memoir of Dr. Beatrice Tinsley, Astronomer* (New York: APS, 1986), 67, 75. On Garmany, see Benjamin F. Shearer and Barbara S. Shearer, eds., *Notable Women in the Physical Sciences: A Biographical Directory* (Westport, CT: Greenwood, 1997), 117–21, quotation from 120). In the 1980s the AAS established an international award for "exceptionally creative or innovative" contributions, open to members of both sexes, in Tinsley's memory. As of 2010 it had been awarded to one woman (S. Jocelyn Bell Burnell, of England) and fifteen men. www.aas.org/grants/awards.html.

13. Nina Roscher, "Women Chemists," *CHEMTECH* 6 (Dec. 1976): 738–43; Molly Gleiser, "The Garvan Women," *Journal of Chemical Education* 62 (Dec. 1985): 1065–68. Caserio was also intrigued that so many had denied that discrimination had been a factor in their careers, for she was sure that it had been. Marjorie C. Caserio, "Pro," *Women Chemists,* Mar. 1976, 3. On Caserio, see Nancy Allee, "Marjorie C. Caserio (1929–)," in Shearer and Shearer, *Notable Women in the Physical Sciences,* 46–51. One additional, unmentioned advantage of winning awards like the Garvan Medal is that it increases one's chances of being included in biographical dictionaries like this one.

14. Jean'ne M. Shreeve, "Con," *Women Chemists,* Mar. 1976, 3–4. On Shreeve, see Cassandra S. Gissendanner, "Jean'ne Marie Shreeve (1933–)," in Shearer and Shearer, *Notable Women in the Physical Sciences,* 367–73. Anna Harrison was similarly conflicted on women's issues and admittedly always felt uneasy about the women-only Garvan Medal.

15. In 1981 the WCC prepared a booklet on past Garvan Medalists. (Maureen Chan, "Atlanta Meeting Highlights," *Women Chemists,* Aug. 1981, 2.) There is a list of past winners on the ACS website, http://portal.acs.org/portal/acs/org. Because so few women were being nominated for ACS awards, the WCC began about 2005 to encourage efforts to submit more, holding two sessions on the subject at the society's spring 2008 meeting. Patricia Ann Mabrouk, "Everything You Ever Wanted to Know About Awards," *Analytical and Bioanalytical Chemistry* 391 (Aug. 2008): 2373–76. In 2009, of the seventy-five persons winning awards from the ACS, eleven to thirteen had identifiably female names. Linda Wang, "ACS 2009 National Award Winners," *C&EN* 86 (Sept. 8, 2008): 56–58.

16. The story behind the Federal Woman's Award can be followed in the *Washington Post,* now available electronically in the ProQuest Historical Newspapers database, and in *Women in Action,* the newsletter of the Federal Women's Program.

17. See www.aaauw.org/education/fga/awards/aa.cfm. In 1972 the AAUW created a second prize, its Recognition Award for Emerging Scholars, which honored thirty-six young women, twenty-three of whom were scientists, engineers, or social scientists, be-

fore being discontinued in 2007. "AAUW Recognition Award for Emerging Scholars," www.aauw.org/educaion/fga/awards/raes.cfm.

18. See the AWM, CSWP, and AWIS websites: http://sites.google.com/site/awm math; www.aps.org/about/governance/committees/cswp/index.cfm; and www.awis.org. Also of interest was the wide publicity accorded these separate women's prizes in the mainstream scientific media. For example, the AWM's Satter Prize was written up in the main mathematical journals with a picture of the winner, as if it were a prize given to members of both sexes by the American Mathematical Society. Thus a prize won by a member of a protected group was not marginalized and ignored but brought before the group as a whole as an award worthy of everyone's notice. On the origins of the Satter Prize, see Joan S. Birman, née Joan Sylvia Lyttle, "Ruth Lyttle Satter," *AWM Newsletter* 20 (Mar.–Apr. 1990): 3–4.

19. The Weizmann Women & Science Award was last given in 2006. See http://en.wikipedia.org/wiki/Weizmann_Women_%26_Science_Award. There are related items in boxes 58, 63, and 64 of the Maxine Singer Papers, Manuscript Division, Library of Congress.

20. Maria Mitchell Association web page, www.mmo.org/category.php?cat_id=14.

21. On the L'Oréal-UNESCO Awards, see the booklet *Beating the Odds: Remarkable Women in Science,* by Laura Bonetta and Julie P. Clayton (n.p., 2008). L'Oréal's foundation also provides fellowships for women graduate students in the United States and abroad and supports several programs for girls in science.

22. See "The Birth of an Award," *BenchMarks,* Dec. 17, 2004, available online at www.rockefeller.edu/benchmarks/benchmarks_121704_c.php; Nicholas Wade, "3 Share Nobel Prize in Medicine for Studies of the Brain," *NYT,* Oct. 10, 2000, A22; and Claudia Dreifus, "He Turned His Nobel into a Prize for Women," ibid., Sept. 26, 2006, F5, F8.

23. Rossiter, *Women Scientists in America: Before Affirmative Action, 1940–1972,* 353–54.

24. Jody Jacobs, "ARCS Man of Science Is a Woman," *Los Angeles Times,* June 25, 1974, pt. 4, C2; Frances Russell Kay, "Woman is 'Man of Science,' Gets ARCS 'Star' Award," unidentified clipping, n.d., and Margie Daniels Oster (who renamed the award), "A[chievement] R[ewards for] C[ollege] S[cientists] Presents Its First 'Woman of Science' Award," *Evening Outlook,* June 27, 1974, both in box 81, Dixy Lee Ray Papers, Library and Archives, Hoover Institution, Stanford University, Stanford, CA. Similar gender mixups involved other prominent women scientists, including the computer expert Grace Murray Hopper, the political scientist Donna Shalala, and the aeronautical engineer Sheila Widnall.

25. "Women of the Year: Great Changes, New Chances, Tough Choices," *Time,* Jan. 5, 1976, 11.

26. Sandra Lee Stuart, comp., *Who Won What When: The Record Book of Winners* (Secaucus, NJ: Lyle Stuart, 1980), 105–6; "Ladies' Home Journal Women of the Year, 1974," *Ladies' Home Journal,* n.d., 81–83; photos, script, letters of congratulation, and critical commentary by TV columnist Bob MacKenzie, "KPIX Dumps Milt Kahn," *Oakland Tribune,* Apr. 10, 1974, in box 83 of Dixy Lee Ray Papers. A live television broadcast was quite stressful, but in this case it was especially so for the moderator, Bess Meyerson Grant, who was scheduled to undergo a hysterectomy the next day. Jennifer

Preston, *Queen Bess: An Unauthorized Biography of Bess Meyerson* (Chicago: Contemporary Books, 1990), 103.

27. Rosalyn Yalow, "Thank You, but No Thank You," *NYT,* June 12, 1978, A19; Lenore Hershey, "Why Woman of the Year Awards Are Not Outdated," ibid., June 20, 1978, A16. This episode is not mentioned in Eugene Straus, *Rosalyn Yalow, Nobel Laureate: Her Life and Work in Medicine; A Biographical Memoir* (New York: Plenum Trade, 1998). For more on Yalow, see *AMWS,* 19th ed., 1995–96, 7:964; the autobiographical "Biomedical Investigation," in *The Joys of Research,* ed. Walter Shropshire Jr. (Washington, DC: Smithsonian Institution Press, 1981), 101–15; idem, "What Being Jewish Means to Me," *NYT,* Mar. 7, 1993, sec. 4, 17; Elizabeth Stone, "A Mme. Curie from the Bronx," *New York Times Magazine,* Apr. 9, 1978; and Carol Kahn, "Rosalyn Yalow, Bronx Housewife: She Cooks, She Cleans, She Wins the Nobel Prize," *Family Health* 10 (June 1978): 24–27. Anna Harrison agreed with Yalow, saying that science was done by people and not by men and women. W. O. Baker to Harrison, June 16, 1978, in box 43, folder 4, Anna Jane Harrison Papers, Mount Holyoke College Archives, South Hadley, MA.

28. "America's Most Important Women," *Ladies' Home Journal,* Nov. 1988, 47–56, 221–26 (12 of the 100 women were scientists or social scientists); Andrea Gross, "The Ten Most Important Women in Medicine," ibid., Mar. 1995, 28–32; "The Most Fascinating Women of the Year," ibid., Jan. 1997, 42–47; Myrna Blyth, "100 Most Important Women of the 20th Century," ibid., May 1999, 56–64; "theljh100," ibid., Nov. 1999, 156–80.

29. Kathy A. Svitil, "The Most Important Women in Science," *Discover* 23 (Nov. 2002): 52–57.

30. Rose M. Johnstone and C. P. Lee, "Why So Few Women Bioscientists at the Podium?" *Scientist,* May 4, 1987, 12; Jessica Gurevitch, "Differences in the Proportion of Women to Men Invited to Give Seminars: Is the Old Boy Still Kicking?" *Bulletin of the Ecological America* 69 (Sept. 1988): 15–60, enclosed with Margaret B. Davis to Dr. [*sic*] Erich Bloch, Nov. 28, 1988, in box 1, Women in Science, Margaret Bryan Davis Papers, University Archives, University of Minnesota, Minneapolis; Marcia Clemmitt, "Toughest Federal Science Jobs Elude Women," *Scientist,* Oct. 15, 1990, 13.

31. Arthur A. Daemmrich, Nancy Ryan Gray, and Leah Shaper, eds., *Reflections from the Frontiers, Explorations for the Future: Gordon Research Conferences, 1931–2006* (Philadelphia: Chemical Heritage Foundation, 2006); items in box 28, folder 8, and box 35, Charlotte Friend Papers, Archives, Mount Sinai Medical Center, New York; Frederick Becker to Charlotte Friend, Jan. 26, 1983, box 21, folder 4 (on hiring Margaret Kripke, whom he had met at an earlier Gordon Conference), ibid. See also Phyllis R. Brown, "Women in Analytical Chemistry—Why So Few?" *Trends in Analytical Chemistry* 5, no. 2 (1986): iv.

32. Jennie Dushek, "Female Primatologists Confer—Without Men," *Science* 249 (1990): 1494–95; "Letters: Sexism and Hypocrisy," ibid. 250 (1990): 887.

33. Robin Wilson, "Papers and Pampers: The Challenges of Attending a Scholarly Meeting, Children in Tow," *CHE,* Dec. 13, 2002, A8–A10.

34. Robert A. Harte, "26 July 1973, MEMO TO FILE RE: *Sharon Johnson versus the University of Pittsburgh,*" in box 58, folder 6, Papers of the American Society for Biochemistry and Molecular Biology, Special Collections, Albin O. Kuhn Library and

Gallery, University of Maryland–Baltimore County, Catonsville, MD; items in box 7, folder 5, Legal Aid Loan Applications, 1976–77, Anna Jane Harrison Papers; correspondence and legal documents about Gleiser's loan from the ACS in box 3, folder 3, ACS—Committee on Professional and Member Relations, ibid.

35. See ch. 1, n. 26.

36. Mildred S. Dresselhaus, Judy R. Franz, and Bunny C. Clark, "Interventions to Increase the Participation of Women in Physics," *Science* 263 (1994): 1392–93. See also ch. 6.

37. Felisa A. Smith and Dawn M. Kaufman, "A Quantitative Analysis of the Contributions of Female Mammalogists from 1919 to 1994," *Journal of Mammalogy* 77, no. 3 (1996): 622.

38. Barbara Krohn, "Women Engineers: Extending Their Influence through the Technical Societies," *SWE Newsletter,* Jan.–Feb. 1977, 1–3, which has information on sixteen women rising in the ranks of the engineering societies; Rossiter, *Women Scientists in America: Before Affirmative Action, 1940–1972,* 309; "Women Hold All Top Spots in St. Louis Section," *C&EN* 64 (Aug. 4, 1986): 29; "Do Women Make the Difference?" *Women Chemists,* May 1988, 2.

39. "79 Jan. 3 Background Guidelines for Sept. Symposium in Washington, DC, on Role of Women in Chemistry," computer printout, in box 4 of the Anna Jane Harrison Papers, is a list of female officeholders in the ACS based on Herman Skolnick and Kenneth M. Reese, *A Century of Chemistry* (Washington, DC: ACS, 1976). There is also material there on Harrison's experience with the Division of Chemical Education.

40. Eileen Horn Stanley, "Mary Lowe Good (1931–), Chemist," in Shearer and Shearer, *Notable Women in the Physical Sciences,* 148–53. See also "Mary L. Good, Transcript of an Interview Conducted by James G. Traynham at Little Rock, Arkansas, on 2 June 1998," Chemical Heritage Foundation, Philadelphia. Good merits a full biography.

41. "Former Members of the Board of Directors (Excluding Presidents), 1964–2000," *AAAS Handbook: 2000–01,* 147–48. On Skillern, see Krohn, "Women Engineers"; and Hogan and Sexton, "Women and the American Psychological Association," 630–31.

42. "Support for the Nominations Process," *ASA Footnotes,* Apr. 1994, 8.

43. Numerous folders in boxes 1–5 of the Vera Kistiakowsky Papers, Institute Archives, MIT, Cambridge, MA, document her extensive involvement with the APS in the 1970s and 1980s.

44. Toby A. Appel, Marie M. Cassidy, and M. Elizabeth Tidball, "Women in Physiology," in *History of the American Physiological Society: The First Century, 1887–1987,* ed. John H. Broheck, Orr E. Reynolds, and Toby A. Appel (Bethesda, MD: American Physiological Society, 1987), 381–90.

45. "Who Am I?" *Science* 258 (1992): 1509; "Meetings and Presidents," *AAAS Handbook: 2000–01,* 144–45; Sigma Xi list on its web page, www.sigmaxi.org. See also "The Record," *American Scientist* 74 (1986): 551–52, for a list from 1893 to 1986–87.

46. Mary Ellen Ellsworth, "A History of the Connecticut Academy of Arts and Sciences, 1799–1999," *Transactions of the Connecticut Academy of Arts and Sciences* 55 (1999): 112, 153. The biologist Dorothea Rudnick, of Albertus Magnus College, served for more than forty years as the academy's reliable secretary, retiring in 1987.

47. Michele L. Aldrich, personal communication, Feb. 27, 2002.

48. Rosalyn Yalow, "Reflections of a Non-establishmentarian," *Endocrinology* 106 (1980): 412–14. For its first thirty-six years the Endocrine Society was named the Association for the Study of Internal Secretions. Neena Schwartz became president in 1982. See also Alfred E. Wilhelmi, "The Endocrine Society: Origin, Organization, and Institutions," ibid. 123 (1988): 13, 26–27.

49. Doris Malkin Curtis, "Guest Column—Rewards of Professional Participation," *Bulletin of the Houston Geological Society,* May 1978, 5. See also "Doris M. Curtis, 77; Led Geological Society," *NYT,* May 30, 1991, D20; Dorothy Jung Echols, "Memorial to Doris M. Curtis, 1914–1991," *Geographical Society of America, Memorials* 23 (1993): 175–83; and "Subaru Outstanding Woman in Science Award," www.geosociety.org/aboutus/awards/curtis.htm#past.

50. "Physics Society Elects First Woman President," *Anaheim California Bulletin,* Jan. 31, 1975, clipping in box 2, APS-CSWP, 1973–75, Vera Kistiakowsky Papers; C. S. Wu Papers, Columbiana Room, Low Library, Columbia University, New York; Ursula Allen, "Chien-Shiung Wu (1912–1997)," in Shearer and Shearer, *Notable Women in the Physical Sciences,* 423–29; memorandum, W. W. Havens Jr. to APS Executive/Budget Committee, Sept. 22, 1975, "Proposal for the Committee on Scientific Society Presidents to become incorporated in Washington as a 501C3 or 501C6 association," in box 1, APS Budget 1975, Vera Kistiakowsky Papers.

51. Harrison was on the road on ACS business for from five to fourteen days every month for two years while president-elect and then president. "Activities for 1977" and "Activities for 1978," both in box 9, folder 13, Anna Jane Harrison Papers. See also a note of appreciation, Mathilde J. Kland to Rebecca L. Rawls, Assistant Editor, C&EN, Dec. 16, 1977, box 4, Anna Jane Harrison Papers.

52. "In Memoriam," *Journal of Bacteriology* 173 (Jan. 1991).

53. See the website for the Society of Women Geographers, www.iswg.org/about.html. See also Sarah Lyall, "Women Meet for Accounts of Unknown," *NYT,* May 13, 1991, B3. On Crews, see www.geog.psu/swig/kim_crews.html.

54. The annual lists of newly elected members of the NAS, the NAE, and the IOM are published in *CHE* and appear on their websites. Formerly, *Science* and the *New York Times* also published them. The academies' full memberships were listed in NAS, NAE, IOM, and NRC, *Organization and Members* (Washington, DC: NAS, NAE, and IOM, [1995]), the last year it appeared in print.

55. "NAE Puts the Focus on Women," *Science* 277 (1997): 483.

56. Ibid.; "Engineering Societies Diversity Summit Summary Report," www.aes.org/diversity/summit_report.asp. See also France A. Córdova, "Projects of the National Academies on Women in Science and Engineering," in *Gendered Innovations in Science and Engineering,* ed. Londa Schiebinger (Stanford, CA: Stanford Univ. Press, 2008), 198–211. On Didion, see http://anitaborg.org/about/who-we-are/catherine-didion/. The NRC's CEEWISE is mentioned in chs. 1 and 8.

57. Rosalyn Yalow to Charlotte Friend, May 6, 1976, box 6, folder 10, Letters of Congratulation, Charlotte Friend Papers.

58. Elizabeth M. O'Hern to Charlotte Friend, Oct. 14, 1977, Friend to O'Hern, Oct. 20, 1977, and O'Hern to Friend, Nov. 29, 1977, all in box 4, folder 4, ibid.; Werner Henle

to Friend, Dec. 6, 1977, and Brigitte Henle to Friend, Apr. 26, 1979, box 24, folder 8, ibid.; and Salome Waelsch to Friend, Dec. 10, 1979, box 29, folder 14, ibid. On Waelsch, see *AMWS,* 19th ed., 1995–96, 7:453; and Michael T. Kaufman, "A Jew, a Woman and Still a Scientist," *NYT,* Feb. 6, 1993, I25.

59. Burbridge served a term as head of the astronomy section in late 1980s. In 1977, before she was elected to the academy, Rubin had tried to get Charlotte Moore Sitterly and Cecilia Payne Gaposchkin nominated to no avail. Vera Rubin to Leo Goldberg, Nov. 19, 1977, and Goldberg to Rubin, Nov. 22, 1977, box 9, Vera C. Rubin Papers. On Wasserman, see Barbara J. Love, comp., *Feminists Who Changed America, 1963–1975* (Urbana: Univ. of Illinois Press, 2006), 480.

60. Yvonne Lohes to Maxine Singer, July 2, 1979, box 54, folder 10, NAS, and nomination, Dec. 18, 1998, box 48, folder 4, Chemists—Barton, Jackie, both in Maxine Singer Papers. At some point the physicist Gertrude Goldhaber and others reportedly formed an informal women's committee, but records on this are not yet available. For more on NAS women, see Elga Wasserman, *The Door in the Dream: Conversations with Eminent Women in Science* (Washington, DC: Joseph Henry, 2000), which is based on a survey of 86 women members with 37 phone interviews and 24 written responses. The transcripts have been deposited at the Schlesinger Library, RIAS, Harvard University, Cambridge, MA. The book is arranged by decade of the women's birth and focuses in particular on the struggles of the married women members who, like her, were mothers.

61. Bryce Crawford Jr. to Members of the Academy, Apr. 6, 1984, and enclosure, box 7, folder 7, NAS, 1984, Charlotte Friend Papers. Friend had been working on getting such changes approved since 1978. Charlotte Friend to David Goddard, May 25, 1978, box 7, folder 2, NAS, 1978, ibid. Vera Rubin also fought for an end to sexist communications, using humor to make her point. Vera C. Rubin, letter to the editor, *Physics Today* 31 (Jan. 1978): 1–14.

62. Vera Rubin to Frank Press, Nov. 15, 1990, Rubin to all the other women in the academy, Nov. 19, 1990, and Press to Rubin, Nov. 29, 1990, all in box 18, Press Letter, Vera C. Rubin Papers; Frank Press to Margaret B. Davis, Dec. 19, 1990, box 5, Margaret Bryan Davis Papers. See also Philip Handler, "Women Scientists: Steps in the Right Direction," *Sciences* 18 (Mar. 1978): 6–9.

63. Cécile DeWitt-Morette to Vera Kistiakowsky, Nov. 9, 1992, Kistiakowsky to DeWitt-Morette, Nov. 24, 1992, and related correspondence in box 10, Correspondence, 1990–93, Vera Kistiakowsky Papers.

64. On the Nobel Prizes awarded to women since 2000, see the epilogue.

65. On Yalow, see above, n. 27; Roy Reed, "Feminist Tone Colors Speech By a Laureate," *NYT,* Dec. 11, 1977, 7; and Laurie Honston, "A Bronx Public School Product," ibid., Oct. 14, 1977, 18. See also Rosalyn S. Yalow, "Radioimmunoassay: A Probe for the Fine Structure of Biologic Systems," *Science* 200 (1978): 1236–45; and idem, "Men and Women Are Not the Same," *NYT,* Jan. 31, 1982, E23.

66. On McClintock, see John Nobel Wilford, "A Brilliant Loner in Love with Genetics," *NYT,* Oct. 11, 1983, C7; Lawrence K. Altman, "Long Island Woman Wins Nobel in Medicine," ibid., Oct. 11, 1983, A1, C6; Barbara McClintock, "The Significance of Responses of the Genome to Challenge," *Science* 226 (1984): 792–801; and Gina Bari

Kolata, "Dr. Barbara McClintock, 90, Gene Research Pioneer, Dies," *NYT,* Sept. 4, 1992, A1, D16. See also Evelyn Fox Keller, *A Feeling for the Organism: The Life and Work of Barbara McClintock* (New York: W. H. Freeman, 1983); Nina Fedoroff and David Botstein, *The Dynamic Genome: Barbara McClintock's Ideas in the Century of Genetics* (Cold Spring Harbor, NY: Cold Spring Harbor Press, 1992); and Nathaniel C. Comfort, *The Tangled Field: Barbara McClintock's Search for the Patterns of Genetic Control* (Cambridge, MA: Harvard Univ. Press, 2001).

67. On Levi-Montalcini, see Harold M. Schmeck Jr., "2 Pioneers in Growth of Cells Win Nobel Prize," *NYT,* Oct. 14, 1986, A1, C3; Roberto Suro, "Unraveler of Mysteries: Rita Levi-Montalcini," ibid., C6; Joan Arehart-Treichel, "Rita Levi-Montalcini: The Woman Who Started It All," *Science News,* May 21, 1977, 331; Rita Montalcini, "Reflections on a Scientific Adventure," in *Women Scientists: The Road to Liberation,* ed. Derek Richter (London: Macmillan, 1982), 99–117; and Rita Levi-Montalcini, *In Praise of Imperfection: My Life and Work* (New York: Basic Books, 1988).

68. On Elion, see *AMWS,* 16th ed., 1989, 2:867; Lawrence K. Altman, "Gertrude Elion, Drug Developer, Dies at 81," *NYT,* Feb. 23, 1999, A21; Richard Kent and Brian Huber, "Obituary: Gertrude Belle Elion (1918–99), Pioneer of Drug Discovery," *Nature* 398 (Apr. 1, 1999): 380; Sharon Bertsch McGrayne, "Damn the Torpedos, Full Speed Ahead!" *Science* 296 (2002): 851–52; Gertrude B. Elion, "The Quest for a Cure," *Annual Review of Pharmacology and Toxicology* 33 (1993): 1–23; idem, "Personal Reflections," *Annals of the New York Academy of Sciences* 869 (Apr. 15, 1999): 16–18; and "A Celebration of the Life of Trudy Elion," Mar. 27, 1999, videotape of her memorial service (I thank Jon Elion for a copy).

69. Milton Friedman and Rose D. Friedman, *Two Lucky People: Memoirs* (Chicago: Univ. of Chicago Press, 1998), 232, 573; Anna Jacobson Schwartz, *Money in Historical Perspective* (Chicago: Univ. of Chicago Press, 1987); James Markham, "3 Nobels in Science—Jerome Karle," *NYT,* Oct. 17, 1985, 17.

70. See Chow's web page, http://main.uab.edu/Sites/biochemistry/faculty/primary/6018/; Anthony Flint, "Behind Nobel, A Struggle for Recognition: Some Scientists Say Colleague of Beverly Researcher Deserved a Share of Medical Prize," *Boston Globe,* Nov. 5, 1993, 1; and Victor McElheny, *Watson and DNA: Making a Scientific Revolution* (Cambridge, MA: Perseus, 2003), 209–11. See also Jon Cohen, "The Culture of Credit," *Science* 268 (1995): 1706–11.

71. See the autobiographical Candace B. Pert, *Molecules of Emotion: Why You Feel the Way You Feel* (New York: Scribner, 1997), 48; and "Interview with Susan Solomon, 5 September 1997," by Dale Kellogg, American Meteorological Society/University Corporation for Atmospheric Research Tape Recorded Interview Project, transcript, National Center for Atmospheric Research Archives, Boulder, CO (I thank Diane Rabson for a copy).

72. For winners from 1962 to 2000, see *The National Medal of Science, 2002,* booklet (Washington, DC: NSF, 2001). Subsequent winners are listed on the website of the National Science & Technology Medals Foundation, www.nationalmedals.org/medals/laureates.php.

73. Items in box 3, folder 44, National Medal of Science, C. S. Wu Papers. There are some files on C. S. Wu in box MA-32, National Medal of Science, White House Central

Files, Gerald Ford Presidential Library, Bentley Library, University of Michigan, Ann Arbor; and there are files on the awardees Katherine Esau, Rosalyn Yalow, and Mildred Dresselhaus in the speechwriting materials of the cabinet-affairs officer Douglas Adair at the George H. W. Bush Presidential Library, College Station, TX.

74. See www.ta.doc.gov/Medal/Recipients-2000.htm (listing winners since 1985); Libby Nelson, "U. at Buffalo Professor, a Prolific Inventor, Is Honored by White House," *CHE,* Oct. 23, 2009, A36; and "Interview with Stephanie Kwolek," 21, by Laura Sweeney, Women in Science and Engineering Archives, Special Collections, Iowa State University Library, Ames.

Epilogue · A New Era of Institutional Contrition and "Transformation"

1. Kate Zernike, "MIT Women Win a Fight against Bias; In Rare Move School Admits Discrimination," *Boston Globe,* Mar. 21, 1999, A1; Carey Goldberg, "M.I.T. Acknowledges Bias against Female Professors," *NYT,* Mar. 23, 1999, A1, A16; "A Study on the Status of Women Faculty in Science at MIT," *MIT Faculty Newsletter* 11, no. 4 (Mar. 1999): special edition, also online at http://web.mit.edu/fnl/women/women.html. See also Nancy Hopkins, "MIT and Gender Bias: Following Up on Victory," *CHE,* June 11, 1999, B4–B5; Andrew Lawler, "Tenured Women Battle to Make It Less Lonely at the Top," *Science* 286 (1999): 1272–78; Robin Wilson, "An MIT Professor's Suspicion of Bias Leads to a New Movement for Academic Women," *CHE,* Dec. 3, 1999, A16–A18; "Women Find It's Still a Man's World at MIT," *Nature* 416 (Mar. 28, 2002): 359; "The MIT Success Story: Interview with Nancy Hopkins," by Raquel Diaz-Sprague, *AWIS Magazine* 32 (Winter 2003): 10–15; and "Oral History of Nancy Hopkins, Apr. 13, 2002," by Laura Sweeney, Women in Science and Engineering Archives, Special Collections, Iowa State University Library, Ames.

2. Lawler, "Tenured Women Battle"; Nancy Hopkins, "Experience of Women at the Massachusetts Institute of Technology," in NRC, Commission on Physical Sciences, Mathematics and Applications, Board on Chemical Sciences and Technology, Chemical Sciences Roundtable, *Women in the Chemical Workforce: A Workshop Report to the Chemical Sciences Roundtable* (Washington, DC: National Academy Press, 2000), 110–24. See also Kate Zernike, "The Reluctant Feminist," *NYT,* Apr. 8, 2001, ED34.

3. Hopkins, "MIT and Gender Bias"; Wilson, "MIT Professor's Suspicion of Bias"; Natasha Loder, "US Science Shocked by Revelations of Sexual Discrimination," *Nature* 405 (June 8, 2000): 713–14; "Women Find It's Still a Man's World."

4. Paula Wasley, "Women's Gains in Sciences at MIT Have Stalled, Study Finds," *CHE,* Apr. 28, 2006, A14.

5. The Nobelist Susumu Tonegawa, head of the Picower Institute for Learning and Memory, had reportedly bullied a female who had been offered a post in the biology department into not accepting the job. Marcella Bombardieri and Gareth Cook, "MIT Star Accused by 11 Colleagues," *Boston Globe,* July 15, 2006, A1, A4. Several months later he stepped down from the directorship of the institute but retained his faculty post. Marcella Bombardieri, "MIT Neuroscience Center Head Quits," ibid., Nov. 17, 2006, B1, B7. See also Linda K. Wertheimer, "Tenure at MIT Still Largely a Male Domain," ibid., Dec. 6, 2007, A1.

6. "Women Worse Off (cont.)," editorial, *Nature* 412 (Aug. 30, 2001): 841; Rex Dalton, "Staff Survey Shows Women Feel Out in the Cold at Caltech," ibid., 844; Andrew Lawler, "Caltech Aims for Big Jump in Women Faculty," *Science* 294 (2001): 2066–67; Anneila I. Sargent, "Committee on the Status of Women Faculty at Caltech, Final Report, Dec. 3, 2001," http://diversity.caltech.edu/documents/CSFWFINALREPORT1.pdf; "Reaffirming and Extending Caltech's Commitment to Attract and Retain a Diverse Faculty," http://diversity.caltech.edu/status_of_women.html.

7. Jeffrey Mervis, "High-Level Groups Study Barriers Women Face," *Science* 284 (1999): 727; CWSE, Office of Scientific and Engineering Personnel, NRC, *Who Will Do the Science of the Future? A Symposium on Careers of Women in Science* (Washington, DC: National Academy Press, 2000). See also Natalie Angier, "Women Swell Ranks of Science, But Remain Invisible at the Top," *NYT,* May 21, 1991, C1, C12; Daniel S. Greenberg, "Science: A Man's World," *Baltimore Sun,* Mar. 31, 1992, clipping in box 18, [NAS] Annual Meeting 1992, Vera C. Rubin Papers; Faye Flam, "What Should It Take to Join Science's Most Exclusive Club?" *Science* 256 (1992): 960–61, pointing out that the academy had no written criteria for membership; and Jeffrey Brainard, "Elitism, Excellence, or Both at the National Academy of Sciences? Critics Question Why So Few Female and Minority Scholars Are Elected," *CHE,* May 11, 2001, A24–A26 (cover story).

8. Correspondence in box 64, folder 7, NAS Annual Meeting, Apr. 2002, Maxine Singer Papers; Natalie Angier, "No Parity Yet, but Science Academy Gains More Women," *NYT,* May 6, 2003, F2; Jeffrey Brainard, "National Academy of Sciences Elects 17 Women," *CHE,* May 9, 2003, A26; idem, "19 Women Elected to National Academy," ibid., May 13, 2005, A22; "Women in the National Academy," ibid., June 10, 2005, A8–A10; "New Members," ibid., May 5, 2006, A31; J.B., "National Academy Elects Fewer Women," ibid., May 11, 2007, A36.

9. Jamilah Evelyn, "Princeton Names Its First Female President," *CHE,* May 18, 2002, A32; Scott Smallwood and Karen Birchard, "Top 2 Leaders at Princeton Are Women," ibid., July 20, 2001, A7; Karen W. Arenson, "More Women Taking Leadership Roles at Colleges," *NYT,* July 43, 2002, A1. See also ch. 11, n. 30.

10. On Hockfield, see "Biography, Susan Hockfield, President, MIT," http://web.mit.edu/Hockfield/biography.html. On Gutmann, see Julianne Basinger, "Penn Picks Prez," *CHE,* Jan. 30, 2004, A7.

11. Zernike, "Reluctant Feminist." Ana Marie Cox and Robin Wilson, "Leaders of 9 Universities Pledge to Improve Conditions for Female Scientists," *CHE,* Feb. 9, 2001, A12. There was, however, a dearth of data about department chairs. Debbie A. Niemeier and Cristina González, "Breaking into the Guildmasters' Club: What We Know about Women Science and Engineering Department Chairs at AAU Universities," *NWSA Journal* 16 (Spring 2004): 157–71.

12. Cox and Wilson, "Leaders of 9 Universities Pledge." The nine were Caltech, Harvard, MIT, Princeton, Stanford, UC–Berkeley, the University of Michigan at Ann Arbor, the University of Pennsylvania, and Yale. John Hennessey, Susan Hockfield, and Shirley Tilghman, "Women and Science: The Real Issue," *Boston Globe,* Feb. 12, 2005, A13; Marcella Bombardieri, "3 University Chiefs Chide Summers on Remarks," ibid., A1, A6.

13. Karen W. Arenson, "Uneven Progress Is Found for Women on Princeton Science and Engineering Faculties," *NYT,* Sept. 30, 2003, B5; Robin Wilson, "Duke and

Princeton Will Spend More to Make Female Professors Happy," *CHE*, Oct. 10, 2003, A12; "Report of the Task Force on the Status of Women Faculty in the Natural Science and Engineering at Princeton, Submitted to President Shirley Tilghman on 5/22/03," www.princeton.edu/pr/news/03/q3/0929-science.htm. See also Robin Wilson, "Where the Elite Teach, It's Still a Man's World," *CHE*, Dec. 3, 2004, A8–A14.

14. "Donna J. Nelson, "Transcript of Interview Conducted by Hilary Domush and Leah Webb-Halpern at University of Oklahoma, Norman, Oklahoma, on 21 and 22 July 2008," 41–51, Chemical Heritage Foundation, Philadelphia; Janice R. Long, "Women Chemists Still Rare in Academia," *C&EN* 78 (2000): 56–57; Robin Wilson, "The Chemistry between Women and Science: 'A Culture That Tolerates Discrimination'; Donna J. Nelson," *CHE*, May 26, 2006, A10–A11.

15. Donna J. Nelson, "The Standing of Women in Academia," *Chemical Engineering Progress* 99 (Aug. 2003): 38S–41S; Robin Wilson, "Women Underrepresented in Sciences at Top Research Universities, Study Finds," *CHE*, Jan. 23, 2004, A9; Nelson Diversity Surveys. http://chem.ou.edu/~djn/diversity/top50html.

16. Jacobs had become a forceful feminist. Madeleine Jacobs, "Perspective: Challenges Await Women Chemists in the New Millennium," *C&EN* 76 (Sept. 21, 1998): 43–55; idem, "Perspective: More Ideas for Women Chemists," ibid. 76 (Nov. 9, 1998): 62–66.

17. Debra R. Rolison, "A Title IX Challenge," *C&EN* 78 (Mar. 13, 2000): 5; Robin Wilson, "The Chemistry between Women and Science: 'The Change is Glacial'; Debra R. Rolison," *CHE*, May 26, 2006, A10–A11.

18. Debra R. Rolison, "Title IX for Women in Academic Chemistry: Isn't a Millennium of Affirmative Action for White Men Sufficient?" and "Discussion," in NRC, Commission on Physical Sciences, Mathematics and Applications, Board on Chemical Sciences and Technology, Chemical Sciences Roundtable, *Women in the Chemical Workforce*, 74–93; idem, "Can Title IX Do for Women in Science and Engineering What It Has Done for Women in Sports?" The Back Page, *APS [American Physical Society] News Online*, May 2003; Debra Rolison, "Title IX as Change Strategy for Women in Science and Engineering . . . and What Comes Next," speech at conference "Women, Work and the Academy: Strategies for Responding to 'Post–Civil Rights Era' Gender Discrimination," Barnard Center for Research on Women, Barnard College, New York, Dec. 9–10, 2004, www.barnard.edu/bcrw/womenand work/Rolison.htm. See also Robert Coontz, "Debra Rolison Profile: Small Thinking, Electrified Froth, and the Beauty of a Fine Mess," *Science* 315 (2007): 787.

19. *Title IX and Science: Hearing Before the Subcommittee on Science, Technology, and Space of the Committee on Commerce, Science, and Transportation, United States Senate, 107th Congress, 2nd Session, October 3, 2002* (Washington, DC: GPO, 2005). The proceedings of the July hearings were not published.

20. U.S. Government Accountability Office, *Gender Issues: Women's Participation in the Sciences Has Increased, but Agencies Need to Do More to Ensure Compliance with Title IX*, GAO-04-639, Report to Congressional Requesters, July 2004, available online at www.gao.gov/new.items/d04639.pdf. See also Jeffrey Mervis, "Can Equality in Sports Be Repeated in the Lab?" *Science* 298 (2002): 356 and "Title IX: From Sports to Science and Engineering," *Chemical Engineering Progress* 99 (Aug. 2003): 19.

21. NRC, CWSE, Committee on the Guide to Recruiting and Advancing Women Scientists and Engineers in Academia, *To Recruit and Advance Women Students and Faculty in Science and Engineering* (Washington, DC: National Academies Press, 2006).

22. John Tierney, "A New Frontier for Title IX: Science," *NYT,* July 15, 2008, F1. See also "Donna J. Nelson, Transcript of Interview Conducted by Hilary Domush and Leah Webb-Halpern at University of Oklahoma, Norman, Oklahoma, on 21 and 22 July 2008," 58–59.

23. "Stanford Women Gain from Anonymous Gift," *Nature* 405 (May 11, 2000): 111; http://med.stanford.edu/diversity/recruiting/funding.html. In 2006 the venture capitalist Michelle Clayman gave Stanford $3 million to set up an endowment for its Institute for Research on Women and Gender, which had been established in the 1970s. Its former director Londa Schiebinger, a historian of science, has focused on issues related to the academic hiring of spouses and couples.

24. "Californian University Gets Money for Women," *Nature* 408 (Nov. 9, 2000): 129; "Bequests," *CHE,* Mar. 2, 2001, A33, which reported the sum as $26.5 million. See also Alvin P. Sanoff, "At U. of Southern California, a Support Network Helps Women in Science and Engineering," ibid., Sept. 29, 2006, B8; and Diane Krieger, "The WISE Women of Science," *USC Trojan Family Magazine,* Autumn 2009, 29–37 (I thank Mary Oates for a copy).

25. Sue V. Rosser, *The Science Glass Ceiling: Academic Women Scientists and the Struggle to Succeed* (New York: Routledge, 2004); Sue V. Rosser and Eliesh O'Neil Lane, "A History of Funding for Women's Programs at the National Science Foundation: From Individual POWRE Approaches to the ADVANCE of Institutional Approaches," *Journal of Women and Minorities in Science and Engineering* 8 (2002): 327–46.

26. Jeffrey Mervis, "NSF Searches for the Right Way to Help Women," *Science* 289 (2000): 379–81; idem, "NSF Program Targets Institutional Change," ibid. 291 (2001): 2063–64; "Transforming the Scientific Enterprise: An Interview with Alice Hogan," by Danielle LaVaque-Manty, in *Transforming Science and Engineering: Advancing Academic Women,* ed. Abigail J. Stewart, Janet E. Malley, and Danielle LaVaque-Manty (Ann Arbor: Univ. of Michigan Press, 2007), 21–27.

27. Rex Dalton, "Grants to Help Women Climb Academic Ladder," *Nature* 413 (Oct. 25, 2001): 761; Lisa M. Frehill, Cecily Jeser-Cannavale, and Janet E. Malley, "Measuring Outcomes: Intermediate Indicators of Institutional Transformation," in Stewart, Malley, and LaVaque-Manty, *Transforming Science and Engineering,* 298–317; Virginia Valian, *Why So Slow? The Advancement of Women* (Cambridge, MA: MIT Press, 1998).

28. Susan Sturm, "Gender Equity as Institutional Transformation," in Stewart, Malley, and LaVaque-Manty, *Transforming Science and Engineering,* 262–80; Danielle LaVaque-Manty and Abigail J. Stewart, "'A Very Scholarly Intervention': Recruiting Women Faculty in Science and Engineering," in *Gendered Innovations in Science and Engineering,* ed. Londa Schiebinger (Stanford, CA: Stanford Univ. Press, 2008), 165–81.

29. Burton Bollag, "Classroom Drama: A U. of Michigan Program Uses Theater to Teach Professors Sensitivity to Women and Minority Groups," *CHE,* July 19, 2005, A12–A14; Matthew Kaplan, Constance E. Cook, and Jeffrey Steiger, "Using Theatre to Stage Instructional and Organizational Transformation," *Change* 38 (May–June 2006): 32–39; Danielle LaVaque-Manty, Jeffrey Steiger, and Abigail J. Stewart, "Interactive

Theater: Raising Issues about the Climate with Science Faculty," in Stewart, Malley, and LaVaque-Manty, *Transforming Science and Engineering*, 204–22.

30. Audrey Williams June, "Colleges Look for New Ways to Help Women in Science," *CHE*, Jan. 29, 2010, A1, A8–A9; Abigail Stewart, e-mail to author, May 14, 2011.

31. Edwin G. Boring, "The Woman Problem," *American Psychologist* 6 (Dec. 1951): 679–82. See also Margaret W. Rossiter, *Women Scientists in America: Before Affirmative Action, 1940–1972* (Baltimore: Johns Hopkins Univ. Press, 1995), 45–48.

32. Yudhijit Bhattacharjee, "Harvard Faculty Decry Widening Gender Gap," *Science* 305 (2004): 1692; Robin Wilson and Piper Fogg, "Female Professors Say Harvard Is Not Granting Tenure to Enough Women," *CHE*, Oct. 1, 2004, A14; Yudhijit Bhattacharjee, "No Meeting of the Minds at Harvard on Women Faculty," *Science* 306 (2004): 389; "Tenure and Gender," *Harvard Magazine*, Jan.–Feb. 2005, 64–68.

33. Arlie Hochschild, *The Second Shift* (New York: Avon Books, 1989).

34. Marcella Bombardieri, "Summers' Remarks on Women Draw Fire," *Boston Globe*, Jan. 17, 2005, A1, B6; idem, "Summers Releases Debated Transcript," ibid., Feb. 18, 2005, A1, B6; Andrew Lawler, "Summers's Comments Draw Attention to Gender, Racial Gaps," *Science* 307 (2005): 492–93. Perhaps the best source of items about the Lawrence Summers affair is the website of the Women in Science and Engineering Leadership Institute at the University of Wisconsin (ironically an ADVANCE site), http://wisesli.engr.wisc.edu/news/Summers.html.

35. Lynn T. Singer et al. and Dorothy G. Swift, in "Letters—Different Opinions on Gender Differences," *CHE*, Feb. 18, 2005, A47. ADVANCE's program officer, Alice Hogan, had been in the group Summers addressed in January 2005. Maureen Byko, "Challenges and Opportunities for Women in Science and Engineering," *J[ournal] O[f] M[etals]* 57 (Apr. 2005). Sara Rimer, "For Women in the Sciences, the Pace of Progress at Top Universities is Slow," *NYT*, Apr. 15, 2005, A15, describes efforts at several institutions that received ADVANCE grants but fails to mention the NSF support.

36. Seventy-nine scientists signed a letter to the editor of *Science* denouncing Summers's remarks. Carol B. Muller et al., "Gender Differences and Performance in Science," *Science* 307 (2005): 1043; Marcella Bombardieri and David Abel, "Summers Gets Vote of No Confidence," *Boston Globe*, Mar. 16, 2005, A1, A20; Sara Rimer, "Professors, in Close Vote, Censure Harvard Leader," *NYT*, Mar. 16, 2005, A15.

37. Marcella Bombardieri, "Summers Sets $50m Women's Initiative," *Boston Globe*, May 17, 2005, A1, A13; Piper Fogg, "Harvard Committees Suggest Steps to Help Women; Summers Pledges $50 million for Effort," *CHE*, May 27, 2005, A8–A9.

38. Marcella Bombardieri, "Harvard Appoints Professor to Senior Diversity Post," *Boston Globe*, July 21, 2005, B4; "Diversity Director," *Harvard Magazine*, Sept.–Oct. 2005, 56–58; Harvard University, Office of the Provost, Faculty Development and Diversity, "End of Year Report, June 2006," online at www.faculty.harvard.edu.

39. Lawrence Summers, "Letter to the Harvard Community," Feb. 21, 2006, http://president.harvard.edu/speeches/2006/0221_summers.html; Robin Wilson and Paul Fain, "Lawrence Summers Quits as Harvard President in Advance of New No-Confidence Vote; Derek Bok to Step In," *CHE*, Feb. 21, 2006, online at http://chronicle.com/article/Lawrence-Summers-Quits-as/118596.

40. "Exit Gray, Enter Keohane," *Harvard Magazine*, Jan.–Feb. 2005, 68. Over the

summer Conrad Harper, the only minority member of the Corporation, had resigned after the other members voted Summers a substantial raise. Paul Fain, "Harvard Board Member Steps Down, Citing Clashes with President," *CHE,* Aug. 21, 2005, A29; "I Can No Longer Support the President," *Harvard Magazine,* Sept.–Oct. 2005, 58–59.

41. Marcella Bombardieri and Maria Sacchetti, "Champagne, Cheers Flow at Harvard," *Boston Globe,* Feb. 12, 2007, A1, B4; Robin Wilson, "Harvard's Historic Choice," *CHE,* Feb. 23, 2007, A1, A23.

42. Eliot Marshall "Harvard Puts Science Campus on Slow Track," *Science* 323 (2009): 1157; Yudhijit Bhattacharjee, "Movers: A Catch for Harvard," ibid., 1653; Steve Bradt, "Cherry A. Murray is Named Dean of SEAS," *Harvard University Gazette Online,* Mar. 10, 2009, www.news.harvard.edu/gazette/2009/03.12/99-seas.html.

43. NAS, NAE, and IOM, Committee on Science, Engineering, and Public Policy, Committee on Maximizing the Potential of Women in Academic Science and Engineering, *Beyond Bias and Barriers: Fulfilling the Potential of Women in Academic Science and Engineering* (Washington, DC: National Academies Press, 2007); Cornelia Dean, "Bias is Hurting Women in Science, Panel Reports," *NYT,* Sept. 19, 2006, A22.

44. On Buck, see Lawrence K. Altman, "Unraveling Enigma of Smell Wins Nobel for Two Americans," *NYT,* Oct. 5, 2004, A18; Alison Abbott, "Science of Smell Wins Medicine Nobel," *Nature* 430 (Oct. 7, 2004): 616; "Linda B. Buck—Curriculum Vitae," http://nobelprize.org/medicine/laureates/2004/buck-cv.html; "Linda B. Buck, Ph.D.," www.hhmi.org/research/investigators/buck_bio.html; and Barbara Berg, "Our Nobel Laureates: The Scent of Surprise," http://fhcrc.org/research/nobel/buck/article.html.

45. On Blackburn, see *AMWS,* 19th ed., 1995–96, 1:672; Nicholas Wade, "Three Americans Share Nobel for Work in Cell Biology," *NYT,* Oct. 6, 2009, 12; Gretchen Vogel and Elizabeth Pennisi, "U.S. Researchers Recognized for Work on Telomeres," *Science* 326 (2009): 212–13; and Catherine Brady, *Elizabeth Blackburn and the Story of Telomeres: Deciphering the Ends of DNA* (Cambridge, MA: MIT Press, 2007). On Greider, see http://en.wikipedia.org/wiki/Carol_W_Greider; Brady, *Elizabeth Blackburn and the Story of Telomeres,* ch. 4; and Liza Mundy, "Success is in Her DNA," *Washington Post,* Oct. 20, 2009, C1, which mentions her children and her life as a single mother.

46. On Ostrom, see www.indiana.edu/~workshop/people/ostromcv.htm; *Who's Who in America, 1994,* 48th ed. (New Providence, NJ: Marquis Who's Who, 1993), 2:2611; and Adrian Cho, "Laureates Analyzed Economics Outside Markets," *Science* 326 (2009): 347.

SELECTED BIBLIOGRAPHY

Archival and Manuscript Collections

Alaska

Fairbanks. University of Alaska. Rasmuson Library. Alaska and Polar Regions Collections. Helen L. Atkinson Collection. Oral history of Florence Weber.

California

Berkeley. University of California. Bancroft Library. Florence Nightingale David Papers. Elizabeth L. Scott Papers.

Los Angeles. UCLA Library. Special Collections. Oral histories of Margaret Kivelson and Clara Szego.

San Diego. Archives, Scripps Institution of Oceanography.

Simi Valley. Ronald Reagan Presidential Library.

Stanford. Stanford University. Hoover Institution. Library and Archives. American Council on Education Papers. Dixy Lee Ray Papers.

Yorba Linda. Richard Nixon Presidential Library. National Science Foundation file. Material on women appointees and awardees.

Colorado

Boulder. National Center for Atmospheric Research Archives. Oral histories of Joanne Simpson and Susan Solomon. Materials on NCAR women.

Connecticut

New Haven. Yale University Archives. Dorothy Horstmann Papers.

District of Columbia

American Association for the Advancement of Science Archives. Office of Opportunities Papers. Janet Welsh Brown Files.

Library of Congress. Manuscript Division. Margaret Mead Papers. Vera C. Rubin Papers. Maxine Singer Papers.

National Academy of Sciences. Archives. Records of the Office of Scientific Personnel—Conference on Women in Science and Engineering, General, 1973. Records of the Committee on Participation of Women in the National Research Council, 1974. Papers of the Committee on the Education and Employment of Women in Science and Engineering.

Smithsonian Institution Archives. Videotape interview with Margaret Geller.

Georgia
Atlanta
Emory University. Manuscripts, Archives, and Rare Book Library. Biographical Files. Augusta S. Cooper Papers. Dora Skypek Papers.
Jimmy Carter Presidential Library. Midge Costanza Files.

Hawaii
Honolulu. University of Hawaii at Manoa Library. Special Collections. Papers of the University of Hawaii Commission on the Status of Women.

Illinois
Evanston. Northwestern University Archives. Neena Schwartz Papers.

Iowa
Ames. Iowa State University Library. Special Collections. Women in Science and Engineering Archives. Papers of the Association for Women Geoscientists. Helen Clark Papers. Fann Harding Papers. Darleane Hoffman Papers. Rosalind Marimont Papers. Nina Matheny Roscher Papers. Mary Singleton Papers. Joan Stadler Papers. Rena F. Subotnik Papers. Kathleen M. Desmond Trehanovsky Papers. Betty Vetter Papers. Beatrice Watt Papers. Lee Anne Willson Papers. Oral histories of Anne Briscoe, Jeannette E. Brown, Jeanette Grasselli Brown, Veona Burton, Mildred Cohn, M. Sue Coleman, Helen Davies, Helen Free, Jenny Glusker, Mary Good, Mary Hamilton, Darleane Hoffman, Nancy Hopkins, Madeleine Jouillie, Judith Klinman, Stephanie Kwolek, Phoebe LeBoy, Barbara Low, Joanne Ravel, Nina Matheny Roscher, Frances Seabright, Mary Singleton, and Lucile Smith.

Maryland
Catonsville. University of Maryland–Baltimore County. Albin O. Kuhn Library and Gallery. Special Collections. Papers of the American Association of Biological Chemists. Papers of the American Association of Immunologists. Papers of the American Society for Biochemistry and Molecular Biology. Papers of the American Society for Cell Biology. Papers of the Biophysical Society.
College Park
American Institute of Physics. Center for the History of Physics. Niels Bohr Library. Nancy Roman Papers. Oral histories of Sandra Faber, Elsa Garmire, Cherry Ann Murray, Nancy Roman, and Sidney Wolff.
National Archives and Records Administration. National Science Foundation, Records of the Women in Science Project.

Massachusetts
Cambridge
Harvard University
Institute for Quantitative Social Sciences. Murray Research Archive Database. Records from Project Access and Joan Ruskus's evaluation of the NSF Vis-

iting Professorships for Women Program for Stanford Research Institute. Oral histories of Ruth Hubbard and Ruth Turner.

Radcliffe Institute for Advanced Study. Schlesinger Library. Association for Women in Psychology Papers. Genes and Gender Group Papers. Sharon L. Johnson Papers. Joanne Simpson Papers. Elga Wasserman Collection. Naomi Weisstein Papers. Papers of the Commission on the Status of Women at Harvard University. Oral history of Joanne Simpson.

Harvard University Archives. Ruth Hubbard Papers. Judith N. Shklar Papers.

Massachusetts Institute of Technology (MIT). Institute Archives. Vera Kistiakowsky Papers. Margaret MacVicar Papers. Oral histories of Mildred Dresselhaus, Ellen Henderson, Christine Jones, Vera Kistiakowsky, and Nancy Koppell.

Polaroid Corporation Archives. Clipping files.

Northampton. Smith College Archives. Department records. Honorary Degree Files.

South Hadley. Mount Holyoke College Archives. Afro-American Students/Alumnae. Chemistry Department Records, Series B. Anna Jane Harrison Papers. Honorary Degree Lists.

Wellesley. Wellesley College Archives. Alice Schafer Accession. Papers of the Association for Women in Mathematics. Papers of the President's Office. Science Department Records. "Self-Study Report," 1979. "Wellesley College: Self-Study for Reaccreditation," 1989. Oral histories of Carolyn Shaw Bell, Carolyn Elliott, Phyllis Fleming, Janet Brown Guernsey, Anne Cohen Heller, Elizabeth J. Rock, Alice Turner Schafer, Eleanor Webster, and Dorothea J. Widmayer.

Woods Hole. Marine Biological Laboratory Library. Ruth Sager Papers.

Michigan

Ann Arbor. University of Michigan. Bentley Library. Gerald Ford Presidential Library. White House Central Files on National Medal of Science. Assorted appointment folders in Name File, White House Records.

Midland. Post Street Archives (formerly Dow Chemical Archives). Clipping File.

Minnesota

Minneapolis. University of Minnesota.

Charles Babbage Institute Archives. Papers of the Association for Women in Computing.

University Archives. Margaret Bryan Davis Papers. Phyllis St. Cyr Freier Papers.

New York

Bronxville. Sarah Lawrence College. Archives. Faculty File.

Ithaca. Cornell University.

Kroch Library. Rare and Manuscript Collections. Barbara Hope Cooper Papers. Eleanor Gibson Papers. Cornell Eleven Papers.

Mathematics Library. *AWM Newsletter*s. Other reports.

New York City

Columbia University.

Butler Library. Rare Book and Special Collections Room. Oral histories of Estelle Ramey (two) and Florence Anderson.
Low Library. Columbiana Room. C. S. Wu Papers.
Mount Sinai Medical Center. Archives. Charlotte Friend Papers.
New York Public Library. Schomburg Center for Research in Black Culture. Jewel Plummer Cobb Papers.
Poughkeepsie. Vassar College Library. Special Collections. Jeanne R. Lowe Papers.

North Carolina
Durham. Duke University Archives. Biographical Files. Juanita Kreps File.

Ohio
Oberlin. Oberlin College Archives. Anna Ruth Brummett Papers.

Oklahoma
Norman. University of Oklahoma. Department of Communication. Julian P. Kanter Political Commercial Archive. Constance Morella's videotaped political advertisements.

Pennsylvania
Philadelphia. Chemical Heritage Foundation. Oral histories of Helen Berman, Sally Chapman, Mary Good, Donna J. Nelson, and Elsa Reichmanis.
University Park. Pennsylvania State University. Pattee Library. Historical Collections and Labor Archives. Jessie Bernard Papers.

Texas
Austin. University of Texas. Center for American History. Archives of American Mathematics. Mathematical Association of America Records.
College Station. George H. W. Bush Presidential Library. Office of White House Speechmaking. Speech File. Backup Files of cabinet affairs officer Douglas Adair.
Waco. Baylor University. Carroll Library. Texas Collection. Vivienne Mayes Papers. Oral history of Vivienne Mayes.

Virginia
Fairfax. George Mason University. Special Collections and Archives. Eugenie V. Mielczarek Papers.
Reston. U.S. Geological Survey Library. Issues of *WinGS*.

Washington
Seattle. University of Washington Archives. Thelma Kennedy Papers. Davida Teller Papers.

Wisconsin
Madison. University of Wisconsin. University Archives. Ruth Bleier Papers.

Wyoming
Laramie. University of Wyoming. American Heritage Center. Hertha Sponer Franck Papers. Christina Lochman-Balk Papers. Sara Rhoads Papers.

Biographical Dictionaries

Ambrose, Susan A., Kristin L. Dunkle, Barbara B. Lazarus, Indira Nair, and Deborah A. Harkus, eds. *Journeys of Women in Science and Engineering: No Universal Constraints.* Philadelphia: Temple Univ. Press, 1997.

American Men and Women of Science. Various editors. 13th–16th eds., New York: Bowker, 1976–89; 18th–19th eds., New Providence, NJ: Bowker, 1992–96.

Bailey, Martha J. *American Women in Science 1950 to the Present: A Biographical Dictionary.* Santa Barbara, CA: ABC-CLIO, 1998.

Grinstein, Louise S., Carol A. Biermann, and Rose K. Rose, eds. *Women in the Biological Sciences: A Biobibliographic Sourcebook.* Westport, CT: Greenwood, 1997.

Grinstein, Louise S., Rose K. Rose, and Miriam H. Rafailovich, eds. *Women in Chemistry and Physics: A Biobibliographic Sourcebook.* Westport, CT: Greenwood, 1993.

Henrion, Claudia. *Women in Mathematics: The Addition of Difference.* Bloomington: Indiana Univ. Press, 1997.

Jordan, Diann. *Sisters in Science: Conversations with Black Women Scientists on Race, Gender, and Their Passion for Science.* West Lafayette, IN: Purdue Univ. Press, 2006.

Love, Barbara J., comp. *Feminists Who Changed America, 1963–1975.* Urbana: Univ. of Illinois Press, 2006.

Morrow, Charlene, and Teri Perl, eds. *Notable Women in Mathematics: A Biographical Dictionary.* Westport, CT: Greenwood, 1998.

Shearer, Benjamin F., and Barbara S. Shearer, eds. *Notable Women in the Life Sciences: A Biographical Dictionary.* Westport, CT: Greenwood, 1996.

———. *Notable Women in the Physical Sciences: A Biographical Dictionary.* Westport, CT: Greenwood, 1997.

Sicherman, Barbara, and Carol Hurd Green, eds. *Notable American Women: The Modern Period.* Cambridge, MA: Harvard Univ. Press, 1980.

Warren, Wini. *Black Women Scientists in the United States.* Bloomington: Indiana Univ. Press, 1999.

Other Reference Works

Aldrich, Michele L., and Paula Quick Hall, comps. *Programs in Science, Mathematics and Engineering for Women in the United States, 1966–1978.* AAAS Publication 80-11. Washington, DC: AAAS, 1980.

"Contributions of 20th Century Women to Physics." http://cwp.library.ucla.edu.

Davies, Doris C. *Idealism at Work: AAUW Educational Foundation Programs, 1967–1981.* Billings, MT: AAUW Educational Foundation, 1981.

Kiger, Joseph C., ed. *Research Institutions and Learned Societies.* Westport, CT: Greenwood, 1982.

Leach, Alicia E., and Michele Aldrich, comps. *Associations and Committees of or for Women in Science, Engineering, Mathematics and Medicine.* AAAS Publication 84-6. Washington, DC: AAAS, 1984.

"A National Analysis of Diversity in Science and Engineering Faculties at Research Universities." Comps. Donna J. Nelson et al., Washington, DC: National Organization for Women, 2003. Revised thereafter. Also available at http://now.org/issues/diverse/diversity%5Freport.pdf.

Nelson Diversity Surveys. http://chem.ou.edu/~djn/diversity/top50html.

Ohles, John F., and Shirley M. Ohles, comps. *Private Colleges and Universities.* 2 vols. Westport, CT: Greenwood, 1982.

Tanselle, G. Thomas, Peter F. Kardon, and Eunice R. Schwager, eds. *The John Simon Guggenheim Memorial Foundation, 1925–2000: A Seventy-Fifth Anniversary Record.* New York: John Simon Guggenheim Memorial Foundation, 2001.

Biographies and Autobiographies

Ajzenberg-Selove, Fay. *A Matter of Choices: Memoirs of a Female Physicist.* New Brunswick, NJ: Rutgers Univ. Press, 1994.

Albers, Donald J., Gerald L. Alexanderson, and Constance Reid. *More Mathematical People: Contemporary Conversations.* Boston: Harcourt Brace Jovanovich, 1990.

Bateson, Mary Catherine. *Composing a Life.* New York: Atlantic Monthly, 1989.

Bell, Terrel H. *The Thirteenth Man: A Reagan Cabinet Memoir.* New York: Free Press, 1988.

Bingham, Clara, and Laura Leedy Gansler. *Class Action: The Story of Lois Jensen and the Landmark Case That Changed Sexual Harassment Law.* New York: Doubleday, 2002.

Blum, Arlene. *Breaking Trail: A Climbing Life.* New York: Scribner, 2005.

Brady, Catherine. *Elizabeth Blackburn and the Story of Telomeres: Deciphering the Ends of DNA.* Cambridge, MA: MIT Press, 2007.

Byers, Nina, and Gary Williams, eds. *Out of the Shadows: Contributions of Twentieth-Century Women to Physics.* Cambridge: Cambridge Univ. Press, 2006.

Cole, Johnnetta. *Conversations: Straight Talk with America's Sister President.* New York: Doubleday, 1993.

Comfort, Nathaniel C. *The Tangled Field: Barbara McClintock's Search for the Patterns of Genetic Control.* Cambridge, MA: Harvard Univ. Press, 2001.

Cook, Alice H. *A Lifetime of Labor: The Autobiography of Alice H. Cook.* New York: Feminist Press, 1998.

Crane, Kathleen. *Sea Legs: Tales of a Woman Oceanographer.* Boulder, CO: Westview, 2003.

Engel, Celeste G. *Rocks in My Head.* New York: Vantage, 1987.

Fischer, Irene K. *Geodesy? What's That? My Personal Involvement in the Age-Old Quest for the Size and Shape of the Earth.* New York: iUniverse, 2005.

Gardner, David Pierpoint. *Earning My Degree: Memoirs of an American University President.* Berkeley and Los Angeles: Univ. of California Press, 2005.

Goetting, Ann, and Sarah Fenstermaker, eds. *Individual Voices, Collective Visions: Fifty Years of Women in Sociology.* Philadelphia: Temple Univ. Press, 1995.

Guzzo, Louis R. *Is It True What They Say about Dixy? A Biography of Dixy Lee Ray.* Mercer Island, WA: Writing Works, 1980.

Hill, Edward. *My Daughter Beatrice: A Personal Memoir of Dr. Beatrice Tinsley, Astronomer.* New York: American Physical Society, 1986.

Hoffleit, Dorrit. *Misfortunes as Blessings in Disguise: The Story of My Life.* Cambridge, MA: American Association of Variable Star Observers, 2002.

Hopkins, Ann Branigar. *So Ordered: Making Partner the Hard Way.* Amherst: Univ. of Massachusetts Press, 1996.

Howe, Florence, ed. *The Politics of Women's Studies: Testimony from Thirty Founding Mothers.* New York: Feminist Press, 2000.

Levi-Montalcini, Rita. *In Praise of Imperfection: My Life and Work.* New York: Basic Books, 1988.

Lightman, Alan, and Roberta Brawer, eds. *Origins: The Lives and Worlds of Modern Cosmologists.* Cambridge, MA: Harvard Univ. Press, 1990.

McLaughlin, Patsy A., and Sandra Gilchrist. "Women's Contributions to Carcinology." In *History of Carcinology,* ed. F. Truesdale, 165–207. Rotterdam: A. A. Balkema, 1993.

Nies, Judith. *The Girl I Left Behind: A Narrative History of the Sixties.* New York: HarperCollins, 2008.

O'Connell, Agnes N. *Models of Achievement: Reflections of Eminent Women in Psychology.* Vol. 3. Mahwah, NJ: Erlbaum, 2001.

O'Connell, Agnes N., and Nancy Felipe Russo, eds. *Models of Achievement: Reflections of Eminent Women in Psychology.* Hillsdale, NJ: Erlbaum, 1981.

———. *Models of Achievement: Reflections of Eminent Women in Psychology.* Vol. 2. Hillsdale, NJ: Earlbaum, 1988.

Orlans, Kathryn P. Meadow, and Ruth A. Wallace, eds. *Gender and the Academic Experience: Berkeley Women Sociologists.* Lincoln: Univ. of Nebraska Press, 1994.

Pert, Candace B. *Molecules of Emotion: Why You Feel the Way You Feel.* New York: Scribner, 1997.

Rubin, Vera. *Bright Galaxies, Dark Matters.* Woodbury, NY: American Institute of Physics, 1997.

Schwartz, Neena B. *A Lab of My Own.* Amsterdam: Rodopi, 2010.

Straus, Eugene. *Rosalyn Yalow, Nobel Laureate: Her Life and Work in Medicine; A Biographical Memoir.* New York: Plenum Trade, 1998.

Other Useful Published Sources

Abramson, Joan. *The Invisible Woman: Sex Discrimination in the Academic Profession.* San Francisco: Jossey-Bass, 1974.

———. *Old Boys and New Women: The Politics of Sex Discrimination.* New York: Praeger, 1979.

Albelda, Randy. *Economics and Feminism: Disturbances in the Field.* New York: Twayne, 1997.

Appel, Toby A. "Physiology in American Women's Colleges: The Rise and Fall of a Female Subculture." *Isis* 85 (1994): 26–56.

———. *Shaping Biology: The National Science Foundation and American Biological Research, 1945–1975.* Baltimore: Johns Hopkins Univ. Press, 2000.

Attwood, Cynthia L. *Women in Fellowship and Training Programs.* Washington, DC: Association of American Colleges, Nov. 1972.

Baker, Liva. *I'm Radcliffe, Fly Me! The Seven Sisters and the Failure of Women's Education.* New York: Macmillan, 1976.

Bart, Jody, ed. *Women Succeeding in the Sciences: Theories and Practices across Disciplines.* West Lafayette, IN: Purdue Univ. Press, 2000.

Berger, Joseph. *The Young Scientists: America's Future and the Winning of the Westinghouse.* Reading, MA: Addison-Wesley, 1994.

Bix, Amy Sue. "From 'Engineeresses' to 'Girl Engineers' to 'Good Engineers': A History of Women's U.S. Engineering Education." *NWSA Journal* 16 (Spring 2004): 27–49.

Blum, Lenore. "A Brief History of the Association for Women in Mathematics: The Presidents' Perspectives." *Notices of the American Mathematical Society* 38 (Sept. 1991): 738–54.

Briscoe, Anne M. "Phenomenon of the Seventies: The Women's Caucuses." *Signs* 4, no. 1 (1978): 152–58.

Briscoe, Anne M., and Sheila M. Pfafflin, eds. "Expanding the Role of Women in the Sciences." Special issue, *Annals of the New York Academy of Sciences* 323 (1979).

Case, Bettye Anne, and Anne M. Leggett, eds. *Complexities: Women in Mathematics.* Princeton, NJ: Princeton Univ. Press, 2005.

Chamberlain, Mariam K., ed. *Women in Academia: Progress and Prospects.* New York: Russell Sage Foundation, 1988.

Cole, Jonathan. *Fair Science: Women in the Scientific Community.* New York: Free Press, 1979.

Cullen, Vicky. *Down to the Sea for Science: 75 Years of Ocean Research, Education, and Exploration at the Woods Hole Oceanographic Institution.* Woods Hole, MA: WHOI, 2005.

Curtin, Jean M., Geneva Blake, and Christine Cassagnau. "The Climate for Women Graduate Students in Physics." *Journal of Women and Minorities in Science and Engineering* 3 (1997): 95–117.

Daemmrich, Arthur A., Nancy Ryan Gray, and Leah Shaper, eds. *Reflections from the Frontiers, Explorations for the Future: Gordon Research Conferences, 1931–2006.* Philadelphia: Chemical Heritage Foundation, 2006.

Daniell, Ellen. *Every Other Thursday: Stories and Strategies from Successful Women Scientists.* New Haven, CT: Yale Univ. Press, 2006.

Davis, Cinda-Sue, Angela B. Ginorio, Carol S. Hollenshead, Barbara B. Lazarus, and Paula M. Rayman, eds. *The Equity Equation: Fostering the Advancement of Women in the Sciences, Mathematics, and Engineering.* San Francisco: Jossey-Bass, 1996.

Duffy, Elizabeth A., and Idana Goldberg. *Creating a Class: College Admissions and Financial Aid, 1955–1994.* Princeton, NJ: Princeton Univ. Press, 1998.

Einarson, Elaine. *Woods-Working Women: Sexual Integration in the U.S. Forest Service.* Tuscaloosa: Univ. of Alabama Press, 1984.

Farley, Jennie, ed. "Resolving Sex Discrimination Grievances on Campus: Four Perspectives." Mimeograph, distributed by Institute of Labor Relations, Cornell University, Ithaca, NY, [1982?].

Ferber, Marianne A., and Julie A. Nelson, eds. *Feminist Economics Today: Beyond Economic Man.* Chicago: Univ. of Chicago Press, 2003.

Fiffer, Steve. *Tyrannosaurus Sue: The Extraordinary Saga of the Largest, Most Fought Over T. Rex Ever Found.* New York: W. H. Freeman, 2000.

Fox, Lynn H., Linda Brody, and Dianne Tobin, eds. *Women and the Mathematical Mystique.* Baltimore: Johns Hopkins Univ. Press, 1980.

Fox, Mary Frank. "Organizational Environments and Doctoral Degrees Awarded to Women in Science and Engineering Departments." *Women's Studies Quarterly* 28 (Spring–Summer 2000): 47–61.

Freeman, Richard B., Tanwin Chang, and Hanley Chiang. "Supporting 'The Best and Brightest' in Science and Engineering: NSF Graduate Research Fellowships." Working Paper 11623, National Bureau of Economic Research, Cambridge, MA, Sept. 2005.

Gleiser, Molly. "The Garvan Women." *Journal of Chemical Education* 62 (Dec. 1985): 1065–68.

Glover, Judith. *Women and Scientific Employment.* Basingstoke, UK: Macmillan, 2000.

Guy-Sheftall, Beverly, and Jo Moore Stewart. *Spelman: A Centennial Celebration, 1881–1981.* Atlanta: Spelman College, 1981.

Hall, Roberta M., and Bernice R. Sandler. *The Classroom Climate: A Chilly One for Women?* Washington, DC: Association of American Colleges, 1982.

Hartmann, Susan M. *The Other Feminists: Activists in the Liberal Establishment.* New Haven, CT: Yale Univ. Press, 1998.

Harwarth, Irene. *Women's Colleges in the United States: History, Issues, and Challenges.* Washington, DC: GPO, 1997.

Herr, Lois Kathryn. *Women, Power, and the AT&T: Winning Rights in the Workplace.* Boston: Northeastern Univ. Press, 2003.

Hoffleit, Dorrit. "The Maria Mitchell Observatory—For Astronomical Research and Public Enlightenment." *Journal of the American Association of Variable Star Observers* 30 (2001): 62–93.

Humphreys, Sheila M., ed. *Women and Minorities in Science: Strategies for Increasing Participation.* Boulder, CO: Westview, 1982.

Kaufman, Polly Welts. *National Parks and the Woman's Voice: A History.* Albuquerque: Univ. of New Mexico Press, 1996.

———, ed. *The Search for Equity: Women at Brown University, 1891–1991.* Hanover, NH: Brown Univ. Press, 1991.

Kendall, Ellen. *"Peculiar Institutions": An Informal History of the Seven Sister Colleges.* New York: Putnam, 1975.

Kindya, Marta Navia. *Four Decades of the Society of Women Engineers.* N.p.: Society of Women Engineers, 1990.

Kohlstedt, Sally Gregory, and Suzanne M. Fischer. "Unstable Networks among Women in Academe: The Legal Case of Shyamala Rajender." *Centaurus* 51 (2009): 37–62.

Kundsin, Ruth B., ed. "Successful Women in the Sciences: An Analysis of Determinants." Special issue, *Annals of the New York Academy of Sciences* 208 (1973).

LaNoue, George R., and Barbara A. Lee. *Academics in Court: The Consequences of Faculty Discrimination Litigation.* Ann Arbor: Univ. of Michigan Press, 1987.

Leggon, Cheryl, and Willie Pearson Jr. "The Baccalaureate Origins of African American Women Female Ph.D. Scientists." *Journal of Women and Minorities in Science and Engineering* 3 (1997): 213–22.

Long, J. Scott, ed. *From Scarcity to Visibility: Gender Differences in the Careers of Doctoral Scientists and Engineers.* Washington, DC: National Academy Press, 2001.

Margolis, Jane, and Alan Fisher. *Unlocking the Clubhouse: Women in Computing.* Cambridge, MA: MIT Press, 2002.

McQuaid, Kim. "Race, Gender, and Space Exploration: A Chapter in the Social History of the Space Age." *Journal of American Studies* 41, no. 2 (2007): 405–34.

Miller-Bernal, Leslie, and Susan L. Poulson, eds. *Going Coed: Women's Experiences in Formerly Men's Colleges and Universities, 1950–2000.* Nashville: Vanderbilt Univ. Press, 2004.

National Research Council. *Postdoctoral Appointments and Disappointments.* Washington, DC: National Academy Press, 1981.

Nerad, Maresi, and Joseph Cerny. "Postdoctoral Patterns, Career Advancement, and Problems." *Science* 285 (1999): 1533–35.

Niemeier, Debbie A., and Cristina González. "Breaking into the Guildmasters' Club: What We Know about Women Science and Engineering Department Chairs at AAU Universities." *NWSA Journal* 16 (Spring 2004): 157–71.

Pattatucci, Angela M., ed. *Women in Science: Meeting Career Challenges.* Thousand Oaks, CA: SAGE, 1998.

Preston, Anne E. *Leaving Science: Occupational Exit from Scientific Careers.* New York: Russell Sage Foundation, 2004.

Rosenberg, Rosalind. *Changing the Subject: How the Women of Columbia Shaped the Way We Think about Sex and Politics.* New York: Columbia Univ. Press, 2004.

Rossiter, Margaret W. "American Scientific Societies and the Equal Rights Amendment, 1977–1982." In "Women Scholars and Institutions: Proceedings of the International Conference (Prague, June 8–11, 2003)," ed. Sona Strbánová, Ida H. Stamhuis, and Katerina Mojsejová, 101–14. Special issue, *Studies in the History of Science and Humanities, Prague* 13 (2004).

———. *Women Scientists in America: Before Affirmative Action, 1940–1972.* Baltimore: Johns Hopkins Univ. Press, 1995.

———. *Women Scientists in America: Struggles and Strategies to 1940.* Baltimore: Johns Hopkins Univ. Press, 1982.

Rothblum, Esther D., Jacqueline Weinstock, and Jessica F. Morris, eds. *Women in the Antarctic.* New York: Harrington Park, 1998.

Russo, Nancy Felipe, and Marie M. Cassidy. "Women in Science and Technology." In "Women in Washington: Advocates for Public Policy," ed. Irene Tinker, 250–62. Special issue, *Sage Yearbook in Women's Policy Studies* 7 (1983).

Sanz, José Luis. *Starring T. Rex! Dinosaur Mythology and Popular Culture.* Trans. Philip Mason. Bloomington: Indiana Univ. Press, 2002.

Selby, Cecily Cannan, ed. "Women in Science and Engineering: Choices for Success." Special issue, *Annals of the New York Academy of Sciences* 869 (1999).

Sonnert, Gerhard. *Gender Differences in Science Careers: The Project Access Study.* With the assistance of Gerald Holton. New Brunswick, NJ: Rutgers Univ. Press, 1995.

———. *Who Succeeds in Science? The Gender Dimension.* With the assistance of Gerald Holton. New Brunswick, NJ: Rutgers Univ. Press, 1995.

Stewart, Abigail J., Janet E. Malley, and Danielle LaVaque-Manty, eds. *Transforming Science and Engineering: Advancing Academic Women.* Ann Arbor: Univ. of Michigan Press, 2007.

Thomas, Auden D. "Preserving and Strengthening Together: Collective Strategies of U.S. Women's College Presidents." *History of Education Quarterly* 48 (Nov. 2008): 565–89.

Tidball, M. Elizabeth, and Vera Kistiakowsky. "Baccalaureate Origins of American Scientists and Scholars." *Science* 193 (1976): 646–52.

Tidball, M. Elizabeth, Daryl G. Smith, Charles S. Tidball, and Lisa E. Wolf-Wendel. *Taking Women Seriously: Lessons and Legacies for Educating the Majority.* Phoenix: Oryx, 1999.

Tiefer, Leonore. "A Brief History of the Association for Women in Psychology, 1969–1991." *Psychology of Women Quarterly* 15 (1991): 635–49.

Tobias, Sheila. *Overcoming Math Anxiety.* New York: Norton, 1978.

Urry, C. Megan, Laura Danly, Lisa E. Sherbert, and Shireen Gonzaga, eds. *Women at Work: A Meeting on the Status of Women in Astronomy, Held at the Space Telescope Science Institute, September 8–9, 1992.* Baltimore: STSI, [1993].

Valian, Virginia. *Why So Slow? The Advancement of Women.* Cambridge, MA: MIT Press, 1998.

Wasserman, Elga. *The Door in the Dream: Conversations with Eminent Women in Science.* Washington, DC: Joseph Henry, 2000.

Williams, Clarence G. *Technology and the Dream: Reflections on the Black Experience at MIT, 1941–1999.* Cambridge, MA: MIT Press, 2001.

Xie, Yu, and Kimberlee A. Shauman. *Women in Science: Career Processes and Outcomes.* Cambridge, MA: Harvard Univ. Press, 2003.

Yentsch, Clarice M., and Carl J. Sindermann. *The Woman Scientist: Meeting the Challenge for a Successful Career.* New York: Plenum, 1992.

Zuckerman, Harriet, Jonathan R. Cole, and John T. Beckwith, eds. *The Outer Circle: Women in the Scientific Community.* New York: Norton, 1991.

INDEX

The letter *t* following a page number denotes a table.